Study Guide to Accompany

General Chemistry

Jean B. Umland

Prepared by

Kenneth J. Hughes
University of Wisconsin, Oshkosh

West Publishing Company
Minneapolis/St. Paul   New York   Los Angeles   San Francisco

WEST'S COMMITMENT TO THE ENVIRONMENT

In 1906, West Publishing Company began recycling materials left over from the production of books. This began a tradition of efficient and responsible use of resources. Today, up to 95% of our legal books and 70% of our college texts are printed on recycled, acid-free stock. West also recycles nearly 22 million pounds of scrap paper annually—the equivalent of 181,717 trees. Since the 1960s, West has devised ways to capture and recycle waste inks, solvents, oils, and vapors created in the printing process. We also recycle plastics of all kinds, wood, glass, corrugated cardboard, and batteries, and have eliminated the use of styrofoam book packaging. We at West are proud of the longevity and the scope of our commitment to our environment.

Production, Prepress, Printing and Binding by West Publishing Company.

TABLE OF CONTENTS

PREFACE

Take every advantage of the opportunity to improve your understanding of and performance in general chemistry by using this study guide which has been written **specifically for <u>General Chemistry</u>, by Umland**. Since you will be frequently referred to many tables of data and information in the textbook, plan to realize the best results from your studies by using both the textbook and study guide together.

The chapter and section titles in this study guide correspond exactly to those found in the textbook. Each section in a chapter consists of a REVIEW of important concepts, followed by solved SAMPLE PROBLEMS with detailed hints and strategies. At the end of each chapter, you will find multiple-choice Self-Test Questions and additional Practice Problems, all with answers provided.

ACKNOWLEDGEMENTS

Many thanks are in order for putting up with me during the preparation of this study guide.

To Keith Dodson at West Publishing for his patience and editing expertise. To Gretchen Webb-Kummer and Lisa Pleban for their technical and formatting skills. To office co-workers Anne Murphy and Joan Ratchman for their understanding and support. To my colleague Paul Kelter for his considerable encouragement and support without which this writing would not have been possible.

Finally, to Jennifer and Phyllis for their continuing love and support.

READING AND LEARNING SCIENTIFIC MATERIAL

Helen H. Allen

University of Houston - Downtown

READING IS A THINKING PROCESS

There are a number of techniques and strategies that you can use to help make reading and understanding your chemistry book and other science textbooks easier. First, you must realize that reading is a thinking process. You should be **actively** engaged in this process.

Scientific material should be read slowly and more than once. Each chapter should be read **before** it is discussed in class and then again **after** class. You should keep a pencil in your hand so that you can make notes in the margins of your book. Think about each new concept as you read, and try to put it in your own words to see if you really understand it. Read all the examples carefully and work the practice problems as you go through the chapter.

In order to **think** while reading and be **actively** engaged in the learning process, you need to first preview the material to develop a structure for your reading and then activate your prior knowledge. Next you should organize the information and make smart decisions.

PREVIEWING

In order to gather information successfully, you should first preview the whole textbook. This initial preview will only take a few minutes, but it could save you a lot of time in the future. You preview the book by examining the title, author, and date of publication of the book. Next, you look to see if the book has a preface, introduction, or foreword. These sections are found at the front of the book and explain

1

why the author wrote the book. The table of contents gives you a listing of the major topics in the book and the order in which the material is presented. You should further survey the book to determine if the book has a glossary, index, appendix, bibliography, references, or selected readings. These are usually located at the end of the chapter or at the back of the book. For instance, your chemistry book has an index at the back. You can look up words alphabetically and be referred to specific page numbers for information. All these features are designed to help you learn, but you must be aware of these features in order to take advantage of them.

Chapters should also be previewed. First, skim the chapter so that your mind has a framework or structure on which to focus. Reading structure may be illustrated by the analogy of constructing a building. First, the perimeter of the building is staked out, the foundation is poured, and the frame is constructed. The frame provides the basis for the construction of the rest of the building. Learning is similar to this method of construction. By previewing a chapter, you know where the chapter begins, the main points that are covered, and where the chapter ends. You develop a framework for gathering information and you are not operating in a vacuum.

You begin **previewing a chapter** by reading the title, headings, subheadings, bold or italicized print, pictures, graphs or charts, questions, and the summary. After you survey the chapter in this manner you begin reading the chapter in sections. You ask **questions** about each section by turning the headings into questions and **reading** the section to answer your questions. You know from writing that most paragraphs have a main idea or point. If the section you are reading contains five paragraphs, the answer to your question could contain as many as five points -- one for each paragraph. Of course, there may be an introductory paragraph or an elaborative paragraph that does not help answer your question. Once you have the question for a section answered, you should **highlight** or underline the key words that explain the answer. This helps you to focus on what is important and edit out the material that you will not need to learn. You should also work the practice problems to see if you understood what you read. Practice problems are provided with most sections in each chapter of this book to help you check your learning.

To illustrate the process of previewing, turn to Section 1.1 of Chapter 1 of your chemistry book. Examine the heading "Observations and Conclusions." You should mentally form the question -- what are observations and what are conclusions? As you read to answer these questions, you discover that observation involves your senses such as sight, touch, and smell. You note changes or differences through observation. Observations should be consistent in that if you repeated the same experiment, you should see, hear, feel, or smell the same things as would anyone else who was observing the same experiment.

Conclusions are based on observations. They are how you interpret your observations or what you think about your observations. Your conclusions may change as you get more information. Conclusions may also vary from person to person.

After you have mentally answered your questions for a section, you should highlight or underline the key words that answer your question. For "what are observations?", you would highlight the following words: sight, touch, smell, senses, and consistent. For "what are conclusions?", you would highlight the key words: interpret, conclude, explain, change mind with more information, and others may explain differently.

You should work the practice problems as you come to them. This ensures learning by breaking each chapter into small, manageable parts and by requiring your **active** participation.

Once the whole chapter has been read following these steps, you should use the features at the end of each chapter for review and to check your understanding.

A **summary** is given to you for review of terms and major points. This summary also acts as a glossary, since all major terms are defined in the narrative of the summary.

Additional **Practice Problems** at the end of the chapter provide you with extra problems from each section to see if you know the information well enough to transfer it to different situations.

Stop and Test Yourself allows you to see if you have mastered specific skills and ideas presented in the chapter.

Putting Things Together helps you to build on your skills by combining skills and by relating previous chapters to the present one.

Applications check your ability to apply or use the information that you have learned. You are asked to write equations, answer questions, solve problems, and think reflectively.

The next major emphasis in reading with understanding is to interact with the information. You interact with the information by relating the subject matter to your previous knowledge.

PREVIOUS KNOWLEDGE

A major factor in learning is to think about what you already know and relate new information to your prior knowledge. By associating new information to related, already-learned information, you are able to understand the information better and retain it longer.

Sometimes there are differences between your previous knowledge and new information. You should realize that your previous knowledge is composed of opinions, generalizations, and conclusions that you have formed over time, and this information may be outdated or inaccurate. You should be open and objective in considering information involving such differences. Contrasting your misconceptions with correct concepts will help you learn the correct concept.

Using your previous knowledge to form a model for learning -- associating new information, forming questions, reading to answer questions, working practice exercises, reviewing, and using the features at the end of the chapter -- actively engages you in the reading/thinking process. Thus, you are successful in gathering important information.

ORGANIZING INFORMATION

Organizing information from textbooks can enhance your learning. Information may be organized in a number of ways. You might begin by taking written notes from textbook chapters. These written notes should contain the mental questions that you formed for each heading and subheading while reading a chapter. Good methods for recording these questions and answers in note form are the Cornell method of note taking and the use of index cards.

The Cornell method of note taking involves recording the book title, chapter title, and page number at the top of notebook paper. The question for each heading or subheading is written on the far left side of the paper. The answer to the question is written on the right side of the paper with a margin between the question and answer. The next question drops down the page to the end of the answer above and is stated on the far left side again. This method allows you to **fold** the left side of the notebook paper back and have the questions visible but **not** the answers. Thus, you can check yourself on your knowledge of the chapter. An example of this method follows:

CHEMISTRY

Chapter 1
Pages 1 - 49

Question	Answer
What are observations?	a. involve senses - sight, touch, smell, etc. b. notice differences or changes through senses c. should be consistent or the same each time the <u>same</u> experiment is repeated. d. same things should be noticed by different people
What are conclusions?	a. conclusions are how you interpret the observations, what you <u>think</u> happened b. conclusions may change as you get more information or reflect on what happened c. different people may draw different conclusions from observing the same experiment

Notes from a text chapter may also be recorded in question-answer form on index cards. Using index cards creates an excellent study method for learning the information. Questions from the chapter are written on the unlined side of the index card, and the answer is written on the lined back side of the card. This system provides an efficient way of keeping track of your learning. All cards that contain learned information can be grouped in one stack while cards containing information you still need to learn may be placed in a separate stack. Furthermore, with index cards the information may be arranged in a random order. An example of an index card follows:

What are Observations?	a. Observations involve your senses in looking at data elements, experiments, or things. b. Notice changes -- differences c. Observations should be consistent for the same experiment.

Paraphrasing information, outlining information, writing summaries, drawing charts, maps, and webs of information are other excellent ways of organizing information for learning. Some of these techniques will be illustrated and discussed in the subsection "Specific Techniques and Strategies for Learning."

SMART DECISIONS

The next aspect of thinking while reading and being actively engaged in this process involves making smart decisions. Making smart decisions means thinking logically and using your common sense. You should form hypotheses, solve problems, and judge the accuracy of your thinking. Smart decisions also means checking your learning. Are you merely working problems step-by-step according to an example, or do you actually understand the logic involved? Is your hypothesis correct or do you now realize where the error lies? The information you have gathered and organized must be transferrable to other situations and problems if you really understand it. For example, in Chapter 1, Section 1.2, you learn about physical and chemical changes. Properties that do **not** involve substances changing into other substances are called **physical** properties. Properties that involve substances changing into other substances are called **chemical** properties.

If you do not realize that the example in the book of an ice cube melting into water or boiling water becoming steam is a **physical** change, you need to reread that section before continuing to the next section. Practice problems are provided for you with each section. Short answers are in the appendix at the back of your book and detailed solutions are in the Student's Solution Manual. The features of this book progressively help you to build higher level thinking skills. You should be able to analyze your learning as you proceed through a chapter section by section and as you build your knowledge from chapter to chapter. Learning is a sequential, progressive process. If there are gaps in your learning, go back to the textbook, the lecture notes, or the laboratory assignment and seek the answer. Ask for help, but be specific! Analyze what you know and what you do not know. Continue to reevaluate your information -- make smart decisions.

SCIENCE VOCABULARY

Some of you have well developed vocabularies and have formed your own techniques for understanding new words, so you will **not** need to read this section. A quick test of your vocabulary will indicate if you should read this section or not. The following words are taken from the first few chapters in your chemistry book. Choose the best answer.

Sample Vocabulary Test

1. Anhydrous means:

   a. futile
   b. waterless
   c. archaic
   d. arduous

2. Nonspontaneous means:

   a. substances not having constant composition
   b. changes not taking place by themselves
   c. substances not changed by physical properties
   d. undesirable chemical reactions

3. Interface refers to:

   a. chemicals drying to the skin
   b. surface lying between two phases
   c. applications of chemistry
   d. learning of names and formulas

4. Luminous means:

   a. alien
   b. credible
   c. irrevocable
   d. shining

5. Kinetic refers to:

   a. motion
   b. ineptness
   c. relatives
   d. study of fire

6. Macroscopic means:

   a. using a microscope
   b. large red blood cells
   c. unicellular bacteria
   d. seeable without magnifying

7. Exothermic means:

   a. releasing heat
   b. not reacting
   c. outermost layer
   d. exhibiting change

8. Heterogeneous means:

   a. debased
   b. different
   d. constant
   d. complete

9. Monatomic refers to:

   a. one reaction
   b. single-celled organism
   c. single atom
   d. a characteristic

10. Nomenclature refers to:

   a. analogous experimental situations
   b. group of chemicals that affect the nervous system
   c. systematic naming in art or science
   d. reactions enhanced by nocturnal conditions

ANSWERS: 1. b, 2. b, 3. b, 4. d, 5. a, 6. d, 7. a, 8. b, 9. c, 10. c

If you scored 9 or 10 correct, skip this vocabulary section. If you scored 7 or 8, you have a good vocabulary, but you might find some helpful hints in this section. If you scored 6 or below, this section should help you.

Reading science can be made easier if you keep a few strategies in mind. Through the use of structure analysis, context clues, and multiple meanings, you can read and understand new words without interrupting your concentration during the reading process. However, in some cases you will need to look a word up in the dictionary. When you do this, write the word down, practice saying it to yourself, and use the word as often as you can so that it will become part of your vocabulary.

1. Structure Analysis

Structure analysis involves learning to examine word parts (prefixes, roots/stems, and suffixes). Science terms involve multisyllable terms that may seem difficult. But many of these terms are composed of word parts that you already know or that you can easily learn.

The English language is composed of over 500 000 words, and new words are being added every day. However, thousands and thousands of these words contain the same word parts with the same meanings. Thus, a single word part may be the clue to thousands of words. This is similar to the numbers 0, 1, 2, 3, 4, 5, 6, 7, 8, and 9. These numbers can be used to construct an infinite set of numbers. You may use a 3 and a 0 for the two digit number 30 or use nine digits for the number 731 265 489. New words are manufactured in much the same way. Word parts are grouped together to convey certain meanings. Examples from the first few chapters of this book follow:

Prefixes	**Meaning**	**Derivatives**
an, a	without, not	anhydrous - without water; apathy - not feeling
con	together, with	convey - to make known with
de	down, away	deflect - turn away, change the direction of
dia	through, across	diagonal - line that cuts through in a slanting direction
dis	not, opposite of	disinfectant - not infected, used to kill germs; displace - not placed, take the place of; disregard - not pay attention to
du	two	duplicate - have two of, repeat exactly
ex	out, from	explode - blow out or from; expand - to push out, increase
extra	outside, beyond	extrapolate - to extend beyond (as to extrapolate the curve beyond the experimental data)
hetero	different	heterogeneous - having dissimilar elements or parts
im	into, not	immerse - place into; immigrate; impale; implant
inter	between	interface - a surface lying between two phases; interpolate - read between
macro	large, long	macroscopic - large enough to be seen with the naked eye
micro	very small	microscopic - very small

mono	one	monatomic - one atom; monorail - one rail
non	not	nonspontaneous - not spontaneous or not occurring by itself
pre	before	predict - know in advance or before
sub	under, below	submerge - covered, below; subnormal - below normal
uni	one	unique - one of a kind, having no like or equal

Roots	**Meaning**	**Derivatives**
alter	other	alternate - first one and then the other; altercation; alteration
auto	self	automate - run by itself; automatic
cin	ashes	incinerator - to burn in, burn into ashes (use the incinerator for burning)
div	separate	divisor - number by which another number is divided; divide; dividend; divorce
graph	write	graphically - write or draw by means of a diagram
ident	same	identical - exactly the same; identify; identity
lum	light	luminous - bright light; illuminate - make clear
maxima	greatest	maximum - the greatest number
memoria	memory, recall	memorize - learn by heart; remember - recall
metr, meter	measure	metric (the metric system is a system of measuring)
mit	send	emit - send or give off
mov	move	removal - to move away; movement; movable
nomen	name	nomenclature - a system of names; nominate - name someone

numer	number	numerical - having to do with numbers
pel	drive	repel - drive or force away
solidus	solid	solidify - become solid
techne	art, skill	technology - study of skilled area; technical - involving skills; technique
therm	heat	exothermic - releasing heat

Suffixes	**Meaning**	**Derivatives**
age	collective	storage - where many things are stored
age	place of	anchorage - place of anchoring
age	action, process	marriage - the act or process of being married
age	result of	damage - the result of destruction
al	relating to	nonmetals - a group not related to metals
ate (verb)	to make, cause to be	automate - to cause to be self operating
ation	process action, state or quality of, result of	fixation - the state of being fixed, not changing or moving
ectomy	surgical removal	tonsillectomy - to remove the tonsils
ene	double-bonded hydrocarbons composed of	benzene; ethylene; pentene; butene
ic	the nature of, characterized by	diatomic - composed of, the nature of having two atoms
ical	of the nature of, characterized by	analytical - the nature of analyzing
ide	chemical suffix	lanthanide series - one of the rows of elements placed at the bottom of the periodic table to save space

ify	to make, cause to be	classify - to be grouped; solidify - to be solid
ist	one who, person who	chemist - person who studies chemistry
itis	inflammatory disease	tonsillitis - inflamed tonsils
ity	state of	impurity - the state of not being pure; toxicity - state of being toxic
ize, ise	subject to, make, carry on	fertilizer - to make fertile; sterilize; temporize
ous	possessing the qualities of	aqueous - composed of water, having the qualities of water; nonspontaneous - taking place only if someone or something makes it happen
tive	state of	reactive - state of reacting

If you learn to analyze the parts of words, many times you will understand the meaning of the words, and be able to continue your reading without interrupting your concentration.

2.   Context Clues

Context clues in a text may also provide valuable information during the reading process. Context clues help inform the reader what a term or word means in a particular context. Widely used context clues include: punctuation, synonyms, antonyms, examples, experience, and sentences before or after a word written specifically to explain the meaning of the word. Examples of these context clues follow:

CONTEXT CLUES

CLUE	EXAMPLE	EXPLANATION
Punctuation can give the definition of a difficult word.	Elements cannot be decomposed (broken into simpler substances) by chemical reactions.	Parentheses are used to help explain the term decomposed. Other punctuation may be used such as dashes (--), commas (,), and semicolons (;).
Synonyms may be used so that familiar words help the reader to understand unfamiliar words.	Microscopic particles are so **tiny** that we need to magnify them in order to see them.	A synonym for microscopic is tiny. The synonym helps explain the meaning of the word.
Antonyms, or the opposite of a word, may help the reader understand the unfamiliar word.	Heterogeneous mixtures do not have the same properties throughout -- they are **not homogeneous.**	If you know homogeneous means having the same properties throughout, it can help explain its opposite, heterogeneous, which means not having the same properties throughout.
Examples can help one understand the definition of a new word.	American households throw away an average of 9 grams of **dangerous waste** (dead batteries, cosmetics, insecticides, drain cleaners, and paint) per day.	The examples make the definition of dangerous waste easier to understand.
Sentences before and sentences after certain terms help to explain what the term means.	Atoms are **electrically neutral.** They do not have a net electric charge. In an atom, the positive charge of the protons in the nucleus is exactly balanced by outside the nucleus.	The second and third sentences help explain what is meant by **electrically neutral** in the first sentence.
Experience or common sense may give the reader a clue to the meaning of a term or word.	A fire extinguisher was present in the lab while the students experimented with **combustible** elements.	Our common sense tells us that combustible elements must be related to fire or there would be no need for a fire extinguisher.

By being alert and recognizing context clues, the reader discovers the meaning of new terms without having to look them up in the dictionary.

3. Multiple Meanings

A note of caution should be mentioned in regard to chemistry vocabulary words. Many times these words have **different** meanings than the meanings you know from your everyday life.

A few examples are listed below:

In chemistry, **physical** has nothing to do with your body. Physical changes in chemistry refer to changes that do not transform a substance into another substance.

In chemistry, **property** does not refer to things you own. Property refers to how substances are identified.

Laws in chemistry are not rules made by governments but rather summaries of series of observations.

To a chemist, a **solution** is not only the answer to a problem but also a homogeneous mixture.

A **nucleus** in chemistry is not a beginning to which additions are made but rather a small, dense, positively charged particle that's at the center of an atom.

In chemistry, **noble** does not mean high rank; rather, it refers to a group of unreactive gases.

To a chemist, a **period** is a horizontal row in the periodic table, as well as a length of time and the punctuation mark at the end of a sentence.

In your scientific reading, learn to recognize word parts, be alert to context clues, and be aware of multiple meanings of words. These simple procedures can extend your comprehension and concentration by allowing you to continue the reading process without interruption.

MEMORY

To succeed in college, it is helpful to understand memory and methods of improving it. Mainly, you will be interested in getting new ideas and information into long-term memory and then being able to recall this information. Widely used techniques for expanding your memory include: drills or practices, associations, visualizations, and the use of your motor skills. You might illustrate memory techniques in the following way:

MEMORY

↓

<u>Short term</u>

↓

to

↓

<u>Long-term</u>

by

Drill & Practice   **Visualization**   **Motor Skills**   **Associations**

Repetition   Drawings   Writing   Prior knowledge
Flash cards   Mental Images   Seeing   Clue words
Notes   Smelling   Mnemonic devices
                                      Feeling

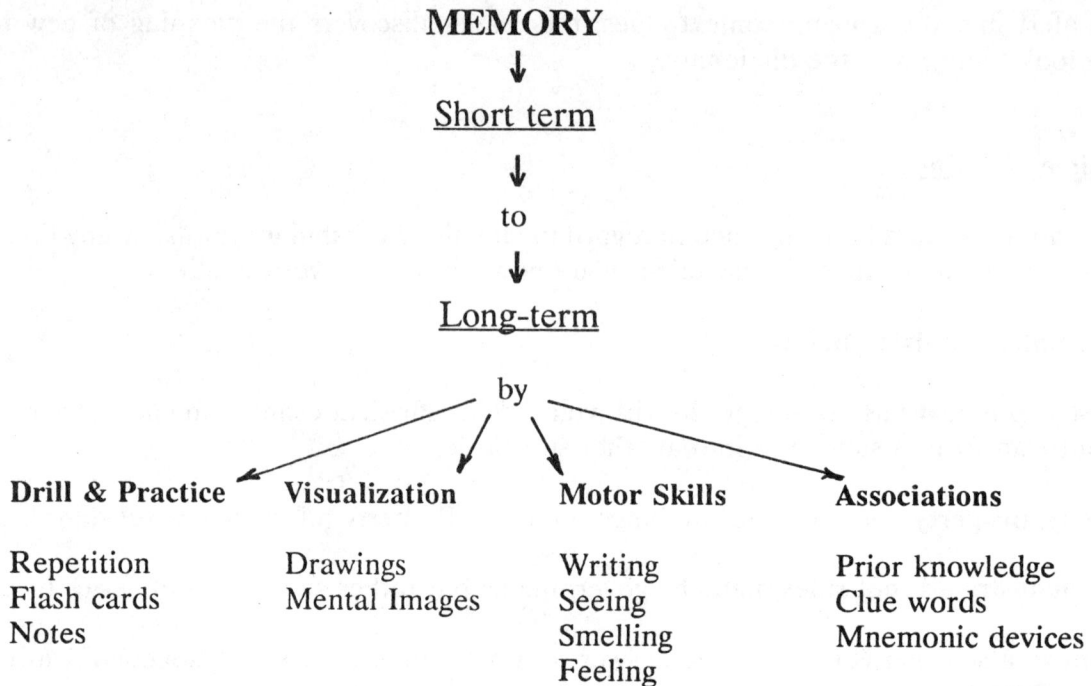

Everyone has a different style of learning. You should determine what strategies work best for you and use them.

1. Drill & Practice

By rehearsing information over and over again, you are able to recall it. You may practice by reciting out loud or by going over information in your mind. It is easier to learn material if you organize it and chunk pieces together. For instance, it is easier to remember your social security number if you recall 455-23-4168 rather than 4-5-5-2-3-4-1-6-8. Making flash cards, reviewing written notes, and practicing answers to questions are effective ways of retaining information.

2. Associations

By relating new information to your **previous knowledge,** you can achieve greater understanding and longer recall. Also by **personalizing** information, you remember it better. You usually remember information about people better than information about things. You will recall more about Abraham Lincoln, the president, than you will about details of his administration. You remember Napoleon long after the names and dates of his battles are forgotten. You can try associating chemical elements or models with people you know to help get data into long-term memory.

Word associations also help to trigger your memory. Word associations are another means of showing connections between ideas. Key words can be listed, and you "brainstorm" to see how many relevant terms you can think to list under the key word. This is a form of grouping or categorizing ideas. Key words will trigger different associations from different people. The more knowledge you have, the more words you will be able to associate with each term.

An example of word associations follows:

Nucleus	Group IVA
small	elements
dense	metals
center	metalloids
positive	nonmetals
charge	periodic table
protons	vertical
atomic number	

Mnemonic devices are another means of making associations and stretching your memory. The names of the Great Lakes may be remembered by using the first letter of each to spell HOMES (Huron, Ontario, Michigan, Erie, and Superior). You may distinguish between a Dromedary camel and a Bactrian camel by turning the initial letter on its side: Bactrian (ⱆ) has two humps while a Dromedary (ⱂ) has only one hump. The spelling of the words stationery and stationary may be distinguished from each other because the one you write on contains the "e" as does the word write. You can make up a number of such devices to help extend your memory. For example, you might use the sentence "He's chosen no one so far cause I balked" to recall symbols for common elements. (H, C, N, O, S, F, Cl, I, and Br).

3. Visualization

You may be able to understand a concept better if you see a picture of it rather than just words describing it. You can draw maps, charts, or diagrams to help you visualize information.

You may also visualize events in your mind. Mental images are usually formed when experiences are repeated, when something out of the ordinary takes place, or when something of particular personal interest occurs. You would recall where you were or what you did if a hurricane or major disaster struck. You would also remember something very personal, such as the birth of your first child.

Since mental images are selective memories of your experience, you may alter what actually happened by incorporating what you have heard from others or by combining a number of similar experiences together. As with all types of memory, you may recall more or fewer events than others.

In science, you may recall general events such as being in the lab. A colorful demonstration or one in which you remember the strong smell of the chemicals may be fixed in your mind. Most likely this mental image is a combination of numerous lab experiences.

4. Motor Skills

Motor skills are important in memory so that abstract information can be reinforced with the physical involvement of the senses. In science, motor skills are involved with observation. In an experiment, you can see and feel the hard ice cube and then you can put your finger in the puddle of water when the ice melts. You can hear and feel steam escaping from a kettle of boiling water. Motor skills are also valuable in remembering scientific material. You **read** the text chapters, you **write** out notes, you **orally** repeat information that you are memorizing, and you **draw** charts or maps of information to help yourself learn.

There is a strong physical element in memory -- you remember better what you do.

Think about **your** memory and employ the techniques that suit you best. This advice is especially true with the next section on specific learning strategies. Many techniques are demonstrated; however, you should **only** be concerned with those that apply to your style of learning and to the type of material you are using.

SPECIFIC TECHNIQUES AND STRATEGIES FOR LEARNING

Learning strategies and techniques can be used to increase your comprehension and promote meaningful understanding. These strategies help you to retain information for a longer period of time, to relate new knowledge to previous knowledge, and to construct new knowledge in ways that make sense and can be remembered. These strategies may be grouped into strategies that expand information, that organize information, and that engage you in monitoring your own learning.

1. Paraphrasing

Paraphrasing is a form of elaboration in which you interpret what you have read by putting it in your own words. You translate the text into your own words for clarity. Paraphrasing involves not only interpreting the words of the text, but also the graphics of the text. You may paraphrase by writing notes in the margin of your text or by writing out notes on notebook paper or index cards. Summarizing text material, solving problems, and answering questions reflectively are other techniques for elaborating and paraphrasing.

An example of paraphrasing graphics could be your interpretation of illustrations that demonstrate a physical change in chemistry. You might look at a picture of an ice cube (water in its solid state) then you might see a picture of the ice cube melting (water in its liquid state). You might paraphrase the pictures by stating that a **physical change** had occurred because the substance merely changed form; it did not turn into a different substance.

Another example of paraphrasing in science might involve your interpretation of the text regarding atoms being electrically neutral. You might say that the nucleus of the atom contains protons, which are positively charged, but the nucleus is surrounded by an equal number of electrons, which are negatively charged. This balance of positive protons with negative electrons makes the atom **electrically neutral.** You may even draw a sketch of an atom to reinforce this concept.

atomic number 6

6 electrons (−)

6 protons (+)
Lithium, Li

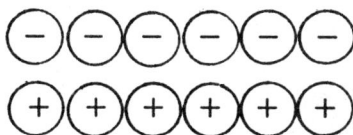

Atom electrically neutral

If you can put material in your own words and reconstruct the concepts from the text, you can master the required learning.

2. Analogies

Constructing analogies may also be considered a form of elaboration. Analogies serve as bridges between what we know and new information. A familiar analogy can make new material clearer and easier to understand. Analogies help activate your previous knowledge and serve as a means of making abstract concepts more concrete and imaginable. For example, you are familiar with cooking. If you were baking cupcakes, the recipe might say, "take 2 and 1/4 cups of flour and 2 eggs (plus other things like sugar, baking powder, butter, milk, and flavoring) to make 30 cupcakes." If you find that you have only one egg, you can still make the cupcakes without going to the store. You can make half a recipe of cupcakes. Since you only have 1 egg, you only need 1 and 1/8 cups of flour, and will only get 15 cupcakes. If there are 3 cups of flour in the bag, you will have 1 and 7/8 cups of flour left.

The cupcake recipe may be used to understand **limiting reactants** in chemistry. When reactions take place, the reactants are **not** usually present in the proportions shown by the equation for the reaction. The quantity of one reactant controls the amount of the product that can be formed. This reactant is called the **limiting reactant** or **limiting reagent**. In the cupcake example, the egg is the limiting reactant. Since the recipe calls for 2 eggs and you only have one egg, you can only make half a recipe. You will only get 15 cupcakes. The egg controls the amount of the product (cupcakes). You will have an excess of flour left.

When a reaction is carried out in chemistry, an excess of one reactant is often present in the reaction mixture. For example, when hydrogen gas burns in air, many more oxygen molecules than hydrogen molecules are present. But, no more than two water molecules can be formed from two molecules of hydrogen gas even if extra oxygen molecules are present.

Hopefully, the analogy of the cupcake recipe and the hydrogen gas burning in air helped you to better understand **limiting reactant** and **excess** of other reactants. By linking the familiar to the unfamiliar, you can better understand the new concept. Such analogies help you to visualize the process, and abstract concepts become more concrete and meaningful.

3. Text Structure

Determining the structure of paragraphs or subsections of text helps you to focus on or determine the major points. An awareness of text structure assists you in forming mental models of the material and in connecting steps.

Text Structure Diagrams

**1. List of Details
or Examples**

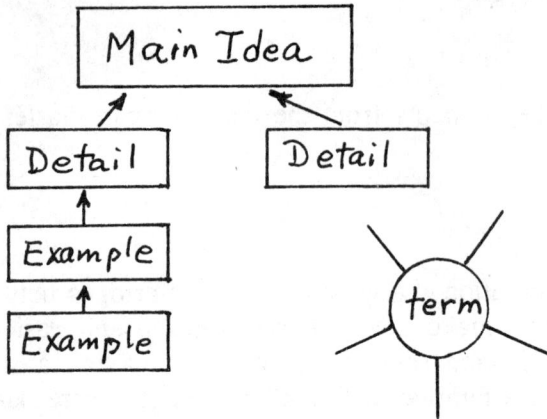

Main Idea

Detail   Detail

Example

Example

term

2. Enumeration

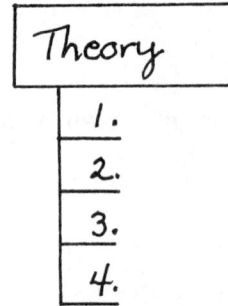

Theory

1.

2.

3.

4.

3. Sequence-Process

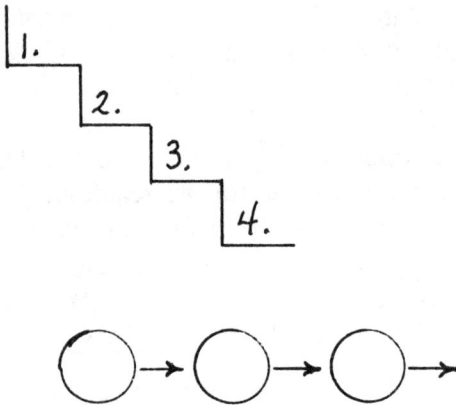

1.

2.

3.

4.

4. Classification

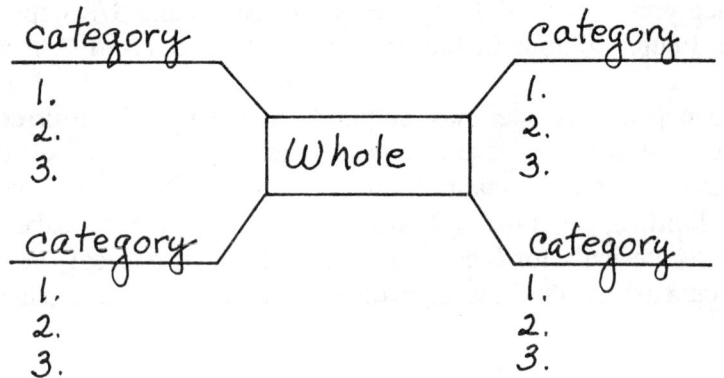

category
1.
2.
3.

category
1.
2.
3.

Whole

category
1.
2.
3.

category
1.
2.
3.

5. Compare and Contrast

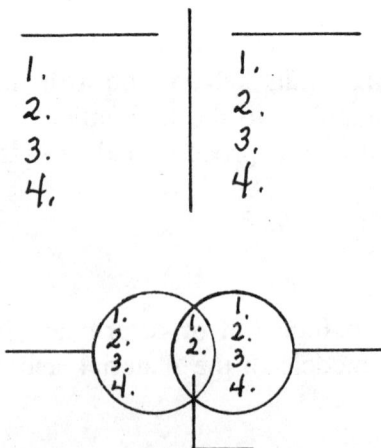

1.
2.
3.
4,

1.
2,
3,
4.

1.
2.
3.
4.

1.
2.
3.
4.

2.

6. Cause and Effect

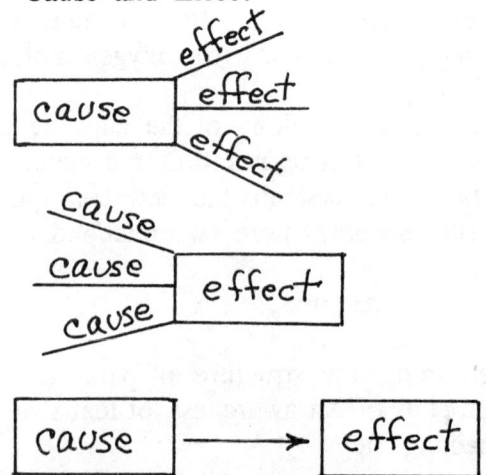

effect

effect

cause

effect

cause

cause   effect

cause

cause ⟶ effect

There are a variety of structures that are used in scientific texts. Some of the most widely used structures include: listing of detail, enumeration, sequence, classification, compare/contrast, and cause and effect.

Listing of detail passages contain a main idea, and the remaining sentences in the paragraph support the main idea by explaining it or extending it. Illustrations and examples may be used for clarification while facts or details are given for extending the meaning. An example of a listing of details of a text structure from your chemistry book follows:

> The nucleus of an atom is a small, dense, positively charged particle at the center of the atom. The positive charge of the nucleus is a result of protons in the nucleus. Protons are small particles with a unit positive charge. The number of protons in the nucleus is called the atomic number of the element. An amber rod rubbed with wool has a positive charge. A leaf or a feather stick to the rod because it is positively "charged."

The main idea of the paragraph is the first sentence which explains that the nucleus has a positive charge. Other sentences give details and examples that support the main idea statement. This example may be shown as follows:

Main idea

The nucleus of an atom is positively charged.

↗ ↖

Detail

Protons in the nucleus are positively charged.

Detail

The atomic number of an element is the number of protons in the nucleus.

↑

Example

Amber rod rubbed with wool has positive charge.

↑

Example

Leaf and feather stick to positively charged rod.

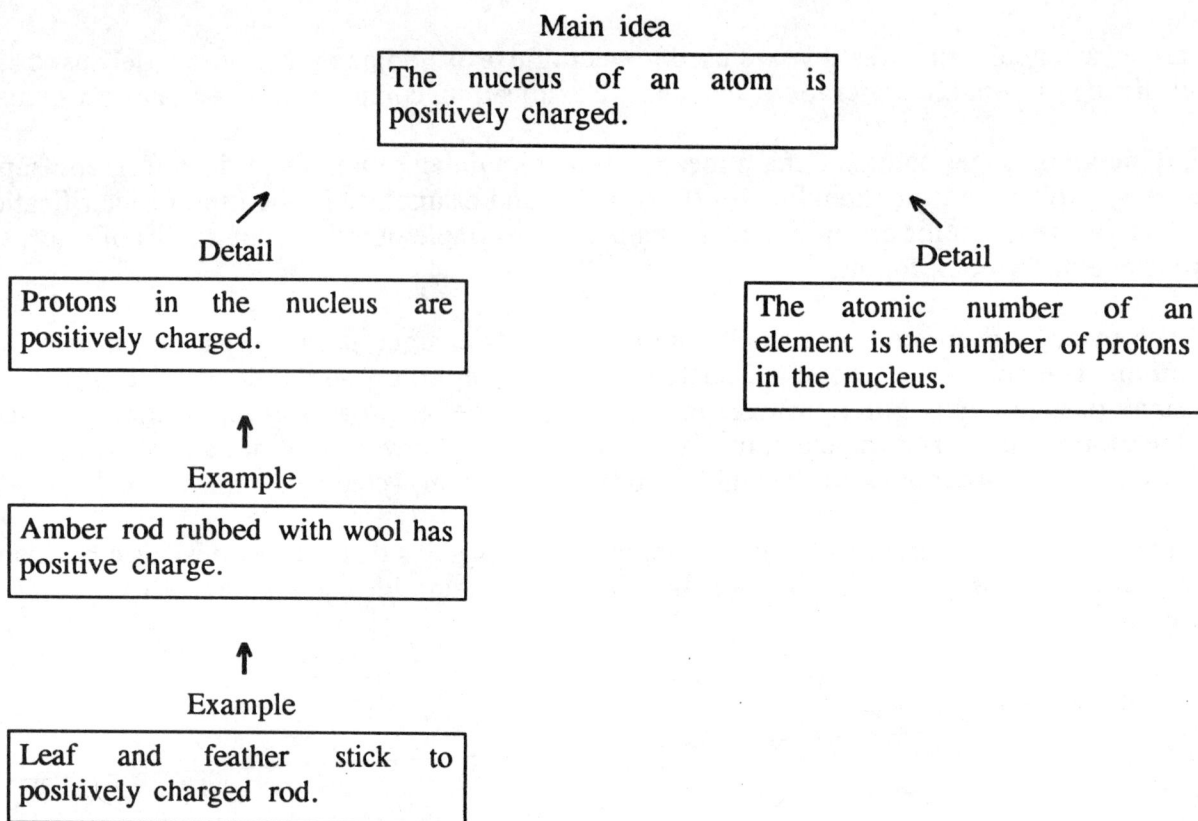

Enumeration structures consist of lists of facts. These lists may be numbered or lettered in a column arrangement or they may appear as a series of facts in paragraph form. You should recognize that this information is important and that you need to be able to recall this list of facts. An example of enumeration follows:

According to the atomic theory suggested by Dalton:

1. All matter is composed of atoms. An atom is the smallest particle of an element that takes part in a chemical reaction.
2. All atoms of a given element are alike; that is, all atoms of gold are the same. Atoms of different elements are different, for example, an atom of copper is lighter than an atom of gold.
3. Compounds are combinations of atoms of more than one element; in a given compound the relative number of each type of atom is always the same. For example, in water, there are always two hydrogen atoms for each oxygen atom.
4. Atoms cannot be created or destroyed. Atoms of one element cannot be changed into atoms of another element by chemical reactions.

Sequence is a text structure that examines events over a period of time or the steps involved in a process.

Sequence may be a narrative of events that proceed in a certain time order, or it may be the steps involved in a particular process. The events or steps **must** be in a particular chronological order. An example from your chemistry book follows:

Summary of steps for writing a chemical equation:

1. Identify reactants and products and write word equation.
2. Write symbols for elements (formulas for elements existing as polyatomic molecules) and formulas for compounds.
3. Balance by changing coefficients in front of symbols and formulas. Do not change formulas or add or remove substances.
4. Check to be sure the same number of each kind of atom is shown on both sides. If coefficients have a common divisor, simplify.
5. Add symbols showing whether substances are solids, liquids, gases, or in aqueous solution.

A chart of writing a chemical equation follows:

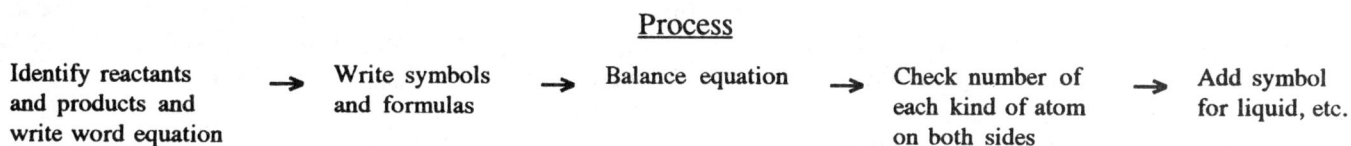

<u>Process</u>

Identify reactants and products and write word equation → Write symbols and formulas → Balance equation → Check number of each kind of atom on both sides → Add symbol for liquid, etc.

Classification structures are those used to group or organize material into separate categories. Each category contains elements or items that share common features or purposes. Organizing by classification helps the scientific reader to group and recall material by certain properties.

An example follows:

Substances can be classified as either elements or compounds. Elements cannot be broken down into simpler substances by chemical reactions. There are 109 known elements. Compounds are substances composed of two or more elements. Compounds can be broken down into simpler substances by decomposition reactions. About eleven million compounds are known.

This example may be drawn as follows:

Substances

Elements   Compounds

↑   ↑

Not broken down by   Composed of two
chemical reactions   or more elements

↑   ↑

109 known elements   Can be broken down into
                                                simpler by decomposition

                                                ↑

                                                11,000,000 known
                                                compounds

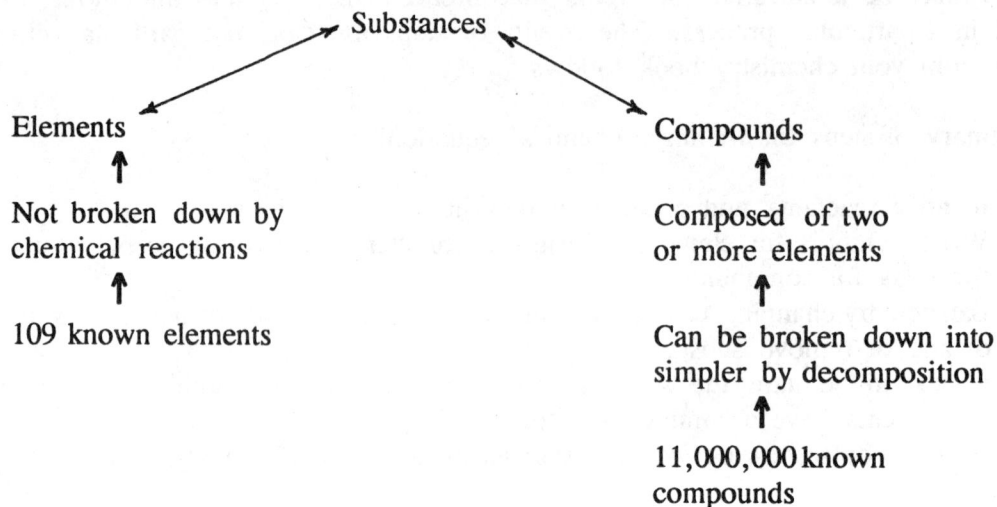

Information may be explained by **comparing and contrasting** people, things, events, or purposes. Comparison recognizes similarities and differences, whereas contrasting emphasizes only differences. You may compare and contrast any number of things, but there must be at least two things.

An example from your chemistry text follows:

> The properties of elements in a group, although similar, are not exactly alike. Of the Group IA metals, lithium has a melting point of 358 °F and reacts slowly with water at room temperature. Sodium has a melting point of 208 °F and reacts rapidly with water at room temperature. Potassium has a melting point of 147 °F and reacts with water faster than sodium does. Rubidium has a melting point of 102 °F and reacts violently with water.

This example may be diagrammed as follows:

Group IA Elements

	Lithium	Sodium	Potassium	Rubidium
Melts	358 °F	208 °F	147 °F	102 °F
Reacts with H_2O at room temperature	Slowly	Rapidly	Faster than sodium	Violently

As you look at the diagram, you can see that the melting points of the elements decrease going down the group, while their reactivity with water increases.

The **cause-effect structure** is used to describe an action or event that is caused by another action or event.

A cause-effect relationship explains why or how something happened and what the result was from that happening. There may be a single cause with a single result, multiple causes with a single result, or a single cause with multiple results.

Your chemistry book describes an experiment in which water is electrolyzed. The water decomposes into hydrogen and oxygen gases. If a burning splint is touched to the mouth of the test tube containing hydrogen gas, there is a loud noise. If a burning splint is inserted into the test tube containing the oxygen, the splint bursts into flame.

The electrolysis of the water **caused** the water to decompose into oxygen and hydrogen.

This example might be diagrammed as follows:

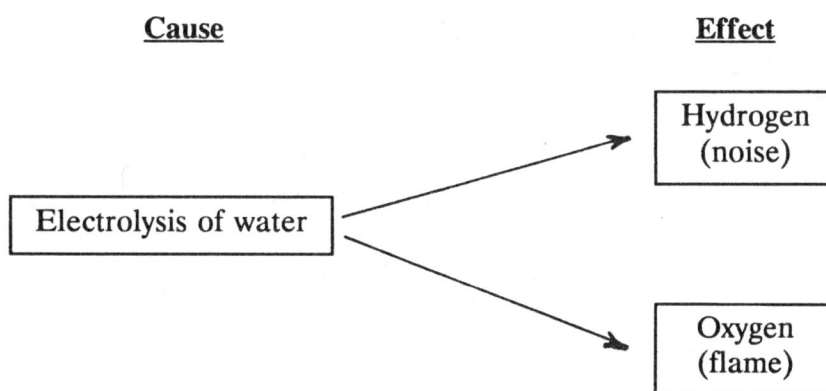

<u>**Cause**</u>   <u>**Effect**</u>

```
                                          ┌─────────────┐
                                    ┌────▶ │  Hydrogen   │
                                    │      │   (noise)   │
┌──────────────────────┐           │      └─────────────┘
│ Electrolysis of water │──────────┤
└──────────────────────┘           │      ┌─────────────┐
                                    └────▶ │   Oxygen    │
                                           │   (flame)   │
                                           └─────────────┘
```

If you recognize the text structure in scientific material, you can then focus on the information that explains the main idea, lists characteristics of the subject, describes steps in a process, separates material into categories, explains how things are similar or different, or shows cause and effect. If you recognize the text structure, you can focus from the beginning on the important parts of the text. This makes you very active in the process of gathering information and answering questions.

Visual aids play a prominent role in learning and recalling scientific material. These aids provide concrete examples that help explain complex information and lengthen the time you can recall information. You may take the information from a text and construct your own visual picture in a number of ways. Different materials lend themselves to different graphics.

4. Concept Mapping

Concept mapping (constructing or organizing your ideas and experiences into meaningful patterns) is an important tool for learning. Information or knowledge maps are made of "concepts" and "linking" words. Your concepts are your ideas, what you believe to be the truth or fact. Linking words show relationships between your ideas or concepts. Concept maps also offer you an opportunity to develop a visual picture of your ideas and how they relate. These visual maps can test your understanding of material and help you to keep track of your learning.

Concept maps are usually arranged with the most general concepts at the top; narrower, more specific concepts are located below. Linking words may connect many of the concepts to one another to show how they are related. This connecting and branching helps you to relate new information to previous knowledge and to expand your information base.

An example of a concept map follows:

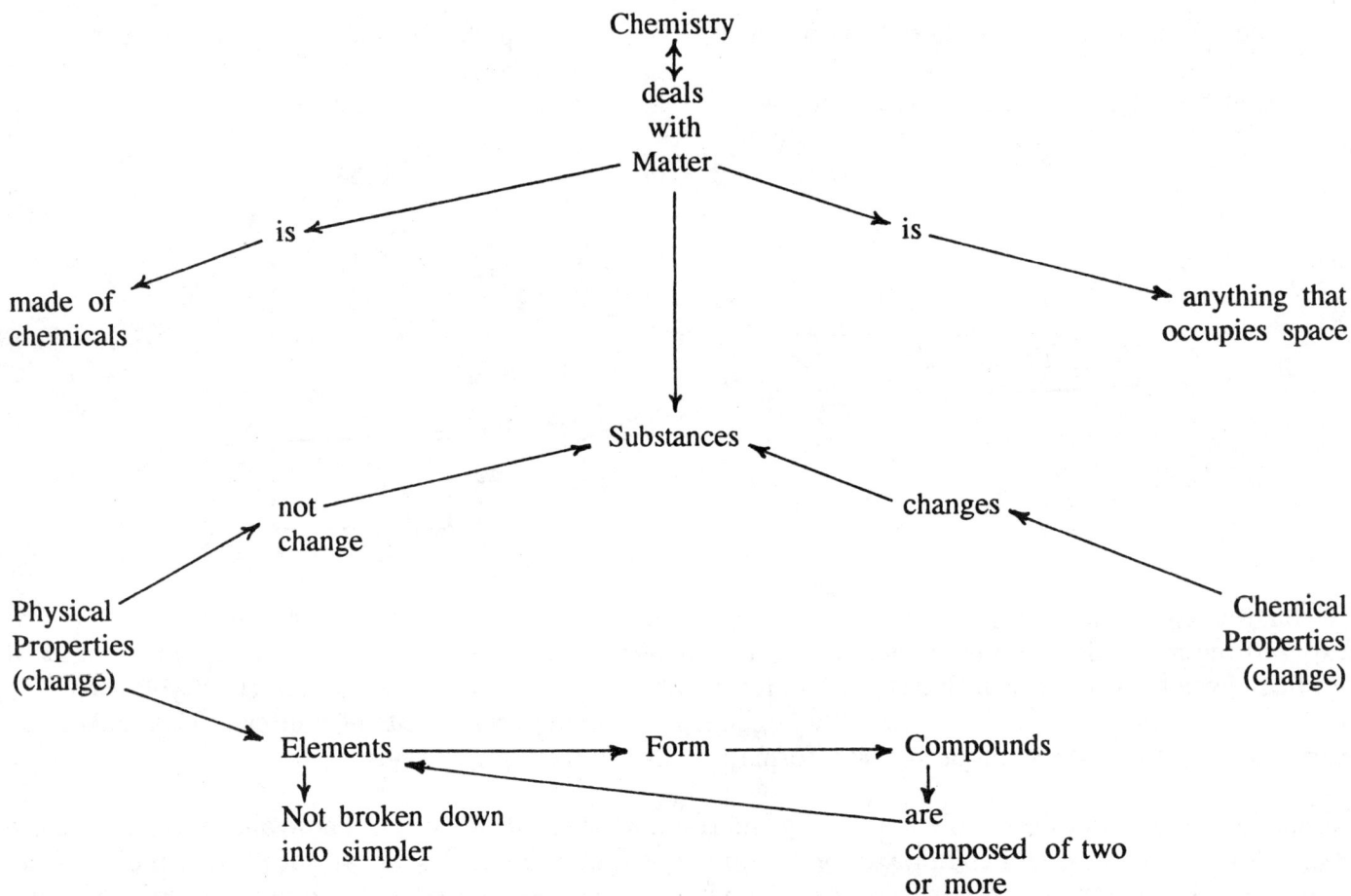

5. Mapping on Cards

Concept mapping on index cards is a useful means of determining how you view a body of information and how you interrelate this information. The object is to create a pattern or visual representation of the various aspects of your information. The general process you should follow is listed below:

- Write key points or concepts on cards.

- Arrange cards so that they make sense to you.

- Rearrange cards until you have the best pattern.

- Connect concepts by linking lines.

- Label types of relationships on connecting lines.

Concept Map of Mixtures

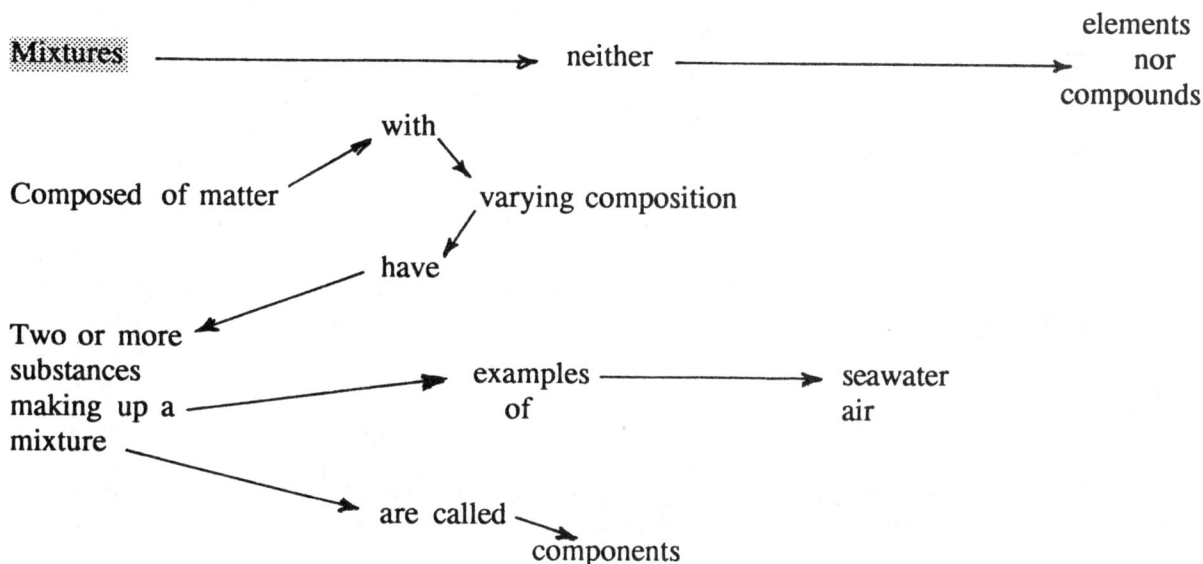

Mixtures ⟶ neither ⟶ elements nor compounds

Composed of matter — with → varying composition

have

Two or more substances making up a mixture

examples ⟶ seawater
of   air

are called → components

6. Semantic Webs

Semantic webs are similar to concept maps. They contain core questions or concepts. Core questions are the center of a web; the web strands are the answers to the core questions and radiate out from the question. Extending from the web strands are the strand supports that are important facts from the reading. Strand ties are details that show relationships between or among strand supports. An example from your book which involves asking a question and reading to answer the questions follows:

Semantic Web

Question

What are observations?

Answer   Answer

Involve your senses   Consistent

                                    Answer   ↓

sight   Noticing changes   When same
↑   or differences   experiment is
touch   repeated
↑
smell
↑
hearing

7.   Venn Diagrams

Venn diagrams are used to examine the relationship between a small number of concepts rather than the understanding of a whole situation. An example follows:

Venn Diagram

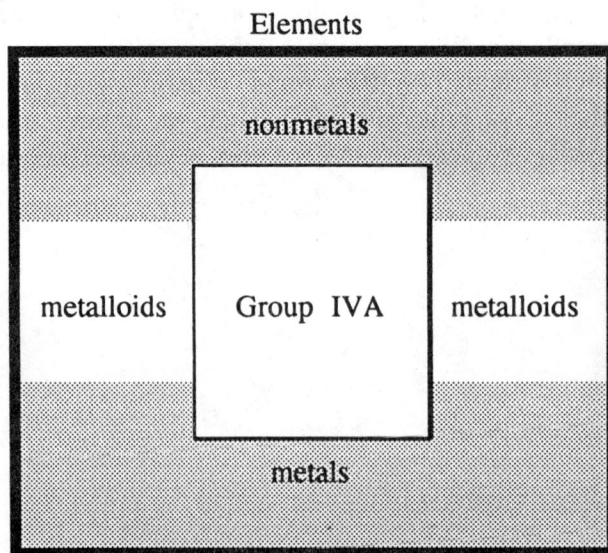

From this diagram, you would infer that Group IVA is a group of elements. You would also infer that elements in Group IVA range from nonmetals to metalloids to metals from top to bottom of the group.

Venn diagrams are a good technique for showing your understanding of the relationship between or among a few concepts.

8. Gowin's Vee

Gowin's Vee helps you understand the structure of knowledge and how you construct new knowledge. Vee diagrams are especially useful with laboratory experiments. They help you to relate the theory and concepts involved with the actual results obtained.

The vee is basically divided into parts. The left side of the vee represents the "thinking" side of knowledge and refers to theories, principles, and concepts. The right side represents the "doing" part of knowledge and refers to value claims, transformations, and records. The center of the vee contains the focus question, and the point of the vee lists the events or objects involved with the learning of the new knowledge and the answering of the focus question.

An Example of a Vee Diagram
from Chemistry

THINKING   **DOING**

Theory:
Dalton's atomic
constant theory

Principles:
Compounds can be
decomposed into simpler
substances by chemical
reactions.
Compounds are
substances composed of
two or more elements.
Wherever and whenever a
compound is found it always
has the same physical and
chemical properties.
Water can be decomposed
into hydrogen gas and
oxygen gas.

Focus Question:
Do compounds
have constant
composition?

Value Claim:
Compounds have
composition.

Knowledge Claim:
Representative samples of
the world's pure water
supply contain
11.19% (by wt.) hydrogen
and 88.81% (by wt.)
oxygen.

Experiment:

Source of H_2O	% by wt. Hydrogen	% by wt. Oxygen
Pacific Ocean	11.18	88.80
Atlantic glacier	11.20	88.82
Lake Louise, Canada	11.10	88.81
Water vapor condensed from office in Japan	11.17	88.80

Events:
4 sources of
pure H_2O from
very different areas.
All contain the same
percent of hydrogen (by wt.)
and oxygen (by wt.) within
experimental error

The Vee diagram helps you understand the law of constant composition by demonstrating that samples of pure water from various sources in the world contain the same percentage (by weight) of hydrogen and oxygen.

Strategies such as paraphrasing, applying analogies, identifying text structures, and constructing visual aids (concept maps, semantic webs, word associations, Venn diagrams, and Gowin's Vee) are excellent tools for expanding your knowledge beyond memorizing to meaningful understanding. You should use the strategies that help **you** put information in your own words, organize your information, and check on your learning.

TEST TAKING

Many students have problems taking tests. Text anxiety causes a person to look at a test and go "blank." Students recall helping classmates study for tests and then the classmate makes a higher score on the test than the one providing the help. Other students report that they studied all the "wrong" things or they have "information overload" and cannot memorize that much material. Excuses for poor test grades also include: bad teacher, not enough time, and the rationalization that you really do not care. Most of these problems produce a sick feeling on the part of the poor test-taker.

Here are some general test-taking techniques that can help you.

General Techniques

General techniques to improve your test-taking ability include the following:

1. Study regularly -- prepare as you go along by reading, taking notes, and constructing learning aids. An athlete does not wait until the night before to get ready for a big game or contest. To be successful, you must plan ahead and work toward your goal.

2. Give yourself the advantage of a good start; be certain of the time, place, and materials required for the exam; arrive with enough time to arrange your working conditions and build a calm, alert attitude. Avoid getting involved in last-minute cram sessions with panicky classmates. Sit at your regular seat if that is possible.

3. Bring all the "tools" that you will need -- pencil, paper, registration form, identification, calculator, and other materials.

4. Approach the test confidently. Take deep breaths, relax, and be calm.

5. Determine the pattern of test questions. If the pattern is easy to hard, be careful not to make careless mistakes or miss questions that you should get right.

6. Associate as many important points as you can in order to fully develop your answer. Write points in margins or on scrap paper as soon as you think of them. Use association to see the relation of one question to another.

7. Do not waste time with emotional reactions to quiz questions. Mentally "talk" to yourself -- stay focused for the period of time that you are taking the test.

8. Do not let lapses of memory produce anxiety or fear; such lapses are normal. Keep working and you will regain your confidence.

9. Pay attention to "clue" words; use the help from the test.

10. Think, avoid hurried answers, guess intelligently.

11. Regulate your speed to the time available, the point value of the question, and your own habits. Allow time to look over your paper at the end. Sometimes a few revisions can add points to your score.

12. Stay until the end of the test and use your time wisely.

These are techniques that should be used when taking tests in general. However, there is more specific advice for multiple choice tests.

Objective Tests

Studying regularly is particularly important for multiple choice tests. Most answers are partially correct. But, you do **not** get partial credit for choosing any answer except the right one.

When taking multiple choice tests, you should:

1. Read the directions carefully, listen to any oral directions or corrections, and check the chalkboard.

2. Glance quickly through the test to plan your time and to see if your test form is complete.

3. Do what you know first. You want to be sure to complete the questions you can answer. Sometimes there are clues within the test that will help you with the more difficult questions. If you have difficulty with some problems or they take too long, skip them and come back if time permits.

4. Read each question carefully. Many students lose points because they do not read the questions correctly.

5. Accept the questions at face value; do not read anything into the question.

6. Give your complete attention to the question on which you are working; do not worry about one question while trying to answer another.

7. Try to supply your own answer before reading the choices provided.

8. If the question involves several steps, be sure you have completed all steps before choosing your answer. Usually answers include choices you might pick if you had only partly answered the question. Don't be tricked!

9. Guess, unless there is a severe penalty.

10. Change your original answer only if you marked in the wrong place or misread the question.

11. When using a separate answer sheet, check frequently to see that you are answering in the properly numbered spaces.

12. Check for omitted questions.

If you learn to relax, focus your attention on the test, and follow most of these strategies, you should become a successful test-taker.

CHAPTER 1

INTRODUCTION

SECTION 1.1   OBSERVATIONS AND CONCLUSIONS

After completing this section, you will be able to do the following:

■ Be able to differentiate between observations and conclusions.

REVIEW

Observations describe what you see, hear, feel, touch, and smell. Since chemistry is a laboratory science based on experiments, careful and complete observations are very important. **Conclusions** are the interpretation of the observations, which may vary from person to person. For example, two people may correctly smell smoke (observation), but may differ as to its source (conclusion). The more observations you make, the better your conclusions will normally be.

SECTION 1.2   PHYSICAL AND CHEMICAL CHANGES

After completing this section, you will be able to do the following:

■ Describe the three states of matter and their relative characteristics;
■ Define and give examples of physical and chemical properties, and physical and chemical changes.

REVIEW

Matter exists in three common "states", "phases", or "forms": **solid, liquid,** and **gas.** [A fourth "state" of matter, **plasma,** only exists at very high temperatures. While plasma comprises most of the stars and interstellar space, it is not common in our daily experience. Therefore, it will not be included in this discussion.] Changes in temperature and/or pressure can change the "state" or "phase" of matter.

Solids, liquids, and gases consist of particles (atoms, ions, or molecules -- discussed in Sections 1.4 and 1.8) which are in constant motion. Whether a substance is a solid, liquid, or gas at a given temperature and pressure depends primarily on two factors:

1. the <u>distance</u> between the particles, and

2. the <u>attraction</u> between the particles.

The particles of a **solid** (like this book) are **closer together** and more **strongly attracted** to each other than the particles of a **liquid** (like the water that we drink) or a **gas** (like the air that we breathe). These properties are what give a solid a definite shape and volume of its own. These characteristics are summarized in Table 1.2A.

Table 1.2A: Characteristics of Solids, Liquids, and Gases

State	Arrangement of particles	Distance between particles	Attraction between particles	Shape[*]	Volume
Solid	Most ordered	Shortest	Greatest	Yes	Yes
Liquid	Intermediate	Intermediate	Intermediate	No	Yes
Gas	Least ordered	Greatest	Least	No	No

[*] Shape means "of its own", i.e., exclusive of the container it may be placed in.

SAMPLE PROBLEM 1.2 A

▶ ▶ Relative to the other two states, why does a gas have neither a shape nor a volume of its own?

Solution:

As shown in Table 1.2A, the particles of a gas are the farthest apart, have the least attraction between each other, and therefore move the fastest and are the most randomly oriented. The particles of a gas are not close enough together to condense or be confined to the volume of the liquid state.

Properties are characteristics which enable us to distinguish one substance from another or recognize an object. **Physical properties** are characteristics which may be observed or measured **without** changing the composition of the material. Size, shape, state/phase, odor, color, electrical conductivity, weight, boiling point, melting point, hardness, luster, and magnetic characteristics are common examples of physical properties. **Chemical properties**, on the other hand, are characteristics involved in a transformation of one substance into another, i.e., matter undergoing **a change in composition**. A chemical property of iron is that it rusts. In the presence of oxygen and moisture, iron metal turns into a substance of different composition: iron oxide --- a "rust-colored" powder, which is called "rust". Burning is another example of a chemical property. When paper burns, the cellulose in the presence of oxygen and heat is transformed into carbon dioxide, water, and soot, all of which have different compositions than paper.

In addition to distinguishing one substance from another by their physical and chemical properties, we also observe that substances undergo physical and chemical **changes**. A **physical change** occurs when there is no change in composition, i.e., only a change in condition or state. Tearing paper, sawing wood, melting snow and evaporation of water are examples of physical changes, because no new substance has been formed. A **chemical change** occurs when a different substance (i.e., one with a different composition) is formed in the process. The resulting substance formed in a chemical change has different **physical and chemical properties** than the starting substance. The rusting of iron and burning of paper described in the previous paragraph are examples of **chemical changes**. All **chemical properties** of a substance describe a **chemical change**. One of the most dramatic examples of a chemical change is that dangerously reactive sodium metal will react with poisonous chlorine gas to produce sodium chloride (common table salt).

SAMPLE PROBLEM 1.2 B

▶▶ Which of the following properties are physical properties and which ones are chemical properties?

burns	odor	luster
conducts heat	freezing point	decomposes
corrodes	non-magnetic	rectangular

Solution:

Although you have not been introduced yet to all the properties described in this example, you should know many of them (if not all) from common usage. The **physical properties** (no different substance formed in the process of measuring or observing the property) are: conducts heat, odor, freezing point, non-magnetic, luster, and rectangular (shape). The **chemical properties** (different substance is formed when this property is observed) are: burns, corrodes, and decomposes.

SECTION 1.3   ELEMENTS, COMPOUNDS, AND MIXTURES

After completing this section, you will be able to do the following:

■ Define and differentiate between elements, compounds and mixtures;

■ Classify material (matter) as either a homogeneous mixture (solution), heterogeneous mixture, element, or compound.

REVIEW

All **material** (or matter) is classified as either a **pure substance** or a **mixture** (see Figure 1.3A). There are two kinds of pure substances: **elements** and **compounds.** There are 109 known **elements,** which are the simplest substances known, i.e., an element **cannot** be broken down into any simpler substance by chemical reactions. A **compound** is composed of two or more elements chemically combined (you can recognize a compound by its formula, which contains the symbols for all the elements in the compound; see Section 1.6 of this study guide).

All materials (matter)

Mixtures — Physical ⇄ Changes — Substances

Homogeneous   Heterogeneous

Compounds ⇄ Chemical Changes (Reactions) — Elements

salt dissolved
in water   iced drink   sodium chloride
(table salt)   sodium &
chlorine

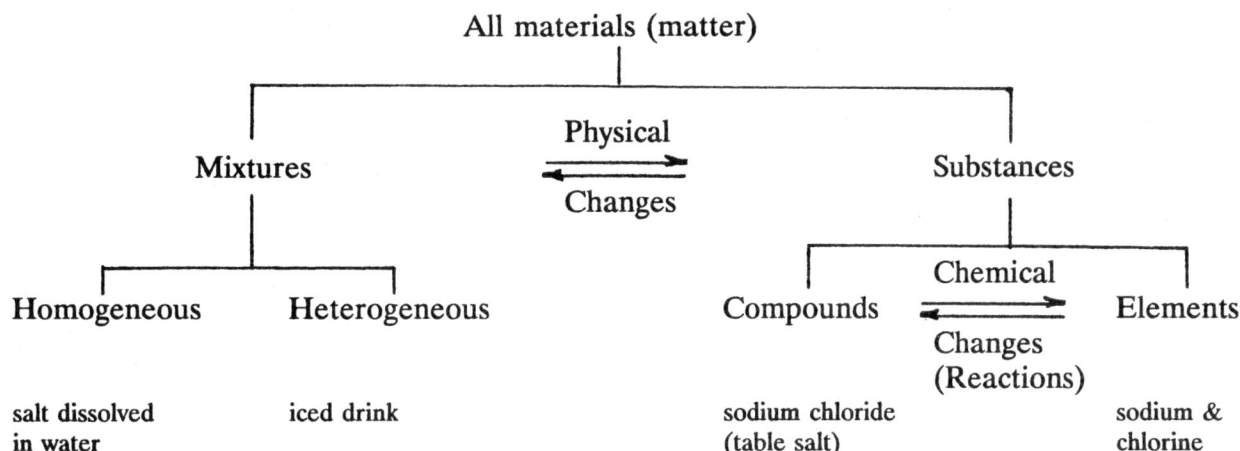

Figure 1.3A   Classification of Materials

If material is not an element or compound, then it must be a **mixture**, which consists of two or more pure substances (elements or compounds) physically mixed together (but **not** chemically combined). There are two kinds of **mixtures: homogeneous** and **heterogeneous.** A **homogeneous mixture** (which is also called a **solution**) is one in which there are no differing parts (i.e., one phase) visible to the naked eye or under a microscope. Solid sugar dissolved in liquid water produces a single phase homogeneous mixture (or solution) of sugar water. A **heterogeneous mixture** has visible differing parts (phases) which can ordinarily be easily seen. Most salad dressings have visible differing parts (one or more solids physically mixed with a liquid) which can be easily seen no matter how thoroughly they are mixed.

There are two major differences between compounds and mixtures. The first is that **compounds** can be decomposed into **elements**, and **elements** can combine to form **compounds**, but only by a **chemical change** (chemical reaction). The compound sodium chloride decomposes into the elements sodium and chlorine, and the elements sodium and chlorine can be combined to form the compound sodium chloride, but only by a **chemical change** (chemical reaction). As a result, the elements sodium and chlorine have **different chemical and physical properties** as uncombined elements than they do when they are in the form of the compound sodium chloride. In contrast, **mixtures** (either heterogeneous or homogeneous) can be separated into elements and/or compounds, and elements and/or compounds can be physically mixed to form a mixture, but only by a **physical change** (no different or new substance is formed). A homogeneous mixture of salt water can be separated into the compounds sodium chloride (table salt) and water by evaporation or boiling, and the compounds sodium chloride and water can be physically mixed to form the homogeneous mixture of salt water. Evaporation, boiling, and dissolving a solid in a liquid are examples of **physical changes**. Since mixtures are formed and separated by **physical changes only**, the components of a mixture (in this case sodium chloride and water) have the **same** physical and chemical properties unmixed as they do when part of a mixture.

The second difference between compounds and mixtures is that the composition of **compounds** is constant (fixed), whereas the composition of **mixtures** may vary. All samples of a **compound** always have the same composition, i.e., the same percent by weight of each element in the compound. This statement is known as the **Law of Constant Composition**. Only a specific amount of sodium metal will chemically combine with a given amount of chlorine gas (using appropriate conditions of temperature, etc.) to produce sodium chloride (with no sodium or chlorine left over). In contrast, **mixtures** have a variable composition by weight because they are formed by a physical change, not a chemical change. That is, differing amounts of salt can be added to a given amount of water to produce salt water of varying composition.

SAMPLE PROBLEM 1.3 A

▸▸ (a) How can a mixture (either homogeneous or heterogeneous) be separated into its components? (b) How can a compound be decomposed into its elements?

Solution:

(a)   Since a mixture is a physical mixing of two or more elements and/or compounds, only a physical change is needed to separate a mixture into its components, e.g., evaporation, boiling, filtration, etc.

(b)   In contrast, the formation of a compound from its elements is a chemical change. Therefore, a compound can be decomposed into its elements only by a chemical change.

SAMPLE PROBLEM 1.3 B

▸▸ Classify each of the following as an element, compound, homogeneous mixture (solution), or heterogeneous mixture:

carbon dioxide gas	cement	oxygen
Kool-Aid®	ice	ocean water
aluminum	vinegar and oil dressing	

Solution:

The elements are (inside front cover of study guide):   oxygen and aluminum.

The compounds are: carbon dioxide gas (consists of the two elements carbon and oxygen); ice (solid water, which consists of the elements hydrogen and oxygen)

The homogeneous mixtures are: Kool-Aid® (one phase visible after preparing mixture); ocean water (assuming clear sample, contains salt dissolved in water; only one phase visible)

The heterogeneous mixtures are: cement (different solid phases visible when dry); vinegar and oil dressing (at least two liquid phases visible no matter how long the bottle is shaken).

SECTION 1.4   ATOMS

After completing this section, you will be able to do the following:

■ Define atom, nucleus, proton, electron, and atomic number;

■ For an atom of any element, given <u>one</u> of the following characteristics, be able to determine the other characteristics using the table of elements:

- name of the element;
- atomic number of the element;
- charge on the nucleus of the atom;
- number of electrons outside the nucleus of the atom.

REVIEW

An **atom** is the smallest particle of an element that takes part in chemical reactions. All atoms of a given element are chemically alike. Atoms can neither be created nor be destroyed by ordinary chemical reactions. An atom can be pictured as an extremely small particle which consists of an even smaller, very dense center called the **nucleus**. The nucleus contains one or more <u>positively</u> ($+1$) charged particles called <u>protons</u>. The number of protons in the nucleus of an atom is called the **atomic number** of the element. Since all atoms of the same element are alike, they will all have the same number of protons in the nucleus, or atomic number. The **atomic number** of every element is given in the table on the inside front cover of this study guide (and the textbook).

Outside the nucleus of the atom are one or more negatively (-1) charged particles called **electrons**. Since **atoms are electrically neutral**, the number of negatively charged electrons outside the nucleus must exactly balance the number of positively charged protons inside the nucleus. Since the atomic number of the element magnesium is 12, all atoms of magnesium have 12 protons inside the nucleus, a charge on the nucleus of $+12$, and (since all atoms of every element must be electrically neutral) 12 electrons (with a total charge of -12) outside the nucleus.

40

SAMPLE PROBLEM 1.4 A

▶▶ Complete the following table from the information provided.

Element	Atomic Number	Charge on Nucleus	No. Electrons Outside Nucleus
Nitrogen			
	27		
Oxygen			
		+35	
Plutonium			
			28
Barium			
	50		

Solution:

HINT:
- You must have a table of the elements like the inside front cover of this study guide or the textbook, which has the name and atomic number of all the elements (plus other information for later use).

- The <u>atomic number</u> = the number of (+1) protons inside the nucleus;
  = the charge on the nucleus;
  = the number of (-1) electrons outside the nucleus.

- Given the name of an element, you can look up its atomic number, which is also the answer to the last two columns.

- Given a number in any of the last three columns (which is automatically the atomic number), you can look up the name of the element. This procedure is a little harder when the elements are arranged alphabetically rather than by increasing atomic number.

Element	Atomic Number	Charge on Nucleus	No. Electrons Outside Nucleus
Nitrogen	7	+7	7
Cobalt	27	+27	27
Oxygen	8	+8	8
Bromine	35	+35	35
Plutonium	94	+94	94
Nickel	28	+28	28
Barium	56	+56	56
Tin	50	+50	50

SECTION 1.6   SYMBOLS

After completing this section, you will be able to do the following:

■ Write the correct symbol from the name of an element, and vice versa.

REVIEW

To write the correct symbol for any element (listed on the inside front cover of this study guide):

- the first letter of the symbol is **always CAPITALIZED**. For most elements, note that the first letter of the name is also the first letter of the symbol. For some elements, the first letter of the symbol (and second letter if there is one) is not found in the English name of the element. In those instances, the symbol has been taken from the Latin or Greek name for that element.

- the second letter (if one is used) is **always lower case**.

The symbol for the element hydrogen is H, **NOT** h. The symbol for antimony, Sb, contains none of the letters in the English name, but is taken from its Latin name *stibium*. Similarly, the symbol for gold, Au, contains letters taken from its Latin name *aurum*. The need to **always** write the **second letter** of a symbol in **lower case** is illustrated by noting that the correct symbol for the element cobalt is Co, **not CO**. A capital letter **anywhere in a formula** (formulas are made up of symbols of elements) **always** indicates the first letter of a different element. CO means a substance containing the elements carbon, C, **and** oxygen, O. CO is the formula for the **compound carbon monoxide**, which is totally different in composition and properties than the **element** cobalt, **Co**. Similarly, the correct symbol for copper is **Cu**, from its Latin name *cu*prum. If the symbol is written as CU, it means a **compound** containing the two elements carbon, C, **and** uranium, U. You must also be able to derive the name of the element from the symbol, particularly for the more common ones used throughout this course.

SAMPLE PROBLEM 1.6 A

▶▶ Complete the following table with the correct symbol or name.

Element	Symbol
Aluminum	
	O
	Ni
Magnesium	
Bismuth	
	N
	K
Tin	
	Pb
	Fe

Solution:

Element	Symbol
Aluminum	Al
Oxygen	O
Nickel	Ni
Magnesium	Mg
Bismuth	Bi
Nitrogen	N
Potassium	K

Tin	Sn
Lead	Pb
Iron	Fe

NOTE: Be very careful to write and use the correct symbols for elements that have similar letters in their name. "N" and "i" are the first two letters of both nickel and nitrogen, but only Ni is the correct symbol for nickel. Magnesium has the symbol Mg, which is frequently confused with manganese, whose symbol is Mn. K is the symbol for potassium (from kalium), not krypton, whose symbol is Kr.

SECTION 1.7   THE PERIODIC TABLE

After completing this section, you will be able to do the following:

■ Given a periodic table and the name or symbol for any element, you will be able to (a) locate it by period and group, (b) identify it as a main group element, transition element, or inner transition element, and (c) classify it as a metal, metalloid, or non-metal.

REVIEW

The periodic table is organized as follows (see the inside front cover of this study guide or the textbook):

- the 109 known elements are arranged in order of **increasing atomic number** reading from left to right, just as you are reading the page of this book. For now, ignore the gaps or open spaces in the periodic table, e.g., between atomic numbers 4 and 5, 12 and 13, etc.

- the **vertical columns** of elements are called **groups** or **families,** with each **group** identified by a number-letter designation, and also a name for some of them. For example, the **Group IIA** elements are Be, Mg, Ca, Sr, Ba, and Ra. Elements in this group are also called the **alkaline earth metals. Group IA** metals are called **alkali metals. Group VIIA** elements are called **halogens. Group 0** elements are called **noble or inert gases.** Elements in the same **group or family** have similar **chemical properties.**

- the **horizontal rows** of elements are called **periods,** with each period identified by a number. The **first period** contains only two elements H and He. The **second period** contains 8 elements, beginning with Na and ending with Ar. **Period 5** has 18 elements in it. **Period 6** has **32** elements in it, because the atomic numbers of the elements in that period run from 55 (Cs) to 86 (Rn). The row of elements called the **lanthanide series** (atomic numbers 57-70) has been removed from **Period 6** and placed at the bottom of the figure to save space; otherwise, the periodic table would be too wide for convenient printing and use. Similarly, the same procedure is used for the **actinide series** (atomic numbers 89-102) in **Period 7**.

- Elements in the **A groups** (IA-VIIA) and **Group 0** are called **main group or representative elements**. Elements in **Groups IIIB, IVB, VB, VIB, VIIB, VIII, IB, IIB** are called **transition elements**. Elements in the **lanthanide and actinide series** are called **inner transition elements**.

- Most of the elements have the properties of **metals** (white spaces in the periodic table). All **metals** are **solids** except for mercury, which is a **liquid**. The elements on the extreme right side of the periodic table do **not** have the properties of metals and are called **non-metals** (diagonal lines in the periodic table). Most of the **non-metals** are **gases** (N, O, F, Cl, and Group 0), some are **solids** (C, P, S, and I), and one is a **liquid** (Br). Separating the **metals** and **non-metals** are the **metalloids** (shaded in the periodic table), which are all **solids** and have properties between the metals and non-metals.

SAMPLE PROBLEM 1.7 A

▶ ▶ Using the periodic table on the inside front cover of this study guide or the textbook, complete the following table for each element given by (a) locating it by group and period, (b) identifying it as a main group element, transition element, or inner transition element, and (c) classifying it as a metal, non-metal, or metalloid.

Element	Atomic Number	Period Number	Group Number & Name (if any)	Main/Trans./ Inner Trans.	Metal/Non-metal/Metalloid
Lithium	3				
Iodine	53				
Silicon	14				
Sodium	11				
Xenon	54				
Iron	26				
Strontium	38				
Fluorine	9				
Silver	47				
Cerium	58				

HINT: • The periodic table will be used for many more concepts to be discussed later in this course. Write on the periodic table (or a copy of it) the information learned in this section (groups, group names, metals, non-metals, metalloids, main group, transition, inner transition) and save it for future use.

Element	Atomic Number	Period Number	Group Number & Name (if any)	Main/Trans./ Inner Trans.	Metal/Non-metal/Metalloid
Lithium	3	2	IA; alkali metal	main	metal
Iodine	53	5	VIIA; halogen	main	non-metal
Silicon	14	3	IVA	main	metalloid
Sodium	11	3	IA; alkali metal	main	metal
Xenon	54	5	0: noble gas	main	non-metal
Iron	26	4	VIII	transition	metal
Strontium	38	5	IIA; alkaline earth metal	main	metal
Fluorine	9	2	VIIA; halogen	main	non-metal
Silver	47	5	IB	transition	metal
Cerium	58	6	no group number	inner transition	metal

NOTE: For inner transition elements, there is no group number normally assigned to them.

SECTION 1.8   MOLECULES AND IONS

After completing this section, you will be able to do the following:

■ Give examples of and differentiate between molecules, simple ions, and polyatomic ions;

■ Given the formula of a molecule, be able to indicate the number of atoms of each element present in the molecule, and vice versa;

■ Given the name or formula of ions forming an ionic compound, be able to write the correct formula for the compound;

■ Given the formula for a compound, be able to predict whether it is a molecular or ionic compound;

■ Given the name of an ionic compound containing polyatomic ions, be able to write the correct formula for the compound by knowing the name, formula, and charge for the polyatomic ions.

REVIEW

Molecules

A **molecule** is the smallest particle of an element or compound that has the chemical properties of the element or compound; **molecules** are electrically **neutral** (no net charge). Compounds that have molecules as unit particles are called **molecular compounds**. Atoms of **non-metals** combine to form **molecules**, which are one of two types:

- **a molecule of a compound** is composed of two or more atoms of at least **two different non-metal elements**. Carbon dioxide is a compound of the two non-metals carbon and oxygen. The smallest particle of the compound carbon dioxide is a **molecule** of carbon dioxide.

- **a molecule of an element** consists of **one or more** atoms of the **same** element. If the element has a single atom as its **unit** (smallest or basic) particle, then a molecule of that element is an atom of that element. For example, the element neon exists of individual neon atoms; a molecule of neon is an atom of neon. The unit particle of some elements has two or more atoms per molecule. **Diatomic** molecules consist of **two** atoms per molecule. Common examples of **elements** with diatomic molecules as their unit particles are hydrogen, oxygen, nitrogen, fluorine, chlorine, bromine, and iodine. [A compound may also have a diatomic molecule as its unit particle.] The unit particle of the element phosphorus is a molecule containing four phosphorus atoms; the unit particle of the element sulfur is a molecule containing eight sulfur atoms.

Formulas are the short-hand notations for representing molecules. The conventions used for writing correct formulas are:

- The formula of a compound consists of the symbols of all the elements in the compound, usually starting with the element closest to the **lower left-hand corner of the periodic table**.

- Each symbol is followed by a **subscript**, which indicates the number of atoms of that element in one molecule of the compound. If there is only one atom of an element in a molecule, the subscript 1 is **not** used.

SAMPLE PROBLEM 1.8 A

▶▶ A molecule of carbon dioxide consists of one carbon atom and two atoms of oxygen. A molecule of carbon monoxide consists of one atom of each element. What are the formulas for carbon dioxide and carbon monoxide?

Solution:

HINT:
- Normally, the **name** of the compound starts with the element that is closest to the lower left-hand corner of the periodic table.

- You will find it extremely useful to memorize the symbols for the most commonly used elements in this course and to be able to locate them quickly in the periodic table.

The formula for carbon dioxide starts with the symbol for the element closest to the lower left-hand corner of the periodic table. First look up the symbols for carbon and oxygen in the inside front cover of the text and locate them in the periodic table. Although both C and O are both in period 2, C is closer to the lower left side of the periodic table. The formula starts with C with **no** subscript, followed by O with the subscript 2.

The correct formula is CO_2. The subscript 1 is **never** used in a formula. No subscript after a symbol **always** means one atom of that element in that molecule.

Since a molecule of carbon monoxide contains only one atom of each element, there are no subscripts in the formula.

The correct formula is CO.

SAMPLE PROBLEM 1.8 B

▶▶ How many atoms of each element are there in (a) one molecule of HF, (b) one molecule of P_2O_5, (c) two molecules of N_2O_4, and (d) six molecules of Cl_2O_7?

Solution:

HINT:
- Determine the number of atoms of each element in **one** molecule, then multiply by the number of molecules.

(a) **No** subscripts means "H_1F_1". Thus, one HF molecule has 1 H atom and 1 F atom.

(b)   One molecule of P_2O_5 has 2 P atoms and 5 O atoms.

(c)   One molecule of N_2O_4 has 2 N atoms and 4 N atoms. **Two** molecules of N_2O_4 will have twice as many atoms of each element, or 4 N atoms and 8 O atoms.

(d)   One molecule of Cl_2O_7 has 2 Cl atoms and 7 O atoms. **Six** molecules of Cl_2O_7 will have six times as many atoms of each element, or 12 Cl atoms and 42 O atoms.

SAMPLE PROBLEM 1.8 C

▶ ▶ Which of the following formulas represent elements and which represent compounds?

(a) NaCl   (b) P_4 (c) SF_6 (d) N_2 (e) CH_4 (f) KNO_3 (g) $HC_2H_3O_2$

Solution:

HINT: • A formula for an element has only **one** symbol in it, regardless of the subscript.

            • A formula for a compound **must** have the symbols of **two** or more elements in it, regardless of the subscripts.

The formulas of **elements** (have symbol of only one element) are: (b) and (d).

The formulas of **compounds** (have symbols of two or more elements) are:

(a), (c), (e), (f), (g).

Ions

Ions are **charged** particles formed by the **transfer of electrons** from one element or combination of elements to another element or combination of elements. Ion formation can be summarized by:

• The metals nearest the left side of the periodic table, Groups IA and IIA, are so reactive because they easily **give up** (transfer) electrons to form **positively charged ions** called **cations**;

• The non-metals in Groups VIA and VIIA are so reactive because they readily **gain** electrons (by transfer) to form **negatively charged ions** called **anions**;

• When atoms of metals in Groups IA and IIA react with atoms of non-metals in Groups VIA and VIIA, enough electrons are transferred to give each atom the **same number of electrons as the noble gas with the closest atomic number**. These charged particles are called **ions**. The electrons given up by the metals to form cations **are transferred to** the non-metals which gain electrons and form anions.

- Oppositely charged ions attract and form **ionic compounds,** which must be electrically **neutral,** i.e., the sum of the positively charged ions must equal the sum of the negatively charged ions.

- The **formula** for an **ionic compound** is written with the **cation** first (which is the element closest to the lower left-hand side of the periodic table, just like molecular compounds), followed by the **anion.** The subscripts in the formula indicate the **lowest ratio** of positive to negative ions which gives an electrically neutral compound. The positive and negative ions are the unit (smallest, basic) particles of **ionic compounds.**

SAMPLE PROBLEM 1.8 D

▶ ▶ Write formulas for (a) a magnesium ion, (b) a fluoride ion, (c) a strontium ion, and (d) an iodide ion.

Solution

HINT:
- To have the **same** number of electrons as atoms of the nearest noble gas elements (Group 0), atoms of:

    - Group IA elements always **lose** 1 electron (same as group number) and form 1+ ions;
    - Group IIA elements always **lose** 2 electrons (same as group number) and form 2+ ions;
    - Al in Group IIIA always **loses** 3 electrons (same as group number) and forms a 3+ ion;
    - Group VIA elements always **gain** 2 electrons and form 2− ions;
    - Group VIIA elements always **gain** 1 electron and form 1− ions.

- The charge on an ion is written as a ^{superscript} with the number always **first,** followed by the charge, e.g. Ca^{2+}.

(a) Start with an atom of magnesium, which has an atomic number of 12 (from the inside front cover of the study guide or textbook). Each **Mg atom** has **12 protons** (abbreviated p; each with a +1 charge) and **12 electrons** (abbreviated e; each with a -1 charge). Then try the following method:

$$Mg \ atom \qquad \rightarrow \qquad Mg^{2+} \ ion \quad + \quad 2e^-$$
$$[12(+1p) + 12(-1e) = 0 \ charge] \qquad [12(+1p) + 10(-1e) = +2 \ charge]$$

Since Mg is in group IIA, a Mg atom needs to give up (transfer to a non-metal) 2 electrons to have the same number of electrons as the nearest noble gas, neon (which has an atomic number of 10, meaning 10 protons and **10 electrons**). The Mg ion formed has 12 protons (+12 charge) and **10 electrons** (−10 charge), for a net charge of −2. The magnesium ion is Mg^{2+}.

(b)   When a non-metal forms an anion, the last syllable of the name of the element is replaced by -ide. The fluoride ion means you start with an atom of fluorine and form a fluoride ion. Fluorine is in group VIIA and has 9 protons and 9 electrons (atomic number is 9). Using the hint above (or the discussion in part (a)), **atoms of group VIIA elements always form 1- ions**, because they only need to gain 1 electron to have the same number of electrons as an atom of the nearest noble gas element. The fluoride ion is F^-.

(c)   Strontium is in group IIA. Atoms of all the elements in group IIA always form **2+** ions, because they give up (transfer) two electrons in order to have the same number of electrons as atoms of the nearest noble gas element. The strontium ion is Sr^{2+}.

(d)   The iodide ion is formed from an atom of iodine. Iodine is in group VIIA. Atoms of all the elements in group VIIA always form **1-** ions, because they gain one electron (by transfer from atoms of metals) in order to have the same number of electrons as atoms of the nearest noble gas element. The iodide ion is I^-.

SAMPLE PROBLEM 1.8 E

▶ ▶ Classify each of the following as either a molecule, a cation, or an anion:

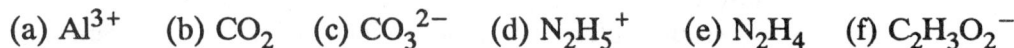

(a) Al^{3+} (b) CO_2 (c) CO_3^{2-} (d) $N_2H_5^+$ (e) N_2H_4 (f) $C_2H_3O_2^-$

Solution:

Cations are **positively** charged ions; **anions** are **negatively** charged ions; **molecules** have an overall charge of **zero** (electrically neutral) and show no charge.

The molecules are the ones with **no** charge: (b) and (e).
The cations are: (a) and (d).
The anions are: (c) and (f).

SAMPLE PROBLEM 1.8 F

▶ ▶ Complete the following table with the correct formula for each ionic compound.

Name	Unit Particles	Formula
sodium iodide	Na^+ and I^-	
strontium fluoride	Sr^{2+} and F^-	
aluminum sulfide	Al^{3+} and S^{2-}	
gallium bromide	Ga^{3+} and Br^-	
potassium sulfide		
magnesium nitride		

Solution:

HINT: In order to write the correct formula for ionic compounds:

- You must know the symbol and charge of each ion. If only the name of the compound is given, you should be able to find the symbol (from the inside front cover of this study guide or the textbook) and determine the charge (see Hint in Sample Problem 1.8 D).

- Combine the **lowest number** of cations and anions so that the overall charge adds up to **zero**. The subscripts are rarely higher than 3.

Name	Unit Particles	Formula
sodium iodide	Na^+ and I^-	NaI $1(+1\ Na) + 1(-1\ I) = 0$
strontium fluoride	Sr^{2+} and F^-	SrF_2 $1(+2\ Sr) + 2(-1\ F) = 0$
aluminum sulfide	Al^{3+} and S^{2-}	Al_2S_3 $2(+3\ Al) + 3(-2\ S) = 0$
gallium bromide	Ga^{3+} and Br^-	$GaBr_3$ $1(+3\ Ga) + 3(-1\ Br) = 0$
potassium sulfide	K^+ (Grp IA) and S^{2-} (Grp VIA)	K_2S $2(+1\ K) + 1(-2\ S) = 0$
magnesium nitride	Mg^{2+} (Grp IIA) and N^{3-} (Grp VA)	Mg_3N_2 $3(+2\ Mg) + 2(-3\ N) = 0$

NOTE:
- Al in Group IIIA always form 3+ ions by giving up 3 electrons in order to have the same number of electrons as atoms of the nearest noble gas element.

- Atoms of **group VA elements form 3- ions** by gaining 3 electrons in order to have the same number of electrons as atoms of the nearest noble gas elements.

Polyatomic Ions

Monatomic ions are formed from one atom (e.g., Na^+, O^{2-}). **Polyatomic ions** are formed from more than one atom (e.g., SO_4^{2-}, OH^-). **The name, formula, and charge of the polyatomic ions in the following table** (which is Table 1.2 in your textbook) **must be memorized in order to write correct formulas for compounds containing these ions.**

Common Polyatomic Ions[*]

Name	Formula	Name	Formula
Acetate ion	$C_2H_3O_2^-$	Nitrate ion	NO_3^-
Ammonium ion	NH_4^+	Perchlorate ion	ClO_4^-
Carbonate ion	CO_3^{2-}	Phosphate ion	PO_4^{3-}
Cyanide ion	CN^-	Silicate ion	SiO_4^{4-}
Hydrogen carbonate ion (Bicarbonate ion)	HCO_3^-	Sulfate ion	SO_4^{2-}
Hydroxide ion	OH^-	Sulfite ion	SO_3^{2-}

[*] A more complete list of anions is given in Appendix B of the textbook.

SAMPLE PROBLEM 1.8 G

▶▶ (a) Write formulas for the hydrogen phosphate ion and the phosphite ion. (b) If the formula for the arsenate ion is AsO_4^{3-}, what are the formulas for the hydrogen arsenate ion and the arsenite ion?

Solution:

HINT:
- To write the formula of a polyatomic ion which starts with the word "hydrogen", start with the formula and charge of the polyatomic ion without the hydrogen. Then place a H at the beginning of the formula and show one less unit of negative charge on the new polyatomic ion.

- To write the formula of a polyatomic ion with the first word ending in "ite", start with the formula and charge of the polyatomic ion that ends in "ate" and decrease the number of oxygen atoms by one. The charge on the ion stays the same.

(a) Using the first hint to write the formula for hydrogen phosphate, start with the formula for phosphate, PO_4^{3-}. Then place a hydrogen at the beginning of the formula and show one less negative charge. The correct formula for the hydrogen phosphate ion is HPO_4^{2-}.

Using the second hint to write the formula for the phosphite ion, start with the formula for the phosphate polyatomic ion, PO_4^{3-}, and decrease the number of oxygen atoms by one. The charge does **not** change. The correct formula for the phosphite ion is PO_3^{3-}.

(b) Similarly, to write the formula for hydrogen arsenate, start with the formula for arsenate, AsO_4^{3-}, add a H to the beginning of the formula, and show one less unit of negative charge. The correct formula for the hydrogen arsenate ion is $HAsO_4^{2-}$.

To write the formula for the arsenite ion, start with the formula for the arsenate ion, AsO_4^{3-}, and decrease the number of oxygen atoms by one. The correct formula for the arsenite ion is AsO_3^{3-}.

SAMPLE PROBLEM 1.8 H

▶▶ Write the formulas for the following ionic compounds using the hints given in Sample Problems 1.8 D, F, and G.

NAME	UNIT PARTICLES	FORMULA
Sodium carbonate		
Calcium cyanide		
Magnesium silicate		
Aluminum hydroxide		
Strontium bicarbonate		
Ammonium acetate		
Rubidium sulfate		
Potassium phosphite		

Solution:

HINT: • If there are two or more of a particular polyatomic ion in the formula of a compound, parentheses must enclose the entire polyatomic ion, followed by a subscript.

NAME	UNIT PARTICLES	FORMULA
Sodium carbonate	Na^+ (Grp IA) and CO_3^{2-} (polyatomic ion)	Na_2CO_3 $2(+1\ Na) + 1(-2\ CO_3) = 0$
Calcium cyanide	Ca^{2+} (Grp IIA) and CN^- (polyatomic ion)	$Ca(CN)_2$ $1(+2\ Ca) + 2(-1\ CN) = 0$
Magnesium silicate	Mg^{2+} (Grp IIA) and SiO_4^{4-} (polyatomic ion)	Mg_2SiO_4 $2(+2\ Mg) + 1(-4\ SiO_4) = 0$
Aluminum hydroxide	Al^{3+} (Grp IIIA) and OH^- (polyatomic ion)	$Al(OH)_3$ $1(+3\ Al) + 3(-1\ OH) = 0$
Strontium bicarbonate	Sr^{2+} (Grp IIA) and HCO_3^- (polyatomic ion)	$Sr(HCO_3)_2$ $1(+2\ Sr) + 2(-1\ HCO_3) = 0$
Ammonium acetate	NH_4^+ (polyatomic ion) and $C_2H_3O_2^-$ (polyatomic ion)	$NH_4C_2H_3O_2$ $1(+1\ NH_4) + 1(-1\ C_2H_3O_2) = 0$
Rubidium sulfate	Rb^+ (Grp IA) and SO_4^{2-} (polyatomic ion)	Rb_2SO_4 $2(+2\ Rb) + 1(-2\ SO_4) = 0$
Potassium phosphite	K^+ (Grp IA) and PO_3^{3-} (polyatomic ion)	K_3PO_3 $3(+1\ K) + 1(-3\ PO_3) = 0$

SECTION 1.9   NAMING INORGANIC COMPOUNDS

After completing this section, you will be able to do the following:

- Given the formula of a compound consisting of a metal and a non-metal **or** two non-metals, write the correct name of the compound, and **vice versa**.

REVIEW

When naming inorganic compounds that consist of a metal and a non-metal, the **same element** appears **first** in both the name and corresponding formula (that is, the element closest to the lower left-hand corner of the periodic table). The second element will be a common non-metal from the following table (which is Table 1.3 in your textbook).

Names and Formulas of Common Monatomic Anions

Name	Formula	Name	Formula
bromide	Br^-	iodide	I^-
chloride	Cl^-	nitride	N^{3-}
fluoride	F^-	oxide	O^{2-}
hydride	H^-	sulfide	S^{2-}

Given the formula of any compound containing two elements that you are expected to name, or the name of any compound containing two elements that you are expected to write the formula for, use the following procedure: If the **first element** in the compound is:

1. a <u>metal</u> from Groups IA, IIA, or Al, **name the compound** by giving the complete name of the metal **first**, followed by the name of the common non-metal with its last syllable changed to **-ide**; to write the **formula from the name**, you need to know the symbols and charges for both elements in the compound (the procedure is identical to that given in Sample Problem 1.8 F).

2. a <u>metal</u> **NOT** in Groups IA, IIA, or Al, **name the compound** by giving the complete name of the metal **first** with the charge on the ion shown as a **roman numeral in parentheses** right **after** it (see Appendix B in the textbook for writing Roman numerals), followed by the name of the common non-metal with its last syllable changed to **-ide**; to write the **formula from the name**, you need to know the symbol and charge for both elements in the compound. The roman numeral after the name of the metal **is** the charge on the metal (add a +), and the charge on the non-metal you should have memorized from the table in this section; then follow the procedure in Sample Problem 1.8 F.

3. a <u>non-metal</u>, **name the compound** by starting with the complete name of the first non-metal in the formula, followed by the name of the second non-metal with its last syllable changed to **-ide**; then, before the name of each element, use a prefix (see Table 1.4 of the textbook) that means the same as the **subscript** for that element (except **mono-** is **NOT** used for the **first** element in the name). To write the **formula from the name**, you need to know the symbol for both elements in the compound; the **subscripts are the numerical values of the prefixes.**

SAMPLE PROBLEM 1.9 A

▶▶ Complete the following table by using the three-step procedure outlined above.

Formula/Name	First Element	Procedure #	Name/Formula
SO_3			
Al_2O_3			
SnS_2			
N_2O			
MnF_2			
Chromium(VI) oxide			
Gold(III) chloride			
Phosphorus pentachloride			
Magnesium oxide			
Dichlorine trioxide			

Solution:

Formula/Name	First Element	Procedure #	Name/Formula
SO_3	non-metal	3	sulfur trioxide (no prefix for 1 atom of **first non-metal**; 3 = tri)
Al_2O_3	metal IA, IIA, Al	1	aluminum oxide (no roman numeral needed for metal in Grps IA, IIA, Al)
SnS_2	metal **NOT** IA, IIA, Al	2	tin(IV) sulfide (roman numeral needed to indicate charge on **metal NOT** from Grps IA, IIA, Al)
N_2O	non-metal	3	dinitrogen monoxide (mono- used for 1 atom of **second** element only)
MnF_2	metal **NOT** IA, IIA, Al	2	manganese(II) fluoride
Chromium(VI) oxide	metal **NOT** IA, IIA, Al	2	CrO_3 1(+6 Cr) + 3(-2 O) = 0
Gold(III) chloride	metal **NOT** IA, IIA, Al	2	$AuCl_3$ 1(+3 Au) + 3(-1 Cl) = 0
Phosphorus pentachloride	non-metal	3	PCl_5 1P; penta = 5 Cl (no prefix means 1 for non-metal)
Magnesium oxide	metal IA, IIA, Al	1	MgO 1(+2 Mg) + 1(-2 O) = 0
Dichlorine trioxide	non-metal	3	Cl_2O_3 di = 2 Cl; tri = 3 O

SECTION 1.10   CHEMICAL EQUATIONS

After completing this section, you will be able to do the following:

■ Given the names or formulas of substances undergoing a chemical change (chemical reaction), be able to (a) identify the reactants and products, and (b) balance the chemical reaction.

REVIEW

When a chemical change occurs, one or more new substances are formed. The substances taking part in this change are described by a **chemical reaction**, consisting of words or (most often) chemical formulas. For example, when carbon (charcoal) completely burns, it reacts with oxygen gas from the air to produce carbon dioxide gas.

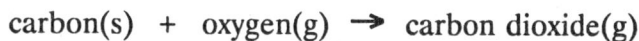

$$\text{carbon(s)} + \text{oxygen(g)} \rightarrow \text{carbon dioxide(g)}$$

You need to be able to convert these words into formulas and write a chemical reaction as shown (remember that oxygen is one of the elements that exists as a diatomic gas; see Section 1.8 of this study guide):

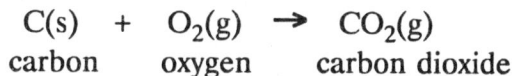

$$\underset{\text{carbon}}{C(s)} + \underset{\text{oxygen}}{O_2(g)} \rightarrow \underset{\text{carbon dioxide}}{CO_2(g)}$$

The **reactants** are the starting substances, C and O_2, which are usually shown on the left-hand side of the reaction. The **product(s)** are the new substance(s) formed in the reaction, CO_2, which are usually shown on the right side. The **arrow** indicates the direction of the change, from reactants to products. The **letters** s, l, g, and aq are used to indicate whether the substance is a solid, liquid, gas, or dissolved in an aqueous (water) solution.

A **balanced chemical reaction** is called a **chemical equation**, which you will always need later in this course before you can do any calculations involving a reaction. A chemical reaction is balanced **only if there are the same number of <u>atoms</u> of each element on both sides of the reaction** (reactant side and product side). That is, matter can be neither created nor destroyed; this will be true only if the masses of all the products **equal** the masses of all the reactants --- which is **always** true **only if the reaction is balanced**.

In the above reaction, **one** atom of C reacts with **one** molecule of O_2 to form **one** molecule of CO_2 (no number or coefficient in front of a formula always means 1). This chemical reaction is already balanced with coefficients of 1 (1 C atom and 2 O atoms on each side of the arrow); therefore, it is a chemical equation. **Remember, atoms (NOT molecules) of each element must be balanced (no matter how complicated the formulas may be).** If the reaction is not balanced with coefficients of one (and most will not), then you balance the reaction **only** by changing the coefficients to higher integers; you **NEVER** change the subscripts of a formula.

60

A summary of the steps for balancing a chemical reaction to obtain a chemical equation are:

1. Write the correct formulas for the reactants and products;
2. Start with the most complicated looking formula and balance the elements in it by changing the coefficients in front of other formulas, **leaving O and H until last. NEVER CHANGE SUBSCRIPTS IN ANY FORMULA;**
3. When balanced, you will have the **lowest, whole-numbered coefficients** that will give you the same number of atoms (NOT molecules) of each element on both sides of the reaction;
4. Add letters s, l, g, aq, to indicate (if known) solid, liquid, gas, or aqueous solution.

SAMPLE PROBLEM 1.10 A

▶▶ Given the equation,

$$3BaCl_2(aq) + 2K_3PO_4(aq) \rightarrow 6KCl(aq) + Ba_3(PO_4)_2(s)$$

(a) list the formulas of the reactants and products; (b) how many atoms of each element appear on each side of the equation?; (c) which substances are solids, liquids, gases, or in aqueous solution?

Solution:

(a) The reactants are the starting materials and are written next to the "tail" of the arrow (which usually is the left side); the products are the resulting materials and are written next to the "head" of the arrow (which is usually the right side). The reactants are $BaCl_2(aq)$, and $K_3PO_4(aq)$, and the products are $KCl(aq)$ and $Ba_3(PO_4)_2(s)$.

(b) To quickly calculate the number of atoms of an element in a formula, multiply the coefficient times the subscript of that element. There are 3 Ba, 6Cl, 6K, 2P, and 8O on each side of the equation. **Be careful when counting atoms in formulas containing two or more polyatomic ions.** $Ba_3(PO_4)_2$, with two PO_4^{3-} polyatomic ions, can be visualized as $Ba_3PO_4PO_4$. To calculate the number of atoms from this polyatomic ion:

coefficient	x	subscript of the element	x	subscript of the polyatomic ion.	= # atoms

For P: 1   x   1   x   2   =   2 P atoms.
For O: 1   x   4   x   2   =   8 O atoms.

(c) There is no (l) or (g) after any formula, so there are no liquids or gases. $Ba_3(PO_4)_2(s)$ is a solid, and $BaCl_2(aq)$, $K_3PO_4(aq)$, and $KCl(aq)$ are in aqueous (water) solution.

SAMPLE PROBLEM 1.10 B

▶▶ Balance each of the following reactions:

(a) $Na_2O(s)$ + $H_2O(l)$ → $NaOH(aq)$

(b) $H_2CO_3(aq)$ + $NH_4OH(aq)$ → $(NH_4)_2CO_3(aq)$ + $H_2O(l)$

(c) $Al_2(SO_4)_3(aq)$ + $Pb(NO_3)_2(aq)$ → $PbSO_4(s)$ + $Al(NO_3)_3(aq)$.

Solution:

HINT: ● Use the 4-step procedure given above.

(a) Step 1: Write down the formulas correctly and leave off the (s), (aq), etc., until the balancing is finished to avoid making errors.

Step 2: Start balancing the atoms of the element Na, always leaving H and O to last.

$$Na_2O + H_2O → 2NaOH$$

Now check H and O; they are also balanced.

Step 3: The reaction is balanced with 2 Na, 2H, and 2O on each side of the equation. The coefficients (1,1,2) are the lowest whole numbers possible.

Step 4: Add the letters s, l, and aq as shown in the statement of the problem to give the complete balanced reaction (chemical equation)

$$Na_2O(s) + H_2O(l) → 2NaOH(aq)$$

(b) Step 2: Start with 1 molecule of what appears to be the most complicated formula, $(NH_4)_2CO_3$, and balance the atoms of N or C (leaving H and O to last). The 2 N on the right require a 2 in front of the NH_4OH on the left to balance the N. The carbons are balanced at 1 on each side.

$$H_2CO_3 + 2NH_4OH \rightarrow (NH_4)_2CO_3 + H_2O$$

\# atoms: 2 H, 3 O + 10 H, 2 O $\rightarrow$ 8 H, 3 O + 2 H, 1 O

There are now 12 H on the left and only 10 H on the right. You can increase the number of H on the **right side** by increasing the coefficient in front of either formula, but try to avoid changing the coefficient in front of a formula in which you have already balanced some of the atoms. (If you change the coefficient to 2 in front of $(NH_4)_2CO_3$, you will have to go back and re-balance the atoms of N and C). Place a 2 in front of the H_2O and the H's will be balanced. Now check the O's; they too are balanced.

Step 3: The reaction is balanced with 2 N, 1 C, 12 H, and 5 O on each side. The coefficients (1,2,1,2) are the lowest whole numbers possible.

Step 4: The balanced reaction (chemical equation) is:

$$H_2CO_3(aq) + 2NH_4OH(aq) \rightarrow (NH_4)_2CO_3(aq) + 2H_2O(l)$$

(c) Step 2: Start with 1 molecule of what appears to be the most complicated formula (it could be any of them in this reaction), and balance the atoms of Al, S, Pb, and N (leaving H and O to last). If we start with one unit of $Al_2(SO_4)_3$, the 2 Al and 3 S will require a 2 in front of the $Al(NO_3)_3$ and a 3 in front of $PbSO_4$ on the right side. Checking for Pb, the 3Pb in $PbSO_4$ will require a 3 in front of $Pb(NO_3)_2$. And now the N's are balanced at 6 on each side. The only element remaining to be checked is O, which is already balanced with 30 O atoms on each side. [By the time you get to checking H and/or O, the coefficients will often already have been determined in the process of balancing the atoms of the other elements.]

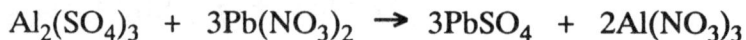

$$Al_2(SO_4)_3 + 3Pb(NO_3)_2 \rightarrow 3PbSO_4 + 2Al(NO_3)_3$$

\# atoms: 12 O   18 O   12 O   18 O

Check to make sure that you agree with these numbers of O atoms. ·

Step 3: The reaction is balanced with 2 Al, 3 S, 3 Pb, 6 N, and 30 O on each side. The coefficients (1,3,3,2) are the lowest whole numbers possible.

Step 4:   The balanced reaction (chemical equation) is:

$$Al_2(SO_4)_3(aq) + 3Pb(NO_3)_2(aq) \rightarrow 3PbSO_4(s) + 2Al(NO_3)_3(aq).$$

SECTION 1.11 PREDICTING REACTIONS

After completing this section, you will be able to do the following:

■   Given a type of reaction for one element, predict a similar reaction for other elements in the same family (group).

REVIEW

Sodium metal reacts with water to form sodium hydroxide and hydrogen gas. The equation is:

$$2Na(s) + 2H_2O(l) \rightarrow 2NaOH(aq) + H_2(g)$$

Since elements in the same group of the periodic table have similar chemical properties, you can predict that a similar equation can be written for $Li(s)$, $K(s)$, $Rb(s)$, and $Cs(s)$.

SAMPLE PROBLEM 1.11 A

▶ ▶ When Mg metal burns, it reacts very rapidly and brightly with oxygen from the air to form solid magnesium oxide. (a) Write the equation for this reaction, and (b) write equations that you would predict to take place when calcium, strontium, and barium metals are burned.

Solution:

(a)   Write the reaction using formulas for the substances given, and balance it.

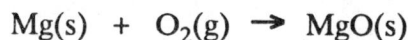

$$Mg(s) + O_2(g) \rightarrow MgO(s)$$

The Mg atoms are balanced. To balance the 2 O atoms on the left will require a 2 in front of MgO. This in turn will require a 2 in front of Mg(s) to balance the Mg atoms. The balanced reaction is:

$$2Mg(s) + O_2(g) \rightarrow 2MgO(s)$$

(b)   For the burning of Mg, the **type reaction** is (where IIA is any group IIA element):

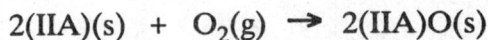

$$2(IIA)(s) + O_2(g) \rightarrow 2(IIA)O(s)$$

This same type reaction can be predicted for Ca, Sr and Ba metals.

$$2Ca(s) + O_2(g) \rightarrow 2CaO(s)$$

$$2Sr(s) + O_2(g) \rightarrow 2SrO(s)$$

$$2Ba(s) + O_2(g) \rightarrow 2BaO(s)$$

SELF-TEST QUESTIONS

Test your understanding of the concepts and skills in this chapter by working through the multiple choice questions **BEFORE** checking the answers. Remember, the information in the tables on polyatomic ions (Section 1.8) and monatomic anions (Section 1.9) must be **memorized**. A periodic table will always be available. The ANSWERS TO SELF-TEST QUESTIONS follow immediately after this section.

1.   Which of the following properties of potassium metal is not a physical property?

A. melting point of 63° C
B. a soft solid substance
C. boiling point of 757° C
D. conducts electricity
E. reacts with water to form hydrogen gas

2.   Which is the only example classified correctly below?

A. sugar dissolved in water (heterogeneous mixture)
B. H_2 (compound)
C. Pb (compound)
D. air (homogeneous mixture)
E. soda water (element)

3. If an atom of an element has an atomic number of 16,

   A. it is an atom of oxygen.
   B. the nucleus has a charge of +16.
   C. it has 8 electrons outside the nucleus.
   D. the atom has an overall charge of +16.
   E. it has 16 protons outside the nucleus.

4. The correct symbol for the element silver is

   A. Si   B. Sl   C. Ag   D. Au   E. Se

5. All of the elements listed have correct symbols except for

   A. strontium (Sr)   B. arsenic (Ar)   C. antimony (Sb)   D. lead (Pb)   E. sodium (Na)

6. In the periodic table,

   A. He is located in period 2 and group 0.
   B. Sn is located in the same period as Pb.
   C. Ca is a non-metal in group IIA.
   D. V and Fe are both transition elements located in period 4.
   E. Br is a non-metal, main group element in group IVA.

7. In three molecules of N_2O_5, there are

   A. 2 N atoms and 5 O atoms.
   B. 3 N atoms and 3 O atoms.
   C. 6 N atoms and 15 O atoms.
   D. 15 N atoms and 6 O atoms.
   E. 7 atoms total of N and O.

8. All of the following formulas represent elements except

   A. Na   B. Cu   C. CO   D. S_8 E. Si

9. The only correct ion is

   A. K^{2+} B. O^- C. Ca^{2+} D. Se^{3-} E. Ba^{3+}

10. The correct formula for lithium bromide is

A. LaBr   B. $LiBr_2$ C. LeBr   D. Li_2Br E. LiBr

11. The correct formula for barium nitrate is

A. $Ba(NO_3)_2$ B. $Ba(NO_2)_2$ C. $B(NO_3)_2$ D. $BaNO_3$ E. Ba_2NO_3

12. The correct name for Cu_2SO_4 is

A. copper sulfate
B. cobalt(II) sulfate
C. copper disulfate
D. copper(I) sulfite
E. copper(I) sulfate

13. The correct formula for dinitrogen pentoxide is

A. NO   B. N_2O_5 C. N_2O_4 D. NiO_5 E. N_5O_2

14. When the following reaction is balanced with the lowest whole numbers possible, the coefficient of O_2 is

$$C_3H_8(g) + O_2(g) \rightarrow CO_2(g) + H_2O(l)$$

A. 1   B. 2   C. 3   D. 4   E. 5

15. When the following reaction is balanced with the lowest whole numbers possible, the coefficient of H_2O is

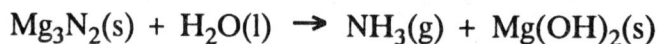

$$Mg_3N_2(s) + H_2O(l) \rightarrow NH_3(g) + Mg(OH)_2(s)$$

A. 3   B. 2   C. 4   D. 1   E. 6

ANSWERS TO SELF-TEST QUESTIONS

1. e	6. d	11. a
2. d	7. c	12. e
3. b	8. c	13. b
4. c	9. c	14. e
5. b	10. e	15. e

PRACTICE PROBLEMS

Test your problem solving skills in this chapter by working through the problems in the space provided **BEFORE** checking with the answers. A periodic table will always be available. The ANSWERS TO PRACTICE PROBLEMS follows immediately after this section.

1. Relative to liquids and gases, why does a solid have both a shape and volume of its own?

2. What is the difference between a physical change and a chemical change?

3. What is the difference between an element, a compound and a mixture? Give an example of each.

68

4.   How many atoms of each element are in (a) four molecules of SO_3, (b) six molecules of N_2O_5, and (c) ten molecules of $C_{12}H_{22}O_{11}$?

5.   Balance each of the following reactions:

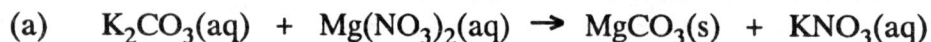

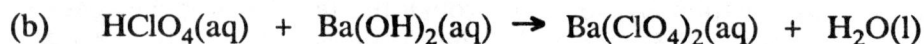

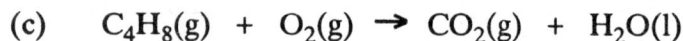

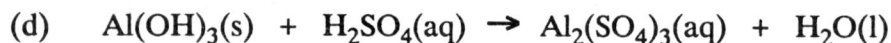

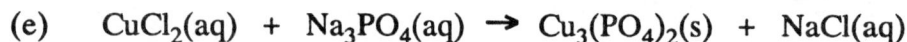

   (a) $K_2CO_3(aq) + Mg(NO_3)_2(aq) \rightarrow MgCO_3(s) + KNO_3(aq)$

   (b) $HClO_4(aq) + Ba(OH)_2(aq) \rightarrow Ba(ClO_4)_2(aq) + H_2O(l)$

   (c) $C_4H_8(g) + O_2(g) \rightarrow CO_2(g) + H_2O(l)$

   (d) $Al(OH)_3(s) + H_2SO_4(aq) \rightarrow Al_2(SO_4)_3(aq) + H_2O(l)$

   (e) $CuCl_2(aq) + Na_3PO_4(aq) \rightarrow Cu_3(PO_4)_2(s) + NaCl(aq)$

ANSWERS TO PRACTICE PROBLEMS

1.   Relative to a liquid and a gas, the particles in a solid are closer together and more strongly attracted to each other. As a result, the particles in a solid vibrate about fixed positions arranged in the most orderly pattern which gives it a shape and a volume.

2.   When a substance undergoes aq physical change, there is no change in composition of the substance (i.e., melting, boiling, evaporation, cutting into smaller pieces, etc.). When a substance undergoes a chemical change, a new substance is formed with different physical and chemical properties than the starting substance (i.e., burning, rusting, decomposing, etc.).

3. An element is the simplest substance known, i.e., cannot be broken down into simpler substances by chemical reactions. Hydrogen, oxygen, carbon, iron, copper, silver and gold are seven examples of the 109 known elements. A compound is composed of two or more elements chemically combined, i.e., the formula of a compound contains the symbols of two or more elements. Examples of compounds are water (H_2O), carbon dioxide (CO_2) and ethyl alcohol (C_2H_5OH). A mixture consists of two or more substances physically mixed together (but not chemically combined); this means components of a mixture can be separated by a physical change. Examples of mixtures are air, salt water, salad dressings, wood, concrete, etc.

4. (a)   4 S atoms and 12 O atoms;

    (b)   12 N atoms and 30 O atoms;

    (c)   120 C atoms, 220 H atoms and 110 O atoms.

5. (a) $K_2CO_3(aq) + Mg(NO_3)_2(aq) \rightarrow MgCO_3(s) + 2KNO_3(aq)$

    (b) $2HClO_4(aq) + Ba(OH)_2(aq) \rightarrow Ba(ClO_4)_2(aq) + 2H_2O(l)$

    (c) $C_4H_8(g) + 6O_2(g) \rightarrow 4CO_2(g) + 4H_2O(l)$

    (d) $2Al(OH)_3(s) + 3H_2SO_4(aq) \rightarrow Al_2(SO_4)_3(aq) + 6H_2O(l)$

    (e) $3CuCl_2(aq) + 2Na_3PO_4(aq) \rightarrow Cu_3(PO_4)_2(s) + 6NaCl(aq)$

CHAPTER 2

MEASUREMENT

SECTION 2.2   CONVERTING UNITS

After completing this section, you will be able to do the following:

■ Write the symbols from the name of SI units, and vice versa;

■ Use SI prefixes to determine the numerical relationship between derived SI units and the base SI unit;

■ Given a measurement in either SI or English units, convert it to other SI or English units.

REVIEW

SI Prefixes

To perform many of the calculations in this course, you will need to **memorize** the information about SI base units and SI prefixes in dark type in Tables 2.1 and 2.2 of the textbook. You will also find that **exponential notation** is a much more convenient way of writing numbers, particularly very large and very small numbers. **See Appendix A.2** for a discussion and examples of using **exponential notation**. Algebraic equations will frequently be used throughout this course for solving numerical problems. See **Appendix A.4** for a discussion and examples of **Solving Equations**.

72

SAMPLE PROBLEM 2.2 A

▶▶ Complete the following table.

Name	Symbol
kilometer	
centigram	
nanosecond	
	pg
	mmol
	kA

Solution:

Using the memorized data from Tables 2.1 and 2.2 in the textbook, replace the name of the prefix with the symbol and vice versa.

Name	Symbol
kilometer	km
centigram	cg
nanosecond	ns
picogram	pg
millimole	mmol
kiloampere	kA

SAMPLE PROBLEM 2.2 B

▶▶ (a) One milligram = ? grams. How many milligrams = 1 gram?
(b) One kilometer = ? meters.
(c) One nanosecond = ? seconds. How many nanoseconds = 1 second?

Solution:

(a) Replace milli- with its numerical value and you have the answer.

$$1 \text{ mg} = 10^{-3} \text{ g.}$$ [1 mg is **smaller** than 1 g.]

To find how many milligrams = 1 gram, you need to solve this equation for 1 gram by dividing both sides of the equation by 10^{-3}. This gives you

$$\frac{1 \text{ mg}}{10^{-3}} = \frac{10^{-3} \text{ g}}{10^{-3}} \quad \text{and} \quad 10^3 \text{ mg} = 1 \text{ g} \qquad \text{[It takes 1000 mg to equal 1 g.]}$$

(b) Replace kilo- with its numerical value. 1 kilometer = 10^3 meters.

(c) Replace nano- with its numerical value.

$$1 \text{ ns} = 10^{-9} \text{ s}$$ [1 ng is much **smaller** than 1 g.]

To find out how many nanoseconds equals 1 second, use the same procedure as in (a).

$$\frac{1 \text{ ns}}{10^{-9}} = \frac{10^{-9} \text{ s}}{10^{-9}} \quad \text{and} \quad 10^9 \text{ ns} = 1 \text{ s} \qquad \text{[It takes 1 000 000 000 ns to equal 1 s.]}$$

Unit conversion factors

A **measurement** contains both a number and a unit, e.g., 5.4 miles, 2.1 seconds, 2.5 gallons, and 17.6 kilograms. A **unit conversion factor** is an equality which allows you to convert from one measurement unit (SI or English) to another measurement unit (SI or English). Using Table 2.2 in the textbook, you can derive all the **SI to SI** conversion factors needed. On the inside back cover of the textbook, you will find the common **English to English** and some **SI to English** conversion factors. Remember that an equality means both sides of the equation are **equal**. For example, 1 foot = 12 inches. This equality does **NOT** say that 1 = 12. What it does say is that 1 **foot** = 12 **inches**, where each inch is 1/12 of a foot.

SAMPLE PROBLEM 2.2 C

▶▶ Convert 25 g to mg. Which is larger, 1 g or 1 mg?

Solution:

HINT:

> As a **general problem-solving** method (which can be used repeatedly throughout this course),
>
> 1. Identify the **unit of what you are looking for** and the **measurement given,** and state the information in the form of a question (in writing at first, mentally only as you gain confidence).
>
> 2. Start the solution to the problem by converting the question into an equation.
>
> 3. Use conversion factors to obtain the unit that you are trying to find.

Step 1:   How many **mg** are equal to **25 g?**

Step 2:   ? (How many) mg = (are equal to) 25 g, or removing the words,

   ? mg = 25 g x (times one or more conversion factors to complete the problem).

Step 3:   From Table 2.2, 1 mg = 10^{-3} g. This is an equality, which also says 10^{-3} g = 1 mg. This equality gives you a unit conversion factor which can be used in one of two ways:

$$\frac{1 \text{ mg}}{10^{-3} \text{ g}} \qquad \text{or} \qquad \frac{10^{-3} \text{ g}}{1 \text{ mg}}$$

Think of the line separating the top and bottom of the conversion factor as an = sign. In this problem we want to convert g to mg. **Always use the conversion factor that has the unit you want to cancel on the bottom.** In this case g will cancel and you are left with mg, just what you are trying to find. Our solution now becomes

$$? \text{ mg} = 25 \text{ g} \times \frac{1 \text{ mg}}{10^{-3} \text{ g}} = 25\ 000 \text{ mg} = 2.5 \times 10^{4} \text{ mg}$$

CHECK: Does this answer make sense? Yes, because 1 mg is only 10^{-3} or 1/1000 of 1 g. 1 gram is larger than 1 mg. It takes 1000 mg to weigh the same as 1 g. Therefore, 25 g will be 25 000 mg.

You will quickly know if you have used the wrong form of the conversion factor. In this case, you would end up with

$$? \text{ mg} = 25 \text{ g} \times \frac{10^{-3} \text{ g}}{1 \text{ mg}} = 2.5 \times 10^{-2} \text{ g}^2/\text{mg}$$

No units would cancel **and they must** if you are to solve for mg.

SAMPLE PROBLEM 2.2 D

▶▶ Convert 4.50 km to mm.

Solution:

Using the general problem-solving approach in Sample Problem 2.2 C,

Step 1:   How many mm are equal to 4.50 km?

Step 2:   ? mm = 4.50 km x ?

Step 3:   From Table 2.2, 1 km = 10^3 m. The correct form of this conversion factor has **km on the bottom** so that it cancels.

$$? \text{ mm} = 4.50 \text{ \st{km}} \times \frac{10^3 \text{ m}}{1 \text{ \st{km}}} \times \text{ ?}$$

So far, we have converted km to m, and you could calculate the number if necessary. But we are trying to find mm. From Table 2.2, 1 mm = 10^{-3} m. The correct form of this conversion factor has **m on the bottom so that it cancels.** Now you have obtained the unit of mm and can calculate the answer.

$$? \text{ mm} = 4.50 \text{ \st{km}} \times \frac{10^3 \text{ \st{m}}}{1 \text{ \st{km}}} \times \frac{1 \text{ mm}}{10^{-3} \text{ \st{m}}} = 4.50 \times 10^6 \text{ mm}$$

CHECK: Does this answer make sense? Yes, because 1 km is 1000 m, and 1 mm is 10^{-3} m or, equivalently, there are 1000 mm in 1 m. The number of mm in 1 km is 1 x 10^6. In 4.50 km, there will be 4.50 x 10^6 mm.

When the SI **base** unit for a quantity is not stated in the problem, usually two conversion factors will be needed (in this example, neither mm nor km are the base unit for length). Another way of looking at it is that the relative size for the units of length in the SI system are: 1 mm < 1 cm < 1 m < 1 km. Since all mm, cm, and km units are related to (derived from) the base unit m, you must convert km to m first, then m to mm.

SAMPLE PROBLEM 2.2 E

▶▶ Convert 5.0×10^3 cm^3 into m^3.

Solution:

Use the general problem-solving approach in Sample Problem 2.2 C.

Step 1: How many m^3 are equal to 5.0×10^3 cm^3?

Step 2: ? m^3 = 5.0×10^3 cm^3 x ?

Step 3: From Table 2.2, 1 cm = 10^{-2} m. This equality will always be valid **as long as you do the same mathematical operation to both sides of the equality (equation).** Thus you can square, cube, take the square root of, etc., any equality as long as it is done to **both sides** of the equality (equation), **including the units.** In this example, squaring both sides of the equality (equation),

$$(1 \text{ cm})^2 = (10^{-2} \text{ m})^2 \qquad \text{or} \qquad 1 \text{ cm}^2 = 10^{-4} \text{ m}^2.$$

And, cubing both sides of the equality (equation),

$$(1 \text{ cm})^3 = (10^{-2} \text{ m})^3 \qquad \text{or} \qquad 1 \text{ cm}^3 = 10^{-6} \text{ m}^3.$$

See Appendix A.2 of the textbook for a discussion and examples of powers and roots of exponential numbers.

The correct form of this conversion factor has **cm^3 on the bottom so that it cancels.** Now you have obtained the unit of m^3 and can calculate the answer.

$$? \text{ m}^3 = 5.0 \times 10^3 \text{ cm}^3 \times \frac{10^{-6} \text{ m}^3}{1 \text{ cm}^3} = 5.0 \times 10^{-3} \text{ m}^3$$

CHECK: Does this answer make sense? Yes, because 1 cm^3 is only 10^{-6} or 1/1 000 000 of a m^3; equivalently, it takes 1 000 000 cm^3 to equal 1 m^3. You are only starting with 5.0×10^3 or 5000 cm^3, which is less than 1 m^3.

SAMPLE PROBLEM 2.2 F

▸▸ Convert 645 g to pounds (lb).

Step 1: How many lb are equal to 645 g?

Step 2: ? lb = 645 g x ?

Step 3: Up until this point, all the examples have dealt with SI units. You also need to be able to convert from SI to English units, and vice versa. This example uses mass --- gram is an SI unit and ounce is an English unit. On the **inside back cover of the textbook**, you will find a number of factors for converting SI to English units. For this example, you need to convert from grams to pounds. Using that conversion factor so that g will cancel, you are left with lb, and can calculate the answer.

$$? \text{ lb} = 645 \text{ g} \times \frac{1 \text{ lb}}{453.6 \text{ g}} = 1.42 \text{ lb}$$

CHECK: Does this answer make sense? Yes, because it takes 453.6 g to equal 1 lb, and 645 g is between 1 and 2 pounds.

SAMPLE PROBLEM 2.2 G

▸▸ Convert 15.5 yards (yd) into cm.

Use the general problem-solving approach in Sample Problem 2.2 C.

Step 1: How many cm are equal to 15.5 yd?

Step 2: ? cm = 15.5 yd x ?

Step 3: Since you are converting units between the two systems (SI and English), use the information on the **inside back cover of the textbook**. There is no conversion factor given that converts length in yd to cm. However, there are two conversion factors given for length that allow you to go from the English to the SI system --- inches (in) to cm, and miles (mi) or feet (ft) to km. Your choices are (a) convert yd to ft, ft to in, and in to cm (three conversion factors); (b) convert yd to ft, ft to km, km to m, and m to cm (four conversion factors); or (c) convert yd to mi, mi to km, km to m, and m to cm (four conversion factors). All choices will give you exactly the same answer. Because choice (a) requires only three conversion factors, we will use that method to solve the problem. Remember, use the form of the conversion factor that will have the unit that you want to cancel **on the bottom**. The solution is

$$? \text{ cm } = 15.5 \text{ yd } \times \frac{3 \text{ ft}}{1 \text{ yd}} \times \frac{12 \text{ in}}{1 \text{ ft}} \times \frac{2.54 \text{ cm}}{1 \text{ in}} = 1.42 \times 10^3 \text{ cm}$$

CHECK: Does this answer make sense? Yes, since 1 yd is larger than 1 ft, which is larger than 1 in, which is larger than 1 cm, the number of cm in 15.5 yd must be considerably greater than 15.5.

SECTION 2.3   UNCERTAINTY IN MEASUREMENT

After completing this section, you will be able to do the following:

■ Given a measurement, indicate the number of significant figures;

■ Given a series of measurements, calculate the median and average.

REVIEW

Whenever a measurement is made, it includes a quantity, a unit, and the uncertainty in making the measurement. In a correctly recorded measurement, the last digit will be estimated or uncertain, and is counted as a **significant figure**. To count **significant digits or figures** in a measurement:

(a) all **nonzero** numbers **are** significant;
(b) **zeros** between **nonzero** numbers **are** significant;
(c) **zeros** at the **end** of a number **to the right of a decimal point are** significant;
(d) zeros at the beginning of a number are **never** significant;
(e) **zeros** at the **end** of a **whole number may** be significant; the uncertainty is removed if the number is always written in exponential form.

When making a series of measurements, you are often asked to calculate the **average** and/or **median** value. If a series of measurements is arranged in order of increasing or decreasing value, the **median** is the **middle one** of an **odd** number of measurements or the **average of the two middle measurements** for an **even** number of measurements. The **average** is obtained by adding all the individual measurements and dividing by the number of measurements.

SAMPLE PROBLEM 2.3 A

▶▶ Complete the following table by indicating how many significant figures there are in each measurement and <u>underlining</u> them.

Measurement	Significant Digits
4.701 cm	
12 064 in	
0.2700 mL	
0.000 06 g	
273.00 lb	
47 300 g	
4.73×10^4 g	
4.730×10^4	
4.7300×10^4	

Solution:

Measurement	Significant Digits
4.701 cm	4; <u>4.701</u>
12 064 in	5; <u>12 064</u>
0.2700 mL	4; 0.<u>2700</u>
0.000 06 g	1; 0.000 0<u>6</u>
273.00 lbs	5; <u>273.00</u>
47 300 g	3-5; <u>47 3</u>??
4.73×10^4 g	3; <u>4.73</u> $\times 10^4$
4.730×10^4	4; <u>4.730</u> $\times 10^4$
4.7300×10^4	5; <u>4.7300</u> $\times 10^4$

NOTE: Whole numbers ending in zero (e.g., 47 300) should **always** be written in exponential form to show the correct number of significant figures in the measurement.

SAMPLE PROBLEM 2.3 B

▶▶ Given the following measurements, what is the median and average for each set?

(a) 5.33 g, 5.34 g, 5.31 g, 5.33 g, 5.30 g.

(b) 43.8 mL, 43.1 mL, 43.5 mL, 43.1 mL.

Solution:

(a) To determine the median, arrange in order of increasing or decreasing value.

$$5.30 \text{ g, } 5.31 \text{ g, } 5.33 \text{ g, } 5.33 \text{ g, } 5.34 \text{ g.}$$

For an **odd** number of measurements, like this example, the median is the middle value. The **median** is 5.33 g.

The average is calculated by adding up all the values and dividing by the number of values (5 in this example).

$$\text{average} = \frac{5.30 + 5.31 + 5.33 + 5.33 + 5.34}{5} = 5.32 \text{ g}$$

NOTE: The median and average may not be the same value.

(b) To determine the median, arrange in order of increasing or decreasing value.

$$43.8 \text{ mL, } 43.5 \text{ mL, } 43.1 \text{ mL, } 43.1 \text{ mL}$$

For an **even** number of measurements, like this example, the **median** is the **average of the two middle values,**

$$\text{median} = \frac{43.5 + 43.1}{2} = 43.3 \text{ mL}$$

The average is calculated by adding up all the values and dividing by the number of values (4 in this example).

$$\text{average} = \frac{43.8 + 43.5 + 43.1 + 43.1}{4} = 43.4 \text{ mL}$$

SECTION 2.4	SIGNIFICANT FIGURES IN CALCULATIONS

After completing this section, you will be able to do the following:

■ Add, subtract, multiply, and divide measurements with the results expressed to the correct number of significant figures.

REVIEW

The answer to a calculation can be no more accurate than the accuracy of the individual measurements used in the calculation. This means that you must know how to round off the answer in any calculation to the correct number of significant figures (s.f.), by using the following procedures:

- For **addition and subtraction**, the number which has the **smallest number of digits to the right of the decimal point** determines the number of significant figures in the answer;

- For **multiplication and division**, the answer should be rounded to the same number of significant figures as there are in the **quantity having the smallest number of significant figures**.

SAMPLE PROBLEM 2.4 A

▶ ▶ Three objects have masses of 26.78 g, 2.4 g, and 0.2 g. (a) How many significant figures does each measurement have? (b) What is the total mass of the three objects, expressed with the correct number of significant figures?

Solution:

(a)
        26.78 g   4 significant figures (s.f.)
         2.4 g   2 significant figures (s.f.)
         0.2 g   1 significant figure (s.f.)

(b)   Sum is   29.38 g
         or   29.4 g 3 s.f.   [See the first procedure in the REVIEW.]

NOTE: For addition and subtraction, the number of significant figures in the answer is **NOT** determined by the measurement having the smallest number of significant figures, 0.2 g with 1 significant figure (which **is** the method used for multiplication and division).

SAMPLE PROBLEM 2.4 B

▶ ▶ The sides of a rectangle (measured with two different rulers) were 2.400 cm and 3.0 cm. Calculate the area of the rectangle, expressed with the correct number of significant figures.

Solution:

HINT: Unlike addition and subtraction, you **can** determine the number of significant figures in your answer before beginning the calculation.

2.400 cm has 4 significant figures and 3.0 cm has 2 significant figures. The answer will have **two** significant figures (see the second procedure before Sample Problem 2.4 A).

Area = length x width = 2.400 cm x 3.0 cm = 7.2000 cm^2 = 7.2 cm^2 (2 s.f.)

SECTION 2.8   DENSITY

After completing this section, you will be able to do the following:

■ Given two of the three quantities density, mass, and volume for any substance, calculate the third quantity.

REVIEW

The density of a material is defined as the mass per unit volume, or $D = M/V$. **Density** is an **intensive property** of a substance, which means that it is **independent** of the quantity of the sample (note that both words start with "in-"). **Mass** and **volume** are **extensive properties** of a substance, which means that they **depend** on the quantity of sample measured.

SAMPLE PROBLEM 2.8 A

▶ ▶ What is the density for a piece of iron which has a mass of 43.2 g and a volume of 5.5 cm^3?

Using the definition for density and substituting the given quantities for mass and volume,

$$D \;=\; \frac{M}{V} \;=\; \frac{43.2 \text{ g Fe}}{5.5 \text{ cm}^3 \text{ Fe}} \;=\; \frac{7.854\ 545\ 5 \text{ g } \cancel{\text{Fe}}}{\text{cm}^3 \; \cancel{\text{Fe}}} \;=\; \frac{7.9 \text{ g}}{\text{cm}^3} \qquad \text{[Only 2 s.f. in the answer.]}$$

SAMPLE PROBLEM 2.8 B

▸▸ What is the mass of a block of aluminum that has measurements of 2.0 cm x 4.0 cm x 8.0 cm? [The density of aluminum is 2.7 g/cm^3.]

HINT: There are two methods that can be used to solve a problem when a formula is given or known: (a) the formula method, or (b) the conversion factor method. Both methods give you identical answers.

Method (a): Solve the density formula for mass and substitute in the values for density and volume. [Rearranging algebraic equations is discussed with examples in Appendix A.4 of the textbook.]

$$D \;=\; \frac{M}{V}$$

and

$$M = D \times V = \frac{2.7 \text{ g Al}}{\text{cm}^3 \text{ Al}} \times (2.0 \times 4.0 \times 8.0) \text{ cm}^3 \text{ Al}$$

$$= \frac{2.7 \text{ g Al}}{\cancel{\text{cm}^3 \text{ Al}}} \times 64 \; \cancel{\text{cm}^3 \text{ Al}} = 172.8 \text{ g} = 1.7 \times 10^2 \text{ g Al}$$

Method (b): Use the three-step procedure described in Sample Problem 2.2 C.

Step 1: How many g Al are equivalent (equal) to a block of Al 2.0 cm x 4.0 cm x 8.0 cm?

Step 2: ? g = 64 cm^3 x ?

Step 3:   You need to convert from a volume in cm³ to mass in g. The other quantity given in the problem, density, always relates mass and volume. The correct form of this conversion factor has **cm³ on the bottom** so that it cancels.

$$? \text{ g} = 64 \text{ cm}^3 \text{ Al } \times \frac{2.7 \text{ g Al}}{\text{cm}^3 \text{ Al}} = 172.8 \text{ g Al} = 1.7 \times 10^2 \text{ g Al}$$

CHECK: Does this answer make sense? Yes, because 1 cm³ of Al has a mass of 2.7 g, so 64 cm³ will be 64 times as much.

SAMPLE PROBLEM 2.8 C

▸▸ What is the volume in cm³ and m³ of a 25 g piece of cork? [Density of cork is 0.24 g/cm³.]

Solution:

Again, you can use the two methods illustrated in the previous example. To rely less on formulas and help you gain more experience and confidence in using conversion factors, we will use the conversion factor method as often as possible.

Step 1:   How many cm³ of cork are equivalent (equal) to 25 g of cork?
[After this example, you should have enough confidence to think through step 1 and start writing with step 2.]

Step 2:   ? cm³ cork = 25 g cork × ?

Step 3:   The relationship between mass in g and volume in cm³ is density, which is given in the problem.

$$? \text{ cm}^3 \text{ cork} = 25 \text{ g cork} \times \frac{1 \text{ cm}^3 \text{ cork}}{0.24 \text{ g cork}} = 104.166 \ 67 \text{ cm}^3 = 1.0 \times 10^2 \text{ cm}^3 \text{ cork}$$

From Table 2.2, 1 cm = 10^{-2} m, which must be cubed to convert the volume from cm³ to m³.

$$? \text{ m}^3 \text{ cork} = 1.0 \times 10^2 \text{ cm}^3 \text{ cork} \times \frac{(10^{-2} \text{ m})^3}{(1 \text{ cm})^3}$$

$$= 1.0 \times 10^2 \text{ cm}^3 \text{ cork} \times \frac{10^{-6} \text{ m}^3}{1 \text{ cm}^3} = 1.0 \times 10^{-4} \text{ m}^3 \text{ cork}$$

CHECK: Do these answers make sense? Yes, because 0.24 g of cork has a volume of only 1 cm³ (roughly the size of a sugar cube). A mass of 25 g will have a much larger volume in cm³. And, 1 cm³ is only 10^{-6} m³, or, equivalently, there are 1 000 000 cm³ equal to 1 m³. Since you only have 100 cm³ of cork, you will have far less than 1 m³.

SECTION 2.9   MEASURING TEMPERATURE

After completing this section, you will be able to do the following:

- Interconvert Celsius and Kelvin temperatures;

- Interconvert Celsius and Fahrenheit temperatures.

REVIEW

An abbreviated form of the equation given in the textbook (2.3) for calculating a temperature in kelvins from degrees Celsius is

K (no degree symbol used) = °C + 273, and, rearranging, °C = K - 273.

The abbreviated form of the equation given in the textbook (2.4) for calculating a temperature in Fahrenheit from Celsius is

°F = 9/5°C + 32, and, rearranging, °C = 5/9 (°F - 32).

SAMPLE PROBLEM 2.9 A

▸▸ Convert 75 °C to K.

Solution:

Using the equation given above,

K = °C + 273 = 75 °C + 273 = 348 K

SAMPLE PROBLEM 2.9 B

▶ ▶ If the temperature in your room is 73 °F, what is it on the Celsius scale?

Solution:

Using the equation given above,

$$°C = 5/9 (°F - 32) = 5/9 (73 - 32) = 5/9 (41) = \boxed{23 °C}$$

SAMPLE PROBLEM 2.9 C

▶ ▶ Liquid nitrogen boils at -196 °C. What is its boiling point on the Fahrenheit scale?

Solution:

Using the equation given above,

$$°F = 9/5°C + 32 = 9/5(-196) + 32 = -353 + 32 = \boxed{-321 °F}$$

SECTION 2.10   MEASURING TIME

After completing this section, you will be able to do the following:

■ Interconvert SI and English units of time.

SAMPLE PROBLEM 2.10 A

▶ ▶ (a) How many seconds are there in 1.0 years (excluding leap year)?   (b) How many milliseconds are there in 15 minutes?

Solution:

Use the general problem-solving method described in Sample Problem 2.2 C.

(a) $? \text{ s } = 1.0 \text{ year} \times \dfrac{365 \text{ day}}{1 \text{ year}} \times \dfrac{24 \text{ h}}{1 \text{ day}} \times \dfrac{60 \text{ min}}{1 \text{ h}} \times \dfrac{60 \text{ s}}{1 \text{ min}} = 3.2 \times 10^7 \text{ s}$

(b) $? \text{ ms } = 15 \text{ min} \times \dfrac{60 \text{ s}}{1 \text{ min}} \times \dfrac{1 \text{ ms}}{10^{-3} \text{ s}} = 9.0 \times 10^5 \text{ ms}$

NOTE: The key is to make sure the unit that you want to **cancel** is in the denominator, and that you write the SI conversion factors correctly, i.e., $1 \text{ ms} = 10^{-3}$ s.

SECTION 2.11   ATOMIC MASSES

After completing this section, you will be able to do the following:

■ Given the mass number and the name or symbol of an atom of an element, determine how many protons, neutrons, and electrons are present in the atom and any ions it forms;

■ Given the isotopic composition of a naturally occurring mixture of isotopes for any element, calculate the atomic mass for that element.

REVIEW

Atoms consist of protons and neutrons inside the nucleus and electrons outside the nucleus. The relative masses of the proton (p), neutron (n), and electron (e) are 1, 1, and 0, respectively. The relative charges on the proton, neutron, and electron are +1, 0, and -1. The **atomic number, Z**, is the number of protons in the nucleus of an atom. The **neutron number, N**, is the number of neutrons in the nucleus of an atom. The **mass number, A**, is the **sum** of the number of protons and neutrons in the nucleus, i.e., $A = Z + N$. The term, **nuclide**, indicates a specific nuclear species by giving the mass number. For example, an atom of oxygen which contains 8 protons and 10 neutrons can be written as oxygen-18 or O-18 (where 18 is the mass number). A nuclide may also be represented by the following notation or symbol (X is an atom of any element), where the **mass number (A)** is always the **superscript** and the **atomic number (Z)** is always the **subscript**:

$$^A_Z X \quad \text{where} \quad A = \#p + \#n \quad \text{and} \quad Z = \#p = \#e.$$

Isotopes are atoms of the same element (same atomic number or number of protons), which differ in the number of neutrons (and therefore have a different mass). Two isotopes of chlorine are Cl-35 and Cl-37, where the 35 and 37 are mass numbers. Atoms of both isotopes have 17 protons (atomic number) in the nucleus, but Cl-35 has **18 neutrons** (35 - 17) in the nucleus and Cl-37 has **20 neutrons** (37 - 17) in the nucleus.

The **atomic mass** for any element (found on the inside front cover of the study guide and textbook) is a weighted average of the naturally occurring isotopes of that element, which means that you must know the atomic masses of the individual isotopes and the percent of each isotope in the mixture. The unit of atomic mass is the **universal atomic mass unit, u (or amu)**, which is defined as 1/12th the mass of an atom of carbon-12 (which by definition has a mass of exactly 12).

SAMPLE PROBLEM 2.11 A

▸▸ For an atom of Cu-63 and an atom of Cu-65, (a) how many protons, neutrons, and electrons are in each? (b) Write the symbol (notation) for each. (c) How many electrons would Cu^{2+} have of each isotope?

Solution:

(a)   All atoms of copper (no matter which isotope) have the same atomic number, Z.

   Cu-63:   Z = 29 = 29 p = 29 e   # n = A - Z = 63 - 29 = 34 n

   Cu-65:   Z = 29 = 29 p = 29 e   # n = A - Z = 65 - 29 = 36 n

(b)
   $^{63}_{29}Cu$ $^{65}_{29}Cu$

(c)   Metal ions are formed by losing **electrons only**; there is **no change** in the number of protons or neutrons. For both isotopes of Cu, the 2+ ions will have (29 - 2) = 27 e.

SAMPLE PROBLEM 2.11 B

▸▸ Naturally occurring oxygen is a mixture of 99.76% O-16 (atomic mass = 15.9949), 0.04% O-17 (atomic mass = 16.9991), and 0.20% O-18 (atomic mass = 17.9992). Calculate the atomic mass of naturally occurring oxygen and compare it with the value given on the inside front cover.

HINT: For each isotope, change the percents to fractions and multiply by the atomic mass.

Atomic mass O = (0.9976 x 15.9949) + (0.0004 x 16.9991) + (0.0020 x 17.9992)

= 15.9 + 0.007 + 0.036 = 16.00

CHECK: Does this make sense? Yes, because over 99% of naturally occurring oxygen is in the form of O-16. Therefore you would expect the atomic mass for the element to be almost the same as the atomic mass for O-16.

The calculated atomic mass of oxygen is almost the same as the value given in the table (15.999). The difference in answers is due to small rounding errors in the intermediate results.

SECTION 2.12   FORMULA MASSES

After completing this section, you will be able to do the following:

■ Given the formula of a compound, calculate the formula mass.

REVIEW

The **formula mass** of a compound is the sum of the atomic masses of the atoms forming the compound.

Formula mass = (# atoms of element A x atomic mass of element A)
+ (# atoms of element B x atomic mass of element B)
+ (# atoms of element C x atomic mass of element C)
+ (etc.).

The **atomic mass** of each element is found on the inside front cover of this study guide and the textbook.

SAMPLE PROBLEM 2.12 A

▶▶ Calculate the formula masses with as many significant figures as possible for (a) H_2SO_4 (b) C_2H_6O (c) Na_2CO_3 (d) $KC_2H_3O_2$ (e) $CuSO_4 \cdot 5H_2O$ (f) $Ba_3(PO_4)_2$.

Solution:

HINT: The atomic masses of the elements are found on the inside front cover of this study guide and the textbook.

(a) In H_2SO_4 there are:

2 H x 1.007 94 u	=	2.015 88 u	[u is the atomic mass unit; see Sec. 2.11]
1 S x 32.066 u	=	32.066 u	
4 O x 15.9994 u	=	63.9976 u	
1 H_2SO_4	=	98.079 48 u	= 98.079 u [3 s.f. after decimal point]

(b) In C_2H_6O there are:

2 C x 12.011 u	=	24.022 u	
6 H x 1.007 94 u	=	6.047 64 u	
1 O x 15.9994 u	=	15.9994 u	
1 C_2H_6O	=	46.069 04 u	= 46.069 u [3 s.f. after decimal point]

(c) In Na_2CO_3 there are:

2 Na x 22.989 768 u =	45.979 536 u	
1 C x 12.011 u	=	12.011 u
3 O x 15.9994 u	=	47.9982 u
1 Na_2CO_3	=	105.988 74 u = 105.989 u

(d) In $KC_2H_3O_2$ there are:

1 K x 39.0983 u	=	39.0983 u
2 C x 12.011 u	=	24.022 u
3 H x 1.007 94 u	=	3.023 82 u
2 O x 15.9994 u	=	31.9988 u
1 $KC_2H_3O_2$	=	98.142 92 u = 98.143 u

(e) In $CuSO_4 \cdot 5H_2O$ there are:

1 Cu x 63.546 u	=	63.546 u
1 S x 32.066 u	=	32.066 u
9 O x 15.9994 u	=	143.9946 u
10 H x 1.007 94 u	=	10.0794 u
1 $CuSO_4 \cdot 5H_2O$	=	249.6860 u = 249.686 u

NOTE: The waters of hydration shown after the dot in a formula are **always** counted as part of the formula mass.

(f) In $Ba_3(PO_4)_2$ there are:

$$
\begin{array}{lll}
3\ Ba \times 137.327\ u & = & 411.981\ u \\
2\ P \times 30.973\ 762\ u & = & 61.947\ 524\ u \\
8\ O \times 15.9994\ u & = & 127.9952\ u \\
\hline
1\ Ba_3(PO_4)_2 & = & 601.923\ 724 \quad = \quad 601.924\ u
\end{array}
$$

SECTION 2.13   AMOUNT OF SUBSTANCE

After completing this section, you will be able to do the following:

■ Given a number of moles of a substance, calculate its mass in grams and the number of unit particles in that sample.

REVIEW

Just as one dozen means 12 and one century means 100, **one mole** (mol) in chemistry is the SI unit that means **Avogadro's number**, or 6.022×10^{23} particles of a substance, which may consist of atoms, ions, or molecules. The **mass** in grams of **one mole** of any substance (often called the **molar mass**) is numerically equal to the formula mass. For example, the **formula mass** of CO_2 is 44.010 u, that is, the mass of 1 CO_2 molecule is 44.010 amu. For a **1.000-mol** sample of CO_2, the following is **always true**:

1.000 mol of CO_2 is 6.022×10^{23} CO_2 molecules, which have a mass of 44.010 g (molar mass).

SAMPLE PROBLEM 2.13 A

▶ ▶ Complete the following table.

Substance	Formula Mass	No. of Moles	Mass of Moles	Number of Particles
Al		1.000		
N_2		1.000		
SO_2		1.000		
SO_2		2.000		
SO_2		0.5000		
MgO		1.000		

Solution:

HINT: See Section 2.12 for calculating formula masses and the discussion at the beginning of this section.

Substance	Formula Mass	No. of Moles	Mass of Moles	Number of Particles
Al	26.981 539 u	1.000	26.98 g	6.022×10^{23} Al atoms
N_2	28.013 48 u	1.000	28.01 g	6.022×10^{23} N_2 molecules 12.04×10^{23} N atoms
SO_2	64.065 u	1.000	64.07 g	6.022×10^{23} SO_2 molecules 6.022×10^{23} S atoms 12.04×10^{23} O atoms
SO_2	64.065 u	2.000	128.1 g	12.04×10^{23} SO_2 molecules 12.04×10^{23} S atoms 24.09×10^{23} O atoms
SO_2	64.065 u	0.5000	32.03 g	3.011×10^{23} SO_2 molecules 3.011×10^{23} S atoms 6.022×10^{23} O atoms
MgO	40.3044 u	1.000	40.30 g	6.022×10^{23} MgO formula units 6.022×10^{23} Mg^{2+} ions 6.022×10^{23} O^{2-} ions

SELF-TEST QUESTIONS

Test your understanding of the concepts and skills in this chapter by working through the multiple choice questions **BEFORE** checking the answers. Remember, the information on SI Units and Prefixes in Tables 2.1 and 2.2 of the textbook must be **memorized**. A periodic table will always be available. The ANSWERS TO SELF-TEST QUESTIONS follow immediately after this section.

1. One millimeter (mm) is

   A. 1 meter.
   B. 10^3 cm.
   C. 10^{-6} μm.
   D. 10^{-6} km.
   E. 10^3 m.

2. The largest unit of mass of the following is

   A. 1 g   B. 1 kg   C. 1 ng   D. 1 dg   E. 1 Mg

3. The mass in kilograms of 2.0×10^4 mg is

   A. 2.0×10^1 B. 2.0×10^{-2} C. 2.0×10^{-6} D. 2.0×10^2 E. 2.0×10^7

4. The length of a 100-yd (3 s.f.) football field is

   A. 5.68×10^{-2} mi   B. 1.50×10^2 ft   C. 3.60×10^2 in   D. 4.50×10^2 ft   E. 1.20×10^3 in

5. The mass in oz of an 8.0-kg weight is

   A. 2.3×10^5 B. 2.8×10^2 C. 2.8×10^{-4} D. 2.3×10^{-1} E. 2.5×10^{-2}

6. Each of the following measurements has the correct number of significant figures <u>except</u>

   A. 165 gal; 3
   B. 2.150×10^6 lb; 4
   C. 0.0140 mL; 2
   D. 0.0041 g; 2
   E. 51.0×10^6 mi; 3

7. Given 10.0 g of each of the following substances, which one will have the largest volume?

   A. magnesium (D = 1.74 g/cm^3)
   B. mercury (D = 13.6 g/cm^3)
   C. silver (D = 10.5 g/cm^3)
   D. wood (D = 0.42 g/cm^3)
   E. water (D = 1.0 g/cm^3)

8. Osmium has the highest density of the naturally occurring elements. Convert its density from 2.20 x 10^4 kg/m^3 to g/cm^3.

   A. 22.0   B. 2.20 x 10^5 C. 2.20 x 10^3 D. 2.20 x 10^7 E. 2.2 x 10^4

9. What temperature on the Fahrenheit scale is equivalent to -10.0 °C?

   A. 50.0 °F   B. -23.3 °F   C. 14.0 °F   D. -12.2 °F   E. 25.0 °F

10. With a mass number (A) of 110, an atom of silver has

   A. 63 protons, 47 neutrons, and 63 electrons.
   B. 47 protons, 47 neutrons, and 63 electrons.
   C. 47 protons, 47 neutrons, and 47 electrons.
   D. 63 protons, 63 neutrons, and 47 electrons.
   E. 47 protons, 63 neutrons, and 47 electrons.

11. If an atom of an element has a mass number (A) of 41 and contains 20 electrons,

   A. the element is Nb.
   B. the atom has 41 neutrons.
   C. the atom has 21 protons.
   D. the element is Ca.
   E. the atom has 21 protons and 20 neutrons.

12. The formula mass for $Pb(NO_3)_2$, to the correct number of significant figures, is

   A. 269.204 94 u   B. 331.2 u   C. 331.209 88 u   D. 331.2099 u   E. 269.2 u

13. The formula mass for $Na_2SO_4 \bullet 10H_2O$, to the correct number of significant figures, is

    A. 322.196 u   B. 142.043 u   C. 322.195 94 u   D. 142.043 14 u   E. 322.2 u

14. The mass in grams of 1.000 mol of P_2O_5 is

    A. 1.000 g   B. 6.022×10^{23} g   C. 141.9 g   D. 42.15×10^{23} g   E. 46.97 g

15. The sample given in the previous question consists of

    A. 1 molecule of P_2O_5.
    B. 6.022×10^{23} atoms of P.
    C. 12.04×10^{23} atoms of O.
    D. 7 atoms of P and O.
    E. 42.15×10^{23} atoms of P and O.

ANSWERS TO SELF-TEST QUESTIONS

1. d	6. c	11. d
2. e	7. d	12. b
3. b	8. a	13. a
4. a	9. c	14. c
5. b	10. e	15. e

PRACTICE PROBLEMS

Test your problem solving skills in this chapter by working through the problems in the space provided **BEFORE** checking with the answers. A periodic table will always be available. The ANSWERS TO PRACTICE PROBLEMS follows immediately after this section.

1. Convert

    (a) 5.50 kg to mg.

    (b) 2.00 L to qt.

    (c) 15.0 km to mi.

    (d) 8.80 kg to lb.

    (e) 6.40 ft to mm.

2. For the three measurements, 13.654 cm, 0.916 cm, and 105.2 cm, (a) indicate the number of significant figures in each measurement, and (b) calculate the sum of the three measurements expressed with the correct number of significant figures.

3. For the two masses, 0.637 25 g and 0.613 g, (a) indicate the number of significant figures in each measurement, and (b) calculate the difference between the two masses expressed to the correct number of significant figures.

4. For an atom of In-113 (indium-113) and an atom of In-116 (indium-116), (a) how many protons, neutrons, and electrons are in each? (b) Write the symbol (notation) for each. (c) How many electrons would In^{3+} have of each isotope?

5. Naturally occurring magnesium is a mixture of 78.70% Mg-24 (atomic mass = 23.9850, 10.13% Mg-25 (atomic mass = 24.9858), and 11.17% Mg-26 (atomic mass = 25.9826). Calculate the atomic mass of naturally occurring magnesium and compare it with the value given on the inside front cover.

ANSWERS TO PRACTICE PROBLEMS

1. (a) $? \text{ mg} = 5.50 \text{ kg} \times \dfrac{10^3 \text{ g}}{1 \text{ kg}} \times \dfrac{10^3 \text{ mg}}{1 \text{ g}} = 5.50 \times 10^6 \text{ mg}$

    (b) $? \text{ qt} = 2.00 \text{ L} \times \dfrac{1 \text{ qt}}{0.946 \text{ L}} = 2.11 \text{ qt}$

    (c) $? \text{ mi} = 15.0 \text{ km} \times \dfrac{1 \text{ mi}}{1.6093 \text{ km}} = 9.32 \text{ mi}$

    (d) $? \text{ lb} = 8.8 \text{ kg} \times \dfrac{2.205 \text{ lb}}{1 \text{ kg}} = 17.6 \text{ lb}$

(e) $? \text{ mm} = 6.40 \text{ ft} \times \dfrac{12 \text{ in}}{1 \text{ ft}} \times \dfrac{2.54 \text{ cm}}{1 \text{ in}} \times \dfrac{10^{-2} \text{ m}}{1 \text{ cm}} \times \dfrac{10^{3} \text{ mm}}{1 \text{ m}} = 1.95 \times 10^{3} \text{ mm}$

OR

$? \text{ mm} = 6.40 \text{ ft} \times \dfrac{1.6093 \text{ km}}{5280 \text{ ft}} \times \dfrac{10^{3} \text{ m}}{1 \text{ km}} \times \dfrac{10^{3} \text{ mm}}{1 \text{ m}} = 1.95 \times 10^{3} \text{ mm}$

2. (a) 13.654 cm (5 s.f.)
       0.916 cm (3 s.f.)
     105.2   cm (4 s.f.)

   (b) _____
        119.770 cm

        119.8 cm [The measurement with the smallest number of digits to the right of the decimal point (105.2 cm with one) determines the number of significant figures in the answer. Therefore, round off the answer to one digit after the decimal point.]

3. (a) 0.637 25 g (5 s.f.)
       −0.613   g (3 s.f.)

   (b) _____
        0.024 25 g

        0.024 g [The measurement 0.613 g has only three significant digits after the decimal point. Therefore, the answer is rounded off to three digits after the decimal point.]

4. (a) In-113: $Z = 49 = 49\text{p} = 49\text{e}$; $\#\text{n} = 113 - 49 = 64\text{n}$

       In-116: $Z = 49 = 49\text{p} = 49\text{e}$; $\#\text{n} = 116 - 49 = 67\text{n}$

   (b) $^{113}_{49}\text{In}$ $^{116}_{49}\text{In}$

   (c) For both isotopes of indium, the 3+ ions will have (49−3 =) 46e.

5. Atomic mass Mg = (0.7870 x 23.9850) + (0.1013 x 24.9858) + (0.1117 x 25.9826) = 24.309.
    The value given on the inside front cover for magnesium is 24.3050.

CHAPTER 3

STOICHIOMETRY

SECTION 3.1   GRAMS, MOLES, AND NUMBER OF PARTICLES

After completing this section, you will be able to do the following:

■ Given **one** of the following quantities for any substance, calculate the other two quantities:

- ■ mass in grams;
- ■ number of moles of particles;
- ■ number of particles.

REVIEW

No matter what quantity of moles, grams, or particles of a substance you are given, you **always** know **two** pieces of information about **one mole** of any substance (Section 2.13):

- ● the mass in grams (from formula mass, using inside front or back cover of this book), and
- ● the number of particles (Avogadro's number, 6.022×10^{23}).

This information provides you with **two invaluable conversion factors** for these and many other applications. The three kinds of calculations that you will need to do are:

1. **Calculate the number of particles from the number of moles of particles, and vice versa.** You will need **one** conversion factor, the **number of particles in one mole**, or 6.022×10^{23} particles/mol.

2. **Calculate the number of grams from the number of moles of particles, and vice versa.** You will need **one** conversion factor, the **mass in grams of one mole of particles,** or g/mol (sometimes called the molar mass), which is numerically equal to the formula mass found on the inside back cover or calculated from the atomic masses on the inside front cover.

3. **Calculate the number of particles from mass in grams, and vice versa.** You will need **both** of the above conversion factors.

The relationships described in these three steps are summarized in the diagram below.

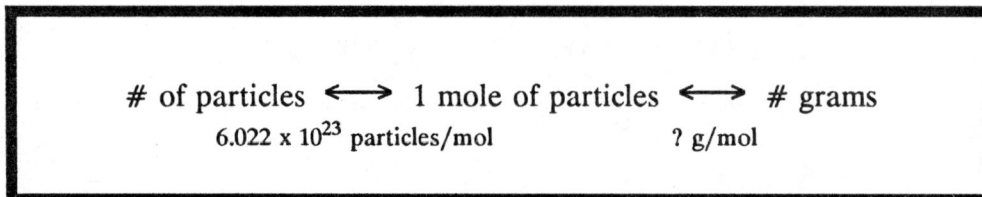

$$\text{\# of particles} \longleftrightarrow \text{1 mole of particles} \longleftrightarrow \text{\# grams}$$
$$6.022 \times 10^{23} \text{ particles/mol} \qquad \text{? g/mol}$$

SAMPLE PROBLEM 3.1 A

▶▶ How many moles of zinc atoms are there in 3.011×10^{23} atoms of zinc?

Solution:

HINT: Use the 3-step, general problem-solving approach given in Sample Problem 2.2 C and used thereafter.

Step 1: How many **moles** of zinc atoms are equivalent to (e.g., are there in, equal to, derived from, etc.) 3.011×10^{23} zinc atoms?

Step 2: ? mol of Zn atoms = 3.011×10^{23} Zn atoms x ?

Step 3: You need to convert **Zn atoms** to **moles of Zn atoms**. [If you were asked to convert Zn atoms to **dozens** of Zn atoms, you would divide by the number in a dozen, 12. The procedure here is the same, except the quantity mole is less familiar.] Regardless of how many atoms of any element you are starting with, you **always** know the number of atoms in **one mole** of that element, 6.022×10^{23}. In other words, one mole (of any formula) = 6.022×10^{23} (of that formula). This equality is a conversion factor which has the two forms:

$$\frac{1 \text{ mol Zn atoms}}{6.022 \times 10^{23} \text{ Zn atoms}} \quad \text{and} \quad \frac{6.022 \times 10^{23} \text{ Zn atoms}}{1 \text{ mol Zn atoms}}$$

Use the form in which the unit that you want to **cancel** is in the denominator. [**Remember**, think of the line separating the top from the bottom as an equal sign.] Choosing the first form will give you what you are trying to find, moles of Zn atoms.

$$? \text{ mol Zn atoms} = 3.011 \times 10^{23} \text{ } \cancel{\text{Zn atoms}} \times \frac{1 \text{ mol Zn atoms}}{6.022 \times 10^{23} \text{ } \cancel{\text{Zn atoms}}}$$

$$= 0.5000 \text{ mol Zn atoms}$$

CHECK: Does this answer make sense? Yes, because 6.022×10^{23} atoms of Zn are 1 mol, and 3.011×10^{23} is **one-half** of 6.022×10^{23}. Thus you only have 0.5000 mol of Zn.

SAMPLE PROBLEM 3.1 B

▶ ▶ What is the weight in grams of 5.0 mol of carbon atoms?

Solution:

HINT: Use the 3-step, general problem-solving approach given in Sample Problem 2.2 C and applied to this chapter beginning with Sample Problem 3.1 A.

Step 1: How many grams of carbon are equivalent to (e.g., are equal to, are the mass of, etc.) 5.0 mol of carbon atoms?

Step 2: $? \text{ g C} = 5.0 \text{ mol of C atoms} \times ?$

Step 3: You need to convert **moles of C atoms** to its **mass in grams**. Regardless of the number of moles of C atoms that you start with, you **always know the mass in grams of one mole of C atoms**. It is numerically equal to the atomic mass found on the inside front cover, 12.011 u. That is, the mass in grams of 1 mol of C atoms is 12.011 g. Or, 1.000 mol of C atoms = (has a mass of) 12.011 g. This equality is a conversion factor which has the two forms:

$$\frac{1 \text{ mol C atoms}}{12.011 \text{ g C}} \quad \text{and} \quad \frac{12.011 \text{ g C}}{1 \text{ mol C atoms}}$$

Use the form in which the unit that you want to **cancel** is in the denominator.

[**Remember,** think of the line separating the top from the bottom as an equal sign.] Choosing the **second** form will give you what you are trying to find, g of C.

$$? \text{ g C} = 5.0 \text{ mol C atoms} \times \frac{12.011 \text{ g C}}{1 \text{ mol C atoms}} = 6.0 \times 10^1 \text{ g C}$$

NOTE: The answer is limited to two significant figures (abbreviated s.f.) by the quantity with the least s.f. (for multiplication and division), 5.0 mol of C atoms.

CHECK: Does this answer make sense? Yes, because 1 mol of C atoms has a mass of 12.011 g, so 5 mol will have a mass five times as great.

SAMPLE PROBLEM 3.1 C

▶▶ Calculate the number of N_2O molecules **and** N atoms in 7.50 g of N_2O.

Solution:

HINT:
- Use the 3-step, general problem-solving approach given in Sample Problem 2.2 C and applied to this chapter beginning with Sample Problem 3.1 A.

- Always calculate the number of **molecules first** (in this case N_2O), then the number of **atoms** of any element in those molecules **second** (N for this example).

Step 1: How many N_2O molecules are equivalent to (e.g., equal to, have a mass of, etc.) 7.50 g of N_2O? [From this point on in this chapter, you should be able to formulate this statement in your head without writing it down.]

Step 2: $? N_2O$ molecules $= 7.50 \text{ g } N_2O \times ?$

Step 3: Without a lot of practice, you will not be able to go directly from grams of N_2O to N_2O molecules. If you look back at the concept diagram immediately preceding Sample Problem 3.1 A, you are starting at the right hand side (g) and will need the two standard conversion factors (used in the two previous examples) derived from the mass of and number of particles in **1 mol**. For this example, the conversion factors are:

$$\frac{1 \text{ mol } N_2O \text{ molecules}}{44.0129 \text{ g } N_2O} \quad \text{and} \quad \frac{1 \text{ mol } N_2O \text{ molecules}}{6.022 \times 10^{23} \text{ } N_2O \text{ molecules}}$$

To convert grams to moles, you will need to use the first conversion factor as shown, so that grams will be in the denominator and cancel. To convert moles to molecules, you will need to use the alternate form of the second conversion factor so that moles will be in the denominator and cancel. Then you will have the units that you need, N_2O molecules.

$$? \text{ } N_2O \text{ molecules} = 7.50 \text{ g } N_2O \times \frac{1 \text{ mol } N_2O \text{ molecules}}{44.0129 \text{ g } N_2O} \times \frac{6.022 \times 10^{23} \text{ } N_2O \text{ molecules}}{1 \text{ mol } N_2O \text{ molecules}}$$

$$= 1.03 \times 10^{23} \text{ } N_2O \text{ molecules} \qquad [3 \text{ s.f. required by } 7.50 \text{ g } N_2O]$$

CHECK: Does this answer make sense? Yes, because 1 mol of N_2O has a mass of ~44 g and you only have 7.5 g or about 1/6 of a mole. Thus you will have only 1/6 of Avogadro's number of molecules, or ~1 x 10^{23}.

Once you have calculated the number of N_2O molecules, you can calculate the number of N atoms by asking yourself, how many N atoms are there in **one** N_2O molecule? The formula for N_2O will give you an equality which is the conversion factor you need, 1 N_2O molecule = (or always consists of, or always contains) 2 N atoms, and

$$\frac{1 \text{ } N_2O \text{ molecule}}{2 \text{ N atoms}} \quad \text{and} \quad \frac{2 \text{ N atoms}}{1 \text{ } N_2O \text{ molecule}}$$

Then use the correct form of the conversion factor so that units of N_2O molecules are in the denominator and cancel, to give you the number of N atoms. **You should be able to quickly calculate the number of atoms of any element in a formula once you know the number of molecules of that formula.**

$$? \text{ atoms N} = 1.03 \times 10^{23} \text{ } N_2O \text{ molecules} \times \frac{2 \text{ N atoms}}{1 \text{ } N_2O \text{ molecule}}$$

$$= 2.06 \times 10^{23} \text{ N atoms}$$

CHECK: Does this answer make sense? Yes, since there are two N atoms in 1 N_2O molecule, there will always be **twice** as many N atoms as N_2O molecules no matter what number of N_2O molecules you are working with.

SECTION 3.3   MORE ABOUT BALANCING EQUATIONS

After completing this section, you will be able to do the following:

■ Given the names or formulas of the reactants and products in a reaction, write a chemical equation (balanced reaction).

REVIEW

A **balanced chemical reaction** is called a **chemical equation**, which you will always need later in this course before you can do any calculations involving a reaction. A chemical reaction is balanced **only if there are the same number of <u>atoms</u> of each element on both sides of the reaction** (reactant side and product side). That is, matter can be neither created nor destroyed. This law is obeyed only if the mass of all the products equals the mass of all the reactants --- which is **always** true **when the reaction is balanced**. See Section 1.10 for a more complete review and procedure for writing and balancing chemical reactions.

SAMPLE PROBLEM 3.3 A

▶ ▶ Balance the following reaction:

$$NH_3(g) \; + \; O_2(g) \; \rightarrow \; N_2(g) \; + \; H_2O(g)$$

Solution:

HINT: Use the 4-step procedure given in Section 1.10 of the study guide.

Step 1: Write down the formulas correctly and leave off the (s), (aq), (g), and (l) until the balancing is finished to avoid making errors. [With experience, you can leave these letters in and reduce the balancing to 3 steps.]

$$NH_3 \; + \; O_2 \; \rightarrow \; N_2 \; + \; H_2O$$

Step 2: Start with 1 molecule of the most complicated looking formula, NH_3, and balance the atoms of N, **always leaving H and O until last** (because they usually appear in so many formulas in the reaction). You cannot produce 2 N on the right side unless you **start** with 2 N on the left side; thus, a coefficient of 2 must be placed in front of NH_3. To balance the other element in NH_3, H, you will need a coefficient of 3 in front of H_2O, so there are 6 atoms of H on each side.

$$2NH_3 \; + \; O_2 \; \rightarrow \; N_2 \; + \; 3H_2O$$

The only element left is O. To produce 3 O on the right, you must start with 3 O on the left side, which means 3/2 O_2 molecules (3 O atoms). The reaction is now balanced (temporarily).

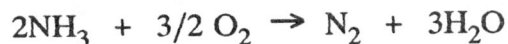

$$2NH_3 + 3/2\,O_2 \rightarrow N_2 + 3H_2O$$

Step 3: When balancing reactions, you can use a fraction of a molecule as an **intermediate** step to the final balanced reaction, since it is the **number of atoms,** NOT molecules, that must be balanced. For this problem, 1 O_2 molecule contains 2 O atoms, so 3/2 O_2 molecules will contain 3 O atoms. However, in practice you cannot use fractions of molecules. A **correctly balanced reaction,** then, is one in which you have the lowest, **whole-numbered** coefficients possible. That **always** means multiplying the entire reaction (i.e., all the coefficients) by the **denominator of the fraction** in order to be left with the lowest, whole-numbered coefficients. For this reaction,

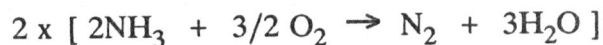

$$2 \times [\ 2NH_3 + 3/2\,O_2 \rightarrow N_2 + 3H_2O\ ]$$

gives you the **correctly balanced reaction** (smallest, whole-numbered coefficients),

$$4NH_3 + 3O_2 \rightarrow 2N_2 + 6H_2O.$$

Step 4: Add the letters describing the state of each substance.

$$4NH_3(g) + 3O_2(g) \rightarrow 2N_2(g) + 6H_2O(g)$$

CHECK: Does this make sense? Yes, because there are the same number of **atoms** (NOT molecules) of each element on each side of the reaction, and the coefficients are the lowest, whole numbers possible (i.e., not all divisible by the same integer other than 1).

SAMPLE PROBLEM 3.3 B

►► Balance the following reaction:

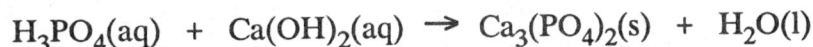

$$H_3PO_4(aq) + Ca(OH)_2(aq) \rightarrow Ca_3(PO_4)_2(s) + H_2O(l)$$

Solution:

HINT: ● Use the 4-step procedure given in Section 1.10 of the study guide.
● If the same polyatomic ion appears on both sides of the reaction, you can balance

the reaction faster by balancing entire polyatomic ions rather than the individual elements in the polyatomic ions.

Step 1: Write down the formulas correctly and leave off the (s), (aq), and (l) until the balancing is finished to avoid making errors. [With experience, you can leave these letters in and reduce the balancing to 3 steps.]

$$H_3PO_4 + Ca(OH)_2 \rightarrow Ca_3(PO_4)_2 + H_2O$$

Step 2: Start with 1 molecule of the most complicated looking formula, $Ca_3(PO_4)_2$, and balance the atoms of Ca and P, **always leaving H and O until last** (because they usually appear in so many formulas in the reaction). You cannot produce 3 Ca on the right side unless you **start** with 3 Ca on the left side; thus, a coefficient of 3 must be placed in front of $Ca(OH)_2$.

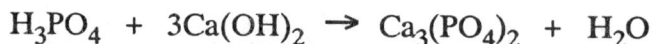

$$H_3PO_4 + 3Ca(OH)_2 \rightarrow Ca_3(PO_4)_2 + H_2O$$

The P atoms can be balanced in one of two ways. **First**, the 2 P atoms on the right side (yes, there are two atoms of P in 1 $Ca_3(PO_4)_2$ molecule) must be balanced by placing a 2 in front of H_3PO_4. Or, the **second method** is to notice that P is part of a polyatomic ion, PO_4^{3-}. When the same polyatomic ion appears on both sides of the reaction, you can balance the P atoms by balancing the polyatomic ions. [Note that this method will also balance the O atoms in the polyatomic ions, but a final check of all O atoms still needs to be made, since O appears in all four formulas in this reaction.] Using either method, the atoms of Ca and P are balanced, but H and O are not. [See Sample Problem 1.10 A for an example on how to calculate the number of atoms of an element in a formula, including polyatomic ions.].

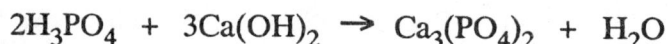

$$2H_3PO_4 + 3Ca(OH)_2 \rightarrow Ca_3(PO_4)_2 + H_2O$$

atoms: 6 H, 8 O + 6 H, 6 O → 8 O + 2 H, 1 O

There are now 12 H on the left side and only 2 H on the right side. Balance the number of H atoms at 12 on each side by using a coefficient of 6 in front of H_2O.

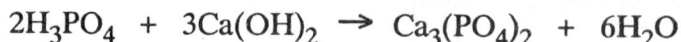

$$2H_3PO_4 + 3Ca(OH)_2 \rightarrow Ca_3(PO_4)_2 + 6H_2O$$

Finally, check the O atoms; they are already balanced at 14 O on each side.

Step 3: The reaction is balanced with 2 P, 3 Ca, 12 H, and 14 O on each side of the reaction. The coefficients (2,3,1,6) are the lowest whole numbers possible, since they are not all divisible by any integer other than 1.

Step 4:   Add the letters aq, s, and l as shown in the statement of the problem.

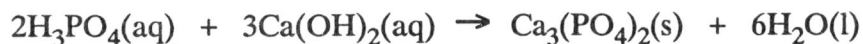

$$2H_3PO_4(aq) \ + \ 3Ca(OH)_2(aq) \ \rightarrow \ Ca_3(PO_4)_2(s) \ + \ 6H_2O(l)$$

SECTION 3.4   EQUATIONS ON A MACROSCOPIC SCALE

After completing this section, you will be able to do the following:

- Interpret the coefficients in a chemical equation (balanced reaction).

REVIEW

You have learned how to balance chemical reactions with formulas representing atoms and molecules by changing the coefficients in front of the formulas (Sections 1.10 and 3.3). In the real world, we can't work with individual atoms and molecules, but must instead use a much larger quantity. If you multiply all the coefficients in a balanced reaction by 6.022×10^{23}, then the coefficients will also indicate numbers of **moles of molecules** as well. Thus, the coefficients in any balanced reaction (chemical equation) should be interpreted as **molecules OR moles of molecules**. You will know which interpretation to use by the type of problem you are asked to solve. If you are asked to **balance** a reaction, it is generally easier if the coefficients are interpreted as **molecules**. On the other hand, if you are asked to **calculate quantities of a reactant or product** in a reaction (especially to weigh out in the laboratory), then the coefficients **must** be interpreted as **moles of molecules**. These relationships, in moles, between reactants and products will give you the conversion factors necessary for finding the solution to the problem (see Sample Problem 3.4 B and Section 3.5).

SAMPLE PROBLEM 3.4 A

▶▶ Interpret the chemical equation (balanced reaction),

$$4NH_3(g) \ + \ 5O_2(g) \ \rightarrow \ 4NO(g) \ + \ 6H_2O(g).$$

Solution:

HINT: The most useful interpretation of the coefficients in **all chemical equations (balanced reactions)** is that they indicate numbers of **molecules** and **moles of molecules**.

The correct interpretation of the reaction is:

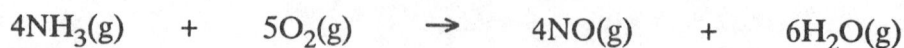

$$4NH_3(g) \quad + \quad 5O_2(g) \quad \rightarrow \quad 4NO(g) \quad + \quad 6H_2O(g)$$

4 NH_3 molecules + 5 O_2 molecules → 4 NO molecules + 6 H_2O molecules

4 mol NH_3 molecules + 5 mol O_2 molecules → 4 mol NO molecules + 6 mol H_2O molecules

The first interpretation says that 4 NH_3 molecules will completely react with 5 O_2 molecules to produce 4 NO molecules and 6 H_2O molecules. If you multiply each of these coefficients by 6.022×10^{23}, then $(4 \times 6.022 \times 10^{23})$ or 4 mol of NH_3 molecules will completely react with $(5 \times 6.022 \times 10^{23})$ or 5 mol of O_2 molecules to produce $(4 \times 6.022 \times 10^{23})$ or 4 mol of NO molecules and $(6 \times 6.022 \times 10^{23})$ or 6 mol of H_2O molecules. The great advantage of interpreting the coefficients as number of moles is that we can always **calculate the mass in grams** of 1 mol of any reactant or product (i.e., numerically equal to the formula mass).

SAMPLE PROBLEM 3.4 B

▶▶ Given the chemical equation (balanced reaction),

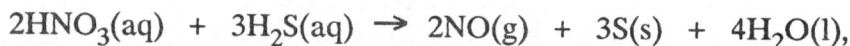

$$2HNO_3(aq) + 3H_2S(aq) \rightarrow 2NO(g) + 3S(s) + 4H_2O(l),$$

(a) How many moles of H_2S are required to react with 4 mol of HNO_3? (b) How many moles of NO can be formed from 1 mol of H_2S, assuming enough HNO_3 is available? (c) Assuming enough H_2S is available, how many moles of HNO_3 will be required to produce 1 mol of NO?

Solution:

HINT:
- Use the 3-step, general problem-solving approach given in Sample Problem 2.2 C and applied to this chapter beginning with Sample Problem 3.1 A.

- Until you have gained enough confidence in interpreting chemical equations (balanced reactions), write the interpretation of the coefficients in moles under each of the reactants and products. This information will give you the necessary conversion factors for solving the problem. **You must know how to correctly interpret a chemical equation (balanced reaction) in terms of moles of molecules before you will be able to perform any calculations with that reaction.**

The correct interpretation of the chemical equation is

$$2HNO_3(aq) + 3H_2S(aq) \rightarrow 2NO(g) + 3S(s) + 4H_2O(l),$$

2 mol   + 3 mol $\rightarrow$ 2 mol   +   3 mol   + 4 mol

(a)   Starting with Step 2 of the general problem-solving approach referenced in the HINT,

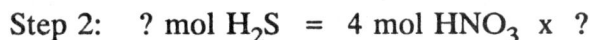

Step 2:   ? mol H_2S = 4 mol HNO_3 x ?

Step 3:   The only relationship between HNO_3 and H_2S is the number of moles from the balanced reaction. This relationship gives us the **equality** 2 mol HNO_3 = (or requires) 3 mol H_2S, which is a **conversion factor** with the two forms

$$\frac{2 \text{ mol } HNO_3 \text{ molecules}}{3 \text{ mol } H_2S \text{ molecules}} \quad \text{and} \quad \frac{3 \text{ mol } H_2S \text{ molecules}}{2 \text{ mol } HNO_3 \text{ molecules}}$$

Use the conversion factor derived from the coefficients in the balanced reaction so that the units of mol HNO_3 are in the denominator, and cancel.

$$? \text{ mol } H_2S = 4 \text{ mol } HNO_3 \text{ x } \frac{3 \text{ mol } H_2S}{2 \text{ mol } HNO_3} = 6 \text{ mol } H_2S$$

CHECK: Does this make sense? Yes! If 2 mol of HNO_3 require 3 mol of H_2S to react (1 and 1/2 times as much, or 50% more), then 4 mol of HNO_3 will require 50% more of H_2S, or 6 mol of H_2S. **To avoid careless mistakes, always take the time to use the conversion factors which come directly from the coefficients in the balanced reaction.**

(b)   Again, starting with Step 2 of the general problem-solving approach referenced in the HINT,

Step 2:   ? mol NO = 1 mol H_2S x ?

Step 3:   The only relationship between H_2S and NO is the number of moles from the balanced reaction. This relationship gives us the **equality** 3 mol H_2S = (or forms, or produces) 2 mol NO, which is a **conversion factor** with the two forms

$$\frac{3 \text{ mol } H_2S \text{ molecules}}{2 \text{ mol NO molecules}} \quad \text{and} \quad \frac{2 \text{ mol NO molecules}}{3 \text{ mol } H_2S \text{ molecules}}$$

Use the conversion factor derived from the coefficients in the balanced reaction so that the units of mol H_2S are in the denominator, and cancel.

$$? \text{ mol NO} = 1 \text{ mol } H_2S \times \frac{2 \text{ mol NO}}{3 \text{ mol } H_2S} = 2/3 \text{ mol NO}$$

CHECK: Does this make sense? Yes! If 3 mol of H_2S will form 2 mol of NO (2/3 as much), then 1 mol of H_2S will form 2/3 as much, or 2/3 (0.67) mol of NO.

(c)   Again, starting with Step 2 of the general problem-solving approach referenced in the HINT,

Step 2: $? \text{ mol } HNO_3 = 1 \text{ mol NO} \times ?$

Step 3:   The only relationship between NO and HNO_3 is the number of moles from the balance reaction. This relationship gives us the **equality** 2 mol NO = (are formed from, or produced by) 2 mol HNO_3, which is a **conversion factor** with the two forms

$$\frac{2 \text{ mol NO molecules}}{2 \text{ mol } HNO_3 \text{ molecules}} \quad \text{and} \quad \frac{2 \text{ mol } HNO_3 \text{ molecules}}{2 \text{ mol NO molecules}}$$

Use the conversion factor derived from the coefficients in the balanced reaction so that the units of mol NO are in the denominator, and cancel.

$$? \text{ mol } HNO_3 = 1 \text{ mol NO} \times \frac{2 \text{ mol } HNO_3}{2 \text{ mol NO}} = 1 \text{ mol } HNO_3$$

CHECK: Does this make sense? Yes! If 2 mol of NO are formed from 2 mol of HNO_3 (same quantity), then 1 mol of NO will require the same amount, or 1 mol of HNO_3.

SECTION 3.5   MASS RELATIONSHIPS IN CHEMICAL REACTIONS

After completing this section, you will be able to do the following:

- Given the names or formulas of reactants and products in a chemical reaction and the quantity of any reactant or product, calculate the quantities of any other reactant needed or product formed in the reaction.

REVIEW

Stoichiometry is the study of quantitative relationships between substances undergoing chemical changes, i.e., chemical reactions. In Section 3.1, you learned that **1 mol** of any formula contains 6.022×10^{23} units of that formula **and** has a mass in grams numerically equal to the formula mass. In Sections 3.3 and 3.4, you learned how to balance reactions and interpret the coefficients of a balanced reaction as numbers of molecules **and** moles of molecules. Now we will combine these concepts to show you how to perform some of the most important quantitative calculations in chemistry. That is, given a quantity (grams or moles) of any reactant or product in a reaction, calculate the quantities (in moles or grams) of any other reactant or product in that reaction.

SAMPLE PROBLEM 3.5 A

▶ ▶ A 130.0 g sample of silver(I) oxide, Ag_2O, is decomposed by heating according to the reaction,

$$Ag_2O(s) \rightarrow Ag(s) + O_2(g).$$

Calculate the mass in grams of (a) Ag formed, and (b) O_2 formed.

Solution:

HINT:

> Use the following procedure for performing all calculations involving mass and/or moles in any chemical reaction.
>
> 1. **Balance the reaction** and, underneath it, interpret the coefficients in terms of moles, indicating the formula masses of needed formulas. Above the reaction, indicate the information given and sought. Sometimes you will be given a chemical equation (i.e., balanced reaction), but many times, you will not. **Always check to make sure the reaction is balanced before proceeding.** An unbalanced reaction will usually give you incorrect or meaningless information.
>
> 2. **Convert the given quantity from grams to moles** by using the numerical value of the formula mass.
>
> 3. **Convert moles of the given quantity to moles of the quantity sought** by using the mole relationships from the balanced reaction.
>
> 4. **Convert the quantity sought from moles to grams** by using the numerical value of the formula mass.

(a)   Step 1:   The first step is critical to the rest of the procedure. The **only** way you can relate quantities of one formula to another formula in a chemical reaction is to use the coefficients (representing numbers of moles) from the **balanced reaction**. In this problem, the reaction is **not** balanced. Balance the reaction (see Sample Problems in Sections 1.10 and 3.3) and write the information indicated in this step.

Problem:	130.0 g		(a) ? g		(b) ? g
Equation:	$2Ag_2O(s)$	$\rightarrow$	$4Ag(s)$	$+$	$O_2(g)$.
Formula mass, u:	231.7358		107.8682		31.9988
Coefficients:	2 mol		4 mol		1 mol

Steps 2-4:   These steps indicate the conversion factors needed for the general problem-solving approach given in Sample Problem 2.2 C and applied to this chapter beginning with Sample Problem 3.1 A. **With practice, you should be able to combine these three steps as shown in the examples in**

this section.

$$? \text{ g Ag} \quad = \quad 130.0 \text{ g Ag}_2\text{O} \; \times \; \frac{1 \text{ mol Ag}_2\text{O}}{231.7358 \text{ g Ag}_2\text{O}} \; \times \; \frac{4 \text{ mol Ag}}{2 \text{ mol Ag}_2\text{O}} \; \times \; \frac{107.8682 \text{ g Ag}}{1 \text{ mol Ag}}$$

$$= \quad 121.0 \text{ g Ag} \qquad\qquad\qquad [4 \text{ s.f. limited by } 130.0 \text{ g Ag}_2\text{O}]$$

Step 2 converts grams of Ag_2O to moles of Ag_2O. You must have the given quantity (Ag_2O) converted to moles before you can do step 3. [If the given quantity is already in moles, then you don't need step 2.] Step 3 converts moles of Ag_2O (what is given) to moles of Ag (what is sought) by using the only relationship available between these two substances --- the coefficients interpreted as number of moles. Step 4 converts moles of what is sought (Ag) to the corresponding mass in grams. [If you are seeking the answer in moles, you don't need step 4.]

CHECK: Does this answer make sense? Yes! According to the balanced reaction, ~463 g of Ag_2O (2 mol Ag_2O × 231.7358 g/mol Ag_2O) will **always** produce ~431 g of Ag (4 mol Ag × 107.8682 g/mol Ag), a quantity slightly less than the starting mass of Ag_2O. Therefore, 130.0 g of Ag_2O should form a slightly smaller mass of Ag metal, or 121.0 g Ag.

(b)   Steps 2-4:   Using the information given above and below the balanced reaction in part (a), step 1,

$$? \text{ g O}_2 \quad = \quad 130.0 \text{ g Ag}_2\text{O} \; \times \; \frac{1 \text{ mol Ag}_2\text{O}}{231.7358 \text{ g Ag}_2\text{O}} \; \times \; \frac{1 \text{ mol O}_2}{2 \text{ mol Ag}_2\text{O}} \; \times \; \frac{31.9988 \text{ g O}_2}{1 \text{ mol O}_2}$$

$$= \quad 8.975 \text{ g O}_2$$

CHECK: Does this answer make sense? Yes! According to the balanced reaction, ~463 g of Ag_2O (2 mol Ag_2O × 231.7358 g/mol Ag_2O) will **always** produce ~32 g of O_2 (1 mol × 31.9988 g/mol O_2), a little less than 10% of the starting mass of Ag_2O. Therefore, 130.0 g of Ag_2O should form a little less than 10% by mass of O_2 gas, or 8.975 g O_2.

114

SAMPLE PROBLEM 3.5 B

▶▶ The reaction for the preparation of pure iron by the thermite process is

$$Al(s) + Fe_2O_3(s) \rightarrow Al_2O_3(s) + Fe(s)$$

(a) How many grams of Al are needed to completely react with 60.0 g of Fe_2O_3? (b) How many grams of Fe_2O_3 are needed to form 250.0 g of Fe, assuming excess Al metal? (c) How many moles of Al_2O_3 will be produced from 125.0 g of Al, assuming an excess of Fe_2O_3?

Solution:

HINT: Use the same procedure given as the HINT in Sample Problem 3.5 A.

(a)   Step 1:   In this problem, the reaction is **not** balanced. Balance the reaction (see Sample Problems in Sections 1.10 and 3.3) and write the information needed. Note that for part (a), the formula masses and coefficients in moles are only needed for Al and Fe_2O_3. However, you must always have a **balanced reaction** to obtain the correct coefficients.

Problem: ? g   60.0 g

Equation: 2Al(s)   + $Fe_2O_3(s) \rightarrow Al_2O_3(s)$ + 2Fe(s)

Formula mass, u: 26.981 539   159.692

Coefficients: 2 mol   1 mol

Steps 2-4:   These steps indicate the conversion factors needed for the general problem-solving approach given in Sample Problem 2.2 C and applied to this chapter beginning with Sample Problem 3.1 A. **With practice, you should be able to combine these three steps as shown in the examples in this section.**

$$? \text{ g Al} = 60.0 \text{ g Fe}_2\text{O}_3 \times \frac{1 \text{ mol Fe}_2\text{O}_3}{159.692 \text{ g Fe}_2\text{O}_3} \times \frac{2 \text{ mol Al}}{1 \text{ mol Fe}_2\text{O}_3} \times \frac{26.981\ 539 \text{ g Al}}{1 \text{ mol Al}}$$

$$= 20.3 \text{ g Al} \qquad\qquad [3 \text{ s.f. limited by } 60.0 \text{ g Fe}_2\text{O}_3]$$

(b)   Step 1:   Using the balanced reaction from part (a),

Problem:	? g		250.0 g
Equation:	2Al(s) +	$Fe_2O_3(s)$ →	$Al_2O_3(s)$ + 2Fe(s)
Formula mass, u:		159.692	55.847
Coefficients:		1 mol	2 mol

Steps 2-4:   **With practice, you should be able to combine these three steps as shown.**

$$? g\ Fe_2O_3 \ =\ 250.0\ \cancel{g\ Fe} \times \frac{1\ \cancel{mol\ Fe}}{55.847\ \cancel{g\ Fe}} \times \frac{1\ \cancel{mol\ Fe_2O_3}}{2\ \cancel{mol\ Fe}} \times \frac{159.692\ g\ Fe_2O_3}{1\ \cancel{mol\ Fe_2O_3}}$$

$$= \ 357.4\ g\ Fe_2O_3 \qquad\qquad \text{[4 s.f. limited by 250.0 g Fe]}$$

(c)   Step 1:   Using the balanced reaction from part (a),

Problem:	125.0 g		? mol
Equation:	2Al(s) +	$Fe_2O_3(s)$ →	$Al_2O_3(s)$ + 2Fe(s)
Formula mass, u:	26.981 539		101.9613
Coefficients:	2 mol		1 mol

   NOTE: You are calculating **moles** of Al_2O_3, not grams.

Steps 2-4:   **With practice, you should be able to combine these three steps as shown.**

$$? mol\ Al_2O_3 \ =\ 125.0\ \cancel{g\ Al} \times \frac{1\ \cancel{mol\ Al}}{26.981\ 539\ \cancel{g\ Al}} \times \frac{1\ \cancel{mol\ Al_2O_3}}{2\ \cancel{mol\ Al}}$$

$$= \ 2.316\ mol\ Al_2O_3 \qquad\qquad \text{[4 s.f. limited by 125.0 g Al]}$$

CHECK: Does this make sense? Yes! The balanced reaction says that it always takes 2 mol of Al to form 1 mol of Al_2O_3 (1/2 as many moles). The starting quantity of Al is ~4.5 mol (125.0 g Al/26.981 539 g/mol Al), which will form approximately 1/2 that number of moles of Al_2O_3, or 2.316 mol Al_2O_3.

116

SECTION 3.6   LIMITING REACTANTS

After completing this section, you will be able to do the following:

■ Given quantities of two or more reactants in a reaction, determine which reactant is the limiting reactant and calculate the excess quantity of the other reactant(s).

REVIEW

In the real world, the reactants are not usually present in the proportions required by the coefficients in the balanced reaction. The quantity of the one reactant that determines how much of the products are formed is called the **limiting reactant**. All other reactants are in excess. **You must be given quantities of two or more reactants in order to have a limiting reactant.** For example, if you have 10 flashlight batteries and 10 empty flashlight cases (no batteries), how many flashlights can you assemble? The "reaction" would be

Problem: 10 flashlight cases + 10 batteries → ? flashlights

"Reaction":   1 flashlight case + 2 batteries → 1 flashlight

The 10 cases **could** make 10 flashlights (1 case/flashlight) if there were enough batteries (2 batteries/flashlight or 20 batteries). However, with 10 batteries, you can make only 5 flashlights (2 batteries/flashlight) or 1/2 the number of flashlights possible. Thus the number of batteries is the **"limiting reactant"**, with 5 flashlight cases in excess (left over). **The quantity of the product(s) formed can be no greater than that allowed by the limiting reactant.**

SAMPLE PROBLEM 3.6 A

▶▶ A mixture of 15.0 g of aluminum and 20.0 g of sulfur is made to react to form aluminum sulfide according to the reaction,

$$Al(s) + S(s) → Al_2S_3(s)$$

(a) Which reactant is the limiting reactant? (b) How many grams of Al_2S_3 can be formed? (c) How many grams of the other reactant are left over (in excess)?

Solution:

HINT: For each reactant, calculate the number of moles and divide by the coefficient in the balanced reaction. The lowest number obtained is the limiting reactant.

(a)   First, you **must always** use a balanced chemical reaction (chemical equation).

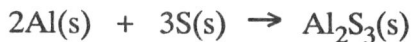

$$2Al(s) \ + \ 3S(s) \ \rightarrow \ Al_2S_3(s)$$

Then for each reactant, calculate the number of moles available and the **fraction** (or multiple) it is of the coefficient in the balanced reaction (called the "recipe" in the textbook). The **lowest fraction** is the **limiting reactant**.

$$? \text{ mol Al} \ = \ 15.0 \text{ g Al} \ \times \ \frac{1 \text{ mol Al}}{26.981\ 539 \text{ g Al}} \ = \ 0.556 \text{ mol Al}$$

$$? \text{ fraction Al} \ = \ \frac{0.556 \text{ mol Al}}{2 \text{ mol Al}} \ = \ 0.278$$

$$? \text{ mol S} \ = \ 20.0 \text{ g S} \ \times \ \frac{1 \text{ mol S}}{32.066 \text{ g S}} \ = \ 0.624 \text{ mol S}$$

$$? \text{ fraction S} \ = \ \frac{0.624 \text{ mol S}}{3 \text{ mol S}} \ = \ 0.208$$

The **limiting reactant** is the reactant with the **smallest fraction, sulfur**. Some aluminum will be left over; thus, there is excess aluminum.

(b)   To calculate the number of grams of Al_2S_3, set up the problem exactly the same as in Sample Problems 3.5 A & B, remembering that the amount of sulfur given will determine how much Al_2S_3 can be produced.

Problem:		20.0 g		? g
Equation:	$2Al(s)$ +	$3S(s)$	$\rightarrow$	$Al_2S_3(s)$
Formula mass, u:		32.066		150.161
Coefficients:		3 mol		1 mol

$$? \text{ mol Al}_2\text{S}_3 \ = \ 20.0 \ \cancel{\text{g S}} \ \times \ \frac{1 \ \cancel{\text{mol S}}}{32.066 \ \cancel{\text{g S}}} \ \times \ \frac{1 \ \cancel{\text{mol Al}_2\text{S}_3}}{3 \ \cancel{\text{mol S}}} \ \times \ \frac{150.161 \ \text{g Al}_2\text{S}_3}{1 \ \cancel{\text{mol Al}_2\text{S}_3}}$$

$$= \ 31.2 \ \text{g Al}_2\text{S}_3 \qquad\qquad \text{[3 s.f. determined by 20.0 g S]}$$

(c)   Some of the other reactant, Al, will be needed to react with all the limiting reactant, S. After the reaction is over, the leftover (unreacted or excess) Al will be the difference between the starting amount of Al and the amount used in the reaction:

$$\text{excess g Al} \ = \ (15.0 \text{ g Al}) - (\text{g Al used in the reaction}).$$

You can calculate the g Al used in the reaction two ways: from the limiting reactant S or the quantity of Al_2S_3 formed in part (b). Both methods are correct and give you the same answer. Choose the method that is easiest for you. Arbitrarily choosing the first method,

$$? \text{ g Al used} \ = \ 20.0 \ \cancel{\text{g S}} \ \times \ \frac{1 \ \cancel{\text{mol S}}}{32.066 \ \cancel{\text{g S}}} \ \times \ \frac{2 \ \cancel{\text{mol Al}}}{3 \ \cancel{\text{mol S}}} \ \times \ \frac{26.981 \ 539 \ \text{g Al}}{1 \ \cancel{\text{mol Al}}}$$

$$= \ 11.2 \text{ g Al} \qquad\qquad \text{[3 s.f. determined by 20.0 g S]}$$

$$? \text{ excess g Al} \ = \ 15.0 \text{ g Al} - 11.2 \text{ g Al} \ = \ 3.8 \text{ g Al}$$

SECTION 3.7   THEORETICAL, ACTUAL, AND PERCENT YIELDS

After completing this section, you will be able to do the following:

■ Calculate the theoretical and percent yield for a chemical reaction.

REVIEW

In Section 3.5, you were given a quantity of a reactant and asked to calculate the amount of product(s) formed. These calculations assumed that the maximum amount of product was formed under ideal conditions, which is called the **theoretical yield**. Normally, the amount of product formed is actually **less** than the **theoretical yield** (hence called the **actual yield**) due to impure starting materials, experimental errors, or unwanted **side reactions** (other reactions that consume some of the reactant). The **percent yield** of a reaction is the actual yield divided by the theoretical yield converted to percent:

$$\% \text{ yield} = \frac{\text{actual yield}}{\text{theoretical yield}} \times 100\%.$$

SAMPLE PROBLEM 3.7 A

▶▶ Given the following reaction, (a) what is the theoretical yield of Cu_2S from 75.0 g of Cu with an excess of S? (b) If 80.0 g of Cu_2S are actually obtained, what is the percent yield?

$$Cu(s) + S(s) \rightarrow Cu_2S(s)$$

Solution:

HINT: The amount of product calculated from a balanced reaction is always the theoretical yield.

(a)   The amount of product calculated in Sample Problem 3.5 A is the theoretical yield for that reaction. Using an identical procedure, you always start with a **balanced reaction and a table of data.**

Problem:	75.0 g			? g (theoretical yield)
Equation:	$2Cu(s)$	+	$S(s)$ →	$Cu_2S(s)$
Formula mass, u:	63.546			159.158
Coefficients:	2 mol			1 mol

$$? \text{ g } Cu_2S = 75.0 \text{ g Cu} \times \frac{1 \text{ mol Cu}}{63.546 \text{ g Cu}} \times \frac{1 \text{ mol Cu}_2S}{2 \text{ mol Cu}} \times \frac{159.158 \text{ g } Cu_2S}{1 \text{ mol Cu}_2S}$$

$$= 93.9 \text{ g } Cu_2S \qquad \qquad [3 \text{ s.f. determined by } 75.0 \text{ g Cu}]$$

(b)   The percent yield is

$$\% \text{ yield} = \frac{\text{actual yield}}{\text{theoretical yield}} \times 100\% = \frac{80.0 \text{ g } Cu_2S}{93.9 \text{ g } Cu_2S} \times 100\% = 85.2 \%$$

SECTION 3.8   QUANTITATIVE ANALYSIS

After completing this section, you will be able to do the following:

■ Calculate percent by mass of a substance in a compound that reacts quantitatively.

REVIEW

Quantitative analysis is finding out how much of a particular substance (usually an element) there is in a given sample. Given the mass of the sample and enough information to find the mass of one or more substances in the sample, you can calculate the **percent by mass** of each substance in the sample by

$$\text{Percent by mass of A in sample} = \frac{\text{mass A in sample}}{\text{total mass of sample}} \times 100\%,$$

where A is any substance in the sample.

SAMPLE PROBLEM 3.8 A

▶▶ If a 0.4352-g sample of a copper oxide is heated and decomposes into 0.3477 g of copper and an unknown amount of oxygen gas, (a) calculate the percent by mass of Cu, and (b) the percent by mass of oxygen in the sample.

Solution:

You aren't given enough information to write a balanced reaction (the copper oxide could be CuO or Cu_2O) nor do you need one for these quantitative reactions. However, it is often helpful to write down a general reaction and show the data given, in order to get a better picture of what is happening. In this problem,

Problem: 0.4352 g   0.3477 g

Reaction: $Cu_xO(s)$ → $Cu(s)$ + $O_2(g)$

(a)   Since you already know the mass of the sample and the mass of the Cu, you can calculate the percent by mass of Cu using the equation above:

$$\% \text{ of Cu in sample} = \frac{\text{g Cu}}{\text{g copper oxide}} \times 100\% = \frac{0.3477 \text{ g Cu}}{0.4352 \text{ g copper oxide}} \times 100\%$$

$$= 79.89\% \text{ Cu}$$

(b)   There are two ways to calculate the percent by mass of oxygen in the copper oxide. **First**, since there are only two elements in the sample, the percent Cu and percent O_2 must add up to 100%. Therefore, the percent O_2 = 100.00% - 79.89% = 20.11% O_2. This is the fastest method.

The other method takes a little longer but is a good check to show that your percent O_2 is correct. Since the total mass of the products **always** equals the total mass of the reactants, it is easy to quickly determine the mass of the oxygen gas produced, (0.4352 g copper oxide) - (0.3477 g Cu) = 0.0875 g O_2. The percent by mass of O_2 can then be calculated in the same way as copper:

$$\% \text{ of } O_2 \text{ in sample} = \frac{\text{g } O_2}{\text{g copper oxide}} \times 100\% = \frac{0.0875 \text{ g } O_2}{0.4352 \text{ g copper oxide}} \times 100\%$$

$$= 20.1\% \ O_2$$

Both methods give you nearly identical answers. Note that the answer found using the second method has one less s.f. because of the 0.0875 g 0_2.

SAMPLE PROBLEM 3.8 B

▶▶ A 0.1465-g sample of an organic compound containing only C, H, and S underwent combustion to quantitatively yield 0.2075 g CO_2, 0.1278 g H_2O, and an unknown amount of SO_2. What is the percent C, H, and S in the compound?

Solution:

Although you don't need to write a chemical reaction to work the problem, it often helps to see what is happening. When a substance undergoes combustion, it reacts with O_2 from the air. The general reaction for this problem is

Problem: 0.1465 g   0.2075 g   0.1278 g

Reaction: $C_xH_yS_z$ + O_2 → CO_2 + H_2O + SO_2

where x, y, and z indicate unknown subscripts in the formula for the starting compound. Starting with C, the percent by mass is

$$\text{Percent by mass of C in sample} = \frac{\text{g C}}{\text{g of sample}} \times 100\% = \frac{\text{g C}}{0.1465 \text{ g}} \times 100\%$$

However, the mass of C is not given; it will have to be calculated. **All the C in the 0.1465-g sample was quantitatively converted** to C in CO_2. To find out the mass of C in 0.2075 g of

CO_2, remember that we always know the mass in grams of 1 mol of CO_2 (44.010 g) and that 1 mol of CO_2 **always** contains 1 mol of C, with a mass of 12.011 g. This information always provides us with an equality or conversion factor, 44.010 g CO_2 = (or always contains) 12.011 g C. Then use the form of the conversion factor so that the proper units cancel.

$$? \text{ g C in } CO_2 = \text{ g } CO_2 \times \frac{\text{mass in g of 1 mol C}}{\text{mass in g of 1 mol } CO_2} = 0.2075 \text{ g } \cancel{CO_2} \times \frac{12.011 \text{ g C}}{44.010 \text{ g } \cancel{CO_2}}$$

$$= 0.056 \, 63 \text{ g C}$$

This is the same mass of C as in the starting compound of C, H, and S. Now you can calculate the percent by mass of C.

$$\% \text{ C} = \frac{\text{g C}}{\text{g sample}} \times 100\% = \frac{0.056 \, 63 \text{ g C}}{0.1465 \text{ g sample}} \times 100\% = 38.66\% \text{ C}$$

In a similar manner, the mass of H in H_2O is calculated by using the equality and conversion factor, 18.0153 g H_2O = (always contains) 2.0159 g H. The mass of H in H_2O and percent by mass of H in the sample are

$$? \text{ g H in } H_2O = \text{ g } H_2O \times \frac{\text{mass in g of 2 mol H}}{\text{mass in g of 1 mol } H_2O} = 0.1278 \text{ g } \cancel{H_2O} \times \frac{2.0159 \text{ g H}}{18.0153 \text{ g } \cancel{H_2O}}$$

$$= 0.014 \, 30 \text{ g H}$$

This is the same mass of H as in the starting compound of C, H, and S. The percent by mass of H is

$$\% \text{ H} = \frac{\text{g H}}{\text{g sample}} \times 100\% = \frac{0.014 \, 30 \text{ g H}}{0.1465 \text{ g sample}} \times 100\% = 9.761\% \text{ H}$$

Since there are only three elements present and you know the percentages of two of them, you can find the percent S by difference. That is,

$$\% \text{ S} = 100\% - (\% \text{ C} + \% \text{ H}) = 100.00\% - (38.66\% \text{ C} + 9.761\% \text{ H}) = 51.58\% \text{ S}$$

Again, you can also calculate the percent S by first finding the mass of S in the sample (you know the masses of C and H) and then using the percent by mass equation.

SECTION 3.9   EMPIRICAL FORMULAS FROM PERCENT COMPOSITION

After completing this section, you will be able to do the following:

■ Calculate empirical formulas from percent composition.

REVIEW

The **empirical formula** shows the lowest whole-numbered ratio of atoms of each element in the formula. For example, NO_2 is the empirical formula for nitrogen dioxide, because the subscripts are the lowest whole-numbered ratio (not divisible by any integer except 1). Dinitrogen tetroxide, N_2O_4, has two N atoms and 4 O atoms in each molecule. Both subscripts are divisible by two, so this is **not** the empirical formula for dinitrogen tetroxide. The **empirical formula** of dinitrogen tetroxide is NO_2. [We will see in Section 3.10 that N_2O_4 is the **molecular formula** for this compound.] Empirical formulas can be calculated from molecular formulas or, more commonly, from percent composition of a compound.

SAMPLE PROBLEM 3.9 A

▸ ▸ What are the empirical formulas for the compounds (a) P_4O_{10} (b) $C_{25}H_{50}$ (c) H_2SO_4 (d) $C_{12}H_{22}O_{11}$?

Solution:

(a)   Since both subscripts are divisible by 2, the empirical formula is P_2O_5.
(b)   Since both subscripts are divisible by 25, the empirical formula is CH_2.
(c)   The empirical formula is the formula given, H_2SO_4, because the subscripts are not divisible by any integer except 1.
(d)   The empirical formula is the formula given, $C_{12}H_{22}O_{11}$, for the same reason as in (c).

SAMPLE PROBLEM 3.9 B

▸ ▸ The composition of a compound containing Fe and Cl has been found to be 44.06% Fe (by mass). What is the empirical formula?

Solution:

HINT: 1. Since you are working with percentages, assume you have an exactly 100-g sample of the compound.

2. Calculate the number of moles of each element.
3. Divide the number of moles of each element by the **smallest** number of moles to obtain the lowest ratio of moles, which will also be the lowest ratio of atoms.

Step 1: First, you will need to determine the percentage of Cl, which is 100.00% - 44.06% = 55.94%. Then, in a 100-g sample, you will have 44.06 g of Fe and 55.94 g of Cl.

Step 2: Calculate the number of moles of Fe and Cl as you did in Section 3.1 of this book.

$$? \text{ mol Fe} = 44.06 \text{ g Fe} \times \frac{1 \text{ mol Fe}}{55.847 \text{ g Fe}} = 0.7889 \text{ mol Fe}$$

$$? \text{ mol Cl} = 55.94 \text{ g Cl} \times \frac{1 \text{ mol Cl}}{35.4527 \text{ g Cl}} = 1.578 \text{ mol Cl}$$

Step 3: Divide both numbers of moles by the smallest number of moles to obtain the simplest ratio of moles **and** atoms (since there are 6.022×10^{23} atoms in 1 mol).

$$\text{Fe} = \frac{0.7889}{0.7889} = 1.000 \qquad\qquad \text{Cl} = \frac{1.578}{0.7889} = 2.000$$

The simplest ratio of moles and atoms is 1.000 Fe to 2.000 Cl. Rounding to the nearest whole number, the empirical formula is $FeCl_2$.

SAMPLE PROBLEM 3.9 C

▶▶ The composition of a compound containing Cd, P, and O, is 63.94% Cd and 11.80% P (by mass). What is the empirical formula for this compound?

Solution:

Use the same procedure as in Sample Problem 3.9 B.

Step 1: First, you will need to determine the percentage of oxygen in the compound. Since you know the percentages of the other two elements and all three percentages must add up to 100%, the % O = 100.00% - (63.94% + 11.80%) = 24.26% O. Then, in a 100-g sample of the compound, you will have 63.94 g Cd, 11.80 g P, and 24.26 g O.

Step 2: Calculate the number of moles of Cd, P, and O using the method shown in Section 3.1 of this book.

$$? \text{ mol Cd} = 63.94 \text{ g Cd} \times \frac{1 \text{ mol Cd}}{112.411 \text{ g Cd}} = 0.5688 \text{ mol Cd}$$

$$? \text{ mol P} = 11.80 \text{ g P} \times \frac{1 \text{ mol P}}{30.973\ 762 \text{ g P}} = 0.3810 \text{ mol P}$$

$$? \text{ mol O} = 24.26 \text{ g O} \times \frac{1 \text{ mol O}}{15.9994 \text{ g O}} = 1.516 \text{ mol O}$$

Step 3: Divide all three numbers of moles by the smallest number of moles to obtain the simplest ratio of moles **and** atoms (since there are 6.022×10^{23} atoms in 1 mol).

$$Cd = \frac{0.5688}{0.3810} = 1.493 \qquad P = \frac{0.3810}{0.3810} = 1.000 \qquad O = \frac{1.516}{0.3810} = 3.979$$

The simplest ratio of moles and atoms is 1.493 Cd to 1.000 P to 3.979 O. Since whole numbers are needed and Cd is half-way between 1 and 2, multiply all the ratios by 2 and round off to the nearest whole number. The empirical formula is $Cd_3P_2O_8$.

SECTION 3.10   MOLECULAR AND STRUCTURAL FORMULAS

After completing this section, you will be able to do the following:

- Knowing the empirical formula and molecular mass for a compound, find the molecular formula;

- Given the condensed structural formula for a compound, write the structural formula, molecular formula, and empirical formula, and calculate the molecular mass.

REVIEW

As defined in Section 3.9, the **empirical formula** of a compound shows the lowest whole-numbered ratio of atoms of each element in the formula. The **molecular formula** shows **all** the atoms of each element in a **molecule**. The **molecular mass** is the mass in u's of one molecule. The **molecular formula** can be found if you know the **empirical formula and the molecular mass**, since

$$\text{molecular formula} = n \times \text{empirical formula, and } n = \frac{\text{molecular mass}}{\text{empirical unit mass}},$$

where n is an integer. If n = 1, then the empirical and molecular formulas are **identical**. If n = 2 or higher, then the molecular formula is a multiple of the empirical formula. For these calculations, the molecular mass need only be known to the nearest whole number.

Two other kinds of formulas are used to show how atoms are joined together to form molecules: a **structural formula** and a **condensed structural formula**. A **structural formula** uses lines to represent bonds that hold atoms together in molecules. A **condensed structural formula** is a structural formula **without** the bonds (lines) shown. The following example, normal butane, will illustrate the **four kinds of formulas** you need to know --- **molecular, empirical, structural, and condensed structural**.

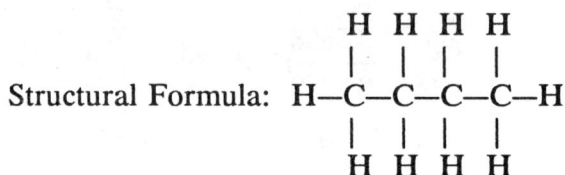

Molecular Formula: C_4H_{10}

Structural Formula:
$$\begin{array}{ccccc} & H & H & H & H \\ & | & | & | & | \\ H- & C- & C- & C- & C-H \\ & | & | & | & | \\ & H & H & H & H \end{array}$$

Empirical Formula: C_2H_5

Condensed Formula: $CH_3CH_2CH_2CH_3$

Note that (a) the molecular formula is 2 times the empirical formula, or n = 2, (b) from the structural formula the C atom forms four bonds and the H atom 1 bond, and (c) you can calculate the **molecular mass** (58.123 u) from all formulas **except** the empirical formula.

SAMPLE PROBLEM 3.10 A

▶▶ What is the molecular formula of a compound that has (a) the empirical formula CH_2 and a molecular mass of 28, and (b) the empirical formula NO_2 and a molecular mass of 92?

Solution:

Using the formulas given above, determine n from the molecular and empirical masses, and then the molecular formula.

(a) $\quad n = \dfrac{\text{molecular mass}}{\text{empirical mass}} = \dfrac{28 \text{ u}}{14 \text{ u}} = 2$

$\quad$ molecular formula $= n \times$ empirical formula $= 2 \times CH_2 = C_2H_4$

(b) $\quad n = \dfrac{\text{molecular mass}}{\text{empirical mass}} = \dfrac{92 \text{ u}}{46 \text{ u}} = 2$

$$\text{molecular formula} \; = \; n \; x \; \text{empirical formula} \; = \; 2 \; x \; NO_2 \; = \; N_2O_4$$

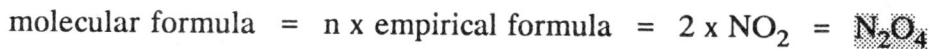

SAMPLE PROBLEM 3.10 B

▶▶ For each of the following condensed structural formulas, write the structural and molecular formulas: (a) CH_3CH_2OH (b) $CH_3CH_2CHClCH_2CH_3$ (c) $CH_3(CH_2)_4CH_2Br$

Solution:

HINT:
- In compounds, C has 4 bonds, O has 2 bonds, and H and Group VIIA elements have 1 bond.
- Start the structural formula by writing down the elements that form **more than 1 bond**, as they appear from left to right in the condensed structural formula. Then add the adjacent elements that form only one bond to complete the structure.

(a) The elements that form more than 1 bond are C and O. Start the formula with

$$C-C-O$$

Then add the atoms of H adjacent to the C and O atoms as shown in the condensed formula, which must agree with the first point in the HINT.

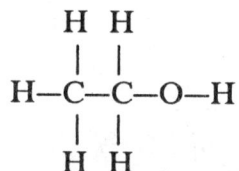

```
      H  H
      |  |
  H—C—C—O—H
      |  |
      H  H
```

The molecular formula can be written from either the condensed structural or complete structural formulas. For this compound, the **molecular formula is** C_2H_6O.

(b) The only element that forms more than 1 bond is C. Start the structural formula with

$$C-C-C-C-C$$

Then add the atoms of H and Cl adjacent to the C atoms in the condensed formula.

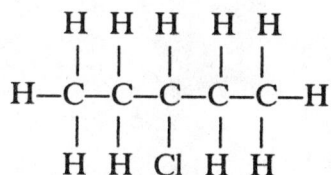

```
      H  H  H  H  H
      |  |  |  |  |
  H—C—C—C—C—C—H
      |  |  |  |  |
      H  H  Cl H  H
```

The **molecular formula** is $C_5H_{11}Cl$.

(c)   The only element that forms more than 1 bond is C. The grouping "$(CH_2)_4$" is an even further condensed method of writing "$CH_2CH_2CH_2CH_2$", to save space. Start the structural formula with

$$C-C-C-C-C-C$$

Then add the atoms of H and Br adjacent to the C atoms in the condensed formula.

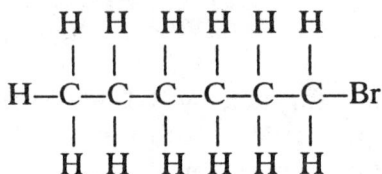

$$
\begin{array}{ccccccccccccc}
 & H & & H & & H & & H & & H & & H & \\
 & | & & | & & | & & | & & | & & | & \\
H- & C & - & C & - & C & - & C & - & C & - & C & -Br \\
 & | & & | & & | & & | & & | & & | & \\
 & H & & H & & H & & H & & H & & H &
\end{array}
$$

The **molecular formula** is $C_6H_{13}Br$.

CHECK: Do these answers make sense? Yes, because the atoms having 2 or more bonds (C and O) are bonded together in the same order as the condensed formula, and each C has 4 bonds, each O has 2 bonds, and each H, Cl, and Br has 1 bond.

SECTION 3.11   PERCENT COMPOSITION FROM FORMULAS

After completing this section, you will be able to do the following:

■ Calculate percent composition from the empirical or molecular formula of a compound.

REVIEW

Since the molecular formula of a compound is a multiple of the empirical formula (Section 3.10), you can calculate the **percent composition** of any element in the compound from **either the empirical or molecular formula.**

SAMPLE PROBLEM 3.11 A

▶▶ Acetic acid has the molecular formula $C_2H_4O_2$ and the empirical formula CH_2O. Calculate the percent by mass of each element (percent composition) in both formulas, to one decimal place.

Solution:

The percent by mass of any element, A, in a molecular formula is

$$\text{percent by mass A} = \frac{\text{total mass A}}{\text{molecular mass}} \times 100\%$$

For the molecular formula, $C_2H_4O_2$, the percent composition is (to one decimal place)

$$\% \, C = \frac{2 \times \text{atomic mass C}}{\text{molecular mass } C_2H_4O_2} \times 100\% = \frac{24.02 \, u}{60.05 \, u} \times 100\% = 40.0\% \, C$$

$$\% \, H = \frac{4 \times \text{atomic mass H}}{\text{molecular mass } C_2H_4O_2} \times 100\% = \frac{4.04 \, u}{60.05 \, u} \times 100\% = 6.7\% \, H$$

$$\% \, O = \frac{2 \times \text{atomic mass O}}{\text{molecular mass } C_2H_4O_2} \times 100\% = \frac{31.98 \, u}{60.05 \, u} \times 100\% = 53.3\% \, O$$

NOTE: Atomic and molecular masses were used to the second decimal place so that they could be rounded off in the final answer to one decimal place specified in the problem.

For the empirical formula, CH_2O, the percent composition will be **identical**, since both the number of C, H, and O atoms and the empirical formula mass are 1/2 of what they are in the molecular formula. You can calculate the percent composition of the empirical formula the same way that was used for the molecular formula to verify this statement.

SAMPLE PROBLEM 3.11 B

▶▶ A compound was analyzed and found to have a percent composition of 30.5% N and 69.5% O. Is the compound N_2O, N_2O_3, N_2O_4, or N_2O_5?

Solution:

Calculate the percent composition (to one decimal place) of each of the possible formulas (as shown in Sample Problem 3.11 A) to find out which formula matches the percent composition found in the analysis.

N_2O: $\%\ N\ =\ \dfrac{2 \times 14.01\ \text{u}}{44.01\ \text{u}}\ \times\ 100\%\ =\ 63.7\%\ N$

        $\%\ O\ =\ \dfrac{16.00\ \text{u}}{44.01\ \text{u}}\ \times\ 100\%\ =\ 36.4\%\ 0$

N_2O_3: $\%\ N\ =\ \dfrac{2 \times 14.01\ \text{u}}{76.01\ \text{u}}\ \times\ 100\%\ =\ 36.9\%\ N$

        $\%\ O\ =\ \dfrac{3 \times 16.00\ \text{u}}{76.01\ \text{u}}\ \times\ 100\%\ =\ 63.1\%\ O$

N_2O_4: $\%\ N\ =\ \dfrac{2 \times 14.01\ \text{u}}{92.01\ \text{u}}\ \times\ 100\%\ =\ 30.5\%\ N$

        $\%\ O\ =\ \dfrac{4 \times 16.00\ \text{u}}{92.01\ \text{u}}\ \times\ 100\%\ =\ 69.6\%\ O$

N_2O_5: $\%\ N\ =\ \dfrac{2 \times 14.01\ \text{u}}{108.01\ \text{u}}\ \times\ 100\%\ =\ 25.9\%\ N$

        $\%\ O\ =\ \dfrac{5 \times 16.00\ \text{u}}{108.01\ \text{u}}\ \times\ 100\%\ =\ 74.1\%\ O$

The percent composition that best matches the analysis is N_2O_4.

SELF-TEST QUESTIONS

Test your understanding of the concepts and skills in this chapter by working through the multiple choice questions **BEFORE** checking the answers. A periodic table will always be available. The ANSWERS TO SELF-TEST QUESTIONS follow immediately after this section.

1. Which is the largest quantity of chromium atoms?

    A. 4 dozen   B. 75 million   C. 2 gross   D. 1 mole   E. 500 billion

2. Exactly one mole of methane molecules, CH_4,

    A. has a mass of 12.011 g.
    B. contains 2.409×10^{24} H atoms.
    C. has a mass of 16.042 u.
    D. contains 16 molecules of CH_4
    E. is correctly described by more than one of the above.

3. A sample containing 1.51×10^{23} SO_3 molecules

    A. has a mass of 20.1 g.
    B. is 3.99 mol of SO_3 molecules.
    C. is 0.502 mol of SO_3 molecules.
    D. has a mass of 80.064 g.
    E. is 2.51 mol of SO_3 molecules.

4. The number of Cl atoms present in 0.00750 mol of PCl_5 molecules is

    A. 4.52×10^{21} B. 3.75×10^{-2} C. 2.26×10^{22} D. 2.71×10^{22} E. 5.00

5. When the following reaction is balanced with the lowest, whole-numbered coefficients, the coefficient of AgCl is

    $$AgNO_3(aq) \ + \ AlCl_3(aq) \ \rightarrow \ AgCl(s) \ + \ Al(NO_3)_3(aq)$$

    A. 1   B. 2   C. 3   D. 4   E. 5

6. In order to balance the following reaction, the lowest whole-numbered coefficient is

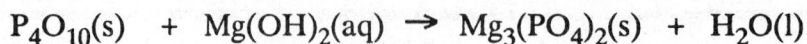

$$P_4O_{10}(s) + Mg(OH)_2(aq) \rightarrow Mg_3(PO_4)_2(s) + H_2O(l)$$

A. 3 for $Mg(OH)_2$.
B. 4 for $Mg_3(PO_4)_2$.
C. 3 for H_2O.
D. 1 for $Mg_3(PO_4)_2$.
E. 6 for $Mg(OH)_2$.

7. How many moles of oxygen can be prepared by the decomposition of 4.000 mol of HgO according to the following reaction?

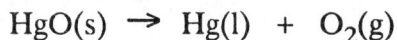

$$HgO(s) \rightarrow Hg(l) + O_2(g)$$

A. 4.000 mol
B. 2.000 mol
C. 8.000 mol
D. 1.000 mol
E. 0.5000 mol

8. Using the reaction given in the previous problem, a sample of HgO is heated to form 2.90 g of O_2. How many grams of Hg metal are formed during that same reaction?

A. 36.4 g   B. 2.90 g   C. 18.2 g   D. 1.45 g   E. 9.10 g

9. How many grams of F_2 will be required (assume excess Fe) to prepare 169.3 g of FeF_3 according to the following reaction?

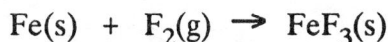

$$Fe(s) + F_2(g) \rightarrow FeF_3(s)$$

A. 57.01 g   B. 42.76 g   C. 38.01 g   D. 85.51 g   E. 19.00 g

10. When 34.5 g of Na react with excess water according to the following reaction, it will

$$Na(s) + H_2O(l) \rightarrow NaOH(aq) + H_2(g)$$

A. produce 40.0 g of NaOH.
B. require 2.00 mols of H_2O.
C. produce 1.51 g of H_2.
D. require 36.0 g of H_2O.
E. produce 1.00 mol of NaOH.

11. How many grams of $BaCl_2$ (in excess $Al_2(SO_4)_3$) are needed to produce 50.0 g of $AlCl_3$ according to the following chemical equation (i.e., balanced reaction)?

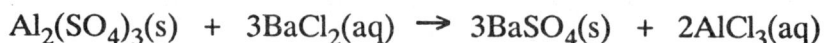

$$Al_2(SO_4)_3(s) + 3BaCl_2(aq) \rightarrow 3BaSO_4(s) + 2AlCl_3(aq)$$

A. 78.1 g   B. 52.1 g   C. 48.0 g   D. 32.0 g   E. 117 g

12. Starting with 5.00 g of Li and 10.0 g of O_2, how many grams of Li_2O can be formed from the following reaction?

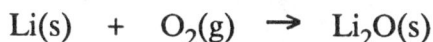

$$Li(s) + O_2(g) \rightarrow Li_2O(s)$$

A. 18.7 g   B. 10.8 g   C. 21.5 g   D. 43.1 g   E. 2.50 g

13. A 0.0850-g sample of a hydrocarbon (contains only carbon and hydrogen) was burned to yield 0.267 g CO_2 and 0.109 g H_2O. The percent by mass of C and H in this compound is

A. 85.7% C and 14.3% H.
B. 71.0% C and 29.0% H.
C. 92.8% C and 7.2% H.
D. 77.4% C and 22.6% H.
E. 80.5% C and 19.5% H.

14. All of the following compounds have different empirical and molecular formulas except

A. $C_{10}H_{22}$ B. Al_2Cl_6 C. $C_{15}H_{30}O_2$ D. P_4O_{10} E. $C_2H_4O_2$

15. The composition of a compound containing only Mg and N is 72.24% Mg by mass. The empirical formula for this compound is

A. MgN
B. Mg_2N_3
C. MgN_2
D. Mg_3N
E. Mg_3N_2

16. The molecular formula of a compound that has the empirical formula P_2O_5 and a molecular mass of 284 is

A. P_2O_5
B. $PO_{2.5}$
C. P_4O_5
D. P_4O_{10}
E. P_2O_{10}

17. For the compound, $CH_3(CH_2)_2OH$,

A. the empirical formula is CH_2O.
B. the molecular formula is C_3H_8O.
C. the complete structural formula is shown in the statement of the problem.
D. the empirical formula is CH_3O.
E. the molecular formula cannot be determined.

18. The percent composition (percent by mass of each element) of $Mg(NO_3)_2$ to two decimal places is

A. 16.39% Mg, 18.89% N, 64.72% O.
B. 28.16% Mg, 16.23% N, 55.61% O.
C. 44.75% Mg, 25.79% N, 29.46% O.
D. 18.10% Mg, 10.43% N, 71.47% O.
E. 12.25% Mg, 13.76% N, 73.99% O.

ANSWERS TO SELF-TEST QUESTIONS

1. d	7. b	13. a
2. b	8. a	14. c
3. a	9. d	15. e
4. c	10. c	16. d
5. c	11. e	17. b
6. e	12. b	18. a

PRACTICE PROBLEMS

Test your problem solving skills in this chapter by working through the problems in the space provided **BEFORE** checking with the answers. A periodic table will always be available. The ANSWERS TO PRACTICE PROBLEMS follows immediately after this section.

1. What is the mass in grams of 0.400 mol of $Ca(OH)_2$?

2. Balance the following reaction with the lowest, whole-numbered coefficients,

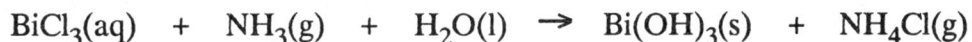

$$BiCl_3(aq) + NH_3(g) + H_2O(l) \rightarrow Bi(OH)_3(s) + NH_4Cl(g)$$

3. How many moles of H_2SO_4 will be needed to completely react with 0.15 mol of NaOH according to the following reaction?

$$NaOH(aq) + H_2SO_4(aq) \rightarrow Na_2SO_4(aq) + H_2O(l)$$

4. Using the reaction in the previous problem, how many grams of NaOH are needed to react with excess H_2SO_4 to prepare 60.0 g of Na_2SO_4?

5. Starting with 27.0 g of Al and 27.0 g of I_2, how many grams of Al_2I_6 can be prepared according to the following reaction?

$$Al(s) + I_2(s) \rightarrow Al_2I_6(s)$$

6. A 0.4000-g sample of an organic compound containing only C, H, and O underwent combustion to quantitatively yield 0.9765 g CO_2 and 0.3997 g H_2O. What is the percent C, H, and O in the compound?

7. The composition of a compound containing C, H, and O is 40.92% C, 4.579% H, and 54.50% O (all mass percent). (a) How many moles of each element are present in a 100.0-g sample of the compound? (b) What is the empirical formula for the compound?

8. What is the complete structural formula for the compound $CH_3(CH_2)_6CH_2Cl$?

9. A compound containing only Cl and O was analyzed and found to contain 38.77% Cl by mass. Is the compound Cl_2O, Cl_2O_3, Cl_2O_5, or Cl_2O_7?

ANSWERS TO PRACTICE PROBLEMS

1. 29.6 g $Ca(OH)_2$

2. 1,3,3 → 1,3

3. 0.075 mol H_2SO_4

4. 33.8 g NaOH

5. 28.9 g Al_2I_6 [$2Al + 3I_2 \rightarrow Al_2I_6$; limiting reactant is I_2]

6. 66.63% C; 11.18% H; 22.2% O

7. (a) 3.407 mol C, 4.543 mol H, 3.406 mol O   (b) $C_3H_4O_3$

8.
$$
\begin{array}{c}
\text{H H H H H H H H} \\
| \ | \ | \ | \ | \ | \ | \ | \\
\text{H—C—C—C—C—C—C—C—C—Cl} \\
| \ | \ | \ | \ | \ | \ | \ | \\
\text{H H H H H H H H}
\end{array}
$$

9. Cl_2O_7

CHAPTER 4

REACTIONS IN SOLUTION

SECTION 4.1   SOME IMPORTANT DEFINITIONS

After completing this section, you will be able to do the following:

- Define solutions as unsaturated, saturated, or supersaturated, and as dilute or concentrated;

- Given the solubility of a substance, classify a solution of it as unsaturated, saturated, or supersaturated.

REVIEW

There are two kinds of mixtures, homogeneous and heterogeneous, both of which can be separated into their components by a **physical change** (Section 1.3). Homogeneous mixtures are more commonly called **solutions**. Unlike compounds, **solutions** can be prepared in varying proportions of the **solvent** and one or more **solutes**. In most instances, the substance present in greatest quantity is called the **solvent**, while the substance(s) present in lesser quantity is the **solute**. For example, vinegar is a solution containing 3-5% acetic acid by weight and the rest of the solution is water; acetic acid is the **solute** and water is the **solvent**. The **solvent** may be a solid (with solid, liquid, or gas solute), liquid (with solid, liquid, or gas solute), or gas (with solid, liquid, or gas solute), resulting in nine combinations of the three phases or nine kinds of solutions. The most important solutions in general chemistry are those in which water is the solvent mixed with a solid or liquid solute.

A **saturated** solution contains as much solute as will dissolve at a given temperature in the presence of undissolved solute. The amount of the solute that dissolves to produce a saturated solution is called the solubility of the solute. For example, the solubility of $PbCl_2(s)$ is 0.99 g per 100 mL of water at 20 °C (Table 4.1 of the textbook). This means that you will have a **saturated** solution of $PbCl_2$ after 0.99 g of solid $PbCl_2$ has been dissolved in 100 mL of water at 20 °C. Solubility of solids and liquids generally increases with temperature, as shown by the fact that $PbCl_2$ has a solubility of 3.34 g per 100 mL of water at 100 °C. An **unsaturated** solution has less dissolved solute than a saturated solution, and a **supersaturated** solution contains more solute than a saturated solution. **Supersaturated** solutions are so unstable that adding a tiny crystal of solute or even sound vibrations will cause additional solute to **precipitate** from the solution until the solution is only saturated.

The **concentration** of a solution is the amount of solute dissolved in a given quantity of solvent or solution. The terms **dilute** and **concentrated** refer to solutions of relatively **low** and **high** concentrations, respectively.

SAMPLE PROBLEM 4.1 A

▶▶ Potassium nitrate, KNO_3, has a solubility of 11.0 g per 100 mL H_2O at 0 °C. (a) How many grams of KNO_3 would you have to dissolve in 100 mL of water at 0 °C to prepare a saturated solution, an unsaturated solution, and a supersaturated solution? (b) Which substance is the solute and which substance is the solvent in each of these solutions?

Solution:

(a)   Since the solubility of a substance, by definition, is the amount of the substance that dissolves in a given quantity of solvent to produce a saturated solution, exactly 11.0 g of KNO_3 would have to be dissolved in 100 mL of water at 0 °C to prepare a **saturated** solution, with additional KNO_3 solid on the bottom of the container. An **unsaturated** solution of KNO_3 will have less than the solubility, or less than 11.0 g of KNO_3 dissolved in 100 mL of water at 0 °C, e.g., 5.2 g, 0.0055 g, 10.9 g, etc. A **supersaturated** solution of KNO_3 will have more than the solubility, or more than 11.0 g of KNO_3 dissolved in 100 ml of water at 0 °C, e.g., 11.1 g, 12.4 g, etc.

(b)   In all three solutions there will be far more H_2O than KNO_3. Using the definitions given above, the solvent is H_2O, and the solute is KNO_3.

SECTION 4.2   ELECTROLYTES

After completing this section, you will be able to do the following:

■ Define strong, weak, and non-electrolytes in terms of their conductivity;

■ Classify electrolytes as acids, bases, or salts;

REVIEW

Electrolytes are compounds that conduct an electrical current when dissolved or melted, by the formation **and** movement of ions. The greater the number of ions formed, the higher the **conductivity** (ability to carry an electrical current) of that electrolyte. In solution, **strong electrolytes** have a **high conductivity** because most of the solute molecules have ionized (formed ions). This is true only if the strong electrolyte is **soluble; insoluble** strong electrolytes will have a **low** conductivity because, even in a saturated solution, the number of ions formed will be too small. In a solution of a soluble **weak electrolyte**, the **conductivity** will be **low** because most of the solute molecules have **not** ionized (relatively few ions formed). A solution of a **nonelectrolyte** conducts **no electrical current** because essentially no ions are present.

Electrolytes are compounds classified as either **acids, bases, or salts**. Acids and bases are classified as strong or weak (electrolytes) by the number of ions formed in solution. Soluble salts are almost always strong electrolytes. The simplest definition of an **acid** is a compound that increases the concentration of hydrogen ions (H^+) when dissolved in water. The unit particles of acids are molecules; non-metals form acids. The normal convention for writing molecular formulas of acids that contain H is to **begin** the formula with the number of H atoms that form H^+ ions when that acid is dissolved in water. For example, the formula for phosphoric acid, H_3PO_4, tells you that each molecule forms 3 H^+ in solution,

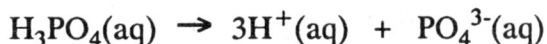

$$H_3PO_4(aq) \rightarrow 3H^+(aq) + PO_4^{3-}(aq)$$

However, the formula for formic acid, $HCHO_2$, **begins** with only 1 H, which tells you that each molecule forms only 1 H^+ in solution,

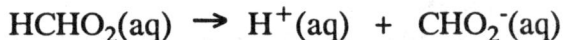

$$HCHO_2(aq) \rightarrow H^+(aq) + CHO_2^-(aq)$$

The H atoms at the beginning of the formulas of these two acids are bonded to O atoms; these bonds are broken by water molecules to form H^+ ions . H atoms **not** at the beginning of formulas are bonded to atoms whose bonds are not easily broken in water (the second H in $HCHO_2$ is bonded to a C atom).

The simplest definition of a **base** is a compound that increases the concentration of hydroxide ions (OH^-) when dissolved in water. The unit particles of bases are normally ions; metals form bases. The convention for writing formulas of bases that contain the OH^- ion is the same as that used in writing the formulas of ionic compounds.

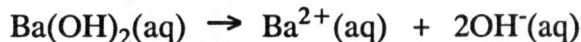

$$Ba(OH)_2(aq) \rightarrow Ba^{2+}(aq) + 2OH^-(aq)$$

The common compound ammonia, NH_3, is a base but has no OH^- ion in its formula. Ammonia undergoes an ionization reaction when dissolved in water.

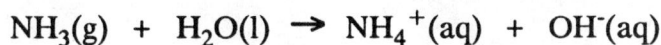

$$NH_3(g) + H_2O(l) \rightarrow NH_4^+(aq) + OH^-(aq)$$

The N atom pulls a H^+ ion away from a water molecule, leaving an OH^- ion as one of the products. This same reaction is typical of many compounds which contain the element nitrogen, N, and function as bases.

Salts are neither acids nor bases, but rather are compounds of metals (or polyatomic cations) and nonmetals (including polyatomic ions, **except OH^- and O^{2-}**). Metal hydroxides and oxides are normally classified as **bases**. Examples of salts are $NaCl$, $CuCl_2$, K_2SO_4, NH_4NO_3, Na_3PO_4, etc. The word salt is a generic name that includes many more compounds than ordinary table salt, $NaCl$. When an acid reacts with a base, a salt and water are formed, or **neutralization** occurs:

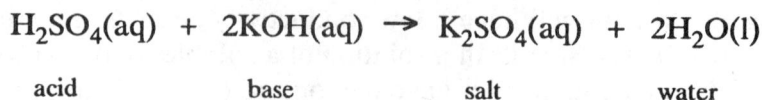

$$H_2SO_4(aq) + 2KOH(aq) \rightarrow K_2SO_4(aq) + 2H_2O(l)$$

$$\text{acid} \qquad\qquad \text{base} \qquad\qquad \text{salt} \qquad\qquad \text{water}$$

That is, if the acid and base are mixed in the proportions shown by the chemical equation (balanced reaction), the H_2SO_4 will completely **neutralize** the KOH, leaving only the products K_2SO_4 (a salt) and water. The cation of the salt, K^+, comes from the base, and the anion of the salt, SO_4^{2-}, comes from the acid.

SAMPLE PROBLEM 4.2 A

▶▶ In each of the following formulas, how many H atoms form H^+ ions in aqueous solutions? (a) H_2SO_4 (b) NaOH (c) H_3AsO_4 (d) $HC_4H_7O_2$ (e) C_6H_6

Solution:

Only the H atoms at the **beginning** of the formula ionize to form H^+ ions in water solution. Therefore, the only formulas in this problem that form H^+ ions in water are (a) 2 H atoms, (c) 3 H atoms, and (d) 1 H atom.

SAMPLE PROBLEM 4.2 B

▶▶ Classify the following compounds as acids, bases, or salts: (a) $HClO_4$ (b) $(NH_4)_2CO_3$ (c) $Sr(OH)_2$ (d) $HBrO_3$ (e) ZnO (f) HN_3 (g) CH_3NH_2 (h) $AgNO_3$.

Solution:

Use the definitions given in the REVIEW section.

Acids have a formula which begins with H:

(a) $HClO_4$, (d) $HBrO_3$, and (f) HN_3.

NOTE: HN_3 is an acid; NH_3 is a base.

Bases normally have a formula which **starts** with a metal (or NH_4^+) and **ends** with OH or O:

(c) $Sr(OH)_2$, (e) ZnO, and (g) CH_3NH_2.

NOTE: The N in CH_3NH_2 attracts a H^+ from H_2O just like the N in NH_3, i.e.,

$CH_3NH_2(aq) + H_2O(l) \rightarrow CH_3NH_3^+(aq) + OH^-(aq)$.

Salts have a formula which starts with a metal (or NH_4^+) and **ends** with a non-metal (or polyatomic anion, except OH^- and O^{2-}):

(b) $(NH_4)_2CO_3$, and (h) $AgNO_3$.

SAMPLE PROBLEM 4.2 C

▶ ▶ Write the formulas **and** names of each of the following: (a) two sodium salts of sulfuric acid, (b) calcium salt of acetic acid, (c) two potassium salts of sulfurous acid, (d) three ammonium salts of phosphoric acid.

Solution:

The neutralization reaction given in the REVIEW shows that the cation of the salt comes from the base and the anion of the salt comes from the acid. In each case, the metal ion (cation) is given. Then, one at a time, replace each of the H atoms at the **beginning** of the formula of the acid by a metal ion. [See examples in Table 4.2 of the textbook.]

(a) Na^+ and H_2SO_4 — $NaHSO_4$ — Sodium hydrogen sulfate
Na_2SO_4 — Sodium sulfate

(b) Ca^{2+} and $HC_2H_3O_2$ — $Ca(C_2H_3O_2)_2$ — Calcium acetate

(c) K^+ and H_2SO_3 — $KHSO_3$ — Potassium hydrogen sulfite
K_2SO_3 — Potassium sulfite

(d) NH_4^+ and H_3PO_4 — $NH_4H_2PO_4$ — Ammonium dihydrogen phosphate
$(NH_4)_2HPO_4$ — Ammonium hydrogen phosphate
$(NH_4)_3PO_4$ — Ammonium phosphate

144

SAMPLE PROBLEM 4.2 D

▶▶ Write a chemical equation (balanced reaction) for the neutralization in aqueous solution of (a) perchloric acid with potassium hydroxide, and (b) carbonic acid (both hydrogen ions react) with sodium hydroxide.

Solution:

Every neutralization reaction has the following reactants and products, regardless of their formula:

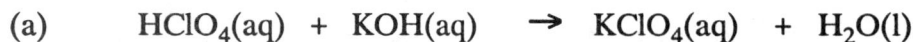

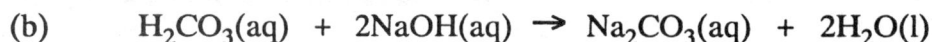

$$Acid \quad + \quad Base \quad \rightarrow \quad Salt \quad + \quad Water$$

(a) $\quad HClO_4(aq) + KOH(aq) \rightarrow KClO_4(aq) + H_2O(l)$

(b) $\quad H_2CO_3(aq) + 2NaOH(aq) \rightarrow Na_2CO_3(aq) + 2H_2O(l)$

SECTION 4.3   REACTIONS BETWEEN IONS IN SOLUTION

After completing this section, you will be able to do the following:

■ Classify a compound as a strong electrolyte, weak electrolyte, or a non-electrolyte;

■ Classify a compound as soluble, slightly soluble, or insoluble;

■ Predict whether a reaction will take place between ions in solution.

REVIEW

To **classify a compound as a strong, weak, or non-electrolyte,** you will need to know the four summary statements in Table 4.3 of the textbook, which collectively will allow you to make decisions for hundreds of compounds. If a compound is not covered in this table, it is a non-electrolyte. To **classify a compound as soluble, slightly soluble, or insoluble,** you will need to know how to interpret and apply the solubility rules in Table 4.4 of the textbook. Again, knowledge of the rules in this table will allow you to make decisions on solubility or formation of a precipitate for hundreds of compounds.

Using the information on strong and weak electrolytes from Table 4.3 and the solubility rules from Table 4.4, you can now predict that when solutions of electrolytes are mixed, a reaction will take place **only if one or more products are insoluble or a weak electrolyte.** Otherwise, no reaction occurs.

SAMPLE PROBLEM 4.3 A

▶▶ Classify each of the following compounds as a strong electrolyte, weak electrolyte, or non-electrolyte: (a) NH_3 (b) HCl (c) NaOH (d) $MgSO_4$ (e) H_2SO_3 (f) H_3PO_4 (g) K_2CO_3 (h) C_2H_6O (ethyl alcohol).

Solution:

HINT: You need to know (memorize) the information in Table 4.3.

The classifications and applicable summary statements from Table 4.3 are:

Strong electrolytes:	(b) HCl (#2)	(c) NaOH (#3)	(d) $MgSO_4$ (#4)
	(g) K_2CO_3 (#4)		
Weak electrolytes:	(a) NH_3 (#3)	(e) H_2SO_3 (#2)	(f) H_3PO_4 (#2)
Non-electrolyte:	(h) C_2H_6O (not acid, base, or salt; must be non-electrolyte)		

SAMPLE PROBLEM 4.3 B

▶▶ Classify each of the following compounds as soluble, slightly soluble, or insoluble: (a) $HClO_3$ (b) $V(NO_3)_3$ (c) Cr_2S_3 (d) $PbCl_2$ (e) $(NH_4)_2CO_3$ (f) $SO_2(g)$ (g) CuI_2 (h) AgCl

Solution:

HINT: You will need to know the information in Table 4.4 of the textbook.

The classifications and applicable rules from Table 4.4 are:

Soluble:	(a) $HClO_3$ (#1)	(b) $V(NO_3)_3$ (#3)
	(e) $(NH_4)_2CO_3$ (#2 & 8)	(g) CuI_2 (#4)
Insoluble:	(c) Cr_2S_3 (#6)	(h) AgCl (#4)
Slightly soluble:	(d) $PbCl_2$ (#4)	(f) SO_2 (#9)

SAMPLE PROBLEM 4.3 C

▸▸ Predict whether a reaction will take place when each of the following pairs of aqueous solutions are mixed. If a reaction occurs, write its equation (balanced reaction). (a) potassium carbonate and hydrochloric acid, (b) potassium phosphate and ammonium chloride, (c) bismuth(III) chloride and sodium sulfide, (d) calcium nitrate and sodium acetate, (e) potassium hydroxide and iron(III) chloride.

Solution:

HINT: A reaction will occur if a weak electrolyte and/or an insoluble (or slightly soluble) substance **is formed** (see Tables 4.3 and 4.4 in textbook).

(a) The two compounds given are soluble and will produce the following four ions:

$$K^+, CO_3^{2-} \text{ and } H^+, Cl^-$$

The only other ion K^+ would be attracted to is the other negative ion, Cl^-. Similarly, the only other ion that H^+ would be attracted to is the oppositely charged CO_3^{2-}. The two **possible new products** are KCl and H_2CO_3. KCl is a salt which is always a strong electrolyte. H_2CO_3 is a weak acid which always breaks down immediately into CO_2 (a slightly soluble gas) and H_2O (a weak electrolyte). The following reaction will occur because of the formation of a weak electrolyte (H_2O) and a slightly soluble gas (CO_2):

$$K_2CO_3(aq) + 2HCl(aq) \rightarrow 2KCl(aq) + H_2O(l) + CO_2(g)$$

(b) Let's shorten the process for the remaining parts of the problem.

Ions present initially: $K^+, PO_4^{3-}, NH_4^+, Cl^-$

Possible new products: KCl and $(NH_4)_3PO_4$

Any insoluble or weak electrolytes? No! [See solubility rules 2,4,8.]

Any chemical reaction? No!

(c) Ions present initially: $Bi^{3+}, Cl^-, Na^+, S^{2-}$

Possible new products: Bi_2S_3 and NaCl

Any insoluble or weak electrolytes? Yes! Bi_2S_3 is insoluble. [See solubility rule 6.]

Any chemical reaction?   Yes!

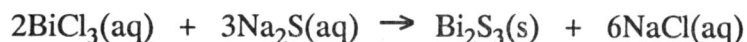

$$2BiCl_3(aq) + 3Na_2S(aq) \rightarrow Bi_2S_3(s) + 6NaCl(aq)$$

(d)   Ions present initially: Ca^{2+}, NO_3^-, Na^+, $C_2H_3O_2^-$

Possible new products: $Ca(C_2H_3O_2)_2$ and $NaNO_3$

Any insoluble or weak electrolytes? No!   [See solubility rules 2 and 3.]

Any chemical reaction?   No!

(e)   Ions present initially: K^+, OH^-, Fe^{3+}, Cl^-

Possible new products: KCl and $Fe(OH)_3$

Any insoluble or weak electrolytes? Yes! $Fe(OH)_3$ is insoluble.   [Solubility rule 7.]

Any chemical reaction?   Yes!

$$3KOH(aq) + FeCl_3(aq) \rightarrow Fe(OH)_3(s) + 3KCl(aq)$$

SECTION 4.4   IONIC EQUATIONS

After completing this section, you will be able to do the following:

■ Write a molecular equation, complete ionic equation, and net ionic equation for a reaction between strong electrolytes.

REVIEW

The chemical equations (balanced reactions) written so far in aqueous solution show all compounds in the form of molecules; these are called **molecular equations**.   Strong electrolytes, however, exist as ions in solution, not molecules. The equation that shows substances **as they exist in aqueous solution** --- strong electrolytes as ions and weak electrolytes and insoluble substances as molecules --- is called a **complete ionic equation**. A shorter form of the complete ionic equation, showing only the species that **take part in the reaction**, is called the **net ionic equation**.

For example, when nitric acid is added to an aqueous solution of sodium carbonate, the three kinds of equations are:

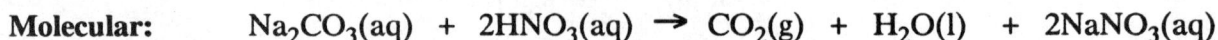

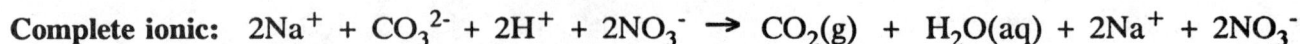

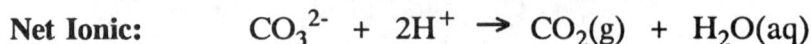

Molecular: $\qquad Na_2CO_3(aq) + 2HNO_3(aq) \rightarrow CO_2(g) + H_2O(l) + 2NaNO_3(aq)$

Complete ionic: $2Na^+ + CO_3^{2-} + 2H^+ + 2NO_3^- \rightarrow CO_2(g) + H_2O(aq) + 2Na^+ + 2NO_3^-$

Net Ionic: $\qquad CO_3^{2-} + 2H^+ \rightarrow CO_2(g) + H_2O(aq)$

Note that from this point on, the (aq) is not shown after the **ions** in **ionic equations only** to save time in writing them. Ions like H^+ and NO_3^-, which do **not** take part in the reaction (they have identical formulas and charges on both sides of the equation), are called **spectator ions**. They "cancel" each other out just like in an algebraic equation. These reactions between ions in solution to form insoluble substances and weak electrolytes are sometimes called **metathesis** or **double replacement reactions**.

SAMPLE PROBLEM 4.4 A

▸▸ (a) Write the molecular, complete ionic, and net ionic equations for the reaction that occurs when solutions of potassium acetate and hydrochloric acid are mixed. (b) Write the molecular equation for two other reactions that have the same net ionic equation.

Solution:

(a)   Using the procedure discussed in Section 4.3, the reaction will occur because the weak electrolyte acetic acid, $HC_2H_3O_2$, is one of the two products formed. The molecular equation (balanced reaction) is

$$KC_2H_3O_2(aq) + HCl(aq) \rightarrow HC_2H_3O_2(aq) + KCl(aq)$$

All salts ($KC_2H_3O_2$) and the acid HCl are strong electrolytes (see Table 4.3 in textbook) and exist as ions in aqueous solution. The complete ionic equation is

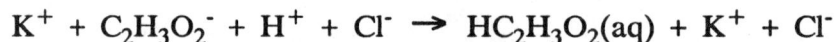

$$K^+ + C_2H_3O_2^- + H^+ + Cl^- \rightarrow HC_2H_3O_2(aq) + K^+ + Cl^-$$

The K^+ and Cl^- ions are called spectator ions because they appear on both sides of the reaction but do not take part in the reaction (i.e., both neither come from nor form a weak electrolyte or insoluble substance). They are important only when the source of the $C_2H_3O_2^-$ and H^+ ions is needed (e.g., when measuring starting quantities in the laboratory). Like an algebraic equation, identical spectator ions "cancel" each other out to produce the net ionic equation

$$C_2H_3O_2^- + H^+ \rightarrow HC_2H_3O_2(aq)$$

(b) Other molecular equations with the **same** net ionic reaction would require (a) any other metal **salt** of the $C_2H_3O_2^-$ ion, and (b) any other acid which is listed as a strong electrolyte in Table 4.3. Two such examples are

$$Cu(C_2H_3O_2)_2(aq) + 2HBr(aq) \rightarrow 2HC_2H_3O_2(aq) + CuBr_2(aq)$$

$$NaC_2H_3O_2(aq) + HNO_3(aq) \rightarrow HC_2H_3O_2(aq) + NaNO_3(aq)$$

SAMPLE PROBLEM 4.4 B

▶▶ Write the complete and net ionic equations for any reactions that occur in Sample Problem 4.3 C.

Solution:

Following the procedure used in Sample Problem 4.4 A,

(a) **Molecular:** $K_2CO_3(aq) + 2HCl(aq) \rightarrow 2KCl(aq) + H_2O(l) + CO_2(g)$

Complete Ionic: $2K^+ + CO_3^{2-} + 2H^+ + 2Cl^- \rightarrow 2K^+ + 2Cl^- + H_2O(l) + CO_2(g)$

Net Ionic: $CO_3^{2-} + 2H^+ \rightarrow H_2O(l) + CO_2(g)$

(b) No Reaction!

(c) **Molecular:** $2BiCl_3(aq) + 3Na_2S(aq) \rightarrow Bi_2S_3(s) + 6NaCl(aq)$

Complete Ionic: $2Bi^{3+} + 6Cl^- + 6Na^+ + 3S^{2-} \rightarrow Bi_2S_3(s) + 6Na^+ + 6Cl^-$

Net Ionic: $2Bi^{3+} + 3S^{2-} \rightarrow Bi_2S_3(s)$

(d) No Reaction!

(e) **Molecular:** $\quad 3KOH(aq) + FeCl_3(aq) \rightarrow Fe(OH)_3(s) + 3KCl(aq)$

Complete Ionic: $3K^+ + 3OH^- + Fe^{3+} + 3Cl^- \rightarrow Fe(OH)_3(s) + 3K^+ + 3Cl^-$

Net Ionic: $\quad Fe^{3+} + 3OH^- \rightarrow Fe(OH)_3(s)$

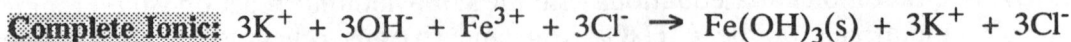

SECTION 4.5   SINGLE REPLACEMENT REACTIONS

After completing this section, you will be able to do the following:

■ State the general trends in reactivity by groups and periods in the periodic table for metals and non-metals;

■ Given the Activity Series for Metals, predict whether reactions will occur between metals;

■ Predict whether reactions will occur between non-metals.

REVIEW

Section 4.4 introduced **double replacement reactions**, where the ions in the two reactant compounds appear to "exchange partners" if a weak electrolyte or insoluble substance is formed. A second very important class of reactions is the **single replacement reaction**, where **one element** takes the place of another element in a compound. That is, the reactants are an element and a compound. Carrying out a series of these single replacement reactions between an element and a compound reveals the following trends in reactivity (ease with which a reaction occurs):

● For **metals in Groups IA and IIA**, reactivity **increases** from right to left in a period and down a group; in other words, reactivity **increases** towards the lower left corner of the periodic table (e.g., K is more reactive than Ca because it is to the left of Ca in Period 4; Rb is more reactive than K because it is lower in Group IA than K).

● For the **transition metals**, reactivity **decreases** down a group (column), but there is **no trend** across a period.

● For **non-metals**, reactivity **increases** from left to right in a period **and** up a group; in other words, reactivity increases towards the upper right corner of the periodic table (e.g., Cl is more reactive than S because it is to the right of S in Period 3, and Cl is more reactive than Br because it is higher than Br in Group VIIA).

The reactivity of many Group IA and IIA metals as well as transition metals is summarized in an **Activity Series** (or reactivity series), found in Table 4.5 (textbook). Referring to this table, the first three elements

(K, Na, Ca) are so reactive that they react immediately with the solvent water. Starting with Mg, each metal will react with dilute acids and replace any element **below** it in salt solutions. For example, Al metal will react with a solution of $Sn(NO_3)_2$ to produce $Al(NO_3)_3$ and Sn metal, but Al metal will not react with Mg metal **above** it in the activity series.

SAMPLE PROBLEM 4.5 A

▶▶ Predict whether a reaction will occur between the following substances; if a reaction occurs, write the molecular, complete ionic, and net ionic equations. (a) barium metal and water, (b) nickel and dilute hydrochloric acid, (c) silver metal and dilute hydrochloric acid, (d) iron metal and chromium(III) chloride solution, (e) lead metal and silver nitrate solution, (f) gold metal and silver nitrate solution.

Solution:

HINT: Refer to the trends above and Table 4.5 in the textbook.

(a) Referring to Table 4.5, barium metal is not listed, but Ca metal in Group IIA is listed and does react with water. The trend for Group IA and IIA elements is that their reactivity increases down a group. Since Ba is lower than Ca in Group IIA, Ba will be more reactive than Ca and will also react with water. The equations are:

Molecular: $Ba(s) + 2H_2O(l) \rightarrow Ba(OH)_2(aq) + H_2(g)$

Complete Ionic: $Ba(s) + 2H_2O(l) \rightarrow Ba^{2+} + 2OH^- + H_2(g)$

Net Ionic: [Same as complete ionic equation; no spectator ions.]

NOTE: The ionic reactions must be balanced with respect to the number of atoms of each element (mass) **and** charge. In the **complete ionic equation**, the net charge is **always balanced at zero** on each side of the reaction if the reaction is correctly balanced.

(b) From Table 4.5, Ni does react with dilute acids. The equations are:

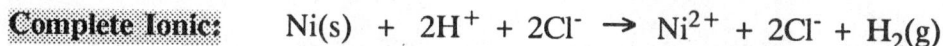

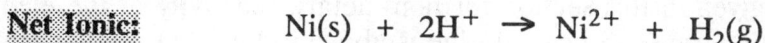

Molecular: $Ni(s) + 2HCl(aq) \rightarrow NiCl_2(aq) + H_2(g)$

Complete Ionic: $Ni(s) + 2H^+ + 2Cl^- \rightarrow Ni^{2+} + 2Cl^- + H_2(g)$

Net Ionic: $Ni(s) + 2H^+ \rightarrow Ni^{2+} + H_2(g)$

NOTE: Again, the ionic reactions must be balanced with respect to number of atoms of each element (mass) **and** charge. In the **complete ionic equation**, the net charge is balanced at zero on each side of the reaction. In the **net ionic equation**, the net charge is balanced at +2 on each side of the reaction.

(c) From Table 4.5, Ag metal **does not** react with dilute acids. No reaction!

(d) From Table 4.5, Fe metal will only react with salt solutions of metals **below** it. Since Cr is **above** Fe, no reaction!

(e) From Table 4.5, Pb metal will react with salt solutions of any metal **below** it. Silver is below Pb; therefore a reaction will occur. The equations are:

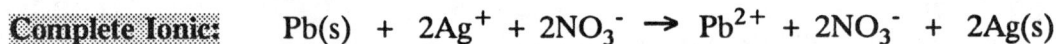

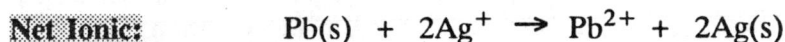

Molecular: $$Pb(s) + 2AgNO_3(aq) \rightarrow Pb(NO_3)_2(aq) + 2Ag(s)$$

Complete Ionic: $$Pb(s) + 2Ag^+ + 2NO_3^- \rightarrow Pb^{2+} + 2NO_3^- + 2Ag(s)$$

Net Ionic: $$Pb(s) + 2Ag^+ \rightarrow Pb^{2+} + 2Ag(s)$$

NOTE: In the **complete ionic equation**, the net charge is balanced at zero on each side of the reaction. In the **net ionic equation**, the net charge is balanced at +2 on each side of the reaction.

(f) From Table 4.5, Au metal is the **least** reactive of the series; it will not react with the salt solutions of any element **above** it. No reaction!

SAMPLE PROBLEM 4.5 B

▶ ▶ Predict whether a reaction will occur between the following substances; if a reaction occurs, write the molecular, complete ionic, and net ionic equations. (a) chlorine water with potassium iodide solution (iodine is a water-insoluble solid), (b) solid iodine and potassium bromide solution.

Solution:

(a) Both chlorine and iodine are non-metals in Group VIIA of the periodic table. From the trends given in this section for non-metals, reactivity of the elements increases as you go up the group. Since chlorine is above iodine, that means aqueous solutions of the **element** Cl_2 (or chlorine water) will react with salt solutions of iodine (or salt solutions of any other element below chlorine). The equations are:

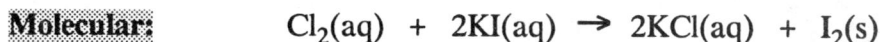

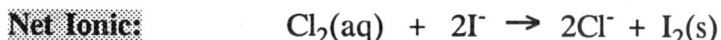

Molecular: $Cl_2(aq) + 2KI(aq) \rightarrow 2KCl(aq) + I_2(s)$

Complete Ionic: $Cl_2(aq) + 2K^+ + 2I^- \rightarrow 2K^+ + 2Cl^- + I_2(s)$

Net Ionic: $Cl_2(aq) + 2I^- \rightarrow 2Cl^- + I_2(s)$

NOTE: In the **complete ionic equation**, the net charge is balanced at zero on each side of the reaction. In the **net ionic equation**, the net charge is balanced at -2 on each side of the reaction.

(b)   Both iodine and bromine are elements in Group VIIA. According to the trend for this group, the element iodine will react with salts of any element **below** it, but **not above it**. Therefore, there will be no reaction!

SECTION 4.6   SOME USES OF REACTIONS IN SOLUTION

After completing this section, you will be able to do the following:

■ Determine whether an insoluble compound will dissolve by adding another reactant to it;

■ Outline a method for the synthesis of a compound.

REVIEW

As we have already seen (Section 4.3), you can predict that when solutions of electrolytes are mixed, a reaction will take place **only if one or more products are insoluble or a weak electrolyte**. Similarly, if one of the reactants is an insoluble compound, it will dissolve (a reaction will take place) **only if one or more of the products is a weak electrolyte, insoluble, or slightly soluble.**

Synthesis is the formation or making of a compound using a variety of reactions, such as double and single replacement reactions already studied (Sections 4.4 and 4.5). If the compound to be synthesized is **insoluble, a double replacement reaction** is often the desirable method. On the other hand, if the compound to be made is **soluble**, the synthesis is more difficult and often requires a **single replacement reaction**.

SAMPLE PROBLEM 4.6 A

▶▶ Answer the following questions and explain your answers. (a) Is $NiCO_3(s)$ soluble in $HNO_3(aq)$? (b) Is $Al(OH)_3(s)$ soluble in $HCl(aq)$? (c) Is $PbI_2(s)$ soluble in $HNO_3(aq)$?

Solution:

(a) H^+ from $HNO_3(aq)$ will react with CO_3^{2-} from $NiCO_3$ to form the weak electrolyte H_2O and the slightly soluble gas (weak electrolyte) $CO_2(g)$,

$$NiCO_3(s) + 2HNO_3(aq) \rightarrow Ni(NO_3)_2(aq) + CO_2(g) + H_2O(l)$$

Nickel carbonate is soluble in nitric acid.

(b) H^+ from $HCl(aq)$ will react with OH^- from $Al(OH)_3$ to form the weak electrolyte H_2O,

$$Al(OH)_3(s) + 3HCl(aq) \rightarrow AlCl_3(aq) + 3H_2O(l)$$

Aluminum hydroxide is soluble in hydrochloric acid.

(c)   The two possible products are the soluble, strong electrolytes $HI(aq)$ and the salt $Pb(NO_3)_2(aq)$. A reaction will occur only if a **weak electrolyte, insoluble, or slightly soluble product is formed**. Therefore, no reaction will occur. Lead iodide is not soluble in nitric acid.

SAMPLE PROBLEM 4.6 B

▶▶ Explain how to synthesize each of the following compounds, and write a molecular equation: (a) CuS (b) $Al(NO_3)_3$ (c) $ZnCO_3$.

Solution:

(a)   CuS is insoluble (Table 4.4). Generally, a double replacement reaction works best to synthesize an insoluble compound. The two reactants and the other product will have to be soluble. One of the reactants must supply the Cu^{2+} ion, and the other must supply the S^{2-} ion. From Table 4.4, possibilities for the **soluble** reactant supplying the Cu^{2+} ion are $Cu(NO_3)_2(aq)$, $CuCl_2(aq)$, $Cu(C_2H_3O_2)_2(aq)$, $CuSO_4(aq)$, $CuBr_2(aq)$, and $CuI_2(aq)$. The possibilities for the soluble reactant supplying the S^{2-} ion are $Na_2S(aq)$,

$K_2S(aq)$, $(NH_4)_2S(aq)$, $MgS(aq)$, $CaS(aq)$, etc. Using the first reactant in each case, the molecular reaction would be

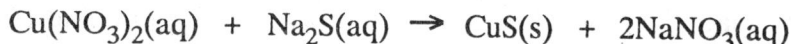

$$Cu(NO_3)_2(aq) + Na_2S(aq) \rightarrow CuS(s) + 2NaNO_3(aq)$$

Carry out the reaction in the proportions shown by the coefficients, separate the CuS(s) from the other product by filtration, and then wash and dry.

(b)   Aluminum nitrate is soluble (Table 4.4). Often, a single replacement reaction works best for synthesizing a soluble compound. One method is to start with Al metal and react it with the salt of a metal below it in the Activity Series (Table 4.5). An even easier method is to react Al metal with a dilute acid and produce H_2 gas as the other product,

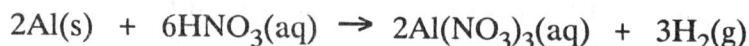

$$2Al(s) + 6HNO_3(aq) \rightarrow 2Al(NO_3)_3(aq) + 3H_2(g)$$

Use an excess of Al metal so that the HNO_3 will be the limiting reactant (and will be entirely used up during the reaction). After the reaction is complete, there will only be excess solid Al (which can be separated by filtration) and a solution of $Al(NO_3)_3$ (the H_2 gas will have bubbled off into the air). The soluble salt, $Al(NO_3)_3$, can be separated from the aqueous solution by evaporation.

(c)   Zinc carbonate is insoluble (Table 4.4). Generally, a double replacement reaction works best to synthesize an insoluble compound. The two reactants and the other product will have to be soluble. One of the reactants must supply the Zn^{2+} ion, and the other must supply the CO_3^{2-} ion. From Table 4.4, possibilities for the **soluble** reactant supplying the Zn^{2+} ion are $Zn(NO_3)_2(aq)$, $ZnCl_2(aq)$, $Zn(C_2H_3O_2)_2(aq)$, $ZnSO_4(aq)$, $ZnBr_2(aq)$, and $ZnI_2(aq)$. The possibilities for the soluble reactant supplying the CO_3^{2-} ion are $Na_2CO_3(aq)$, $K_2CO_3(aq)$, $(NH_4)_2CO_3(aq)$, etc. Using the second reactant in each case, the molecular reaction would be

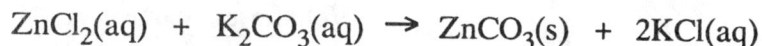

$$ZnCl_2(aq) + K_2CO_3(aq) \rightarrow ZnCO_3(s) + 2KCl(aq)$$

Carry out the reaction in the proportions shown by the coefficients, separate the $ZnCO_3(s)$ from the other product by filtration, and then wash and dry.

SECTION 4.7   CONCENTRATION EXPRESSED AS PERCENT

After completing this section, you will be able to do the following:

- Calculate the volume of solute necessary to prepare a solution of known volume and concentration in volume percent;

- Calculate the mass of solute and solvent necessary to prepare a solution of known mass and concentration in mass percent;

- Calculate the concentration of a solution in parts per million (ppm) and parts per billion (ppb).

REVIEW

Since a solution is a homogeneous mixture (Section 1.3), it can be prepared with differing amounts of solute and solvent. This means a solution can have a variable composition or **concentration**, depending upon the amount of solute and solvent mixed together. Outside scientific laboratories, concentration of solutions is often expressed in **percent**. However, additional information is needed to clarify which kind of percent --- **% by volume** or **% by mass**. For a given solution, the **volume %** and **mass %** are NOT the same number. Masses are additive no matter what the solute and solvent are, but volumes are NOT additive. [Will a cup of sugar added to a cup of dry cereal fill a two-cup container?] Concentrations of mass % and volume % for component A are defined as

$$\text{Mass \% A} = \frac{\text{Mass component A}}{\text{Total mass solution}} \times 100$$

$$\text{Volume \% A} = \frac{\text{Volume component A}}{\text{Total volume solution}} \times 100$$

When concentrations are very small (as in measuring air pollutants, insecticides, etc.), similar units used are **parts per million (ppm)** and **parts per billion (ppb)**, either one **by volume or by mass**. The corresponding formulas (by volume) are

$$\text{ppm (volume)} = \frac{\text{Volume component A}}{\text{Total volume solution}} \times 10^6$$

$$\text{ppb (volume)} = \frac{\text{Volume component A}}{\text{Total volume solution}} \times 10^9$$

All of the concentration units in this section are called **physical units**, because they do not require the identity or formula of the solute to perform any of the calculations. [Section 4.8 takes up the concentration unit of molarity, which is called a **chemical unit** because you will need to know the formula of the solute in order to do many of the calculations.]

SAMPLE PROBLEM 4.7 A

▸▸ Write directions for preparing 175 mL of a 0.20% by volume solution of isopropyl alcohol in water.

HINT: ● Use the 3-step, general problem-solving approach given in Sample Problem 2.2 C and used throughout the rest of Chapter 2 and in Chapter 3.

● When given more than one piece of information initially, start with the given piece of information that has the **same number of units** as the information you are trying to find.

Step 1: To start the problem, you will need to calculate the number of milliliters of the solute, isopropyl alcohol, before you can add water to it and prepare the solution. The equivalency statement or question that needs to be asked is:

How many milliliters of isopropyl alcohol are equivalent to (i.e., equal to, needed to prepare) 175 mL of a 0.20% by volume solution?

Step 2: Which equation do you use?

? mL isopropyl alcohol (solute) = 175 mL solution x ?

OR   ? mL isopropyl alcohol (solute) = 0.20% by volume solution x ?

See the **second HINT** at the beginning of this problem. You are given two quantities in the problem --- 175 mL (which has 1 unit, volume in mL) and 0.20% by volume solution (which has 2 units, volume in mL of solute and volume in mL of solution, since 0.20% by volume means 0.20 mL of isopropyl alcohol per 100 mL of solution). **All methods of expressing concentrations of solutions (mass percent, volume percent, ppm, ppb, molarity, etc.) have TWO measurement units: one in the numerator and one in the denominator.**

Since you are looking for a quantity (mL solution) that has only **1 unit,** you must start with the given quantity that has only 1 unit --- 175 mL. [This strategy works for many other applications as well.] The correct setup for this problem is

? mL isopropyl alcohol (solute) = 175 mL solution x ?

Step 3: The connecting link between volume of solution and volume of solute is % by volume. **All concentration units are equalities or conversion factors.** In this example, a 0.20% by volume solution **means** that 0.20 mL isopropyl alcohol = (or is present in every) 100-mL solution. The two forms of this conversion factor are:

$$\frac{0.20 \text{ mL isopropyl alcohol}}{100 \text{ mL solution}} \quad \text{and} \quad \frac{100 \text{ mL solution}}{0.20 \text{ mL isopropyl alcohol}}$$

[**Remember**, think of the line separating the top from the bottom as an equal sign.] Then use the form of the conversion factor in which the unit that you want to **cancel** (mL solution) is in the **denominator**. Choosing the **first** form will give you the unit that you are trying to find, mL isopropyl alcohol.

$$? \text{ mL isopropyl alcohol (solute)} = 175 \text{ mL solution} \times \frac{0.20 \text{ mL isopropyl alcohol}}{100 \text{ mL solution}}$$

$$= 0.35 \text{ mL isopropyl alcohol}$$

Measure 0.35 mL of isopropyl alcohol into a 200-mL graduated cylinder. Add water until the liquid level is at 175 mL, and mix. You will now have 175 mL of a 0.20% by volume solution of isopropyl alcohol.

NOTE: We do not need the identity or formula mass of the solute. We could measure out 0.35 mL of **any liquid solute** (soluble in water), add water to the 175 total volume level, and prepare 175 mL of a 0.20% by volume solution of any solute.

SAMPLE PROBLEM 4.7 B

▶▶ What is the concentration in parts per million (ppm) of the isopropyl alcohol solution in Sample Problem 4.7 A?

Solution:

Using the definition of ppm given above for the 0.20% by volume solution of isopropyl alcohol,

$$\text{ppm (vol)} = \frac{\text{volume of solute}}{\text{volume of solution}} \times 10^6 = \frac{0.20 \text{ mL isopropyl alcohol}}{100 \text{ mL solution}} \times 10^6$$

$$= 2.0 \times 10^3 \text{ ppm}$$

NOTE: The label on the container with this isopropyl alcohol solution could read:

CONCENTRATION:	0.20% by volume isopropyl alcohol
	2.0×10^3 ppm isopropyl alcohol

They both are absolutely correct! They are simply two different ways of expressing the concentration of the same solution.

SAMPLE PROBLEM 4.7 C

▸▸ Write directions for preparing 200.0 g of aqueous solution containing 15.0% by mass NaOH.

Solution:

Use the HINTS and procedure shown in Sample Problem 4.7 A.

Step 1: To prepare a solution with concentration in mass percent, you will need to calculate grams of solute and grams of solvent, since

$$g \text{ solution} = g \text{ solute} + g \text{ solvent}$$

For this problem, you will need to calculate the grams of NaOH (solute) and grams of H_2O (solvent), since

$$200.0 \text{ g solution} = g \text{ NaOH} + g \text{ H}_2\text{O}$$

Always calculate grams of solute first (from the mass percent and mass of solution), then use this equation to calculate grams of H_2O. To calculate grams of NaOH needed to prepare the solution, ask the question:

How many grams of NaOH are equivalent to (required to prepare) 200.0 g of 15.0% by mass NaOH solution?

Step 2: Which equation do you use?

$$? \text{ g NaOH} = 200.0 \text{ g solution} \times ?$$

OR $\quad ? \text{ g NaOH} = 15.0\%$ by mass solution $\times$?

See the procedure for selecting the correct setup in Part 2 of Sample Problem 4.7 A. Since you are looking for only 1 unit in your answer (g NaOH), select the piece of given information with 1 unit also, 200.0 g solution (percent by mass and all other concentrations always have **2** units in their definition). The correct setup for this problem is

$$? \text{ g NaOH} = 200.0 \text{ g solution} \times ?$$

Step 3: The connecting link between grams of solution and grams of solute is percent by mass. A 15.0% by mass NaOH solution **means** that 15.0 g NaOH = (or is dissolved in every) 100 g of solution. The two forms of this conversion factor are:

$$\frac{15.0 \text{ g NaOH}}{100 \text{ g solution}} \quad \text{and} \quad \frac{100 \text{ g solution}}{15.0 \text{ g NaOH}}$$

[**Remember**, think of the line separating the top from the bottom as an equal sign.] Then use the form of the conversion factor in which the unit that you want to **cancel** (g solution) is in the **denominator**. Choosing the **first** form will give you the unit that you are trying to find, g NaOH.

$$? \text{ g NaOH} = 200.0 \text{ g solution} \times \frac{15.0 \text{ g NaOH}}{100 \text{ g solution}} = 30.0 \times 10^1 \text{ g NaOH}$$

The number of grams of water is found from the equation in Step 1,

$$200.0 \text{ g solution} = 30.0 \text{ g NaOH} + \text{g H}_2\text{O}$$

$$? \text{ g H}_2\text{O} = 170.0 \text{ g}$$

Measure out 30.0 g of NaOH and dissolve it in 170.0 g of water. You will now have 200.0 g of a 15.0% by mass solution of NaOH.

SECTION 4.8   MOLARITY

After completing this section, you will be able to do the following:

- Given two of the three quantities for a solution --- molarity, grams or moles of solute, and volume --- calculate the third quantity;

- Prepare a dilute solution from a concentrated solution;

- Given the percent by mass of a solution, convert it to molarity;

- Given different volumes and concentrations of the same solution, calculate the molarity of the mixture.

REVIEW

Percent by mass and percent by volume are called **physical units** of concentration because the identity and formula mass of the solute are not required. In the scientific world, stoichiometric calculations use chemical equations where the formulas are related to each other through the coefficients standing for moles. **Molarity** is a **chemical unit** of concentration because it requires that the identity and formula mass of the solute be known. You will find that **molarity** is the most common method of expressing concentration of a solution. It is very important that you learn how to work with and prepare solutions involving molarity.

Molarity is defined as the number of moles of solute per liter of solution,

$$\text{molarity, M} = \frac{\text{number of moles of solute}}{\text{volume of solution in liters}}$$

Since the quantity of solute is measured in moles, remember that you know the mass in grams of **one** mole of any formula --- numerically equal to the formula mass (Section 3.1, point 2 in the REVIEW, and Sample Problem 3.1 B of this book). Molarity also requires that the volume of the solution be in liters. Often the volume is given in mL, which can be converted to L by multiplying by the conversion factor 1 L/1000 mL. Mathematically, this is equivalent to moving the decimal point 3 places to the **left** to convert mL to L, e.g., 425 mL = 0.425 L. Writing down the volume initially in liters will save set up and calculation time. Of course, to convert a volume in L to mL, move the decimal point three places to the **right**, e.g., 0.750 L = 750 mL.

SAMPLE PROBLEM 4.8 A

▸▸ What is the molarity of a solution prepared by dissolving 26.5 g of $Cu(NO_3)_2$ in water and diluting to 6.50×10^2 mL?

Solution:

HINT: See Sample Problems 4.7 A & C for the application of the 3-step general problem-solving approach to concentrations of solutions.

Step 1: How many moles of $Cu(NO_3)_2$ per liter of solution (**not** just the word, molarity, or its symbol, M) are equivalent to (equal to) 26.5 g of $Cu(NO_3)_2$ in 6.50×10^2 mL of solution? [Or, since this statement is a little awkward when calculating the concentration unit itself, use the definition of molarity. Step 2 is the same regardless of the choice.]

Step 2: $? \dfrac{\text{mol } Cu(NO_3)_2}{\text{L solution}} \text{ (M)} = \dfrac{26.5 \text{ g } Cu(NO_3)_2}{6.50 \times 10^2 \text{ mL solution}} \times ?$

Molarity (and all expressions of concentration) has **two** units, moles of solute in the numerator, and liters of solution in the denominator. Therefore, you must start with two units of given information --- some quantity of solute in the numerator, and some quantity of solution in the denominator.

Step 3: Your goal is to convert grams of $Cu(NO_3)_2$ to moles of $Cu(NO_3)_2$, and mL of solution to L of solution. The conversions can be done in either order. You always know the mass in grams of **1 mole** of any formula; it is numerically equal to the formula mass. One mole of $Cu(NO_3)_2$ has a mass of 187.556 g. And, 1 L = 1000 mL. Using these two conversion factors so that the units of g and mL cancel,

$$? \frac{mol\ Cu(NO_3)_2}{L\ solution}\ (M) = \frac{26.5\ g\ \cancel{Cu(NO_3)_2}}{6.50 \times 10^2\ \cancel{mL\ soln.}} \times \frac{1000\ \cancel{mL\ soln.}}{1\ L\ soln.} \times \frac{1\ mol\ Cu(NO_3)_2}{187.556\ g\ \cancel{Cu(NO_3)_2}}$$

$$= \frac{0.217\ mol\ Cu(NO_3)_2}{L\ soln.} = \boxed{0.217\ M}$$

You can shorten the writing process by converting the volume from mL to L (see REVIEW in this section) in your head before beginning the calculation. Of course, the molarity is the same.

$$? \frac{mol\ Cu(NO_3)_2}{L\ solution}\ (M) = \frac{26.5\ g\ \cancel{Cu(NO_3)_2}}{0.650\ L\ solution} \times \frac{1\ mol\ Cu(NO_3)_2}{187.556\ g\ \cancel{Cu(NO_3)_2}}$$

$$= \frac{0.217\ mol\ Cu(NO_3)_2}{L\ soln.} = \boxed{0.217\ M}$$

NOTE: This shorter method will be used from this point on.

CHECK: Does this answer make sense? Yes! Approximately 0.15 mol of $Cu(NO_3)_2$ is dissolved in 2/3 of a liter, so 1 L of solution will contain 50% more, or ~0.22 mol $Cu(NO_3)_2$.

SAMPLE PROBLEM 4.8 B

▶ ▶ How many grams of $AgNO_3$ are needed to prepare 150.0 mL of 2.0×10^{-2} M $AgNO_3$?

Solution:

See Sample Problem 4.8 A.

SAMPLE PROBLEM 4.8 F

▶ ▶ What is the molarity of the solution formed by mixing 45.0 mL of 0.065 M $FeCl_3$ with 60.0 mL of 0.048 M $FeCl_3$?

HINT: Add together the number of **moles** of $FeCl_3$ contributed by each solution and divide by the total volume.

When mixing different quantities of solutions of the same substance, there is no chemical reaction occurring. To find the molarity of the mixture, divide the **total number of moles** of $FeCl_3$ by the **total volume**.

$$? \quad \frac{\text{\# mol } FeCl_3 \text{ (total)}}{\text{\# L soln. (total)}} = \frac{\text{\# mol } FeCl_3 \text{ (0.065 M soln.)} + \text{\# mol } FeCl_3 \text{ (0.048 M soln.)}}{\text{\# L soln. (0.065 M soln.)} + \text{\# L soln. (0.048 M soln.)}}$$

You **cannot** average the molarities of the individual solutions to obtain the molarity of the mixture, because you are using different volumes.

To calculate the number of **moles** of a substance in solution, **always** multiply the M times the V in liters, # mol solute = M (mol/L) x V (L). **You will use this equation frequently throughout this course.** For each of the two solutions in this problem, the number of moles of $FeCl_3$ is

$$? \text{ mol } FeCl_3 \text{ (0.065 M soln.)} = 0.045 \text{ } \cancel{\text{L soln.}} \text{ x } \frac{0.065 \text{ mol } FeCl_3}{1 \text{ } \cancel{\text{L soln.}}}$$

$$= 2.93 \text{ x } 10^{-3} \text{ mol } FeCl_3$$

$$? \text{ mol } FeCl_3 \text{ (0.048 M soln.)} = 0.060 \text{ } \cancel{\text{L soln.}} \text{ x } \frac{0.048 \text{ mol } FeCl_3}{1 \text{ } \cancel{\text{L soln.}}}$$

$$= 2.88 \text{ x } 10^{-3} \text{ mol } FeCl_3$$

Substituting into the equation at the beginning of this problem, the molarity of the mixture is

$$? \quad \frac{\text{\# mol FeCl}_3 \text{ (total)}}{\text{\# L soln. (total)}} = \frac{2.93 \times 10^{-3} \text{ mol FeCl}_3 + 2.88 \times 10^{-3} \text{ mol FeCl}_3}{0.045 \text{ L soln.} + 0.060 \text{ L soln.}}$$

$$= \frac{5.81 \times 10^{-3} \text{ mol FeCl}_3}{1.05 \times 10^{-1} \text{ L soln.}} = 5.5 \times 10^{-2} \text{ M} = 0.055 \text{ M}$$

CHECK: Does this make sense? Yes, because the molarity of the mixture has to be between the two initial concentrations and closer to the molarity of the solution which contributed the greater volume. The final molarity, 0.055 M, is between 0.065 M and 0.048 M and is closer to the molarity of the solution which contributed the greater volume, 60.0 mL of 0.048 M FeCl$_3$.

SECTION 4.9   STOICHIOMETRY OF REACTIONS IN SOLUTION

After completing this section, you will be able to do the following:

■ Given the quantity of any reactant or product in a reaction in grams, or molarity and volume (if a solution), calculate the quantities of any other reactant needed or product formed.

REVIEW

You have already learned to carry out quantitative calculations in chemical equations where one given quantity was in grams or moles (Section 3.5), and when two given quantities were given (limiting reactant).

After learning about molarity (Section 4.8), you now know another way to provide information in a reaction when using solutions --- the molarity and volume of a solution. Multiplying molarity times volume in liters **always** gives you the number of moles of solute, which can be readily used for quantitative calculations in equations.

SAMPLE PROBLEM 4.9 A

▶ ▶ How many mL of 0.0852 M HCl are needed to neutralize 25.5 mL of 0.0648 M Ba(OH)$_2$, according to the reaction,

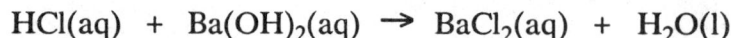

$$\text{HCl(aq)} + \text{Ba(OH)}_2\text{(aq)} \rightarrow \text{BaCl}_2\text{(aq)} + \text{H}_2\text{O(l)}$$

Solution:

HINT: Refer to the 4-step HINT given at the beginning of Sample Problem 3.5 A for solving stoichiometry problems, and to any of the Sample Problems in Section 3.5.

Step 1: **Check to make sure the reaction is balanced and write the information given and needed.** This reaction is **NOT** balanced. The balanced reaction and appropriate information is

Problem:	? mL	25.5 mL
	0.0852 M	0.0648 M

Equation: $2HCl(aq) + Ba(OH)_2(aq) \rightarrow BaCl_2(aq) + 2H_2O(l)$

Coefficients:	2 mol	1 mol

The formula mass is not needed for a substance in solution if **both** the molarity and volume of the solution are known.

Step 2: **Convert the given quantity to moles.** If the given quantity is a solution of known molarity and volume, convert to moles (always) by multiplying the molarity times the volume. The given or starting quantity in this problem is $Ba(OH)_2$, because we know both the molarity and volume of the solution. Converting the volume to liters initially,

$$? \text{ mol } Ba(OH)_2 = 0.0255 \text{ L } Ba(OH)_2 \text{ soln. } \times \frac{0.0648 \text{ mol } Ba(OH)_2}{1 \text{ L } Ba(OH)_2 \text{ soln.}} = 1.652 \times 10^{-3} \text{ mol } Ba(OH)_2 \text{ used.}$$

Step 3: **Convert moles of $Ba(OH)_2$ used to moles of HCl needed,** using the coefficients from the **balanced** reaction,

$$? \text{ mol HCl } = 1.652 \times 10^{-3} \text{ mol } Ba(OH)_2 \times \frac{2 \text{ mol HCl}}{1 \text{ mol } Ba(OH)_2} = 3.304 \times 10^{-3} \text{ mol HCl}$$

Step 4: **Convert moles of HCl needed to mL of HCl solution,** using the molarity of the HCl.

$$? \text{ mL HCl soln. } = 3.304 \times 10^{-3} \text{ mol HCl } \times \frac{1 \text{ L HCl soln.}}{0.0852 \text{ mol HCl}} \times \frac{1000 \text{ mL HCl soln.}}{1 \text{ L HCl soln.}} = 38.8 \text{ mL HCl soln.}$$

With practice, you should be able to combine steps 2-4 and eliminate the calculation of answers for intermediate steps,

$$\text{? mL HCl soln.} = 0.0255 \text{ L Ba(OH)}_2 \text{ soln.} \times \frac{0.0648 \text{ mol Ba(OH)}_2}{1 \text{ L Ba(OH)}_2 \text{ soln.}} \times \frac{2 \text{ mol HCl}}{1 \text{ mol Ba(OH)}_2} \times \frac{1 \text{ L HCl soln.}}{0.0852 \text{ mol HCl}} \times \frac{1000 \text{ mL soln.}}{1 \text{ L soln.}}$$

$$= 38.8 \text{ mL HCl soln.}$$

CHECK: Does this answer make sense? Yes. The concentration of the HCl solution is about 4/3 the concentration of the $Ba(OH)_2$ solution. One mole of $Ba(OH)_2$ requires 2 mol of HCl. Therefore, 3/4 as much HCl solution times 2 (3/4 x 25.5 mL x 2 ~ 38.2 mL) will be needed.

SAMPLE PROBLEM 4.9 B

▶ ▶ If 8.5 g of solid tin is treated with 200.0 mL of 0.500 M HCl according to the reaction given, (a) how many grams of hydrogen gas will be formed? (b) How much of which reactant will be left after the reaction is over?

$$\text{Sn(s)} + \text{HCl(aq)} \rightarrow \text{SnCl}_2\text{(aq)} + \text{H}_2\text{(g)}$$

Solution:

HINT: This is a limiting reactant problem.

(a) See Sample Problem 4.9 A.

Step 1: This reaction is **NOT** balanced. The balanced reaction and appropriate information is

Problem:	8.5 g	200.0 mL 0.500 M			? g
Equation:	Sn(s)	+ 2HCl(aq)	$\rightarrow$	SnCl$_2$(aq) +	H$_2$(g)
Formula mass, u:	118.710				2.015 88
Coefficients:	2 mol	1 mol			1 mol

Step 2: Which reactant determines how much hydrogen gas can be produced? Whenever you are given the quantities of **two** reactants (that is, enough information to calculate number of moles of each reactant), it is a **limiting reactant problem** (see Section 3.6 and Sample Problem 3.6 A in this study guide). For each reactant, calculate the number of moles and divide by its coefficient in the balanced reaction. The lowest number obtained is the limiting reactant.

$$? \text{ mol Sn } = 8.5 \text{ g Sn} \times \frac{1 \text{ mol Sn}}{118.710 \text{ g Sn}} = 0.0716 \text{ mol Sn}$$

$$? \text{ fraction Sn } = \frac{0.0716 \text{ mol Sn}}{1 \text{ mol Sn}} = 0.0716$$

$$? \text{ mol HCl } = 0.2000 \text{ L soln.} \times \frac{0.500 \text{ mol HCl}}{1 \text{ L soln.}} = 0.1000 \text{ mol HCl}$$

$$? \text{ fraction HCl } = \frac{0.1000 \text{ mol HCl}}{2 \text{ mol HCl}} = 0.05000$$

Since $0.05000 < 0.0716$, HCl is the limiting reactant.

Step 3-4: The quantity of H_2 gas that will be formed is

$$? \text{ g } H_2 = 0.1000 \text{ mol HCl} \times \frac{1 \text{ mol } H_2}{2 \text{ mol HCl}} \times \frac{2.015 \, 88 \text{ g } H_2}{1 \text{ mol } H_2} = 0.101 \text{ g } H_2$$

> **NOTE:** One extra significant figure (s.f.) was carried through this multiple-part problem until the answer for part (a) was obtained. The 3 s.f. in the answer are determined by the 3 s.f. in the 0.500 M HCl.

(b)   If HCl is the limiting reactant, the other reactant, Sn, is present in excess. First calculate the amount of tin that reacted with all the HCl, then the amount of tin left over (in excess).

$$? \text{ g Sn } = 0.1000 \text{ mol HCl} \times \frac{1 \text{ mol Sn}}{2 \text{ mol HCl}} \times \frac{118.710 \text{ g Sn}}{1 \text{ mol Sn}} = 5.936 \text{ g Sn}$$

$$? \text{ g Sn left over } = 8.5 \text{ g Sn present} - 5.936 \text{ g Sn reacted } = 2.6 \text{ g Sn left over}.$$

SECTION 4.10   TITRATION

After completing this section, you will be able to do the following:

- Given the molarity and volume of a standard solution used to titrate a specific volume of an unknown, calculate the molarity of the unknown solution and the mass in grams and mass percent of the sample of unknown.

REVIEW

Titration is a method of determining the concentration of an unknown solution by adding just enough of a **standard** solution (whose concentration is known) to complete the reaction. The point in the titration at which the addition of **standard** solution stops, is called the **endpoint**. The endpoint is signaled by a sudden change in color (typically) due to the addition of a small quantity of an **indicator** at the beginning of the titration. Once the volume of **standard solution** needed to reach the endpoint is known, enough information is available to calculate the molarity of the unknown solution and the mass of solute in the sample of unknown. If the density of the unknown solution is available, the mass percent of the sample of unknown can also be calculated.

SAMPLE PROBLEM 4.10 A

▶▶ A 25.00-mL sample of nitric acid required 18.75 mL of 0.150 M KOH to reach the endpoint of a titration. The density of the nitric acid was 1.0071 g/mL. (a) What was the molarity of the nitric acid? (b) How many grams of nitric acid did the 25.00-mL sample contain? (c) What was the mass % nitric acid in the sample? The reaction for the titration is

$$HNO_3(aq) \ + \ KOH(aq) \ \rightarrow \ KNO_3(aq) \ + \ H_2O(l)$$

Solution

HINT: Use a procedure similar to that in Sample Problem 4.9 A, except leave steps 2-4 **uncombined** when solving for the molarity of a solution in a reaction.

(a)   Step 1:   **Check to make sure the reaction is balanced and write the information given and needed.** This reaction is balanced. The balanced reaction and appropriate information is

Problem: ? M   0.150 M
                      25.00 mL   18.75 mL

Equation: $HNO_3(aq)$ + $KOH(aq)$ $\rightarrow$ $KNO_3(aq)$ + $H_2O(l)$

Coefficients: 1 mol   1 mol

The formula mass is not needed for a substance in solution if **both** the molarity and volume of the solution are known.

       Step 2:   **Convert M and mL of standard KOH to moles of KOH.**

$$? \text{ mol KOH} \ = \ 0.018 \ 75 \ \text{L KOH soln.} \ \times \ \frac{0.150 \text{ mol KOH}}{1 \text{ L KOH soln.}} \ = \ 0.002 \ 81 \text{ mol KOH}$$

Step 3: **Convert moles of KOH to moles of HNO₃.**

$$? \text{ mol HNO}_3 = 0.002\ 81 \text{ mol KOH} \times \frac{1 \text{ mol HNO}_3}{1 \text{ mol KOH}} = 0.002\ 81 \text{ mol HNO}_3$$

Step 4: **Convert moles and mL of HNO₃ to M.**

$$? \text{ M HNO}_3 = \frac{0.002\ 81 \text{ mol HNO}_3}{0.025\ 00 \text{ L soln.}} = \frac{0.112 \text{ mol HNO}_3}{1 \text{ L soln.}} = 0.112 \text{ M}$$

(b) From part (a), step **3**, you calculated the number of moles of HNO₃ present in the 25.00-mL sample. By now you should be able to easily convert moles to grams of HNO₃ in the sample. The formula mass of HNO₃ is 63.0129 u.

$$? \text{ g HNO}_3 \text{ in sample} = 0.002\ 81 \text{ mol HNO}_3 \times \frac{63.0129 \text{ g HNO}_3}{1 \text{ mol HNO}_3} = 0.177 \text{ g HNO}_3$$

(c) To calculate the mass percent of the sample, you need to know the mass of HNO₃ in the sample (0.177 g from part b), and the total mass of the 25.00 mL sample. You can calculate the **total mass of the sample** of solution (which means mass of HNO₃ **and** H₂O) from the density (given) and volume of the sample (see Sample Problem 2.8 B).

$$? \text{ g HNO}_3 \text{ soln.} = 25.00 \text{ mL HNO}_3 \times \frac{1.0071 \text{ g HNO}_3 \text{ soln.}}{1 \text{ mL HNO}_3} = 25.18 \text{ g HNO}_3 \text{ soln.}$$

The mass percent of the sample is

$$? \text{ mass \% HNO}_3 = \frac{\text{mass solute}}{\text{total mass soln.}} \times 100 = \frac{0.177 \text{ g HNO}_3}{25.18 \text{ g soln.}} \times 100 = 0.703\%$$

SELF-TEST QUESTIONS

Test your understanding of the concepts and skills in this chapter by working through the multiple choice questions **BEFORE** checking the answers. A periodic table will always be available. The ANSWERS TO SELF-TEST QUESTIONS follow immediately after this section.

1. When 10.0 g of oxalic acid is dissolved in 100.0 g of water,

   A. water is the solute.
   B. a homogeneous mixture is produced.
   C. oxalic acid is the solvent.
   D. 110.0 g of solution is produced.
   E. more than one of the above is correct.

2. The solubility of a substance is the amount of a solute that dissolves to produce

   A. an unsaturated solution.
   B. a dilute solution.
   C. a saturated solution.
   D. a concentrated solution.
   E. a supersaturated solution.

3. The number of hydrogens that form hydrogen ions in aqueous solution is indicated correctly for each formula **except** for

   A. $HC_2H_3O_2$; 4 H

   B. H_3AsO_4; 3 H

   C. $H_3C_6H_5O_7$; 3 H

   D. H_2SeO_4; 2 H

   E. $HCHO_2$; 1 H

4. The only salt listed below is

   A. CuO B. $(NH_4)_2SO_4$ C. $H_2C_2O_4$ D. NH_3 E. $HC_2H_3O_2$

5. All the following compounds are classified correctly as a strong or weak electrolyte **except**

   A. $HC_2H_3O_2$ (weak)

   B. RbOH (strong)

   C. $(NH_4)_2CO_3$ (strong)

   D. HNO_2 (strong)

   E. HBrO (weak)

6. The only insoluble compound listed below is

   A. $KC_2H_3O_2$ B. $Ni(NO_3)_2$ C. $Al(OH)_3$ D. $FeCl_3$ E. Na_3PO_4

7. When aqueous solutions of sodium phosphate and silver nitrate are mixed, the net ionic equation is

   A. $Na_3PO_4(aq) + 3AgNO_3(aq) \rightarrow Ag_3PO_4(s) + 3NaNO_3(aq)$

   B. $Ag^+ + PO_4^- \rightarrow AgPO_4(s)$

   C. $3Na^+ + PO_4^{3-} + 3Ag^+ + 3NO_3^- \rightarrow Ag_3PO_4(s) + 3Na^+ + 3NO_3^-$

   D. $Na^+ + NO_3^- \rightarrow NaNO_3(aq)$

   E. $3Ag^+ + PO_4^{3-} \rightarrow Ag_3PO_4(s)$

8. Solid zinc sulfite is soluble in

   A. NaCl(aq)   B. HNO_3(aq)   C. KNO_3(aq)   D. $NaC_2H_3O_2$(aq)   E. none of those given.

9. Exactly fifty grams of a 15% by mass KOH solution contains

   A. 7.5 g KOH   B. 35 g of soln.   C. 43 g KOH   D. 35 g H_2O E. 15 g KOH

10. A 100.0-mL blood sample was found to contain 0.090 g of glucose (formula mass = 180.0 u). The molarity of the solution is

A. 0.050 M   B. 0.00050 M   C. 0.0050 M   D. 9.0 M   E. 0.0090 M

11. The number of grams of H_2SO_4 required to prepare 200.0 mL of a 0.400 M H_2SO_4 solution is

A. 15.7 g   B. 78.5 g   C. 196 g   D. 7.85 g   E. 12.5 g

12. If 47 g of $MgCl_2$ are dissolved in enough water to prepare 250 mL of solution, the molarity is

A. 8.0 M   B. 0.50 M   C. 0.25 M   D. 4.0 M   E. 2.0 M

13. How many mL of 0.60 M NaOH are needed to prepare 300.0 mL of a 0.40 M NaOH solution?

A. 4.5×10^2 mL

B. 2.0×10^2 mL

C. 4.0×10^2 mL

D. 72 mL

E. 1.2×10^2 mL

14. If 30.0 mL of a 1.8 M $Ni(NO_3)_2$ solution is diluted to 150.0 mL, the concentration of the dilute solution is

A. 0.36 M   B. 0.18 M   C. 9.0 M   D. 0.54 M   E. 1.8 M

15. The number of mL of 0.2510 M $HC_2H_3O_2$ needed to neutralize 35.60 mL of 0.1816 M KOH is

$$HC_2H_3O_2(aq) + KOH(aq) \rightarrow KC_2H_3O_2(aq) + H_2O(l)$$

A. 6.465 mL   B. 49.20 mL   C. 25.76 mL   D. 30.54 mL   E. 18.72 mL

ANSWERS TO SELF-TEST QUESTIONS

1. e [both b and d true]	6. c	11. d
2. c	7. e	12. e
3. a	8. b	13. b
4. b	9. a	14. a
5. d	10. c	15. c

PRACTICE PROBLEMS

Test your problem-solving skills in this chapter by working through the problems in the space provided **BEFORE** checking the answers. A periodic table will always be available. The ANSWERS TO PRACTICE PROBLEMS follow immediately after this section.

1. Write the formulas and names of the three potassium salts of boric acid, H_3BO_3 (the anion of boric acid is called the borate ion).

2. Write a chemical equation (balanced reaction) for the neutralization in aqueous solution of (a) chlorous acid, $HClO_2$, with sodium hydroxide; and (b) oxalic acid, $H_2C_2O_4$, with potassium hydroxide (both hydroxide ions react).

3. Using the information in Tables 4.3 and 4.4, predict whether a reaction will take place when each of the following pairs of solutions are mixed. If a reaction occurs, write its equation (balanced reaction). (a) ammonium carbonate and nitric acid (b) magnesium bromide and silver(I) nitrate (c) sodium phosphate and potassium sulfate.

4.   Write the complete and net ionic equations for any reactions that occur in the previous problem.

5.   Using the information in Table 4.5, predict whether a reaction will occur between the following substances. If a reaction occurs, write the molecular, complete ionic, and net ionic equations.
(a) iron metal and water (b) magnesium metal and calcium nitrate solution (c) nickel metal and tin(II) chloride solution (d) magnesium metal and copper(II) chloride solution.

6.   Explain how to synthesize $Mg_3(PO_4)_2$, including a molecular equation.

7.   How would you prepare 500.0 g of a 25.0% by mass solution of KCl in water?

8.   What is the molarity of concentrated nitric acid, if it is 70.0% by mass HNO_3 and has a density of 1.42 g/mL?

9.   How much water should be added to 1.0×10^2 mL of 1.0 M HCl to prepare a 0.40 M HCl solution?

10. If 24.3 g of solid iron is treated with 2.000 L of 0.4850 M HCl according to the reaction given, (a) how many grams of hydrogen gas will be formed? (b) What mass of iron and volume in liters of HCl(aq) will be left after the reaction is over?

$$Fe(s) \; + \; HCl(aq) \; \rightarrow \; FeCl_2(aq) \; + \; H_2(g)$$

11. A 21.50-mL sample of hydrochloric acid required 28.52 mL of a 0.700 M NaOH solution to reach the endpoint in a titration. The density of the hydrochloric acid was 1.0339 g/mL. (a) What was the molarity of the hydrochloric acid? (b) How many grams of hydrochloric acid did the 21.50-mL sample contain? (c) What was the mass percent hydrochloric acid in the sample? The equation for the titration is

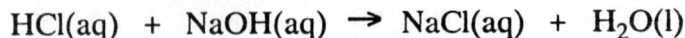

$$HCl(aq) \ + \ NaOH(aq) \ \rightarrow \ NaCl(aq) \ + \ H_2O(l)$$

ANSWERS TO PRACTICE PROBLEMS

1. KH_2BO_3 Potassium dihydrogen borate
          K_2HBO_3 Potassium hydrogen borate
          K_3BO_3 Potassium borate

2.   Acid   +   Base $\rightarrow$ Salt   +   Water

     (a) $HClO_2(aq) \ + \ NaOH(aq) \ \rightarrow \ NaClO_2(aq) \ + \ H_2O(l)$

     (b) $H_2C_2O_4(aq) \ + \ 2KOH(aq) \ \rightarrow \ K_2C_2O_4(aq) \ + \ 2H_2O(l)$

3.   (a) $(NH_4)_2CO_3(aq) \ + \ 2HNO_3(aq) \ \rightarrow \ 2NH_4NO_3(aq) \ + \ CO_2(g) \ + \ H_2O(l)$

     (b) $MgBr_2(aq) \ + \ 2AgNO_3(aq) \ \rightarrow \ 2AgBr(s) \ + \ Mg(NO_3)_2(aq)$

     (c) $Na_3PO_4(aq) \ + \ K_2SO_4(aq) \ \rightarrow$ No Reaction (N.R.)

4.   **(a)**   Complete Ionic: $2NH_4^+ + CO_3^{2-} + 2H^+ + 2NO_3^- \rightarrow 2NH_4^+ + 2NO_3^- + CO_2(g) + H_2O(l)$

           Net Ionic: $CO_3^{2-} + 2H^+ \rightarrow CO_2(g) + H_2O(l)$

     (b)   Complete Ionic: $Mg^{2+} + 2Br^- + 2Ag^+ + 2NO_3^- \rightarrow 2AgBr(s) + Mg^{2+} + 2NO_3^-$

           Net Ionic: $2Br^- + 2Ag^+ \rightarrow 2AgBr(s)$

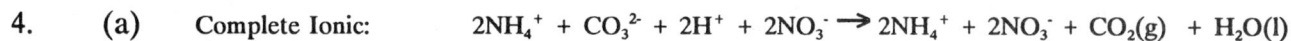

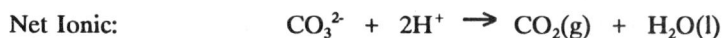

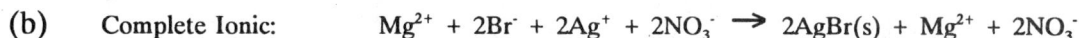

5.   **(a)**   No reaction! Fe metal is not reactive enough (see Table 4.5).

     (b)   No reaction! Mg will only react with salt solutions of metals **below** it in Table 4.5. Ca is **above** Mg.

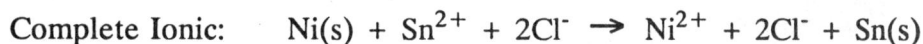

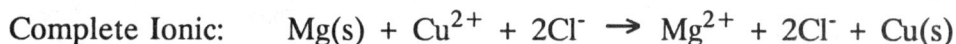

     (c)   Molecular: $Ni(s) + SnCl_2(aq) \rightarrow NiCl_2(aq) + Sn(s)$

           Complete Ionic: $Ni(s) + Sn^{2+} + 2Cl^- \rightarrow Ni^{2+} + 2Cl^- + Sn(s)$

           Net Ionic: $Ni(s) + Sn^{2+} \rightarrow Ni^{2+} + Sn(s)$

     (d)   Molecular: $Mg(s) + CuCl_2(aq) \rightarrow MgCl_2(aq) + Cu(s)$

           Complete Ionic: $Mg(s) + Cu^{2+} + 2Cl^- \rightarrow Mg^{2+} + 2Cl^- + Cu(s)$

           Net Ionic: $Mg(s) + Cu^{2+} \rightarrow Mg^{2+} + Cu(s)$

6.   Magnesium phosphate is insoluble (Table 4.4), so a double replacement reaction is best to synthesize this compound. See Sample Problem 4.6 B, parts (a) and (c), for method. Possibilities for soluble magnesium salt are nitrate, chloride, acetate, sulfate, bromide, or iodide. Possibilities for soluble phosphate are ammonium, potassium, sodium, and phosphoric acid. Using the first example in each case,

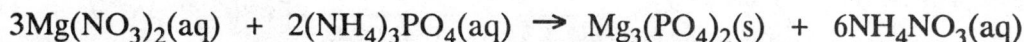

$$3Mg(NO_3)_2(aq) + 2(NH_4)_3PO_4(aq) \rightarrow Mg_3(PO_4)_2(s) + 6NH_4NO_3(aq)$$

Carry out the reaction in the proportions shown by the coefficients and separate $Mg_3(PO_4)_2(s)$ from the other product by filtration. Then wash and dry.

7.
$$? \text{ g KCl } = 500.0 \text{ g soln. } \times \frac{25.0 \text{ g KCl}}{100 \text{ g soln.}} = 125 \text{ g KCl}$$

$$? \text{ g H}_2\text{O } = 500.0 \text{ g soln} - 125 \text{ g KCl } = 375 \text{ g H}_2\text{O}$$

Measure out 125 g KCl and dissolve in 375 g H_2O.

8.
$$? \frac{\text{mol HNO}_3}{\text{L soln}} \text{ (M) } = \frac{70.0 \text{ g HNO}_3}{100 \text{ g soln.}} \times \frac{1.42 \text{ g soln.}}{1 \text{ mL soln.}} \times \frac{1000 \text{ mL soln.}}{1 \text{ L soln.}} \times \frac{1 \text{ mol HNO}_3}{63.0129 \text{ g HNO}_3} = 15.8 \text{ M HNO}_3$$

9.
$$V(\text{H}_2\text{O}) = V_{ad} - V_{bd} = V_{ad} - 1.0 \times 10^2 \text{ mL}$$

$$M_{bd} \times V_{bd} = M_{ad} \times V_{ad}$$

$$1.0 \text{ M} \times (1.0 \times 10^2 \text{ mL}) = 0.40 \text{ M} \times V_{ad} \qquad\qquad V_{ad} = 2.5 \times 10^2 \text{ mL}$$

$$V(\text{H}_2\text{O}) = 2.5 \times 10^2 \text{ mL} - 1.0 \times 10^2 \text{ mL} = 1.5 \times 10^2 \text{ mL}$$

10. Limiting reactant problem.

(a)
$$? \text{ mol Fe present } = 24.3 \text{ g Fe } \times \frac{1 \text{ mol Fe}}{55.847 \text{ g Fe}} = 0.435 \text{ mol Fe}$$

$$? \text{ fraction Fe } = \frac{0.435 \text{ mol Fe}}{1 \text{ mol Fe}} = 0.435$$

$$? \text{ mol HCl present } = 2 \text{ L soln. } \times \frac{0.4850 \text{ mol HCl}}{1 \text{ L soln.}} = 0.9700 \text{ mol HCl}$$

$$? \text{ fraction HCl } = \frac{0.9700 \text{ mol HCl}}{2 \text{ mol HCl}} = 0.4850$$

$0.435 < 0.4850$, Fe is limiting reactant.

$$? \text{ g H}_2 = 0.435 \text{ mol Fe } \times \frac{1 \text{ mol H}_2}{1 \text{ mol Fe}} \times \frac{2.015 \, 88 \text{ g H}_2}{1 \text{ mol H}_2} = 0.877 \text{ g H}_2$$

(b) Since iron is the limiting reactant, there will be zero grams of iron left after the reaction is over.

$$? \text{ L HCl reacted } = 0.435 \text{ \st{mol Fe} } \times \frac{2 \text{ \st{mol HCl}}}{1 \text{ \st{mol Fe}}} \times \frac{1 \text{ L soln.}}{0.4850 \text{ \st{mol HCl}}} = 1.79 \text{ L soln.}$$

Excess HCl soln. = 2.000 L soln. - 1.79 L soln. = 0.21 L

= 210 mL HCl soln. left over.

11. (a)

$$? \text{ mol NaOH } = 0.028\ 52 \text{ \st{L soln.} } \times \frac{0.700 \text{ mol NaOH}}{\text{\st{L soln.}}} = 0.0200 \text{ mol NaOH}$$

$$? \text{ mol HCl } = 0.0200 \text{ \st{mol NaOH} } \times \frac{1 \text{ mol HCl}}{1 \text{ \st{mol NaOH}}} = 0.0200 \text{ mol HCl}$$

$$? \text{ M HCl } = \frac{0.0200 \text{ mol HCl}}{0.02150 \text{ L soln.}} = 0.0929 \text{ M}$$

(b)

$$? \text{ g HCl } = 0.0200 \text{ \st{mol HCl} } \times \frac{36.461 \text{ g HCl}}{1 \text{ \st{mol HCl}}} = 0.728 \text{ g HCl in sample}$$

(c)

$$? \text{ g soln. } = 21.50 \text{ \st{mL soln.} } \times \frac{1.0339 \text{ g HCl soln.}}{1 \text{ \st{mL soln.}}} = 22.23 \text{ g HCl soln.}$$

$$? \text{ mass \% HCl } = \frac{0.728 \text{ g HCl}}{22.23 \text{ g HCl soln.}} \times 100 = 3.27\%$$

CHAPTER 5

GASES

SECTION 5.1   PRESSURE

After completing this section, you will be able to do the following:

- Define pressure and understand how a barometer works;

- Interconvert pressure units of atmosphere, Pascal, torr and mmHg.

REVIEW

Pressure is defined as **force per unit area** (force/unit area). In chemistry, the pressure of the earth's atmosphere is normally measured with a mercury barometer, which consists of a column of mercury sealed in a tube with the open end under the surface of a pool of mercury. The pressure of the column of mercury equals the pressure of the atmosphere pushing down on that same pool of mercury. If the atmospheric pressure decreases, then the pressure of the column of mercury must also decrease correspondingly, which means the column of mercury will decrease in height. If the atmospheric pressure increases, then the column of mercury will rise until its pressure equals that of the atmosphere. Pressure is measured in units of atmosphere (atm), torr, mmHg, or the SI unit of pascal (Pa).

SAMPLE PROBLEM 5.1 A

▶▶ Arrange the following units of pressure in order of increasing size: 1 atm, 1 Pa, 1 torr, 1 mmHg.

Solution:

HINT: Derive the necessary information from the conversion factors:

$$1 \text{ atm} = 1.013 \ 25 \times 10^5 \text{ Pa} = 760 \text{ torr} = 760 \text{ mmHg}$$

The atm is the largest unit and the Pa is the smallest, with torr and mmHg in between. The units of pressure in order of increasing size are

$$1 \text{ Pa} < 1 \text{ mmHg} = 1 \text{ torr} < 1 \text{ atm.}$$

SAMPLE PROBLEM 5.1 B

▶▶ Convert (a) 685.0 mmHg to atm (b) 5.432 x 10⁴ Pa to torr (c) 0.750 atm to torr (d) 715.5 mmHg to Pa.

Solution:

HINT: Derive all the conversion factors needed from the information given in the HINT of Sample Problem 5.1 A.

(a) Use the correct form of the conversion factor for the equality given between mmHg and atm, 1 atm = 760 mmHg, in the general problem-solving technique used in previous chapters.

$$? \text{ atm} = 685.0 \text{ mmHg} \times \frac{1 \text{ atm}}{760 \text{ mmHg}} = 0.9013 \text{ atm}$$

(b) Similarly, use the given equality between Pa and torr, 1.013 25 x 10⁵ Pa = 760 torr.

$$? \text{ torr} = 5.432 \times 10^4 \text{ Pa} \times \frac{760 \text{ torr}}{1.013 \ 25 \times 10^5 \text{ Pa}} = 407.4 \text{ torr}$$

(c) Similarly, use the given equality between atm and torr, 1 atm = 760 torr.

$$? \text{ torr} = 0.750 \; \text{atm} \; \times \; \frac{760 \text{ torr}}{1 \; \text{atm}} \; = \; 570 \text{ torr}$$

(d)   Similarly, use the given equality between mmHg and Pa, 760 mmHg = 1.013 25 x 10^5 Pa.

$$? \text{ Pa} = 715.5 \; \text{mmHg} \; \times \; \frac{1.013 \; 25 \times 10^5 \text{ Pa}}{760 \; \text{mmHg}} \; = \; 9.539 \times 10^4 \text{ Pa}$$

SECTION 5.2   RELATION BETWEEN PRESSURE AND VOLUME OF A GAS

After completing this section, you will be able to do the following:

■ Use Boyle's law to calculate the new pressure or volume of a gas at constant temperature.

REVIEW

The relationship between pressure and volume of a gas **at constant temperature** is known as Boyle's law, which states that the volume of a gas is inversely proportional to the pressure of the gas, at constant temperature. That is,

$$P = \frac{a \text{ constant}}{V} \quad \text{or} \quad V = \frac{a \text{ constant}}{P} \quad \text{or} \quad PV = a \text{ constant}$$

In other words, if the volume of a gas decreases, the pressure must increase accordingly so that the product of P x V is always constant, at constant temperature. Similarly, if the volume of a gas increases, the pressure must decrease accordingly at constant temperature. To calculate the pressure or volume of a gas at constant temperature, you will need to use the following equation derived from Boyle's law,

$$P_1 \times V_1 = P_2 \times V_2$$

where P_1 is the pressure at volume V_1, and P_2 is the pressure at volume V_2. Any units of pressure and volume may be used as long as they are the **same units** throughout the problem. Given three of the variables, you can solve for the fourth variable.

SAMPLE PROBLEM 5.2 A

▶▶ If the volume of a sample of gas is doubled, what happens to the pressure?

Solution:

You need to assign values to P_1, V_1, P_2, and V_2, where the subscript $_1$ means the initial (or first or old) set of variables and the subscript $_2$ means the final (or second or new) set of variables. Although no specific pressures or volumes are given, you can assign the **simplest** values possible and pick the units you want (as long as they are the same throughout the problem). The simplest value for V_1 is 1 mL (unit arbitrary); then V_2 must be 2 mL since the volume is doubled (or you could have used 2 mL and 4 mL, or 0.33 L and 0.66 L, or 150 mL and 300 mL, for V_1 and V_2, respectively). For P_1, assign the simplest value possible, 1 mmHg (or Pa or torr or atm). Then see what happens to the pressure -- that is, what is P_2?

Solve for P_2 from the equation given above, or remember what happens to the pressure of a gas when the volume increases (the pressure decreases). In either case, you arrive at the same answer.

$$P_2 = P_1 \times \frac{V_1}{V_2} = 1 \text{ mmHg} \times \frac{1 \text{ mL}}{2 \text{ mL}} = 0.5 \text{ mm Hg}$$

CHECK: Does this answer make sense? Yes, because the pressure **always** decreases when the volume of a gas increases. More specifically, if the volume of a gas doubles, the pressure **must** be one-half as great, at the same temperature. The volume doubled from 1 mL to 2 mL, and the pressure became one-half the initial value, 1 atm to 0.5 atm. If the volume had tripled, the pressure would be one-third as great. Similarly, if the pressure of a gas is four times as great as the initial pressure, then the volume becomes one-fourth of the initial volume.

SAMPLE PROBLEM 5.2 B

▶▶ If a sample of a gas has a volume of 15.7 L at 1.25 atm, what will the volume be at a pressure of 0.652 atm, if the temperature remains constant?

Solution:

See Sample Problem 5.2 A. The values of the variables are

$$P_1 = 1.15 \text{ atm} \quad \text{when} \quad V_1 = 15.7 \text{ L, and}$$
$$P_2 = 0.652 \text{ atm} \quad \text{when} \quad V_2 = \text{? L.}$$

$$V_2 = V_1 \times \frac{P_1}{P_2} = 15.7 \text{ L} \times \frac{1.15 \text{ atm}}{0.652 \text{ atm}} = 27.7 \text{ L}$$

CHECK: Does this answer make sense? Yes! The pressure decreases, so the volume must increase. The initial (old) volume V_1 **must** be multiplied by a fraction (or ratio of pressures) **greater than 1**, which can only mean P_1/P_2. Again, you can either reason out the solution this way to determine whether V_1 is multiplied by P_1/P_2 or P_2/P_1, or else solve for V_2 from the equation.

NOTE: The assignment of the subscripts $_1$ and $_2$ to pressure and volume can be done one of two ways. Both are correct and produce the same answer. The **key** is that the **same subscript** must be assigned to the volume of a gas at a particular pressure. For example, in this problem the alternate assignment of subscripts is

$$P_1 = 0.652 \text{ atm} \quad \text{when} \quad V_1 = ? \text{ L, and}$$
$$P_2 = 1.15 \text{ atm} \quad \text{when} \quad V_2 = 15.7 \text{ L.}$$

The point is that this gas has a pressure of 1.15 atm when the volume is 15.7 L, which means that **this** pressure and volume must be assigned the **same subscript**, P_1 and V_1, or P_2 and V_2, respectively. You are trying to find the volume of this gas when the pressure is 0.652 atm. The same answer of 27.7 L is obtained in both sets of subscript assignments (solving for V_2 in the first case, and solving for V_1 in the second case).

SAMPLE PROBLEM 5.2 C

▶▶ If a sample of a gas has a pressure of 2.16×10^3 Pa when the volume is 376 mL, what will the pressure be when the volume is 147 mL, if the temperature remains constant?

Solution:

See Sample Problem 5.2 B. The values of the variables are

$$P_1 = 2.16 \times 10^3 \text{ Pa} \quad \text{when} \quad V_1 = 376 \text{ mL, and}$$
$$P_2 = ? \text{ Pa} \quad \quad \text{when} \quad V_2 = 147 \text{ mL.}$$

$$P_2 = P_1 \times \frac{V_1}{V_2} = 2.16 \times 10^3 \text{ Pa} \times \frac{376 \text{ mL}}{147 \text{ mL}} = 5.52 \times 10^3 \text{ Pa}$$

CHECK: Again, does this answer make sense? Absolutely! If the volume of a gas decreases, the inverse must happen to the pressure -- that is, the pressure must increase. The new volume decreases to almost one-third the original volume, so the pressure is almost tripled.

SECTION 5.3 RELATION BETWEEN VOLUME AND TEMPERATURE OF A GAS

After completing this section, you will be able to do the following:

■ Use Charles's law to calculate the new volume or temperature of a gas at constant pressure.

REVIEW

The relationship between volume and temperature of a gas **at constant pressure** is known as Charles's law, which states that the volume of a gas is directly proportional to the Kelvin or absolute temperature of the gas, at constant pressure. That is,

$$V = \text{a constant} \times T \quad \text{or} \quad \frac{V}{T} = \text{a constant}$$

In other words, if the temperature of a gas decreases, the volume of the gas must decrease proportionately so that the quotient V/T is always constant, at constant pressure. Similarly, if the temperature of a gas increases, the volume must increase proportionately at constant pressure. To calculate the volume or temperature of a gas at constant pressure, you will need to use the following equation derived from Charles's law,

$$\frac{V_1}{T_1} = \frac{V_2}{T_2}$$

where V_1 is the volume at temperature T_1, and V_2 is the volume at temperature T_2. Although any unit of volume may be used as long as it is the **same unit** throughout the problem, temperature **must** be in Kelvin. Given three of the variables, you can solve for the fourth variable.

SAMPLE PROBLEM 5.3 A

▶ ▶ If a sample of a gas doubles in volume at constant pressure, what happens to the temperature?

See Sample Problem 5.2 A. You need to assign values to V_1, T_1, V_2, and T_2, where the subscript $_1$ means the initial (or first or old) set of variables and the subscript $_2$ means the final (or second or new) set of variables. Although no specific volumes are given, you can assign the **simplest** values possible and pick the unit you want (as long as it is the same throughout the problem). The temperature **must** be Kelvin (**no degree symbol used**). When a temperature is given in °C, you **must** convert it to K by adding 273.2 °C before doing any calculation. The simplest value for V_1 is 1 L (unit arbitrary); then V_2 must be 2 L, since the volume is doubled (or you could have used 2 mL and 4 mL, or 0.33 L and 0.66 L, or 150 mL and 300 mL, for V_1 and V_2, respectively). For T_1, assign any simple value **in Kelvin**, as long as it is **above 0.0 K**. For example, use 100 K for T_1. Then see what happens to the temperature -- that is, what is T_2?

Solve for T_2 from the equation given above or, better yet, remember what happens to the temperature of a gas when the volume increases according to Charles's law (the temperature also increases). In either case, you arrive at the same answer.

$$T_2 = T_1 \times \frac{V_2}{V_1} = 100 \text{ K} \times \frac{2 \text{ L}}{1 \text{ L}} = 200 \text{ K}$$

CHECK: Does this answer make sense? Yes, because the temperature **always** increases as the volume of a gas increases, at constant pressure. If the volume of a gas doubles, the Kelvin temperature **must** double, when the pressure remains constant. The volume doubled from 1 L to 2 L, and the temperature doubled from 100 K to 200 K. If the volume had tripled, the Kelvin temperature would have tripled also. Similarly, if the volume of a gas becomes one-third its original volume, the Kelvin temperature becomes one-third its original value also.

SAMPLE PROBLEM 5.3 B

▶▶ If a sample of gas has a volume of 237 mL at 15.0 °C, what will the volume be when the temperature becomes -20.0 °C, if the pressure remains constant?

See Sample Problem 5.3 A. The values of the variables are

$$V_1 = 237 \text{ mL} \quad \text{when} \quad T_1 = 15.0 \text{ °C} = 288.2 \text{ K, and}$$
$$V_2 = ? \text{ mL} \quad \text{when} \quad T_2 = -20.0 \text{ °C} = 253.2 \text{ K.}$$

Remember, Charles's law only is valid for temperature in **Kelvin**, NOT °C (K = °C + 273.2). You can solve this equation by solving the above formula for V_2 or, better yet, remember what happens to the volume of a gas as the temperature decreases (the volume also decreases). You always find V_2 by starting with V_1 and multiplying by a ratio of temperatures. Since the final volume (V_2) must be less than V_1, the correct ratio of temperatures must be **less** than 1,

$$V_2 = 237 \text{ mL} \times \frac{253.2 \text{ K}}{288.2 \text{ K}} = 208 \text{ mL}$$

CHECK: Does this answer make sense? Yes! The temperature decreases, so the volume must decrease. The initial (old) volume V_1 **must** be multiplied by a fraction (or ratio of temperatures) **less than 1**, which can only mean 253.2 K/288.2 K.

SAMPLE PROBLEM 5.3 C

▶▶ At 125 °C, the volume of a gas is 14.7 L. In order for the volume of the gas to double, what does the temperature have to become, if the pressure remains constant?

Solution:

If you are reasoning through your solutions to these gas law problems rather than memorizing or rearranging formulas, you should be able to come up with the answer immediately --- **double the Kelvin equivalent of 125 °C!** Or, if you find it more helpful to assign subscripts to the variables first and then set up the solution,

$$V_1 = 14.7 \text{ L} \quad \text{when} \quad T_1 = 125 \text{ °C} = 398 \text{ K, and}$$
$$V_2 = 29.4 \text{ L} \quad \text{when} \quad T_2 = ? \text{ K.}$$

Since the volume doubles (you could have used 1 L and 2 L, etc., for V_1 and V_2, respectively), the temperature in Kelvin must double. Or, to find V_2, you must multiply V_1 by a fraction (or ratio of volumes) **greater than 1**, or 29.4 L/14.7 L,

$$T_2 = 398 \text{ K} \times \frac{29.4 \text{ L}}{14.7 \text{ L}} = 796 \text{ K}$$

CHECK: Does this answer make sense? By now, you should agree that it does. At constant pressure, the Kelvin temperature of a gas increases in direct proportion to the increase in volume of the gas.

SECTION 5.4   STANDARD TEMPERATURE AND PRESSURE

After completing this section, you will be able to do the following:

- ■ Convert the observed volume of a gas to the volume at standard temperature and pressure (STP), and vice versa, by applying both Boyle's and Charles's laws.

REVIEW

The **standard temperature and pressure (STP)** of a gas is defined as 0 °C or 273.15 K and 760 mmHg (or 1 atm or 1.013 25 x 10^5 Pa). By applying both Boyle's and Charles's laws, you can convert the observed volume of a gas at non-standard conditions to the volume at STP, and vice versa.

SAMPLE PROBLEM 5.4 A

▶ ▶ A sample of gas has a volume of 714 mL at 40.0 °C and 1.25 atm. What is the volume of the gas at STP?

Solution:

Remember that STP means 0 °C = 273.2 K, and 1.0 atm = 760 mmHg. The assignment of variables is

$$V_1 = 714 \text{ mL} \qquad P_1 = 1.25 \text{ atm} \qquad T_1 = 40.0 °C = 313.2 \text{ K}$$
$$V_2 = ? \text{ mL} \qquad P_2 = 1.00 \text{ atm} \qquad T_2 = 0.00 °C = 273.2 \text{ K}$$

The final or new volume of the gas (V_2) is determined by the change in pressure and temperature. Reason through that the volume of the gas (a) **increases** due to a decrease in pressure (Boyle's law) and (b) **decreases** due to a decrease in temperature (Charles's law). This means that the ratio of pressures must be >1 (1.25 atm/1.00 atm), and the ratio of temperatures must be <1 (273.2 K/313.2 K). Again, this reasoning process avoids memorizing a combined equation of Boyle's and Charles's laws. The solution is

$$V_2 = 714 \text{ mL} \times \frac{1.25 \text{ atm}}{1.00 \text{ atm}} \times \frac{273.2 \text{ K}}{313.2 \text{ K}} = 779 \text{ mL}$$

SAMPLE PROBLEM 5.4 B

▶▶ A sample of gas has a volume of 0.625 L at STP. What is the volume of the gas at 75.0 °C and 805 mmHg?

Solution:

See Sample Problem 5.4 A. The assignment of variables is

$$V_1 = 0.625 \text{ L} \qquad P_1 = 760 \text{ mmHg} \qquad T_1 = 0.00 \text{ °C} = 273.2 \text{ K}$$
$$V_2 = ? \text{ L} \qquad P_2 = 805 \text{ mmHg} \qquad T_2 = 75.0 \text{ °C} = 348.2 \text{ K}$$

The final or new volume of the gas (V_2) is determined by the change in pressure and temperature. Reason through that the volume of the gas (a) **decreases** due to an increase in pressure (Boyle's law), and (b) **increases** due to an increase in temperature (Charles' law). That means the ratio of pressures must be <1 (760 mmHg/805 mmHg) and the ratio of temperatures must be >1 (348.2 K/273.2 K). The solution is

$$V_2 \;=\; 0.625 \text{ L} \;\times\; \frac{760 \text{ mmHg}}{805 \text{ mmHg}} \;\times\; \frac{348.2 \text{ K}}{273.2 \text{ K}} \;=\; 0.752 \text{ L}$$

SECTION 5.5	GAY-LUSSAC'S LAW OF COMBINING VOLUMES AND AVOGADRO'S LAW

After completing this section, you will be able to do the following:

■ Perform stoichiometric calculations on chemical equations involving gases where the coefficients represent volumes of gases.

REVIEW

Previously (Section 3.4), you learned that the coefficients in a chemical equation (balanced reaction) stand for numbers of molecules or numbers of moles of molecules. If the reactants and products in the reactions are **gases**, the coefficients also represent **volumes of gases**, since the volume of a gas at **constant pressure and temperature** is directly proportional to the number of molecules of the gas. This conclusion arises from the application of Gay-Lussac's law of combining volumes (at constant temperature and pressure, the volumes of gases involved in chemical reactions are in the ratio of small whole numbers) **and** Avogadro's law (the volume of a gas at constant temperature and pressure is directly proportional to the number of molecules of the gas). Thus, a correct interpretation of the following equation, where the volumes of gases are measured at constant temperature and pressure, is

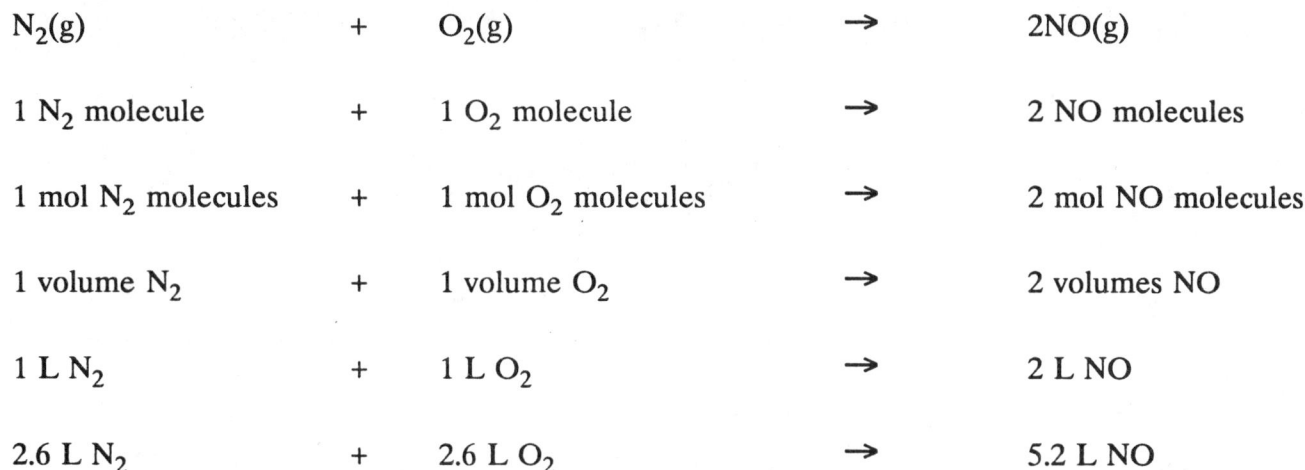

$N_2(g)$	+	$O_2(g)$	$\rightarrow$			$2NO(g)$
1 N_2 molecule	+	1 O_2 molecule	$\rightarrow$			2 NO molecules
1 mol N_2 molecules	+	1 mol O_2 molecules	$\rightarrow$			2 mol NO molecules
1 volume N_2	+	1 volume O_2	$\rightarrow$			2 volumes NO
1 L N_2	+	1 L O_2	$\rightarrow$			2 L NO
2.6 L N_2	+	2.6 L O_2	$\rightarrow$			5.2 L NO

Any volume unit can be used, as long as it is the same unit throughout the reaction; the volumes must be measured (used) **at constant temperature and pressure.** The volume of gas measured at STP that contains **1 mol** of gas is called the **STP molar volume,** which has been found to be 22.4 L.

SAMPLE PROBLEM 5.5 A

▶ ▶ When water vapor is formed, (a) how many liters of hydrogen gas are required to react with 3.7 L of $O_2(g)$? (b) How many liters of $H_2O(g)$ are formed in part (a), assuming the reaction is quantitative (i.e., goes to completion)? (Assume all volumes are measured at the same temperature and pressure.)

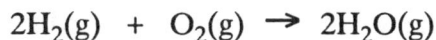

$$2H_2(g) + O_2(g) \rightarrow 2H_2O(g)$$

Solution:

First, you should **always** start any stoichiometric calculation with a balanced reaction (chemical equation). If you are working with gases at constant temperature and pressure, the quickest solution is found by interpreting the coefficients as volumes of gases. Set up the problem in a manner similar to Sample Problem 3.5 A, and solve it using the same general problem-solving procedure. Again, the coefficients yield the necessary conversion factors.

Problem:	(a) ? L	3.7 L		(b) ? L
Equation:	$2H_2(g)$ +	$O_2(g)$	$\rightarrow$	$2H_2O(g)$
Coefficients:	2 L	1 L		2 L

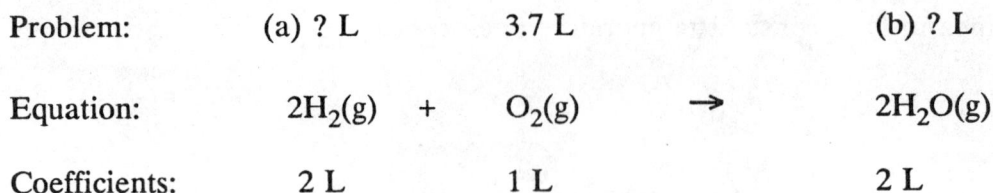

(a) $\quad ? \text{ L } H_2 = 3.7 \text{ L } O_2 \times \dfrac{2 \text{ L } H_2}{1 \text{ L } O_2} = $ 7.4 L $H_2(g)$

(b) $\quad ? \text{ L } H_2O(g)$ formed = # L $H_2(g)$ reacted = 7.4 L $H_2(g)$ = 7.4 L $H_2O(g)$

SAMPLE PROBLEM 5.5 B

►► If 5.4×10^{22} molecules of $O_2(g)$ occupy 18.4 L, (a) what volume will 5.4×10^{22} molecules of $H_2(g)$ occupy if both gases are at the same temperature and pressure? (b) What volume will 2.7×10^{22} molecules of $O_2(g)$ occupy (still at the same temperature and pressure)?

Solution:

(a) The 5.4×10^{22} molecules of H_2 **or any other gas at the same temperature and pressure** will occupy the same volume, 18.4 L. This results directly from Avogadro's law --- the volume of a gas at constant temperature and pressure is directly proportional to the number of molecules. There is so much empty space in gases (which is why they have the lowest densities) that only the **number** of molecules (NOT size or kind) determines the volume of the gas.

(b) At the same temperature and pressure as in part (a), Avogadro's law states that the volume of a gas is directly proportional to the number of molecules of the gas. If the number of molecules doubles (at the same T and P), the volume doubles. If the number of molecules is half as great, the volume is half as great. In this problem,

$$? \text{ L } O_2 = 2.7 \times 10^{22} \text{ } O_2 \text{ molecules} \times \frac{18.4 \text{ L}}{5.4 \times 10^{22} \text{ } O_2 \text{ molecules}} = 9.2 \text{ L } O_2$$

| SECTION 5.6 | THE IDEAL GAS EQUATION |
| SECTION 5.7 | USING THE IDEAL GAS EQUATION TO SOLVE PROBLEMS |

After completing this section, you will be able to do the following:

■ Use the ideal gas equation, $PV = nRT$, to calculate (a) P, V, n, or T of a gas given three of the four quantities, (b) V of a gas formed in a chemical reaction from a given mass of a reactant, (c) the density of a gas, and (d) the molecular mass of a gas.

REVIEW

Although Boyle's law, Charles's law, and Avogadro's law deal with certain properties of gases under specific conditions, the **most useful equation** to understand and master when dealing with gases is the **ideal gas equation, $PV = nRT$**, where P = pressure, V = volume, n = number of moles of a gas, R = the gas constant (whose value depends on the units used for P and V), and T = temperature (**always in kelvin**). You must be able to apply the ideal gas equation to the four applications given above and illustrated in the following sample problems.

SAMPLE PROBLEM 5.7 A

▶▶ What is the pressure of 1.20 mol of O_2 gas that have a volume of 17.4 L at 20.4 °C?

Solution:

Of all the gas laws discussed so far in this chapter, only one has number of moles in it --- the ideal gas law, or ideal gas equation. Start with $PV = nRT$, solve for pressure, P, and substitute the values given for n, R, and T, remembering that the correct value of R from Table 5.3 in the textbook is the one in which the units of R agree with the units of P and V.

$$P = \frac{nRT}{V} = \frac{(1.20 \text{ mol})(0.082\ 06 \text{ atm} \cdot \text{L} \cdot \text{mol}^{-1} \cdot \text{K}^{-1})(293.6 \text{ K})}{17.4 \text{ L}} = 1.66 \text{ atm}$$

NOTE: Only when the correct value and units of R are used will all the units cancel except what you are looking for, pressure in atm.

SAMPLE PROBLEM 5.7 B

▶▶ At what temperature will 15.8 g of CO_2 gas have a pressure of 950.0 mmHg and a volume of 27.6 L?

Solution:

See Sample Problem 5.7 A. The ideal gas equation must be used, because the quantity of gas is given, i.e., grams. The mass in grams will need to be converted to moles as you have done many times before,

$$n = \frac{m}{MM} = \frac{15.8 \text{ g}}{44.010 \text{ g/mol}} = 0.359 \text{ mol},$$

where m is the mass in grams and MM is the molecular mass in g/mol (mass in grams of 1 mol of the gas). Then solve the ideal gas equation for T and substitute in the values for P, V, n, and R, making sure that the units of R include mmHg for P and L for V (Table 5.3 in textbook).

$$T = \frac{PV}{nR} = \frac{(950.0 \text{ mmHg})(27.6 \text{ L})}{(0.359 \text{ mol})(62.36 \text{ mmHg} \cdot \text{L} \cdot \text{mol}^{-1} \cdot \text{K}^{-1})} = 1170 \text{ K}$$

SAMPLE PROBLEM 5.7 C

▶▶ What volume of O_2 gas at 0.950 atm and 40.0 °C will be formed from the decomposition of 45.8 g of CaO(s)?

Solution:

You are given the pressure and temperature of the O_2 gas. Before you can find the volume of the O_2 gas from the ideal gas equation, you will need to know how many moles of O_2 gas are produced by the chemical equation. See Sample Problem 3.5 A of this book for the procedure.

Problem:	45.8 g			? mol

Equation: $\qquad$ 2CaO(s) $\rightarrow$ 2Ca(s) $+$ O$_2$(g).

Formula mass, u: $\quad$ 56.077

Coefficients:	2 mol	2 mol	1 mol

$$? \text{ mol O}_2 = 45.8 \text{ g } \cancel{\text{CaO}} \times \frac{1 \text{ mol } \cancel{\text{CaO}}}{56.077 \text{ g } \cancel{\text{CaO}}} \times \frac{1 \text{ mol O}_2}{2 \text{ mol } \cancel{\text{CaO}}} = 0.408 \text{ mol O}_2$$

Now find the volume of the gas at the conditions given.

$$V = \frac{nRT}{P} = \frac{(0.408 \text{ mol})(0.082\ 06 \text{ atm} \cdot \text{L} \cdot \text{mol}^{-1} \cdot \text{K}^{-1})(313.2 \text{ K})}{0.950 \text{ atm}} = 11.0 \text{ L}$$

SAMPLE PROBLEM 5.7 D

▶▶ What is the density of chlorine gas (Cl$_2$) at 10.5 °C and 2.7 atm?

Solution:

You can use the two-step method of finding (1) the volume of the gas from the ideal gas equation and then (2) the density from the density equation (this method is shown in Sample Problem 5.8 of the textbook), OR you can solve the ideal gas equation directly for density by using a slightly different form of the ideal gas equation (which will be used in this problem and Sample Problem 5.7 E). Since the number of moles, n, of any substance is the mass, m, in grams divided by the molecular mass, MM (see Sample Problem 5.7 B), make this substitution in the ideal gas equation,

$$PV = nRT = \frac{m}{MM} RT$$

Then solve the ideal gas equation for density and complete the problem.

$$D = \frac{m}{V} = \frac{P \cdot MM}{RT} = \frac{(2.7 \text{ atm})(70.9054 \text{ g} \cdot \text{mol}^{-1})}{(0.082\ 06 \text{ atm} \cdot \text{L} \cdot \text{mol}^{-1} \cdot \text{K}^{-1})(283.7 \text{ K})} = 8.2 \text{ g} \cdot \text{L}^{-1}$$

NOTE: Remember, the formula mass of Cl$_2$ from the inside back cover is 70.9054 u/molecule. The molecular mass (MM) of Cl_2 is the mass in grams of one mole of Cl_2, or 70.9054 g/mol.

SAMPLE PROBLEM 5.7 E

▶▶ What is the molecular mass (MM) of a 1.37-g sample of gas which occupies a volume of 2.04 L at 27.6 °C and 724.3 mmHg?

Solution:

You can use the two-step method of finding (1) the number of moles of the gas from the ideal gas equation under the conditions given and then (2) the mass of one mole of the gas, which is the molecular mass (this method is shown in Sample Problem 5.9 of the textbook), OR you can solve the ideal gas equation directly for molecular mass (mass in grams of 1 mol) by using the **same** variation of the ideal gas equation derived in Sample Problem 5.7 D. The molecular mass (MM) of the gas is

$$? \frac{g}{mol} (MM) = \frac{mRT}{PV} = \frac{(1.37 \text{ g})(62.36 \text{ mmHg•L•mol}^{-1}\text{•K}^{-1})(300.8 \text{ K})}{(724.3 \text{ mmHg})(2.04 \text{ L})} = 17.4 \text{ g•mol}^{-1}$$

SECTION 5.8   DALTON'S LAW OF PARTIAL PRESSURES

After completing this section, you will be able to do the following:

■ Calculate the partial pressure of a gas in a mixture of gases;

■ Calculate the dry volume of a gas collected over water.

REVIEW

According to **Dalton's law of partial pressures**, the total pressure of a mixture of gases is equal to the sum of the partial pressures of the individual gases in the mixture,

$$P_{total} = P_{gas\ 1} + P_{gas\ 2} + P_{gas\ 3} + ...$$

When a gas is collected over water, the pressure measured is that of the gas wet,

$$P_{total} = P_{gas} + P_{water\ vapor}$$

The pressure of the gas dry (p_{gas}) is found by subtracting the vapor pressure of water at the temperature of the experiment ($p_{water\ vapor}$ from Table 5.4 in the textbook) from the total pressure, p_{total}. Then the volume of the dry gas can be calculated.

SAMPLE PROBLEM 5.8 A

▶▶ (a) What is the partial pressure of each gas in a mixture of 0.65 mol of hydrogen gas and 1.15 mol of helium gas in a 0.500-L container at 200.0 °C? (b) What is the total pressure of the mixture in (a)?

Solution:

(a)   The partial pressure of each gas is calculated from the ideal gas equation as if no other gas was present. That is, each gas in a mixture occupies the entire volume of the container.

$$P_{hydrogen} = \frac{nRT}{V} = \frac{(0.65\ mol)(0.082\ 06\ atm{\bullet}L{\bullet}mol^{-1}{\bullet}K^{-1})(473.2\ K)}{0.500\ L} = 5.0 \times 10^1\ atm$$

$$P_{helium} = \frac{nRT}{V} = \frac{(1.15\ mol)(0.082\ 06\ atm{\bullet}L{\bullet}mol^{-1}{\bullet}K^{-1})(473.2\ K)}{0.500\ L} = 8.93 \times 10^1\ atm$$

(b)   The total pressure of the mixture is the sum of the partial pressures,

$$P_{total} = P_{hydrogen} + P_{helium} = 5.0 \times 10^1\ L + 8.93 \times 10^1\ L = 1.39 \times 10^2\ L$$

SAMPLE PROBLEM 5.8 B

▶▶ In a student experiment, 0.850 L of nitrogen gas was collected over water (wet gas) at 22.1 °C and a barometric pressure of 738.3 mmHg. Calculate the volume the nitrogen gas would have occupied if it had been dry at the same barometric pressure and temperature.

Solution:

First find the partial pressure of the nitrogen gas in the nitrogen + water vapor mixture, since water vapor is **always** present when a gas is collected over water. The vapor pressure of water at various temperatures is given in Table 5.4 of the textbook. The vapor pressure of water at

22.1 °C is not given, but the vapor pressures at the next closest temperatures (22.0 °C and 22.2 °C) are given. For 22.1 °C, calculate the vapor pressure half-way in between, $(19.827 + 20.070)/2 = 19.949$ mmHg. The partial pressure of nitrogen is

$$P_{nitrogen} = P_{total} - P_{water\ vapor} = 738.3\ mmHg - 19.949\ mmHg = 718.4\ mmHg$$

In the presence of water vapor, the pressure of the nitrogen gas is 718.4 mmHg. If no water vapor were present in this experiment (i.e., the nitrogen gas was collected by some other method), the pressure of the dry nitrogen would have been equal to the barometric pressure, 738.3 mmHg. To summarize the information,

nitrogen, wet	V = 0.850 L	P = 718.4 mmHg	T = 22.1 °C
nitrogen, dry	V = ? L	P = 738.3 mmHg	T = 22.1 °C

Since the temperature remains constant, the volume of the dry nitrogen is equal to the volume of the wet nitrogen times a ratio of pressures **less than 1** (since when the pressure increases from 718.4 to 738.3, the volume must decrease proportionately),

$$V_{dry\ nitrogen} = 0.850\ L \times \frac{718.4\ \cancel{mmHg}}{738.3\ \cancel{mmHg}} = 0.827\ L$$

SECTION 5.9   GRAHAM'S LAW OF THE DIFFUSION RATES OF GASES

After completing this section, you will be able to do the following:

■ Use Graham's law to compare rates of diffusion of gases.

REVIEW

Gases spread out or **diffuse** to fill the container they are confined to, but not at the same rate. **Graham's law of diffusion** states that, under the same conditions, the rates of diffusion of different gases are inversely proportional to the square roots of their molecular masses,

$$\frac{\text{Rate of diffusion of gas 1}}{\text{Rate of diffusion of gas 2}} = \frac{\sqrt{MM\ gas\ 2}}{\sqrt{MM\ gas\ 1}}$$

In other words, the smaller the molecular mass (MM) of the gas molecule, the faster it will diffuse.

SAMPLE PROBLEM 5.9 A

▶▶ Molecules of which gas will diffuse faster: O_2, He, or H_2?

Solution:

The lighter the molecule (smallest molecular mass, MM), the faster it will diffuse. The molecular masses of O_2, He and H_2 from the inside front and back covers are 31.9988 g/mol, 4.002 602 g/mol, and 2.015 88 g/mol, respectively. Since H_2 has the smallest molecular mass, H_2 molecules will diffuse the fastest, with He molecules next, and O_2 molecules diffusing the slowest because they are the heaviest.

SAMPLE PROBLEM 5.9 B

▶▶ How many times as fast would NO(g) diffuse as NO_2(g)?

Solution:

Use Graham's law of diffusion equation given above. Since you are comparing NO to NO_2, the equation is easiest to work with if the rate of NO is in the numerator,

$$\frac{\text{Rate of NO(g)}}{\text{Rate of NO}_2\text{(g)}} = \frac{\sqrt{46.0055 \text{ g/mol}}}{\sqrt{30.0061 \text{ g/mol}}} = 1.238\ 23 \text{ times as fast}$$

The equation says that the rate of diffusion of NO gas is about 1.24 times faster than the rate of diffusion of NO_2 gas.

SECTION 5.10   THE KINETIC-MOLECULAR THEORY

After completing this section, you will be able to do the following:

■ Explain the behavior of gases using the kinetic-molecular theory.

REVIEW

The **kinetic-molecular theory** is the explanation for why gases behave the way they do. The theory refers to the properties of an "ideal" gas. The properties of real gases (e.g., H_2, O_2, N_2, etc.) in some instances come close to those of an ideal gas, but no "ideal" gas exists. Nevertheless, the **kinetic-molecular theory** explains reasonably well the behavior of real gases. The major points in the **kinetic-molecular theory** are:

- **Gases consist of small particles called molecules.**
  Molecules of a real gas occupy only about 0.04% of the volume occupied by the gas; the rest of the volume is empty space. Molecules of an **ideal gas** have **no** volume. That is, 0% of the volume of an **ideal gas** is occupied by the gas molecules.

- **The molecules of a gas are constantly moving at very high speeds.**
  At room temperature, molecules are moving at speeds in excess of 1000 miles per hour.

- **Molecules of ideal gases cannot hit each other because they have no "volume."**
  But molecules of real gases do have a volume and collide with each other as well as the walls of the container they are confined to, at very high speeds. Gas molecules striking the inside wall of a container is what is called the pressure of a gas. The more gas molecules pumped into a fixed-volume container at constant temperature, the greater the number of collisions with the inside wall of the container and the higher the pressure.

- **There are no attractive or repulsive forces between the molecules of an ideal gas.**
  Collisions between ideal gas molecules are called **elastic collisions** because the total energy of motion of the molecules remains unchanged. Collisions between molecules of real gases are **inelastic** because some of the kinetic energy of the molecules is converted into attractive and repulsive forces.

- **Energy of motion is called kinetic energy.**
  The **average** kinetic energy of the molecules of a gas is proportional to the Kelvin temperature. At any given temperature, the molecules of **all gases** have the **same average** kinetic energy. The kinetic energy of a gas depends on the mass and velocity of its molecules,

$$\text{Kinetic energy} = \frac{1}{2}mv^2,$$

where m is mass and v is velocity. At a given temperature, all molecules of a particular gas do not have the same velocity (kinetic energy) but are distributed in the shape of a "bell-shaped" curve (see O_2 at 25 °C in Figure 5.16 in the textbook). When the temperature of the O_2 gas is increased to 700 °C, the average speed increases (and therefore kinetic energy), and the speeds of the O_2 molecules are distributed over a wider range.

All the gas laws can be derived from the **kinetic-molecular theory.**

SECTION 5.11   REAL GASES

After completing this section, you will be able to do the following:

■ Use the kinetic-molecular theory to explain under what conditions the behavior of real gases approaches ideal gas behavior.

REVIEW

By graphing the **compressibility factor, PV/nRT,** vs. pressure for real gases, the conditions can be observed under which ideal gas behavior is **approached** (but not reached). From the ideal gas equation, PV=nRT, the compressibility factor of an ideal gas is 1. Graphs of the compressibility factor under varying conditions of pressure and temperature for several real gases are shown in Figures 5.17 (a), (b), and (c) of the textbook. The fact that **real gases** have compressibility factors other than 1 (ideal gas) can be explained by the existence of **attractive forces between molecules** and by the fact that molecules of real gases are not points (like ideal gases) but **have volume.** Remember, in an **ideal gas,** there are **no** attractive forces between molecules, and **0%** of the volume of the gas is occupied by the gas molecules themselves.

The same graphs show that all gases **approach ideal gas behavior** as **pressure approaches zero** and as the **temperature increases.** As **pressure approaches zero,** the volume of a real gas becomes larger, the gas molecules move farther apart and experience weaker attractive forces, and the volume of the gas molecules is a smaller fraction of the total volume of the gas. As the **temperature increases,** the volume of a real gas also increases, the molecules move faster and overcome some of the attractive forces, and there is a decrease in the fraction of the total volume of the gas occupied by the gas molecules. Thus, real gases approach (but never reach) ideal gas behavior at **low pressures and high temperatures.**

SELF-TEST QUESTIONS

Test your understanding of the concepts and skills in this chapter by working through the multiple choice questions **BEFORE** checking the answers. A periodic table will always be available. The ANSWERS TO SELF-TEST QUESTIONS follow immediately after this section.

1. A pressure of 525.0 mmHg is the same as

   A. 0.6908 torr   B. 1.429 atm   C. 760.0 torr   D. 6.999×10^4 Pa   E. 0.525 atm

2. If the pressure of a gas is reduced to one-fourth its original value, the volume of the gas at constant temperature will

   A. remain unchanged.
   B. be four times larger.
   C. be reduced to one-fourth its original volume.
   D. double.
   E. be reduced to one-half its original volume.

3. If a sample of gas has a volume of 475 mL at 20.0 °C, the volume of the gas at 0.0 °C and constant pressure will be

   A. 443 mL   B. 0.00 mL   C. 510 mL   D. 475 mL   E. 462 mL

4. A sample of gas has a volume of 295 mL at 15.0 °C and 480.0 mmHg. The volume of the gas at STP is

   A. 197 mL   B. 443 mL   C. 177 mL   D. 493 mL   E. 295 mL

5. The number of milliliters of $N_2(g)$ formed when 284 mL of NO(g) decomposes is (all volumes measured at the same temperature and pressure)

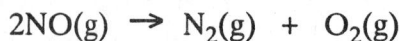

$$2NO(g) \rightarrow N_2(g) + O_2(g)$$

   A. 284 mL   B. 568 mL   C. 71 mL   D. 213 mL   E. 142 mL

6. The volume of 0.250 mol of a gas at 35.7 °C and 1.12 atm is

   A. 573 L   B. 0.654 L   C. 5.66 L   D. 7.10 L   E. 22.4 L

7. The density of HCl gas at 40.0 °C and 685 mmHg is

   A. 1.28 g/L   B. 9.59 g/L   C. 10.0 g/L   D. 0.782 g/L   E. 0.497 g/L

8. If the partial pressures of helium, neon, and argon gases are 264 mmHg, 179 mmHg, and 311 mmHg, respectively, the total pressure of the mixture of gases is

   A. 251 mmHg   B. 754 mmHg   C. 310 mmHg   D. 1063 mmHg   E. 912 mmHg

9. The vapor pressure of water at 28.9 °C is (use Table 5.4 in textbook)

   A. 29.697 mmHg   B. 30.043 mmHg   C. 29.870 mmHg   D. 31.461 mmHg   E. 31.824 mmHg

10. The following gases arranged in order of increasing rate of diffusion are

$$HCl(g),\ NH_3(g),\ F_2(g),\ H_2S(g),\ SO_2(g)$$

   A. $NH_3 < H_2S < F_2 < HCl < SO_2$.
   B. $NH_3 < SO_2 < H_2S < HCl < F_2$.
   C. $H_2S < SO_2 < HCl < NH_3 < F_2$.
   D. $SO_2 < F_2 < HCl < H_2S < NH_3$.
   E. $HCl < NH_3 < H_2S < SO_2 < F_2$.

ANSWERS TO SELF-TEST QUESTIONS

1. d	4. c	7. a
2. b	5. e	8. b
3. a	6. c	9. c
		10. d

PRACTICE PROBLEMS

Test your problem-solving skills in this chapter by working through the problems in the space provided **BEFORE** checking the answers. A periodic table will always be available. The ANSWERS TO PRACTICE PROBLEMS follow immediately after this section.

1. Convert a pressure of 0.910 atm to its equivalent in units of mmHg, torr, and Pa.

2. If a sample of a gas has a volume of 541 mL at a pressure of 376 mmHg, what will the pressure be when the volume is 684 mL, assuming the temperature remains constant?

3. If a sample of a gas has a volume of 0.641 L at 10.0 °C, at what Celsius temperature will the volume of the gas be 0.550 L under constant pressure?

4. A sample of a gas has a volume of 1.55 L at STP. What is the volume of the gas at -50.0 °C and 0.375 atm?

5. What volume of CO_2 gas at 2.75 atm and 75.0 °C will be formed by the decomposition of 64.2 g of $NaHCO_3$?

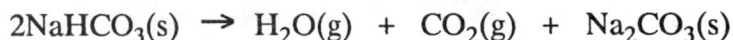

$$2NaHCO_3(s) \rightarrow H_2O(g) + CO_2(g) + Na_2CO_3(s)$$

6.   What is the molecular mass (MM) of a 0.297-g sample of a gas which occupies a volume of 0.176 L at 18.4 °C and 1.48 atm?

7.   (a) What is the partial pressure, in units of atm, of each gas in a mixture of 1.41 mol of oxygen gas and 0.96 mol of chlorine gas in a 2.6-L container at 75 °C? (b) What is the total pressure of the mixture in (a)?

8.   How many times as fast would $CH_4(g)$ diffuse as $NH_3(g)$?

ANSWERS TO PRACTICE PROBLEMS

1.   692 mmHg = 692 torr = 9.22×10^4 Pa

2.   297 mmHg

3.   243.0 K = −30.2 °C

4. $\quad V_2 = 1.55 \text{ L} \times \dfrac{1.000 \text{ atm}}{0.375 \text{ atm}} \times \dfrac{223.2 \text{ K}}{273.2 \text{ K}} = 3.38 \text{ L}$

5. $\quad ? \text{ mol } CO_2 = 64.2 \text{ g NaHCO}_3 \times \dfrac{1 \text{ mol NaHCO}_3}{84.007 \text{ g NaHCO}_3} \times \dfrac{1 \text{ mol } CO_2}{2 \text{ mol NaHCO}_3} = 0.382 \text{ mol } CO_2$

$\quad V = \dfrac{nRT}{P} = \dfrac{(0.382 \text{ mol})(0.082\ 06 \text{ atm}\bullet\text{L}\bullet\text{mol}^{-1}\bullet\text{K}^{-1})(348.2 \text{ K})}{2.75 \text{ atm}} = 3.97 \text{ L } CO_2$

6. $\quad MM = \dfrac{mRT}{PV} = \dfrac{(0.297 \text{ g})(0.082\ 06 \text{ atm}\bullet\text{L}\bullet\text{mol}^{-1}\bullet\text{K}^{-1})(291.6 \text{ K})}{(1.48 \text{ atm})(0.176 \text{ L})} = 27.3 \text{ g}\bullet\text{mol}^{-1}$

7. (a) $\quad p_{oxygen} = \dfrac{nRT}{V} = \dfrac{(1.41 \text{ mol})(0.082\ 06 \text{ atm}\bullet\text{L}\bullet\text{mol}^{-1}\bullet\text{K}^{-1})(348.2 \text{ K})}{2.6 \text{ L}} = 15 \text{ atm}$

$\quad p_{chlorine} = \dfrac{nRT}{V} = \dfrac{(0.96 \text{ mol})(0.082\ 06 \text{ atm}\bullet\text{L}\bullet\text{mol}^{-1}\bullet\text{K}^{-1})(348.2 \text{ K})}{2.6 \text{ L}} = 11 \text{ atm}$

(b) $\quad p_{total} = p_{oxygen} + p_{chlorine} = 15 \text{ atm} + 11 \text{ atm} = 26 \text{ atm}$

8. $\quad \dfrac{\text{Rate of } CH_4}{\text{Rate of } NH_3} = \dfrac{\sqrt{17.030\ 56 \text{ g/mol}}}{\sqrt{16.043 \text{ g/mol}}} = 1.0303 \text{ times as fast}$

CHAPTER 6

CHEMICAL THERMODYNAMICS: THERMOCHEMISTRY

SECTION 6.1   SYSTEM, SURROUNDINGS, AND UNIVERSE

After completing this section, you will be able to do the following:

■ Define thermochemistry, system, surroundings, and universe.

REVIEW

Thermodynamics is concerned with the energy changes that accompany chemical and physical processes. **Thermochemistry** is the part of thermodynamics that deals with the relationship between **chemical reactions and heat.** Two conditions must be satisfied in order for us to actually observe a chemical change (chemical reaction) or physical change. First, the change must be possible from an energy standpoint. And second, the change must take place at an observable or measurable rate. A study of thermodynamics (Chapters 6, 17) satisfies the first condition, and the study of kinetics (Chapter 18) addresses the second condition.

In thermodynamics, you must learn the definitions of frequently used terms. The **universe** consists of the **system** and the **surroundings**,

$$universe = system + surroundings$$

where the **system** is the part of the universe that we are interested in, and the **surroundings** are the rest of the universe. That is, if you are dissolving salt in water, the **system** is the salt and water, and the **surroundings** are the container, atmosphere above the container, the bench top, and everything else that makes up the universe.

SECTION 6.2   WHY CHANGES TAKE PLACE

After completing this section, you will be able to do the following:

■ Explain how energy and disorder determine whether a change is spontaneous or nonspontaneous.

REVIEW

Spontaneous physical and chemical changes take place naturally -- that is, on their own unassisted. When you add solid NaCl to water, the salt slowly dissolves (even when not stirred). This is a **spontaneous** change. **Nonspontaneous** physical and chemical changes do not take place naturally, but rather only occur when assistance is provided (often some form of energy). A salt solution does not separate into solid salt and liquid water on its own. That is a **nonspontaneous** change. However, if the salt solution is heated to boil off the water, the two components can be separated. **Two factors** determine whether a change is **spontaneous** or **nonspontaneous**: energy and **disorder**.

Energy is the ability to do work. The different types of energy are **thermal** (heat), **electrical, radiant** (including light), **chemical, mechanical** (including sound), and **nuclear**. Energy can also be classified as either **kinetic** [energy of motion (Section 5.10)] or **potential** [stored energy]. Energy can be transferred from one object to another, and one form of energy can be changed into another form. A fireworks display is the conversion of stored chemical energy in the chemicals making up the charge into the kinetic energy of the display, heat energy, light energy, and sound energy. At room temperature, most (but not all) **spontaneous changes** give off energy. That is, the system cools down as the surroundings heat up. If the energy given off is in the form of heat (thermal), then the change is called **exothermic**. Many **nonspontaneous changes** will occur if energy is added to the system. That is, the system heats up as the surroundings cool down. If the energy added to the system is in the form of heat, then the change is called **endothermic**. We are familiar with many more spontaneous, exothermic changes (e.g., burning fuels) than nonspontaneous, endothermic changes (e.g., decomposition of numerous compounds). Notable exceptions are the **spontaneous, endothermic changes** of ice to liquid water, and of liquid alcohol to a gas at room temperature. In both cases, the changes occur **spontaneously** but absorb heat from the surroundings (**endothermic**). If you hold an ice cube in your hand, or place a few drops of rubbing alcohol on your arm, you will feel cooler because heat is flowing from your hand/arm to the system. How can this happen?

An **endothermic** change can take place **spontaneously** if the change results in a large enough **increase in disorder** of the system. That is what happens with the melting of the ice cube and evaporation of the alcohol. The term used to measure **disorder** is **entropy**. The higher the disorder, the higher the entropy of the system. If a system consists of a deck of cards, its **entropy** is much lower when the 52 cards are stacked in a pile (less disorder) than when they are dropped on the floor and scatter in all directions (higher disorder).

Even though a change may be **spontaneous**, it may not be observable or measurable because it takes place too slow. For some chemical changes (reactions), a material can be added which makes the reaction occur faster **without being used up** in the reaction. This material is called a **catalyst**.

To predict whether a change will take place, you need to know two things: (1) how the energy and entropy of the system will change, and (2) whether the change will take place at a practical rate.

SAMPLE PROBLEM 6.2 A

▶ ▶ Which of the following changes are spontaneous? (a) toasting bread (b) burning wood in a fireplace (c) cleaning up a messy room (d) going down a water slide.

Solution:

(a) nonspontaneous - you can place the bread in the toaster for a week and it will not toast until heat energy is supplied.

(b) spontaneous - once the wood is lit, the reaction of burning continues by itself.

(c) nonspontaneous - clothes, papers, food, etc., laying all over a room do not get hung up, washed, or picked up by themselves. Energy (human) needs to be expended.

(d) spontaneous - once started, you continue down the slide on your own.

SAMPLE PROBLEM 6.2 B

▶ ▶ Indicate whether the following changes are exothermic or endothermic: (a) burning candle (b) melting wax (c) freezing water

Solution:

(a) exothermic - once the candle is lit, you can feel the heat given off to the surroundings.

(b) endothermic - if you stop applying heat to solid wax, it will no longer melt.

(c) exothermic - ice melts by absorbing heat from the surroundings; the reverse process (water freezing) must give off heat to the surroundings.

SECTION 6.3   TEMPERATURE, THERMAL ENERGY, AND HEAT

After completing this section, you will be able to do the following:

■ Define and know the difference between temperature, thermal energy, and heat.

REVIEW

Temperature is a measure of the hotness or coldness of a substance; it is an **intensive** property, independent of sample size. **Thermal energy** is the energy of motion of the particles (atoms, ions, or molecules) of a substance; it is an **extensive** property, which does depend on sample size. **Heat** is the thermal energy **transfer** that results from a difference in temperature. There is always a net flow of thermal energy from a hot to a cold object.

SECTION 6.5   ENERGY UNITS

After completing this section, you will be able to do the following:

■ Interconvert energy units of joules, kilojoules, calories, and kilocalories.

REVIEW

The SI derived unit of energy is the **joule, J**, which is defined as $1 \text{ kg} \cdot \text{m}^2 \cdot \text{s}^{-2}$. The conversion factor between the more familiar energy unit of the calorie and the joule is

$$1 \text{ calorie (cal)} = 4.184 \text{ J (exactly)}$$

Since both units of energy are frequently used, you must be able to interconvert between them routinely.

SAMPLE PROBLEM 6.5 A

▶▶ Convert (a) 65.0 J to kJ (b) 14.7 kcal to cal (c) 155 J to cal (d) 38.6 kcal to KJ.

Solution:

(a) $? \text{ kJ} = 65.0 \text{ J} \times \dfrac{1 \text{ kJ}}{1000 \text{ J}} = 0.0650 \text{ kJ}$

(b) $? \text{ cal} = 14.7 \text{ kcal} \times \dfrac{1000 \text{ cal}}{1 \text{ kcal}} = 14\ 700 \text{ cal} = 1.47 \times 10^4 \text{ cal}$

(c) $? \text{ cal} = 155 \text{ J} \times \dfrac{1 \text{ cal}}{4.184 \text{ J}} = 37.0 \text{ cal}$

(d) $? \text{ kJ} = 38.6 \text{ kcal} \times \dfrac{1000 \text{ cal}}{1 \text{ kcal}} \times \dfrac{4.184 \text{ J}}{1 \text{ cal}} \times \dfrac{1 \text{ kJ}}{1000 \text{ J}} = 162 \text{ kJ}$

However, you can shorten the process considerably by noticing that you are both multiplying and dividing by 1000. Multiplying the three conversion factors together gives you the very useful conversion factor, 1 kcal = 4.184 kJ, with the identical result,

$? \text{ kJ} = 38.6 \text{ kcal} \times \dfrac{4.184 \text{ kJ}}{1 \text{ kcal}} = 162 \text{ kJ}$

SECTION 6.6   HEAT CAPACITY AND SPECIFIC HEAT

After completing this section, you will be able to do the following:

- Calculate the thermal energy change of a substance from its mass, specific heat, and temperature change.

REVIEW

When an object undergoes a temperature change, the thermal energy change depends on the mass and specific heat of the object and the temperature change it undergoes,

thermal energy change = mass x specific heat x temperature change,

where the **specific heat** of a substance is the amount of thermal energy required to heat one gram of the substance one degree, with units of $J \bullet g^{-1} \bullet °C^{-1}$. A closely related term is the **heat capacity** of a substance, which is the quantity of thermal energy needed to raise the temperature of an object one degree. **Heat capacity** is calculated by multiplying mass times specific heat. The temperature change is always

$$\text{temperature change} = t_{\text{final}} - t_{\text{initial}},$$

which automatically will give you the correct sign for the thermal energy change. **Exothermic** changes will always have a **negative** sign because energy is **lost by the system**. **Endothermic** changes will always have positive signs because energy is **added to the system**.

SAMPLE PROBLEM 6.6 A

▶▶ How many joules must be added to a 25.41-g sample of $Br_2(l)$ to raise the sample temperature from 22.6 °C to 41.7 °C?

Solution:

Using the two equations given above,

temperature change $= t_{final} - t_{initial} = $ 41.7 °C $-$ 22.6 °C $=$ 19.1 °C

thermal energy change = mass x specific heat x temperature change

$=$ 25.41 g x 0.47 $J{\bullet}g^{-1}{\bullet}°C^{-1}$ x 19.1°C $=$ 2.3 x 10^2 J

The positive sign means thermal energy is added to (absorbed by) the system (Br_2), an endothermic change.

SAMPLE PROBLEM 6.6 B

▶▶ What thermal energy change will occur when the temperature of a 56.9-g sample of copper is lowered from 115.0 °C to 35.0 °C? Assume that the specific heat of copper, 0.387 $J{\bullet}g^{-1}{\bullet}°C^{-1}$, is constant over the temperature range.

Solution:

See Sample Problem 6.6 A. Using the same two equations,

temperature change $= t_{final} - t_{initial} = $ 35.0 °C $-$ 115.0 °C $=$ -80.0 °C

thermal energy change = mass x specific heat x temperature change

$=$ 56.9 g x 0.387 $J{\bullet}g^{-1}{\bullet}°C^{-1}$ x (-80.0°C) $=$ -1.76 x 10^3 J

The negative sign means thermal energy is lost (given off) by the system (Cu), an exothermic change.

SECTION 6.7   MEASUREMENT OF THERMAL ENERGY GAINED OR LOST DURING CHANGES

After completing this section, you will be able to do the following:

■ Calculate the heat capacity of a calorimeter and the heat of neutralization of an acid.

REVIEW

Thermal energy changes are measured with a **calorimeter**. In general chemistry laboratories, a calorimeter usually consists of a covered styrofoam cup with thermometer and stirrer inserted. When a warm substance (object or water) is added to a cool substance (typically water) in a calorimeter, and the temperature carefully monitored until there is no longer any change, a final temperature is observed which is between the starting temperatures of the warm and cold substances. Since the Law of Conservation of Energy says that energy can be neither created nor destroyed, the thermal energy given off by the warmer substance must be equal but opposite in sign to the thermal energy gained by the cool substance and the calorimeter itself.

thermal energy change warm substance = − (thermal energy change cool substance + thermal energy gained by calorimeter)

The thermal energy gained by the calorimeter itself (i.e., styrofoam cup, thermometer, and stirrer) is called the **calorimeter constant** or **heat capacity** of the calorimeter. It is usually small, but measurable, in comparison to the thermal energy gained by the cool substance.

Once the heat capacity of the calorimeter has been found, the calorimeter can be used to measure the gain or loss of thermal energy for all kinds of physical and chemical changes, including neutralization reactions. When an acid is added to a base in a calorimeter, the thermal energy given off by the chemical reaction of neutralization is gained by the solution (containing a salt and water when the reaction occurs quantitatively) and the calorimeter.

thermal energy given off = − (thermal energy gained by solution + thermal energy gained by calorimeter)

The thermal energy given off by the neutralization reaction is the **heat of neutralization**. The **heat of neutralization per mole** can then be calculated.

218

SAMPLE PROBLEM 6.7 A

▶▶ After adding 40.0 mL of a 0.75 M H_3PO_4 solution to 90.0 mL of a 1.00 M KOH solution in a calorimeter with a heat capacity of 17.43 J•°C^{-1} (both solutions and calorimeter at 21.5 °C), the two solutions were thoroughly mixed, with the temperature rising to 29.2 °C. What is the heat of neutralization of 1 mol of the base? Assume the densities of the acid and base are 1.00 g/mL, and that the specific heat of the solution after reaction was the same as water, 4.18 J•g^{-1}•°C^{-1}.

Solution:

The change in temperature = t_{final} - $t_{initial}$ = 29.2 °C - 21.5 °C = 7.7 °C.
The thermal energy given off by the reaction of the acid neutralizing the base had to warm up (was gained by) the 130.0 g (90.0 + 40.0) of solution,

thermal energy = − (thermal energy gained + thermal energy gained)
given off   by solution   by calorimeter
by reaction

= - [(130.0 g soln. x 4.18 J•g^{-1}•°C^{-1} x 7.7 °C) + (17.43 J•°C^{-1} x 7.7 °C)]

= - [4200 J + 130 J] = -4300 J

This is the heat of neutralization for 90.0 mL of 1.0 M base. To find the number of moles of base used (or any other substance in solution), always multiply the molarity times the volume in liters,

? mol base = M (mol•L^{-1}) x V (L) = $\frac{1.00 \text{ mol base}}{\text{L soln.}}$ x 0.090 L soln. = 0.090 mol base

The heat of neutralization for one mole of base is

? heat of neutralization = $\frac{-4300 \text{ J}}{0.090 \text{ mol base}}$ = $\frac{-4.8 \times 10^4 \text{ J}}{1 \text{ mol base}}$ = $\frac{-48 \text{ kJ}}{1 \text{ mol base}}$

SECTION 6.8   ENTHALPY

After completing this section, you will be able to do the following:

- Determine whether a reaction is exothermic or endothermic from the change in thermal energy;

- Calculate the amount of thermal energy given off or absorbed during a reaction.

REVIEW

If thermal energy is gained or lost when the system is under **constant pressure** (such as atmospheric pressure in many laboratories, industrial plants, and biological reactions), the thermal energy change is called more specifically the **enthalpy change, ΔH**, where the delta symbol, Δ, means "change in". In Section 6.7, the change in temperature could have been called ΔT. Similarly for ΔH, the change in enthalpy is the difference between the final and initial enthalpies,

$$\Delta H = H_{final} - H_{initial}$$

$\Delta H_{reaction}$, ΔH_{rxn}, and ΔH are used interchangeably to mean the thermal energy change for a reaction taking place under constant pressure. Unless indicated otherwise, all reactions will be taking place under constant pressure. If ΔH is given for a specific type of reaction (neutralization, fusion, vaporization, etc.), the type may be noted by subscripts, such as ΔH_{neut}, ΔH_{fus}, ΔH_{vap}, etc.

Thermal energy given off or gained (absorbed) can be shown in one of two ways in a **thermochemical equation**: by giving the value of ΔH with proper sign, or by showing the thermal energy as a reactant or product. It is **very important** to know what each representation means. Only one representation is used for a given reaction.

Thermal energy change	Sign of ΔH	Location in reaction
Exothermic	−	Product
Endothermic	+	Reactant

For example, 565.98 kJ are given off when carbon monoxide gas reacts with oxygen gas to form carbon dioxide gas. The **thermochemical reaction** (that is, reactants, products, and thermal energy change) can be written as

$$2CO(g) + O_2(g) \rightarrow 2CO_2(g) + 565.98 \text{ kJ}$$

or

$$2CO(g) + O_2(g) \rightarrow 2CO_2(g) \qquad \Delta H = -565.98 \text{ kJ}$$

This reaction is **exothermic**, which means that thermal energy is given off or released, like a product in a reaction. The negative sign of ΔH also indicates that this is an **exothermic reaction**. In either case, the coefficients always represent the specific number of **moles** shown for that reaction. In this reaction, 2 mol of CO gas react with 1 mol of O_2 gas to form the products of 2 mol of CO_2 gas and 565.98 kJ of thermal energy.

SAMPLE PROBLEM 6.8 A

▶▶ For the following reaction,

$$4NH_3(g) + 5O_2(g) \rightarrow 4NO(g) + 6H_2O(l)$$

$\Delta H_{rxn} = -1169.7$ kJ. (a) Is the reaction exothermic or endothermic? Explain! (b) Rewrite the equation showing thermal energy as a reactant or product. (c) How many kJ are gained or given off by the reaction of 43.5 g of $NH_3(g)$ with excess O_2?

Solution:

(a)   **A negative ΔH always means thermal energy is given off by the system, so this is an exothermic reaction.**

(b)   In an <u>exothermic</u> reaction, thermal energy is <u>given off</u> and always shown as a product.

$$4NH_3(g) + 5O_2(g) \rightarrow 4NO(g) + 6H_2O(l) + 1169.7 \text{ kJ}$$

NOTE: When thermal energy is given off by a system (reaction), think of it as a "product" given off to the surroundings, just like NO and H_2O. The mathematical equivalent of the words "given off by the system" is a minus sign in front of the thermal energy, or $\Delta H = -1169.7$ kJ.

(c)   When ΔH is given for a reaction, it is the thermal energy change for the reaction **exactly** as given, with the coefficients representing moles. From the information given in this problem, 1169.7 kJ are given off when **4 mol** of ammonia gas react with **5 mol** of oxygen gas to form **4 mol** of NO gas and **6 mol** of liquid H_2O. This gives us the equality and conversion factor we need to solve the problem, 4 mol NH_3 = (or, react to give a thermal change of) -1169.7 kJ.

$$? \text{ kJ} = 43.5 \text{ g NH}_3 \times \frac{1 \text{ mol NH}_3}{17.030\ 56 \text{ g NH}_3} \times \frac{-1169.7 \text{ kJ}}{4 \text{ mol NH}_3} = -747 \text{ kJ}$$

CHECK: Does this answer make sense? Yes. The 1169.7 kJ are given off when **4 mol** of NH_3 react. The 43.5 g of NH_3 are only ~2.5 mol of NH_3, which will react to give off less than 1169.7 kJ, or 747 kJ.

SAMPLE PROBLEM 6.8 B

▶▶ Methyl alcohol decomposes according to the reaction

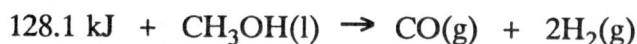

$$128.1 \text{ kJ} + CH_3OH(l) \rightarrow CO(g) + 2H_2(g)$$

(a) Is this reaction exothermic or endothermic? Explain! (b) What is ΔH_{rxn}? (c) How many kJ are gained or given off by the decomposition of 80.0 g of $CH_3OH(l)$?

Solution:

(a) When thermal energy is shown in the thermochemical equation as a **reactant**, the thermal energy is **gained** by the system and the reaction is **always endothermic**.

(b) $\Delta H_{rxn} = 128.1 \text{ kJ}$. The value of ΔH_{rxn} is always <u>positive</u> for <u>endothermic</u> reactions (like all other mathematical values, no sign in front of a number always means it is positive).

(c) See Sample Problem 6.8 A.

$$? \text{ kJ} = 80.0 \text{ g } CH_3OH \times \frac{1 \text{ mol } CH_3OH}{32.042 \text{ g } CH_3OH} \times \frac{128.1 \text{ kJ}}{1 \text{ mol } CH_3OH} = 3.20 \times 10^2 \text{ kJ}$$

CHECK: Does this answer make sense? Definitely! The 128.1 kJ are gained by the system when <u>1 mol</u> of CH_3OH decomposes. The 80.0 g of CH_3OH is ~2.5 mol, which will gain ~2.5 times as much thermal energy, or 320 kJ.

SECTION 6.9   HESS'S LAW

After completing this section, you will be able to do the following:

■ Calculate the enthalpy of reaction, ΔH_{rxn}, from other chemical equations with known ΔH_{rxn}'s;

■ Calculate the enthalpy of reaction, ΔH_{rxn}, from enthalpies of formation, ΔH_f's;

■ Calculate the enthalpy of formation, ΔH_f, from the enthalpy of reaction, ΔH_{rxn}.

REVIEW

The many enthalpy changes that cannot be measured by calorimetry can often be found by using **Hess's law**, which states that the thermal energy given off or absorbed (gained) in a given chemical change (reaction) is the **same** whether the chemical change (reaction) takes place in a single step or in several steps. This is possible because **enthalpy** (and pressure, volume, temperature, etc.) is a property known as a **state function**. That is, it depends only on the **initial** and **final** states of the system, **not on the pathway** between the initial and final states. Thus, if reactants A and B form products K and L, the enthalpy change, ΔH_{rxn}, is **identical** no matter if A and B form K and L in one step, or if A and B react to form C and D, which in turn react to produce E and F, which finally yield K and L. **The multiple steps and their ΔH's are added together to give the overall reaction wanted and its ΔH_{rxn},**

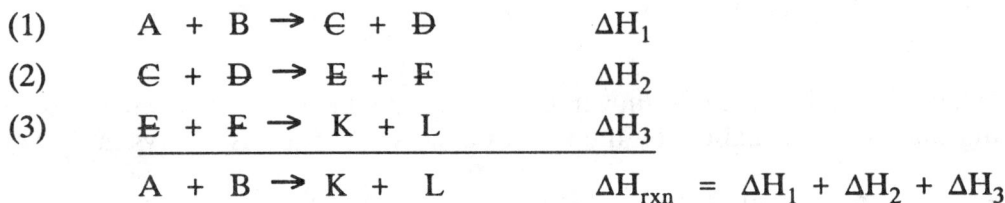

$$
\begin{array}{lll}
(1) & \text{A} + \text{B} \rightarrow \cancel{\text{C}} + \cancel{\text{D}} & \Delta H_1 \\
(2) & \cancel{\text{C}} + \cancel{\text{D}} \rightarrow \cancel{\text{E}} + \cancel{\text{F}} & \Delta H_2 \\
(3) & \cancel{\text{E}} + \cancel{\text{F}} \rightarrow \text{K} + \text{L} & \Delta H_3 \\
\hline
& \text{A} + \text{B} \rightarrow \text{K} + \text{L} & \Delta H_{rxn} = \Delta H_1 + \Delta H_2 + \Delta H_3
\end{array}
$$

The multiple steps must be added correctly so that everything cancels out except the overall reaction wanted (see Sample Problem 6.9 A).

Hess's law becomes particularly useful if all the multiple steps to be added together are **formation reactions**, which are defined as the formation of 1 mol of a compound from **the elements**. The enthalpy change for a formation reaction is called the enthalpy of formation, ΔH_f. If the reactants and products in the formation reactions are in their **standard state**, then the enthalpy change is called the **standard enthalpy of formation, $\Delta H_f°$**. The **standard state** of **solids and liquids** is the pure solid or pure liquid at 1-atm pressure. For gases, the **standard state** is the ideal gas at a pressure of 10^5 Pa. There is no standard temperature for thermodynamics, but many thermodynamic properties, including enthalpies of formation, are usually tabulated at 25 °C. Table 6.2 and Appendix E of the textbook give standard enthalpies of formation, $\Delta H_f°$, for a number of compounds and **non-standard states** of elements. **The standard enthalpies of formation, $\Delta H_f°$, of the elements in their standard states are zero.** With a table of standard enthalpies of formation, you don't need to write out all the formation reactions and add them together to obtain the overall reaction and its enthalpy change, $\Delta H_{rxn}°$, because the procedure is always the same.

$$
\begin{aligned}
\Delta H_{rxn}° &= \text{sum of standard enthalpies} \quad - \quad \text{sum of standard enthalpies} \\
&\quad\ \ \text{of formation of } \textbf{products} \qquad\qquad \text{of formation of } \textbf{reactants} \\
\\
&= \sum \Delta H_f°{}_{\text{products}} \quad - \quad \sum \Delta H_f°{}_{\text{reactants}},
\end{aligned}
$$

where each standard enthalpy of formation, $\Delta H_f°$, must be multiplied by the coefficient of the substance in the chemical equation.

The Greek letter sigma, Σ, means sum. Remember, the Greek letter delta, Δ, in front of any property always means a specific kind of change, **final** (property) − **initial** (property). For example, $\Delta t = t_{final} - t_{initial}$ and $\Delta H_{rxn}^{\circ} = \Delta H_{products \, (final)} - \Delta H_{reactants \, (initial)}$.

This same equation can be used to obtain the standard enthalpy of formation of a compound, ΔH_f°, if ΔH_{rxn}° for a reaction involving the compound can be measured **and** the standard enthalpies of formation of all the other reactants and products are known.

SAMPLE PROBLEM 6.9 A

▸▸ Calculate ΔH_{rxn} for the reaction

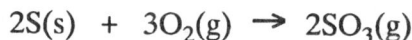

$$2S(s) + 3O_2(g) \rightarrow 2SO_3(g)$$

from the following information,

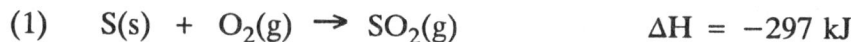

(1) $S(s) + O_2(g) \rightarrow SO_2(g)$ $\Delta H = -297$ kJ

(2) $2SO_3(g) \rightarrow 2SO_2(g) + O_2(g)$ $\Delta H = 198$ kJ

Solution:

HINT:
- Start with the equation whose ΔH you are trying to find and work backwards.
- The equations whose ΔH's you are given must be written (they may need to be reversed, or coefficients multiplied by a whole number or fraction) so that when they are added together, everything cancels **except** the exact equation (including coefficients) that you are trying to find.
- Whatever you do to an equation (reverse it, multiply by a fraction or whole number) affects ΔH. If an equation is reversed, **only the sign** of ΔH is changed. When an equation is multiplied by a fraction or whole number, so is the value of ΔH.

The given thermochemical equations are numbered (1) and (2) for convenience and reference. The equation for which you are trying to find ΔH_{rxn} has 2 mol S(s) on the reactant side. The only given thermochemical equation with S(s) in it is (1), with 1 mol S(s) on the reactant side. Start by multiplying equation (1) by 2 (including ΔH). Then note that the equation you want has 2 mol $SO_3(g)$ as a product. The only given equation with $SO_3(g)$ in it is (2), but $SO_3(g)$ is on the wrong side. To have $SO_3(g)$ on the product (right) side, you will need to **reverse the equation**. Reversing the equation means the sign of ΔH changes. Notice that we have said nothing about $O_2(g)$ in the equation you want, because it appears in both equations (1) and (2). If we have done everything else right, the oxygens will turn out correctly. Let's summarize our data.

224

Action		ΔH, kJ

2 x (1)　　　　$2S(s) + 2O_2(g) \rightarrow \cancel{2SO_2(g)}$　　　　　　2 x (−297) = −594

Reverse (2)　$\cancel{2SO_2(g)} + O_2(g) \rightarrow 2SO_3(g)$　　　　　−(198) = −198

$2S(s) + 3O_2(g) \rightarrow 2SO_3(g)$　　　　　$\Delta H_{rxn} = -792$

Everything cancels, so that we are left with only the **exact** equation (including coefficients) that we want. The ΔH values are added together just like the equations. **Remember, reversing** an equation means that the **sign of ΔH** must change. If a reaction is exothermic in one direction, that same reaction must be endothermic in the reverse direction.

SAMPLE PROBLEM 6.9 B

▶▶ Calculate ΔH_{rxn} for the reaction

$$C_2H_6(g) \rightarrow C_2H_4(g) + H_2(g)$$

from the following information,

(1)　$2C_2H_6(g) + 7O_2(g) \rightarrow 4CO_2(g) + 6H_2O(l)$　　$\Delta H = -3120$ kJ
(2)　$C_2H_4(g) + 3O_2(g) \rightarrow 2CO_2(g) + 2H_2O(l)$　　$\Delta H = -1411$ kJ
(3)　$2H_2(g) + O_2(g) \rightarrow 2H_2O(l)$　　$\Delta H = -572$ kJ

Solution:

See Sample Problem 6.9 A for HINTS and procedure. The equation we want has 1 mol $C_2H_6(g)$ on the left side; equation (1) has 2 mol $C_2H_6(g)$ on the left side, so we multiply equation (1) by 1/2 (including ΔH). The equation we want has 1 mol $C_2H_4(g)$ on the right side; equation (2) has 1 mol $C_2H_4(g)$ on the left side, so we reverse equation (2) and change the sign of ΔH. The equation we want has 1 mol $H_2(g)$ on the right side; equation (3) has 2 mol $H_2(g)$ on the left side, so we reverse equation (3) **and** multiply it by 1/2 (which means change the sign of ΔH and multiply by 1/2). We now are able to add the equations together to find ΔH_{rxn}.

Action		ΔH, kJ
1/2 x (1)	$C_2H_6(g)$ + ~~7/2O$_2$(g)~~ → ~~2CO$_2$(g)~~ + ~~3H$_2$O(l)~~	1/2 x (−3120) = −1560
Rev. (2)	~~2CO$_2$(g)~~ + ~~2H$_2$O(l)~~ → $C_2H_4(g)$ + ~~3O$_2$(g)~~	−(−1411) = 1411
1/2 x Rev. (3)	~~H$_2$O(l)~~ → $H_2(g)$ + ~~1/2 O$_2$(g)~~	−1/2 x (−572) = 286

$$C_2H_6(g) \rightarrow C_2H_4(g) + H_2(g) \qquad \Delta H_{rxn} = 137$$

Everything cancels so that we are left with only the **exact** equation (including coefficients) that we want, and the value of ΔH_{rxn}.

SAMPLE PROBLEM 6.9 C

►► Use standard enthalpies of formation, ΔH_f°, from Appendix E to calculate ΔH_{rxn}° for the reaction

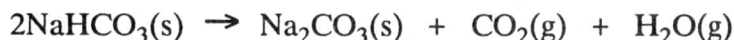

$$2NaHCO_3(s) \rightarrow Na_2CO_3(s) + CO_2(g) + H_2O(g)$$

Solution:

Look up the standard enthalpies of formation in Appendix E.

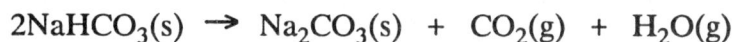

$$2NaHCO_3(s) \rightarrow Na_2CO_3(s) + CO_2(g) + H_2O(g)$$

ΔH_f°, kJ/mol =	−947.7	−1130.9	−393.51	−238.92

Use the equation given in the REVIEW and substitute the values of ΔH_f°. Remember to multiply the ΔH_f° (kJ/mol) for a substance by its coefficient (mol) in the chemical equation. Be very careful to avoid making errors with signs.

$$\Delta H_{rxn}{}^{\circ} = \text{sum of standard enthalpies} \qquad - \qquad \text{sum of standard enthalpies}$$
$$\text{of formation of \textbf{products}} \qquad \text{of formation of \textbf{reactants}}$$

$$= [\Sigma \Delta H_f{}^{\circ}{}_{products}] \qquad\qquad - \qquad [\Sigma \Delta H_f{}^{\circ}{}_{reactants}]$$

$$= [1 \text{ mol } Na_2CO_3 \text{ x } \Delta H_f{}^{\circ}(Na_2CO_3) \ + \ 1 \text{ mol } CO_2 \text{ x } \Delta H_f{}^{\circ}(CO_2) \ + \ 1 \text{ mol } H_2O \text{ x } \Delta H_f{}^{\circ}(H_2O)]$$

$$- [2 \text{ mol } NaHCO_3 \text{ x } \Delta H_f{}^{\circ}(NaHCO_3)]$$

$$= [1 \text{ mol } Na_2CO_3 \text{ x } (-1130.9 \text{ kJ/mol } Na_2CO_3) \ + \ 1 \text{ mol } CO_2 \text{ x } (-393.51 \text{ kJ/mol } CO_2)$$

$$+ \ 1 \text{ mol } H_2O \text{ x } (-238.92 \text{ kJ/mol } H_2O)] - [2 \text{ mol } NaHCO_3 \text{ x } (-947.7 \text{ kJ/mol } NaHCO_3)]$$

$$= [(-1130.9 \text{ kJ}) + (-393.51 \text{ kJ}) + (-238.92 \text{ kJ})] - [(-1895.4 \text{ kJ})]$$

$$= [-1763.33 \text{ kJ}] - [-1895.4 \text{ kJ}] \ = \ 132.1 \text{ kJ}$$

CHECK: Make sure you double check your math, particularly the signs. See Sample Problem 6.9 D for a simplified process.

SAMPLE PROBLEM 6.9 D

▶▶ Use standard enthalpies of formation, $\Delta H_f{}^{\circ}$, from Appendix E to calculate $\Delta H_{rxn}{}^{\circ}$ for the reaction,

$$2CH_3CH_3(g) \ + \ 7O_2(g) \ \rightarrow \ 4CO_2(g) \ + \ 6H_2O(g)$$

Solution:

Look up the standard enthalpies of formation in Appendix E.

$$2CH_3CH_3(g) \ + \ 7O_2(g) \ \rightarrow \ 4CO_2(g) \ + \ 6H_2O(g)$$

| $\Delta H_f{}^{\circ}$, kJ/mol = | -84.67 | 0 | -393.51 | -238.92 |

See Sample Problem 6.9 C for the detailed procedure. If you understand the procedure, then the following simplified format will allow you to arrive at the correct answer more quickly and with less errors.

$$\Delta H_{rxn}^{\circ} = \text{sum of standard enthalpies} \quad - \quad \text{sum of standard enthalpies}$$
$$\text{of formation of } \textbf{products} \qquad \text{of formation of } \textbf{reactants}$$

$$= [\Sigma \Delta H_f^{\circ}{}_{products}] \qquad - \qquad [\Sigma \Delta H_f^{\circ}{}_{reactants}]$$

$$= [4 \text{ mol x } \Delta H_f^{\circ}(CO_2) + 6 \text{ mol x } \Delta H_f^{\circ}(H_2O(g))]$$
$$- [2 \text{ mol x } \Delta H_f^{\circ}(CH_3CH_3) + 7 \text{ mol x } \Delta H_f^{\circ}(O_2)]$$

$$= [4 \text{ mol x } (-393.51 \text{ kJ/mol}) + 6 \text{ mol x } (-238.92 \text{ kJ/mol})]$$
$$- [2 \text{ mol x } (-84.67 \text{ kJ/mol}) + 7 \text{ mol x } (0 \text{ kJ/mol})]$$

$$= [(-1574.04 \text{ kJ}) + (-1433.52 \text{ kJ})] - [(-169.34 \text{ kJ}) + (0 \text{ kJ})]$$

$$= [-3007.56 \text{ kJ}] - [-169.34 \text{ kJ}] = \boxed{-2838.22 \text{ kJ}}$$

CHECK: Again, check your math carefully. Remember, the ΔH_{rxn}° is equal to the sum of the heats of formation (ΔH_f°) of the products minus the reactants, with each ΔH_f° multiplied by the coefficient of that substance in the balanced chemical reaction.

SAMPLE PROBLEM 6.9 E

▶▶ The standard enthalpy of combustion of methanol, $CH_3OH(l)$, is -726.53 kJ/mol. What is ΔH_f° for methanol?

Solution:

The standard enthalpy of combustion, ΔH_{comb}°, is **always** the thermal energy given off when **1 mol** of a substance burns in an excess of oxygen to form $CO_2(g)$ and $H_2O(l)$, with all reactants and products in their standard states. The thermochemical equation for the combustion of methanol under standard conditions at $25^{\circ}C$ is

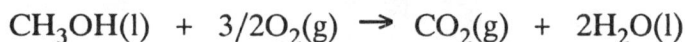

$$CH_3OH(l) + 3/2O_2(g) \rightarrow CO_2(g) + 2H_2O(l)$$

with $\Delta H_{rxn}^{\circ} = \Delta H_{comb}^{\circ} = -726.53$ kJ. You should be able to write out the equation that corresponds to the ΔH_{comb}° for any substance.

This is an application where you know the ΔH_{rxn}° (which more specifically is ΔH_{comb}°) and all the standard heats of formation, ΔH_f°, except methanol.

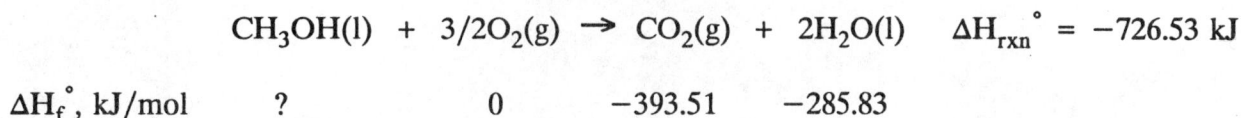

$$CH_3OH(l) + 3/2O_2(g) \rightarrow CO_2(g) + 2H_2O(l) \qquad \Delta H_{rxn}^{\circ} = -726.53 \text{ kJ}$$

| ΔH_f°, kJ/mol | ? | 0 | -393.51 | -285.83 |

228

See Sample Problems 6.9 C & D for the formulas and procedure.

ΔH_{rxn}° = sum of standard enthalpies of formation of **products** − sum of standard enthalpies of formation of **reactants**

$$= [\Sigma \Delta H_f^{\circ} \text{ products}] - [\Sigma \Delta H_f^{\circ} \text{ reactants}]$$

$$= [1 \text{ mol} \times \Delta H_f^{\circ}(CO_2) + 2 \text{ mol} \times \Delta H_f^{\circ}(H_2O(l))]$$
$$- [1 \text{ mol} \times \Delta H_f^{\circ}(CH_3OH) + 3/2 \text{ mol} \times \Delta H_f^{\circ}(O_2)]$$

At this point, you can solve the equation for ΔH_f° of CH_3OH, substitute in the numerical values, and then calculate the answer (as was done in Sample Problem 6.8 in the textbook). However, you may find rearranging this bulky equation somewhat awkward. As an alternative solution, you may want to try substituting in the numerical values at this point, and then simplify the equation before finally solving it for ΔH_f° of methanol. Using the alternative method,

$$-726.53 \text{ kJ} = [1 \text{ mol} \times (-393.51 \text{ kJ/mol}) + 2 \text{ mol} \times (-285.83 \text{ kJ/mol})]$$
$$- [1 \text{ mol} \times \Delta H_f^{\circ}(CH_3OH) + 3/2 \text{ mol} \times (0 \text{ kJ/mol})]$$

$$= [(-393.51 \text{ kJ}) + (-571.66 \text{ kJ})] - [1 \text{ mol} \times \Delta H_f^{\circ}(CH_3OH) + (0 \text{ kJ})]$$

$$= [-965.17 \text{ kJ}] - [1 \text{ mol} \times \Delta H_f^{\circ}(CH_3OH)]$$

Now solve this simplified equation for $\Delta H_f^{\circ}(CH_3OH)$.

$$?\Delta H_f^{\circ}(CH_3OH) = \frac{-965.17 \text{ kJ} + 726.53 \text{ kJ}}{1 \text{ mol } CH_3OH(l)} = -238.64 \text{ kJ/mol } CH_3OH$$

CHECK: Make sure that you check your math carefully.

SELF-TEST QUESTIONS

Test your understanding of the concepts and skills in this chapter by working through the multiple choice questions **BEFORE** checking the answers. Use Appendix E for any heat of formation, ΔH_f°, data needed. A periodic table will always be available. The ANSWERS TO SELF-TEST QUESTIONS follow immediately after this section.

1. All of the following changes are spontaneous <u>except</u>

    A. a balloon bursts when poked with a pin.
    B. the odor of freshly popped popcorn spreads through a room.
    C. a stationary bike rolls up a hill.
    D. puddles of water evaporate.
    E. solid AgCl precipitates from a solution of $AgNO_3$ and $NaCl$.

2. The correct conversion for 76.9 kJ is

    A. 76.9 kcal   B. 0.0769 J   C. 18.4 cal   D. 7.69×10^3 J   E. 1.84×10^4 cal

3. The thermal energy change that occurs when the temperature of 38.1 g of $C_{graphite}$ is raised from 23.5 °C to 250.0 °C is

    A. 6.1×10^3 J   B. 4.4×10^3 J   C. 6.8×10^3 J   D. 6.4×10^2 J   E. 1.4×10^4 J

4. In the following chemical equation, 164 kJ of thermal energy is absorbed (gained). Which of the following statements is true?

$$CH_4(g) \ + \ N_2(g) \ \rightarrow \ HCN(g) \ + \ NH_3(g)$$

    A. The reaction is exothermic.
    B. The thermal energy would be written as a reactant.
    C. $\Delta H_{rxn} = -164$ kJ.
    D. 164 kJ will also be absorbed (gained) by the reverse reaction.
    E. More than one of the above statements is correct.

5. An example of an endothermic reaction is

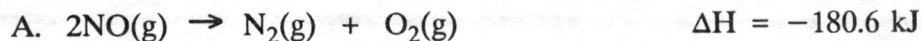

    A. $2NO(g) \rightarrow N_2(g) + O_2(g)$ $\Delta H = -180.6$ kJ

    B. $CaC_2(s) + 2H_2O(l) \rightarrow C_2H_2(g) + Ca(OH)_2(s) + 128$ kJ

    C. $C_2H_4(g) + 3O_2(g) \rightarrow 2CO_2(g) + 2H_2O(l) + 1401$ kJ

    D. $N_2(g) + 3H_2(g) \rightarrow 2NH_3(g) + 91.8$ kJ

    E. not given.

6. The standard enthalpy change, ΔH_{rxn}, for the following reaction is

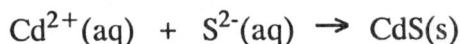

$$Cd^{2+}(aq) + S^{2-}(aq) \rightarrow CdS(s)$$

A. -270.9 kJ   B. -53.0 kJ   C. -204.8 kJ   D. -119.1 kJ   E. -184.2 kJ

7. The standard enthalpy change, ΔH_{rxn}, for the following reaction is

$$2Fe(s) + 3CO_2(g) \rightarrow Fe_2O_3(s) + 3CO(g)$$

A. 24.8 kJ   B. -466.9 kJ   C. -762.3 kJ   D. -541.2 kJ   E. -2336.3 kJ

8. The standard enthalpy change, ΔH_{rxn}, for the following reaction is

$$3NO_2(g) + H_2O(l) \rightarrow 2HNO_3(aq) + NO(g)$$

A. 135.5 kJ   B. -318.8 kJ   C. -138.3 kJ   D. -71.9 kJ   E. -221.5 kJ

9. The standard enthalpy change, ΔH_{rxn}, for the following reaction is

$$CaO(s) + 3C(s) \rightarrow CaC_2(s) + CO(g)$$

A. -805.4 kJ   B. 464.8 kJ   C. 805.4 kJ   D. -464.8 kJ   E. 362.8 kJ

10. The standard enthalpy change, ΔH_{rxn}, for the following reaction is

$$2Na(s) + 2H_2O(l) \rightarrow 2NaOH(s) + H_2(g)$$

A. -377.70 kJ   B. -141.94 kJ   C. -1427.20 kJ   D. -283.88 kJ   E. -494.16 kJ

ANSWERS TO SELF-TEST QUESTIONS

1. c	4. b	7. a
2. e	5. e	8. c
3. a	6. d	9. b
		10. d

PRACTICE PROBLEMS

Test your problem-solving skills in this chapter by working through the problems in the space provided **BEFORE** checking the answers. Use Appendix E for any heat of formation, ΔH_f°, data needed. A periodic table will always be available. The ANSWERS TO PRACTICE PROBLEMS follow immediately after this section.

1.   Convert 19.8 kcal to cal, J, and kJ.

2.   (a) What thermal energy change occurs, in J, when 24.7 g of $H_2O(l)$ lowers its temperature from 67.5 °C to 40.0 °C? (b) Is this an exothermic or endothermic change?

3. For the following reaction, $\Delta H = -233$ kJ. (a) Is the reaction endothermic or exothermic? (b) Rewrite the equation showing thermal energy as a reactant or product. (c) How many kJ are gained or given off by the reaction of 25.2 g of H_2S with excess SO_2?

$$2H_2S(g) + SO_2(g) \rightarrow 3S(s) + 2H_2O(l)$$

4. Calculate ΔH_{rxn} for the reaction

$$2Pb(s) + 2C(s) + 3O_2(g) \rightarrow 2PbCO_3(s)$$

from the following information,

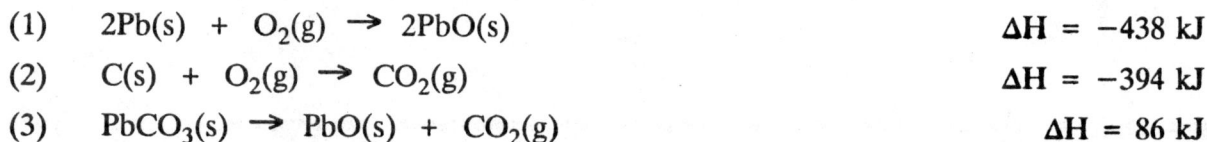

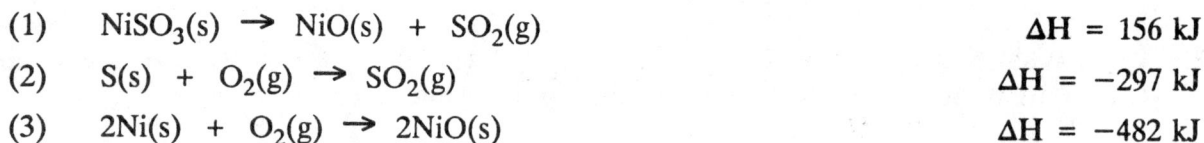

(1)	$2Pb(s) + O_2(g) \rightarrow 2PbO(s)$	$\Delta H = -438$ kJ
(2)	$C(s) + O_2(g) \rightarrow CO_2(g)$	$\Delta H = -394$ kJ
(3)	$PbCO_3(s) \rightarrow PbO(s) + CO_2(g)$	$\Delta H = 86$ kJ

5. Calculate ΔH_{rxn} for the reaction

$$2Ni(s) + 2S(s) + 3O_2(g) \rightarrow 2NiSO_3(s)$$

from the following information,

(1)	$NiSO_3(s) \rightarrow NiO(s) + SO_2(g)$	$\Delta H = 156$ kJ
(2)	$S(s) + O_2(g) \rightarrow SO_2(g)$	$\Delta H = -297$ kJ
(3)	$2Ni(s) + O_2(g) \rightarrow 2NiO(s)$	$\Delta H = -482$ kJ

6. Calculate the standard enthalpy change, $\Delta H_{rxn}°$, for the following reaction,

$$6NH_3(g) \;+\; 8NO_2(g) \;\rightarrow\; 9H_2O(g) \;+\; 7N_2O(g)$$

7. The standard enthalpy of combustion of ethane, $H_2C{=}CH_2(g)$, is -1410.96 kJ. Calculate the standard heat of formation, $\Delta H_f°$, for ethane.

ANSWERS TO PRACTICE PROBLEMS

1. 1.98×10^4 cal $= 8.28 \times 10^4$ J $= 82.8$ kJ

2. (a) -2.84×10^3 J

    (b) negative sign means exothermic; energy has been lost by the system (water) to the surroundings.

3. (a) exothermic; ΔH is negative.

    (b) $2H_2S(g) \;+\; SO_2(g) \;\rightarrow\; 3S(s) \;+\; 2H_2O(l) \;+\; 233$ kJ

    (c) -86.1 kJ given off.

4.   Action   ΔH, kJ

(1) $2Pb(s) + O_2(g) \rightarrow 2PbO(s)$ −438

2 x (2) $2C(s) + 2O_2(g) \rightarrow 2CO_2(g)$ $2 \times (-394) = -788$

2 x Rev.(3) $2PbO(s) + 2CO_2(g) \rightarrow 2PbCO_3(s)$ $-2 \times (86) = -172$
_____   _____

$2Pb(s) + 2C(s) + 3O_2(g) \rightarrow 2PbCO_3(s)$ $\Delta H_{rxn} = -1398$

5.   Action   ΔH, kJ

2 x Rev.(1) $2NiO(s) + 2SO_2(g) \rightarrow 2NiSO_3(s)$ $-2 \times (156) = -312$

2 x (2) $2S(s) + 2O_2(g) \rightarrow 2SO_2(g)$ $2 \times (-297) = -594$

(3) $2Ni(s) + O_2(g) \rightarrow 2NiO(s)$ −482
_____   _____

$2Ni(s) + 2S(s) + 3O_2(g) \rightarrow 2NiSO_3(s)$ $\Delta H_{rxn} = -1388$

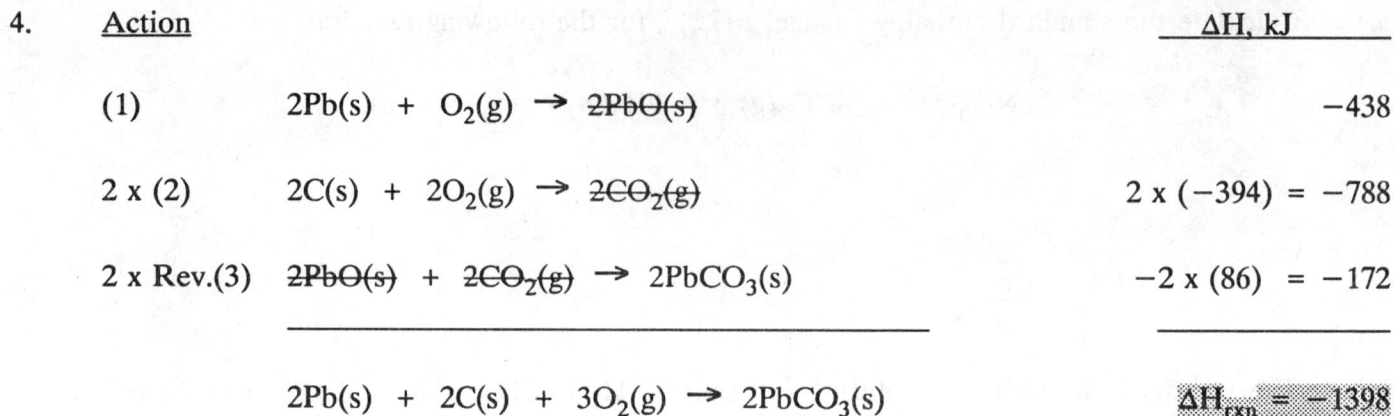

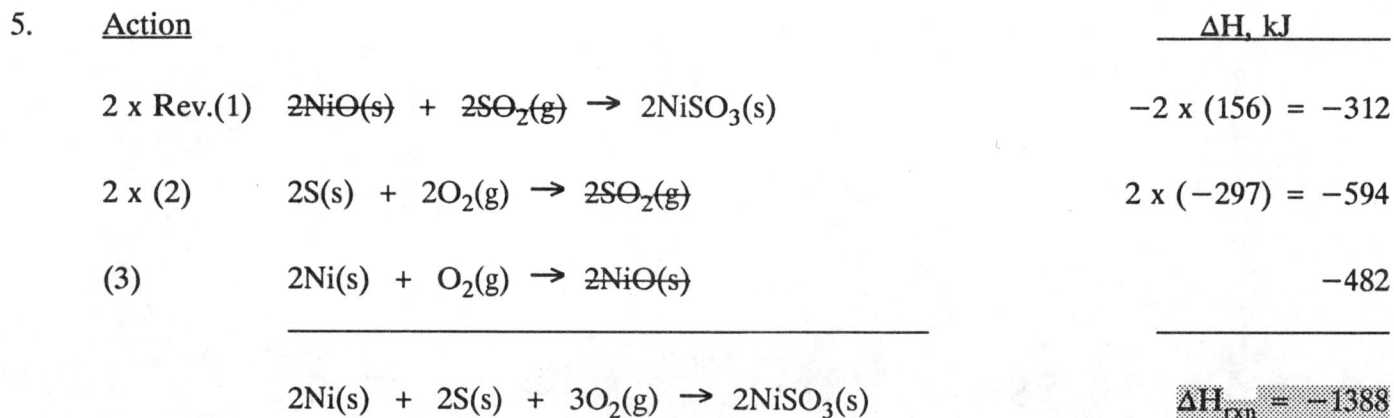

6. ΔH_{rxn}° $=$ $\sum \Delta H_{f\ products}^{\circ}]$ $-$ $[\sum \Delta H_{f\ reactants}^{\circ}]$

$=$ $[9\ \text{mol} \times \Delta H_f^{\circ}(H_2O(g)) + 7\ \text{mol} \times \Delta H_f^{\circ}(N_2O)]$
          $- [6\ \text{mol} \times \Delta H_f^{\circ}(NH_3) + 8\ \text{mol} \times \Delta H_f^{\circ}(NO_2)]$

$=$ $[9\ \text{mol} \times (-238.92\ \text{kJ/mol}) + 7\ \text{mol} \times (82.05\ \text{kJ/mol})]$
          $- [6\ \text{mol} \times (-46.11\ \text{kJ/mol}) + 8\ \text{mol} \times (33.2\ \text{kJ/mol})]$

$=$ $[(-2150.28\ \text{kJ}) + (574.35\ \text{kJ})] - [(-276.66\ \text{kJ}) + (265.6\ \text{kJ})]$

$=$ $[-1575.93\ \text{kJ}] - [-11.06\ \text{kJ}] =$ $-1564.9\ \text{kJ}$

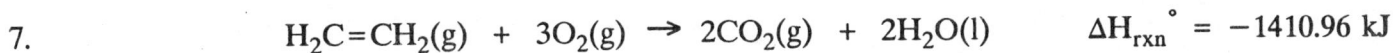

7. $\qquad$ $H_2C=CH_2(g) + 3O_2(g) \rightarrow 2CO_2(g) + 2H_2O(l)$ $\qquad$ $\Delta H_{rxn}^\circ = -1410.96$ kJ

H_f°, kJ/mol $\qquad$? $\qquad$ 0 $\qquad$ -393.51 $\qquad$ -285.83

ΔH_{rxn}° $\quad = \quad$ $\sum \Delta H_{f\ products}^\circ]$ $\qquad - \qquad [\sum \Delta H_{f\ reactants}^\circ]$

$\qquad = \quad$ [2 mol x $\Delta H_f^\circ(CO_2)$ + 2 mol x $\Delta H_f^\circ(H_2O(l))$]
$\qquad\qquad\qquad$ $-$ [1 mol x $\Delta H_f^\circ(H_2C=CH_2)$ + 3 mol x $\Delta H_f^\circ(O_2)$]

-1410.96 kJ $\ = \ $ [2 mol x $(-393.51$ kJ/mol$)$ + 2 mol x $(-285.83$ kJ/mol$)$]
$\qquad\qquad\qquad\quad$ $-$ [1 mol x $\Delta H_f^\circ(H_2C=CH_2)$ + 3 mol x $(0$ kJ/mol$)$]

$\qquad\qquad = \quad$ [-1358.68 kJ] $-$ [1 mol x $\Delta H_f^\circ(H_2C=CH_2)$]

$?\Delta H_f^\circ(H_2C=CH_2) \ = \ \dfrac{-1358.68 \text{ kJ} + 1410.96 \text{ kJ}}{1 \text{ mol } H_2C=CH_2} \ = \ $ 52.28 kJ/mol $H_2C=CH_2$

CHAPTER 7

ATOMIC STRUCTURE

REVIEW

Between 1897 and 1932, physicists carried out a variety of experiments involving physical changes to discover the existence of protons, neutrons, and electrons. The details of these experiments are described in Sections 7.1-7.4 of the textbook. From those discoveries and others, we know that an atom is made up of a small, dense nucleus surrounded by rapidly moving electrons. The nucleus contains the protons and neutrons. A summary of the major characteristics of these atomic particles is as follows:

PARTICLE	RELATIVE CHARGE	RELATIVE MASS
Proton	$+1$	1 u
Neutron	0	1 u
Electron	-1	~ 0 u

These three basic atomic particles are of greatest interest to chemists because the **chemical and physical properties** of the elements and their compounds can be explained by protons, neutrons, and electrons. More specifically, the **chemical properties** of the elements and their compounds depend on the way that the **electrons are arranged around the nucleus** -- that is, on the **electronic structure** of the atom. Much of the information about the electronic structure of atoms comes from observations of the interaction between visible light and atoms which means that you need to know something about waves (Section 7.5 of the textbook) and electromagnetic radiation.

| SECTION 7.6 | ELECTROMAGNETIC RADIATION |

After completing this section, you will be able to do the following:

■ Calculate the frequency of electromagnetic radiation from wavelength and vice versa;

■ Given two of the three quantities (photon energy, frequency, and wavelength), calculate the third quantity.

REVIEW

Electromagnetic radiation is a form of energy that consists of perpendicular electric and magnetic fields that change, at the same time and in phase, with time. All kinds of electromagnetic radiation travel through **empty space (a vacuum)** at the same speed, 2.998×10^8 **m/s**. The SI unit of **frequency** is called the **hertz, Hz,** which equals one cycle (or wave) per second (see Section 7.5 of the textbook for a description of waves, cycles, and frequency)

$$1 \text{ Hz} = \frac{1 \text{ cycle}}{s} = \frac{1}{s} = 1 \text{ s}^{-1}$$

Note that for frequency, it is understood that the unit "cycle" or "wave" is meant but not shown in the numerator. The product of the frequency (ν) and wavelength (λ) of electromagnetic radiation is equal to the speed of light (c),

$$\lambda \nu = c$$

If the speed of light is in m/s, then wavelength must be in meters and frequency in hertz. The higher the frequency of electromagnetic radiation, the shorter the wavelength.

Electromagnetic radiation includes **visible light** as well as other familiar kinds of radiation, from very low frequency **radio waves** to the high-frequency **x-rays** and **gamma rays** (see Figure 7.15 of the textbook for the complete electromagnetic spectrum). The visible part of the electromagnetic spectrum ranges from the low- frequency red color to the high-frequency violet color. In a vacuum (empty space), all wavelengths of electromagnetic radiation are **transmitted** or **passed through**, whether we can "see it" (visible part of spectrum) or not. If matter is in the path of the electromagnetic radiation, different wavelengths are **absorbed** or **taken up**, depending on the kind of matter present. For example, water solutions of copper compounds appear light blue to us because the copper compounds **absorb** wavelengths in the red, orange, and yellow regions of the visible spectrum and **transmit** wavelengths in the green and blue regions. The color that you see is the color of the transmitted wavelengths.

The particles of electromagnetic radiation are called **photons**. The energy of a **photon** is proportional to the **frequency** of the electromagnetic radiation and (using the equation above) inversely proportional to the **wavelength** of the electromagnetic radiation,

$$\text{energy of a photon} = h\nu = \frac{hc}{\lambda},$$

where **h** is called **Planck's constant** (6.626×10^{-34} J•s), **c** is the speed of light in a vacuum (2.998×10^8 m•s^{-1}), ν is the frequency of radiation in s^{-1}, and λ is the wavelength of the radiation in meters. Overall, this equation says that the shorter the wavelength, the higher the frequency and the greater the energy of a photon of electromagnetic radiation. Given two of the three quantities (energy, frequency, and wavelength), you can calculate the third.

Thus, **electromagnetic radiation** has characteristics of both waves (frequency and wavelength) and particles (photons). That is, all the properties of electromagnetic radiation can only be explained by using both a wave model and a particle model.

SAMPLE PROBLEM 7.6 A

▶▶ (a) Arrange the following types of electromagnetic radiation in order of increasing frequency: x-rays, TV waves, microwaves, gamma rays, and visible light. (b) What is the relationship between increasing frequency of the radiation in (a) and wavelength and energy of that same radiation? (c) Arrange the "colors" of the visible spectrum (yellow, orange, green, red, blue, and violet) in order of increasing energy. (d) What is the relationship between increasing energy of the radiation in (c) and frequency and wavelength of the same radiation?

Solution:

HINT: Refer to Figure 7.15 in the textbook.

(a)   The order of increasing frequency is

TV waves < microwaves < visible light < x-rays < gamma rays

(b)   As **frequency increases, energy increases** and **wavelength decreases**.

(c)   The order of increasing energy is

red, orange, yellow, green, blue, violet

(d)   As **energy increases, frequency increases** and **wavelength decreases.**

SAMPLE PROBLEM 7.6 B

▶▶ If an FM radio station broadcasts at 94.3 MHz, (a) what is the wavelength of the radio waves? (b) What is the energy of a single photon of that radiation?

Solution:

(a)   Before using the above equation which relates wavelength, λ, and frequency, ν, the frequency must be converted from MHz to s^{-1},

$$? \, s^{-1} \; = \; 94.3 \; \text{MHz} \; \times \; \frac{10^6 \; \text{Hz}}{1 \; \text{MHz}} \; \times \; \frac{1 \; s^{-1}}{1 \; \text{Hz}} \; = \; 9.43 \times 10^7 \; s^{-1}$$

Now solving the equation for wavelength,

$$? \, \lambda \; = \; \frac{c}{\nu} \; = \; \frac{2.998 \times 10^8 \; \text{m} \bullet s^{-1}}{9.43 \times 10^7 \; s^{-1}} \; = \; 3.18 \; \text{m}$$

(b)   Since you know both the frequency and wavelength of the radiation, you can solve for the energy from either equation given in the REVIEW. The shortest route is to calculate energy from frequency,

$$\text{energy/photon} \; = \; h\nu \; = \; (6.626 \times 10^{-34} \; \text{J} \bullet s)(9.43 \times 10^7 \; s^{-1}) \; = \; 6.25 \times 10^{-26} \; \text{J}$$

SAMPLE PROBLEM 7.6 C

▶▶ If a certain microwave radiation has a wavelength of 1.20 cm, calculate (a) the frequency and (b) the energy, in kJ/mol, of this radiation.

Solution:

See Sample Problem 7.6 B for procedure and formulas.

(a) $\qquad ?\,\nu \;=\; \dfrac{c}{\lambda} \;=\; \dfrac{2.998 \times 10^8 \; \text{m} \bullet \text{s}^{-1}}{1.20 \; \text{cm} \; \times \; \dfrac{10^{-2} \; \text{m}}{1 \; \text{cm}}} \;=\; \dfrac{2.998 \times 10^8 \; \text{m} \bullet \text{s}^{-1}}{1.20 \times 10^{-2} \; \text{m}} = 2.50 \times 10^{10} \; \text{s}^{-1}$

(b) Calculate the energy of **one photon** first using the same procedure as in Sample Problem 7.6 B.

$$\text{energy/photon} \;=\; h\nu \;=\; (6.626 \times 10^{-34} \; \text{J} \bullet \text{s})(2.50 \times 10^{10} \; \text{s}^{-1}) \;=\; 1.66 \times 10^{-23} \; \text{J}$$

Then using Avogadro's number and the conversion factor to kJ, you can calculate the energy in kJ/mol photons,

$$? \; \dfrac{\text{kJ}}{\text{mol photons}} \;=\; \dfrac{1.66 \times 10^{-23} \; \text{J}}{1 \; \text{photon}} \; \times \; \dfrac{6.022 \times 10^{23} \; \text{photons}}{1 \; \text{mol photons}} \; \times \; \dfrac{1 \; \text{kJ}}{10^3 \; \text{J}}$$

$$=\; 1.00 \times 10^{-2} \; \text{kJ/mol photons}$$

SECTION 7.7   THE BOHR MODEL OF THE HYDROGEN ATOM

After completing this section, you will be able to do the following:

■ Calculate the wavelength and energy of spectral lines emitted by hydrogen.

REVIEW

To explain why negatively charged electrons are not pulled into the positively charged nucleus of an atom and why hydrogen atoms only emit certain wavelengths of light, Niels Bohr proposed that the electron moves around the nucleus in one of many possible orbits (similar to the orbits of the planets around the sun). Each orbit has a different energy associated with it. The energy that an electron has in an atom is the energy of the orbit it occupies. Since each orbit has a different (discrete) energy, an electron may have only the energies of the orbits. That is, the energy of an electron in an atom is NOT continuous, but **quantized**. Each orbit was assigned a number, **n**, called the **principal quantum number**, which can have values from 1 (for the energy level closest to the nucleus) to infinity (see Figure 7.23 of the textbook).

An electron in an atom can change energy only by going from one energy level (orbit) to another energy level (orbit), which is called a **transition**. When an electron **falls** from a higher energy level (orbit) to a lower energy level (orbit), energy in the form of electromagnetic radiation is **given off**. The energy given off is equal to the **difference** in energies between the two orbits and produces an **emission** spectral line of corresponding wavelength. If the transition energy is in the visible region of the electromagnetic spectrum, you can see the different color spectral lines (see Figure 7.24 of the textbook). The equation for calculating the wavelength of the **visible** spectral lines emitted by the **hydrogen atom** when an electron falls from a higher energy level (orbit) to n=2 is

$$\frac{1}{\lambda} = 1.097 \times 10^7 \text{ m}^{-1} \left[\frac{1}{2^2} - \frac{1}{n^2} \right] \qquad n = 3, 4, 5, \ldots$$

When an electron is **raised** from a **lower** energy level to a **higher** energy level, electromagnetic radiation is **absorbed** that is equal to the difference in energy level and produces an **absorption** spectral line. For example, the energy emitted when an electron falls from energy level n=3 to n=2, is **identical** to the energy absorbed when an electron is raised from n=2 to n=3 (see Figure 7.24 of textbook).

Although the idea of electrons moving in fixed orbits is no longer valid, Bohr's model is frequently used to introduce **quantum numbers** and **quantized energy states** for electrons in atoms.

SAMPLE PROBLEM 7.7 A

▶▶ (a) Calculate the wavelength in nm of the line in the spectrum of the hydrogen atom for which n=7. (b) Calculate the energy, in kJ/mol, of photons of light with the wavelength calculated in (a).

Solution:

(a) Substituting into the equation given above,

$$\frac{1}{\lambda} = 1.097 \times 10^7 \text{ m}^{-1} \left[\frac{1}{2^2} - \frac{1}{7^2} \right] = 2.519 \times 10^6 \text{ m}^{-1}$$

Solving the equation $1/\lambda = 2.519 \times 10^6 \text{ m}^{-1}$ for λ and converting meters to nanometers,

$$\lambda = \frac{1}{2.519 \times 10^6 \text{ m}^{-1}} = 3.970 \times 10^{-7} \text{ m} \times \frac{1 \text{ nm}}{10^{-9} \text{ m}} = 397.0 \text{ nm}$$

(b) First, calculate the energy of **one photon** using the wavelength in meters from part (a) in the equation given in the REVIEW.

$$\text{energy/photon} = \frac{hc}{\lambda} = \frac{(6.626 \times 10^{-34} \text{ J} \cdot \text{s})(2.998 \times 10^8 \text{ m} \cdot \text{s}^{-1})}{3.970 \times 10^{-7} \text{ m}} = 5.004 \times 10^{-19} \text{ J}$$

Then, using Avogadro's number and the conversion factor to kJ, find the energy in kJ/mol as in Sample Problems 7.6 B and C.

$$? \frac{\text{kJ}}{\text{mol photons}} = \frac{5.004 \times 10^{-19} \text{ J}}{1 \text{ photon}} \times \frac{6.022 \times 10^{23} \text{ photons}}{1 \text{ mol photons}} \times \frac{1 \text{ kJ}}{10^3 \text{ J}}$$

$$= 301.3 \text{ kJ/mol photons}$$

SECTION 7.12 ELECTRON SPIN AND THE PAULI EXCLUSION PRINCIPLE

After completing this section, you will be able to do the following:

■ Determine the number of electrons that occupy a shell and subshell in an atom.

REVIEW

Three quantum numbers are needed to describe an electron in an orbital of an atom:

n   principal quantum number - tells the size and energy of an orbital;
l   angular momentum quantum number - tells the shape of the orbital;
$\mathbf{m_l}$ magnetic quantum number - describes the direction of the orbital in space.

The values that these quantum numbers may take and a table of allowed combinations are described in Section 7.10 and Table 7.1 of the textbook. A fourth quantum number, the **spin quantum number, $\mathbf{m_s}$**, had to be added to account for the spectral lines produced by an electron spinning clockwise or counterclockwise. The **Pauli Exclusion Principle** says that **no two electrons in an atom can have all four quantum numbers alike** (three of the four may be alike, but not all four). As a result, an **orbital** can only hold **two** electrons of opposite spin. The number of electrons that can occupy any shell or subshell can be determined by finding the number of orbitals available (Table 7.1 in the textbook) and multiplying by two electrons/orbital.

SAMPLE PROBLEM 7.12 A

▸▸ (a) How many electrons can occupy the 2p subshell, the 3p subshell, the 6p subshell? (b) How many electrons can occupy the 3d subshell, the 4d subshell, the 7d subshell?

Refer to Table 7.1 in the textbook.

(a)   The 2p, 3p, and 6p subshells each contain **three** orbitals (because m_l has three permissible values for a p subshell). Since each **orbital** can hold a maximum of **two** electrons, the 2p, 3p, and 6p subshells can each hold six electrons. **In fact, all p subshells have three orbitals and therefore can hold a total of six electrons.**

(b)   All d subshells have **five** orbitals (because m_l has five permissible values for a d subshell). Since each **orbital** can hold a maximum of **two** electrons, **each d subshell can hold 10 electrons.**

SAMPLE PROBLEM 7.12 B

▶ ▶ How many electrons can occupy the third shell in an atom?

Solution:

Refer to Table 7.1 in the textbook. Determine the number of orbitals in the shell and multiply by two electrons/orbital. The third shell has **one** s orbital, **three** p orbitals, and **five** d orbitals, for a total of **nine** orbitals. Multiplying by two electrons/orbital, the third shell can hold **18 electrons**. [Alternatively, the number of electrons occupying any **shell** is $2n^2$, where n is the principal quantum (or shell) number.]

SELF-TEST QUESTIONS

Test your understanding of the concepts and skills in this chapter by working through the multiple choice questions **BEFORE** checking the answers. A periodic table will always be available. The ANSWERS TO SELF-TEST QUESTIONS follow immediately after this section.

1. Electromagnetic radiation with a wavelength of 2.54 cm has a frequency of

    A. 1.18×10^8 s^{-1}

    B. 8.47×10^{-11} s^{-1}

    C. 3.00×10^8 s^{-1}

    D. 8.47×10^{-9} s^{-1}

    E. 1.18×10^{10} s^{-1}

2. The energy of a single photon of the radiation in the previous problem (#1) is

    A. 7.82×10^{-24} J   B. 4.71 J   C. 1.18×10^{10} J   D. 4.71×10^{-3} J   E. 5.61×10^{-18} J

3. If a spectrometer uses 60.0-MHz radiation, the wavelength, in meters, of this frequency is

    A. 1.10×10^{-41} m   B. 5.00 m   C. 0.200 m   D. 5.00×10^6 m   E. 6.40 m

4. The energy in kJ/mol of the radiation in the previous problem (#3) is

    A. 3.97×10^{-26} kJ/mol

    B. 2.39×10^{-5} kJ/mol

    C. 4.18×10^4 kJ/mol

    D. 1.20×10^{-4} kJ/mol

    E. 6.75×10^3 kJ/mol

5. In atoms of a particular element, the electron energy level change that will **emit** a spectral line with

the shortest wavelength will be

A. n=3 to n=7.
B. n=5 to n=3.
C. n=6 to n=3.
D. n=4 to n=7.
E. n=7 to n=3.

6. The number of electrons that can occupy a 3d orbital is

A. 10   B. 5   C. 2   D. 18   E. 6

7. The subshell that can hold 6 electrons is

A. p   B. s   C. g   D. f   E. d

8. The number of electrons that can occupy the fourth shell (n=4) is

A. 16   B. 8   C. 7   D. 32   E. 64

9. All of the following can hold a maximum of 2 electrons except

A. 4s subshell   B. 2p subshell   C. 4f orbital   D. 8s subshell   E. 5d orbital

10. Which of the following can hold the most electrons?

A. 4f subshell   B. second shell   C. 6d subshell   D. 5s subshell   E. 8p subshell

ANSWERS TO SELF-TEST QUESTIONS

1. e	4. b	7. a
2. a	5. e	8. d
3. b	6. c	9. b
		10. a

PRACTICE PROBLEMS

Test your problem-solving skills in this chapter by working through the problems in the space provided **BEFORE** checking the answers. A periodic table will always be available. The ANSWERS TO PRACTICE PROBLEMS follow immediately after this section.

1. If electromagnetic radiation has a wavelength of 755 nm, (a) what is the frequency of this radiation? (b) What is the energy in kJ/mol of this radiation?

2. (a) Calculate the wavelength in nm of the line in the spectrum of the hydrogen atom for which $n=8$. (b) Calculate the energy, in kJ/mol, of photons of light with the wavelength calculated in (a).

ANSWERS TO PRACTICE PROBLEMS

1. (a) $? \nu = \dfrac{c}{\lambda} = \dfrac{2.998 \times 10^8 \text{ m} \bullet \text{s}^{-1}}{755 \text{ nm} \times \dfrac{10^{-9} \text{ m}}{1 \text{ nm}}} = \dfrac{2.998 \times 10^8 \text{ m} \bullet \text{s}^{-1}}{755 \times 10^{-9} \text{ m}} = 3.97 \times 10^{14} \text{ s}^{-1}$

(b)   energy/photon $= h\nu = (6.626 \times 10^{-34} \text{ J}\bullet\text{s})(3.97 \times 10^{14} \text{ s}^{-1}) = 2.63 \times 10^{-19} \text{ J}$

$$? \; \frac{\text{kJ}}{\text{mol photons}} = \frac{2.63 \times 10^{-19} \text{ J}}{1 \text{ photon}} \times \frac{6.022 \times 10^{23} \text{ photons}}{1 \text{ mol photons}} \times \frac{1 \text{ kJ}}{10^3 \text{ J}}$$

$$= 158 \text{ kJ/mol photons}$$

2.   (a) $\dfrac{1}{\lambda} = 1.097 \times 10^7 \text{ m}^{-1} \left[\dfrac{1}{2^2} - \dfrac{1}{8^2} \right] = 2.571 \times 10^6 \text{ m}^{-1}$

$$\lambda = \frac{1}{2.571 \times 10^6 \text{ m}^{-1}} = 3.889 \times 10^{-7} \text{ m} \times \frac{1 \text{ nm}}{10^{-9} \text{ m}} = 388.9 \text{ nm}$$

(b)   energy/photon $= \dfrac{hc}{\lambda} = \dfrac{(6.626 \times 10^{-34} \text{ J}\bullet\text{s})(2.998 \times 10^8 \text{ ms}^{-1})}{3.889 \times 10^{-7} \text{ m}} = 5.108 \times 10^{-19} \text{ J}$

$$? \; \frac{\text{kJ}}{\text{mol photons}} = \frac{5.108 \times 10^{-19} \text{ J}}{1 \text{ photon}} \times \frac{6.022 \times 10^{23} \text{ photons}}{1 \text{ mol photons}} \times \frac{1 \text{ kJ}}{10^3 \text{ J}}$$

$$= 307.6 \text{ kJ/mol photons}$$

CHAPTER 8

ELECTRONIC STRUCTURE AND THE PERIODIC TABLE

SECTION 8.1   ELECTRON CONFIGURATIONS

After completing this section, you will be able to do the following:

■ Given an atom of an element in its ground or excited state, (a) sketch the energy level diagram, (b) write the electron configuration, (c) write the abbreviated electron configuration, and (d) write an orbital diagram.

REVIEW

The arrangement of electrons in the orbitals of atoms determines many of the chemical properties of the elements. Remember that an **orbital** is a volume in space where an electron with a particular energy is likely to be found. When the electrons occupy the **lowest energy orbitals** available, the atom is in its **ground state**. If one or more electrons in an atom occupy a higher energy orbital while leaving a lower energy orbital unfilled, the atom is in an **excited state**. There are many excited states possible for an atom of any element. The energy of the orbitals in an atom increases as the principal quantum number, n, increases, although the difference in energy between the same two shells (say 1s and 2s) will vary with the element (see Figure 8.1 in the textbook). **Within a given shell**, the subshells increase in energy in the order s < p < d < f. All of the orbitals in a subshell have the same energy and are called **degenerate orbitals**.

The distribution of electrons in an atom can be shown by using an arrow pointing up (↑) or an arrow pointing down (↓) to represent electrons of opposite spin in an orbital of an energy level diagram. A second and faster method of showing the distribution of electrons in an atom is by writing an **electron**

configuration, which shows how many electrons (as superscripts) occupy each subshell in order of increasing energy. For example, the ground-state **electron configuration** of Li is $1s^2 2s^1$, which shows that the lowest energy 1s orbital contains 2 electrons and the next higher energy 2s orbital has 1 electron. Because **electron configurations** can become quite long for elements with high atomic numbers, an **abbreviated electron configuration** is often used. Using the ground state of Li again, the **abbreviated electron configuration** would be [He] $2s^1$, where the symbol [He] represents the electron configuration of the nearest noble gas with lower atomic number. Thus, [He] means $1s^2$. To show whether any unpaired electrons are present in an atom, a fourth method is used to show the distribution of electrons in an atom; this method is known as an **orbital diagram**. The **orbital diagram** for the Li atom in the ground state is

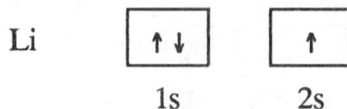

Li $\boxed{\uparrow\downarrow}$ $\boxed{\uparrow}$

1s   2s

or in an abbreviated form,

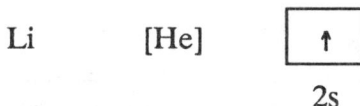

Li   [He] $\boxed{\uparrow}$

2s

where each box represents an orbital. When placing electrons into orbitals with the same energy, **Hund's rule** says that **electrons occupy orbitals singly** before beginning to pair up. Atoms with **unpaired** electrons are **paramagnetic** -- that is, they are weakly attracted into a magnetic field. Atoms with all electrons paired are **diamagnetic** -- that is, they are pushed out of a magnetic field.

SAMPLE PROBLEM 8.1 A

▶▶ For the carbon atom in an excited state, (a) sketch the energy level diagram (including the electrons), and (b) write the electron configuration for the excited state of C in (a).

Solution:

(a)   The atomic number (Z) of carbon is 6 (inside front cover), which means that a carbon atom has 6 electrons around the nucleus. The energy level diagram shows the subshells and orbitals in order of increasing energy. The ground state of a C atom will have all six electrons in the lowest energy level (shell or subshell) possible. The C atom in an excited state will have at least 1 electron in a higher energy orbital than the ground state. These 6 electrons are shown in the ground state and in one of many possible excited states for the carbon atom.

$$\underset{3s}{\underline{\quad}} \qquad\qquad \underset{3s}{\underline{\uparrow}}$$

$$\underset{2s}{\underline{\uparrow\downarrow}} \; \underset{}{\underline{\uparrow}} \; \underset{2p}{\underline{\uparrow}} \; \underline{\quad} \qquad \underset{2s}{\underline{\uparrow\downarrow}} \; \underset{}{\underline{\uparrow}} \; \underset{2p}{\underline{\quad}} \; \underline{\quad}$$

↑
Energy

$$\underset{1s}{\underline{\uparrow\downarrow}} \qquad\qquad\qquad \underset{1s}{\underline{\uparrow\downarrow}}$$

Ground state of C   One excited state of C

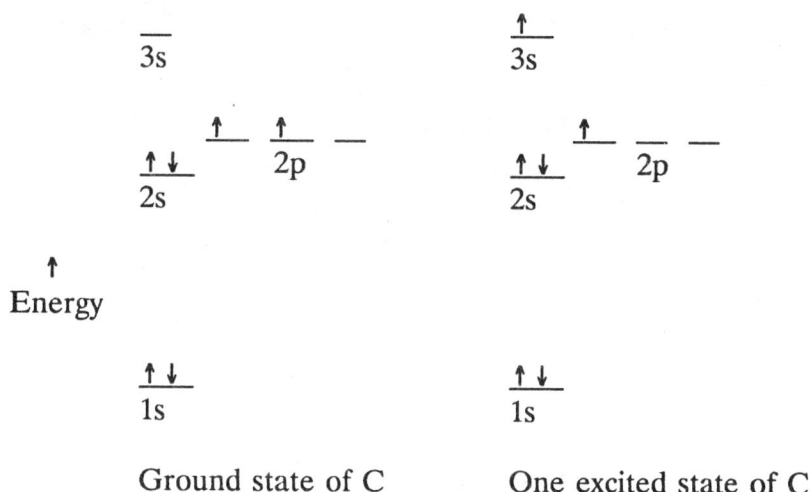

(b)   The electron configuration gives the main shell number (1, 2, 3, ...) followed by the subshell letter (s, p, d, f, ...), with a superscript indicating the number of electrons in the subshell, **in order of increasing energy.**   The electron configuration for the C atom in the excited state shown in (a) is

$$C\ (Z=6) \qquad 1s^2 2s^2 2p^1 3s^1,$$

which is read as two 1s electrons, two 2s electrons, one 2p electron, and one 3s electron. The superscripts are **NOT** exponents.   The superscripts **must add up to the atomic number (Z) 6.** The "(Z=6)" after the symbol for carbon, C, is not actually part of the electron configuration but is shown there as a reminder that the atomic number of an element is also the number of electrons, and that the sum of the superscripts must add up to that number.

SAMPLE PROBLEM 8.1 B

▶▶ For the oxygen atom in the ground state, (a) sketch the energy level diagram, (b) write the electron configuration, (c) write the abbreviated electron configuration, and (d) write the orbital diagram.

Solution:

HINT: All methods of showing the distribution of electrons in an atom (i.e., energy level diagram, electron configuration, abbreviated electron configuration, orbital diagram) must show the orbitals **in order of increasing energy.**

(a)   See Sample Problem 8.1 A (a). The atomic number of O is 8, so an oxygen atom has 8 electrons around the nucleus. In the ground state, the 8 electrons occupy the lowest energy levels available.

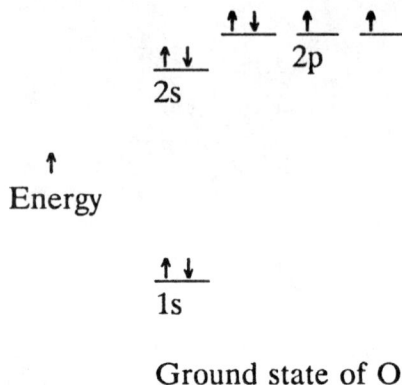

$$\underset{2s}{\uparrow\downarrow} \qquad \underset{2p}{\overline{\uparrow\downarrow}\ \overline{\uparrow}\ \overline{\uparrow}}$$

↑
Energy

$$\underset{1s}{\uparrow\downarrow}$$

Ground state of O

When half-filled, the three 2p orbitals are degenerate -- that is, they have the same energy. The last (eighth) electron could be paired up with an electron in **any** of the three half-filled orbitals. Under those conditions, it is **customary to start pairing electrons in the orbitals in the same way as you read text, from left to right.**

(b)   See Sample Problem 8.1 A (b). The electron configuration of the O atom in the ground state is

$$O\ (Z=8) \qquad 1s^2 2s^2 2p^4$$

Again, the superscripts must add up to the atomic number, 8.

(c)   In the abbreviated electron configuration, the symbol of the **nearest noble gas with lower atomic number** is used to represent the electron configuration up through that noble gas. For the ground state of O, the abbreviated electron configuration is

$$O\ (Z=8) \qquad [\text{He}]\ 2s^2 2p^4,$$

where [He] represents the electron configuration of the atom He, $1s^2$.

(d)   The orbital diagram for an atom shows the distribution of electrons, **including any unpaired electrons,** by using a box for each orbital, arranged horizontally in order of increasing energy. The orbital diagram of O in the ground state is

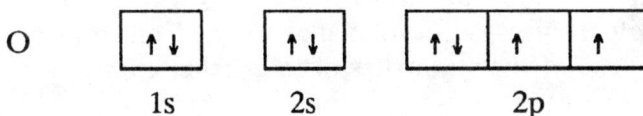

O $\boxed{\uparrow\downarrow}$ $\boxed{\uparrow\downarrow}$ $\boxed{\uparrow\downarrow\ |\ \uparrow\ |\ \uparrow}$

       1s   2s   2p

There are **two** unpaired electrons in an oxygen atom in the ground state; the oxygen atom in the ground state is **paramagnetic**.

SECTION 8.3	USING THE PERIODIC TABLE TO WRITE ELECTRON CONFIGURATIONS

After completing this section, you will be able to do the following:

- For an **atom** of an element in its ground state, (a) write the electron configuration, (b) write the abbreviated electron configuration, (c) write the orbital diagram, and (d) determine the number of unpaired electrons;

- For an **ion** of an element in its ground state, (a) write the electron configuration, and (b) write the abbreviated orbital diagram.

REVIEW

The periodic table is read like a newspaper, from left to right and down the page. That is exactly the arrangement of the elements in order of increasing atomic number, which is also the order of increasing energy of the subshells (orbitals). The **period numbers** in the periodic table are the **principal quantum numbers, n, or the numbers of the main energy shells.** The periodic table can be blocked into **four** sections corresponding to the last electron of an element occupying an s, p, d, or f orbital (see Figure 8.6 of the textbook). Group **IA and IIA** elements (a row or block of 2 elements) always have their last electron occupying an **s subshell** (or s orbital). An s subshell with one s orbital can only hold 2 electrons. Groups **IIIA through VIIA** and Group **0** (a row or block of 6 elements) have their last electrons occupying a **p subshell** (or p orbital) in the same main shell as the immediately preceding ns subshell. A p subshell with its three p orbitals can hold a total of 6 electrons. Groups **IIIB through IIB** (a row or block of 10 elements) have their last electrons occupying a **d subshell** (or d orbital) in the **(n-1)** shell of the ns subshell immediately preceding it. A d subshell with its five orbitals can hold a total of 10 electrons. Elements in the **lanthanide** and **actinide** series (a row or block of 14 elements) shown at the bottom of the periodic table (for convenience in printing) have their last electrons occupying an **f subshell** (or f orbital) in the **(n-2)** shell of the ns subshell immediately preceding it. An f subshell with its seven orbitals can hold a total of 14 electrons.

In an atom or ion, the electrons occupy the orbitals that give the atom or ion the **lowest total energy** (which depends both on orbital energies and on the energy of repulsion between electrons). An electron configuration **in order of increasing energy of the orbitals** (the so-called **aufbau principle,** or building up of electrons) can be readily written with the aid of a **periodic table.** For example, to write the electron configuration of element 118 (or any other element) in the ground state, always start with hydrogen ($Z = 1$) in period 1 ($n = 1$) and "read" the periodic table from left to right by adding 1 electron for each increase in atomic number of 1. When you come to the end of a period (Group 0 element or noble gas), follow the increasing atomic numbers to the beginning of the next row. Every time you see a block of two, six, ten, or fourteen elements in any period (row), electron(s) are always occupying s, p, d, or f subshells, respectively. The step-by-step process for writing the electron configuration of element 118 (yet to be discovered) is

PERIOD (main shell)	CONFIGURATION	GROUP 0 ELEMENT	ATOMIC NUMBER
1 (n=1)	$1s^2$	He	2
2 (n=2)	$2s^2 2p^6$	Ne	10
3 (n=3)	$3s^2 3p^6$	Ar	18
4 (n=4)	$4s^2 3d^{10} 4p^6$	Kr	36
5 (n=5)	$5s^2 4d^{10} 5p^6$	Rn	54
6 (n=6)	$6s^2 4f^{14} 5d^{10} 6p^6$	Xe	86
7 (n=7)	$7s^2 5f^{14} 6d^{10} 7p^6$	?	118

Note that a p subshell is in the same main shell as the s subshell immediately preceding it. Also, a d subshell is in the (n-1) shell of the s subshell immediately preceding it. Similarly, an f subshell is in the (n-2) shell of the s subshell immediately preceding it. The complete electron configuration for element 118 would be

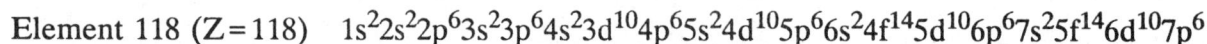

Element 118 (Z=118) $1s^2 2s^2 2p^6 3s^2 3p^6 4s^2 3d^{10} 4p^6 5s^2 4d^{10} 5p^6 6s^2 4f^{14} 5d^{10} 6p^6 7s^2 5f^{14} 6d^{10} 7p^6$

An **alternate** method of writing the electron configuration is to **group the orbitals with the same principal quantum number together.** This method (which may be useful when writing the electron configurations for cations) does **NOT** always follow the order of increasing atomic numbers in the periodic table once you reach potassium, **K,** with atomic number 19. The **alternate** electron configuration for element 118 (with extra spacing between main shells shown here for ease of reading) would be

Element 118 (Z=118) $1s^2$ $2s^2 2p^6$ $3s^2 3p^6 3d^{10}$ $4s^2 4p^6 4d^{10} 4f^{14}$ $5s^2 5p^6 5d^{10} 5f^{14}$ $6s^2 6p^6 6d^{10}$ $7s^2 7p^6$

Once the electron configuration is written from the periodic table for an atom in the ground state, the abbreviated electron configuration and orbital diagrams can be written and the number of unpaired electrons determined (see Section 8.1 REVIEW and Sample Problem 8.1 B).

To write the electron configuration for an ion in the ground state, use the periodic table to write the electron configuration for an **atom** of the element first. Then, for an **anion, add** to the lowest energy orbital(s) available the number of electrons equal to the negative charge on the ion. For a **cation, subtract** from the **highest** (outermost) energy orbital(s) the number of electrons equal to the positive charge on the ion. Once the electron configuration is written for any ion, the abbreviated electron configuration can be written in an identical manner to that used for atoms.

SAMPLE PROBLEM 8.3 A

▶ ▶ For the germanium (Ge) atom in its ground state, (a) write the electron configuration, (b) write the abbreviated electron configuration, (c) write the orbital diagram, and (d) indicate how many unpaired electrons are present.

Solution:

HINT: From this point on, write electron configurations from a periodic table using the procedure given above.

(a) For germanium, the atomic number (Z) is 32, which means that germanium atoms have 32 electrons. The electron configuration must have 32 electrons occupying orbitals in order of increasing energy, which is identical to the order of increasing atomic number of the elements in the periodic table. Using the procedure given above, the electron configuration of germanium is

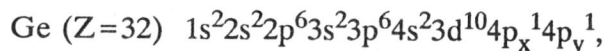

$$Ge \ (Z=32) \quad 1s^2 2s^2 2p^6 3s^2 3p^6 4s^2 3d^{10} 4p_x^{\ 1} 4p_y^{\ 1},$$

where $4p_x^{\ 1} 4p_y^{\ 1}$ indicates that the last two electrons obey Hund's rule and **singly** occupy two separate 4p orbitals (the $4p_x$ and $4p_z$ or $4p_y$ and $4p_z$ orbitals would be equally correct; common practice is to use the orbitals alphabetically). Often, the electron configuration is written as

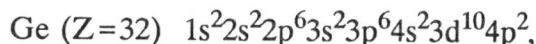

$$Ge \ (Z=32) \quad 1s^2 2s^2 2p^6 3s^2 3p^6 4s^2 3d^{10} 4p^2,$$

where $4p^2$ is **understood** to mean $4p_x^{\ 1} 4p_y^{\ 1}$. **For clarity, we will use this latter representation from this point on.**

The **alternate method** of writing the electron configuration by grouping orbitals with the same principal quantum number together in order of increasing quantum number is

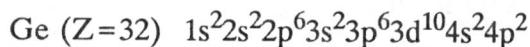

$$Ge \ (Z=32) \quad 1s^2 2s^2 2p^6 3s^2 3p^6 3d^{10} 4s^2 4p^2$$

CHECK: The superscripts must add up to the atomic number 32, no matter which way the electron configuration is written.

(b) The abbreviated electron configuration (see Sample Problem 8.1 B) is

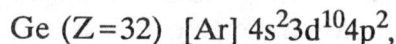

$$Ge \ (Z=32) \quad [Ar] \ 4s^2 3d^{10} 4p^2,$$

or grouping the orbitals with the same quantum number together,

$$\text{Ge } (Z=32) \text{ [Ar] } 3d^{10}4s^24p^2,$$

where [Ar] means $1s^22s^22p^63s^23p^6$.

(c) The abbreviated orbital diagram (see Sample Problem 8.1 B) for the germanium atom in the ground state with the orbitals of the same quantum number grouped together is

Ge [Ar] | ↑↓ | ↑↓ | ↑↓ | ↑↓ | ↑↓ | | ↑↓ | | ↑ | ↑ | |

             3d   4s   4p

(d) The number of unpaired electrons (if any) in an atom can be determined by counting them in the orbital diagram in part (c). There are two unpaired electrons in a germanium atom in the ground state, which makes it **paramagnetic** (see Section 8.1 REVIEW).

SAMPLE PROBLEM 8.3 B

▶▶ For the tungsten (W) atom in its ground state, (a) write the electron configuration, (b) write the abbreviated electron configuration, (c) write the orbital diagram, and (d) indicate how many unpaired electrons are present.

Solution

HINT: See Sample Problem 8.3 A.

(a) For tungsten, the atomic number (Z) is 74, which means tungsten atoms have 74 electrons. The electron configuration must have 74 electrons occupying orbitals in order of increasing energy, which is identical to the order of increasing atomic number of the elements in the periodic table. Using the procedure given above, the electron configuration of tungsten is

$$\text{W } (Z=74) \quad 1s^22s^22p^63s^23p^64s^23d^{10}4p^65s^24d^{10}5p^66s^24f^{14}5d^4,$$

where $5d^4$ is understood to mean the last four electrons **singly** occupy four of the five 5d orbitals (that is, they obey Hund's rule).

The **alternate method** of writing the electron configuration by grouping orbitals with the same principal quantum number together is

$$\text{W } (Z=74) \quad 1s^22s^22p^63s^23p^63d^{10}4s^24p^64d^{10}4f^{14}5s^25p^65d^46s^2$$

CHECK: The superscripts must add up to the atomic number 74, no matter which way the electron configuration is written.

(b) The abbreviated electron configuration is

$$W \ (Z=74) \ \ [Xe] \ 6s^2 4f^{14} 5d^4,$$

or grouping the orbitals with the same quantum number together,

$$W \ (Z=74) \ \ [Xe] \ 4f^{14} 5d^4 6s^2,$$

where [Xe] means $1s^2 2s^2 2p^6 3s^2 3p^6 4s^2 3d^{10} 4p^6 5s^2 4d^{10} 5p^6$.

(c) The abbreviated orbital diagram for the tungsten atom in the ground state with the orbitals of the same quantum number grouped together is

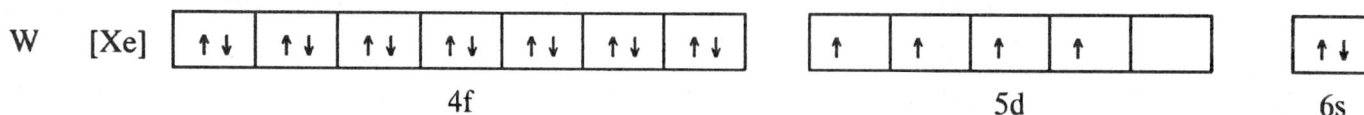

W [Xe]

| ↑↓ | ↑↓ | ↑↓ | ↑↓ | ↑↓ | ↑↓ | ↑↓ | | ↑ | ↑ | ↑ | ↑ | | | ↑↓ |

4f 5d 6s

(d) The number of unpaired electrons (if any) in an atom can be determined by counting them in the orbital diagram in part (c). There are four unpaired electrons in a tungsten atom in the ground state, which makes it **paramagnetic** (see Section 8.1 REVIEW).

SAMPLE PROBLEM 8.3 C

▶▶ For the nitride ion, N^{3-} in the ground state, (a) write the electron configuration, and (b) give the abbreviated orbital diagram.

Solution

(a) Begin by writing the electron configuration for the N **atom** (see Sample Problems 8.3 A & B). Nitrogen has an atomic number of 7, or 7 electrons.

$$N \ (Z=7) \ \ 1s^2 2s^2 2p^3,$$

where $2p^3$ is understood to mean $2p_x^1 2p_y^1 2p_z^1$. The N^{3-} ion has three more electrons than the N atom, or a total of 10 electrons. The last three electrons will occupy the lowest energy orbitals available, which are the three half-filled 2p orbitals of the N atom. The electron configuration of N^{3-} is

$$N^{3-} \quad (10 \text{ electrons}) \quad 1s^2 2s^2 2p^6,$$

which is identical to the electron configuration of the element neon.

(b)   The abbreviated orbital diagram for the nitrogen atom is

N   [He]   [↑↓]   [↑] [↑] [↑]
         2s   2p

The abbreviated orbital diagram for the nitride ion, N^{3-}, is

N^{3-} [He]   [↑↓]   [↑↓] [↑↓] [↑↓]
         2s   2p

Note that the nitrogen atom is paramagnetic with three unpaired electrons, and that the nitride ion is diamagnetic with no unpaired electrons.

SAMPLE PROBLEM 8.3 D

▸▸ For the tin(II) ion, Sn^{2+}, and the tin(IV) ion, Sn^{4+}, in their ground states, (a) write the electron configuration, and (b) give the abbreviated orbital diagram.

Solution:

(a)   For cations, begin by writing the **alternate** electron configuration for the Sn **atom** that has orbitals with the same principal quantum number grouped together and arranged in order of increasing principal quantum number (see Sample Problems 8.3 A & B). Tin has an atomic number of 50, or 50 electrons. Its electron configuration is

$$Sn \ (Z=50) \quad 1s^2 2s^2 2p^6 3s^2 3p^6 3d^{10} 4s^2 4p^6 4d^{10} 5s^2 5p^2,$$

where $5p^2$ is understood to mean $5p_x^{\ 1} 5p_y^{\ 1}$. The abbreviated form of this alternate configuration is

$$Sn \ (Z=50) \quad [Kr] \ 4d^{10} 5s^2 5p^2$$

The Sn^{2+} ion has **two** less electrons than the Sn atom. The electrons **lost** first by an atom to become a cation are those in the **highest energy orbitals**. By using the alternate electron configuration, you start at the right side of the electron configuration (where the highest energy orbitals are automatically shown) and remove enough electrons to give the required positive charge. For a Sn atom to become a Sn^{2+} ion, two electrons will have to be removed from the two half-filled 5p orbitals. The electron configuration for Sn^{2+} is

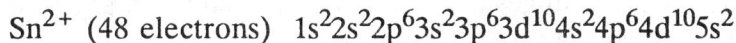

$$Sn^{2+} \text{ (48 electrons)} \quad 1s^2 2s^2 2p^6 3s^2 3p^6 3d^{10} 4s^2 4p^6 4d^{10} 5s^2$$

or, Sn^{2+} (48 electrons) [Kr] $4d^{10} 5s^2$.

The Sn^{4+} ion will have four less electrons than a tin atom (or two less electrons than Sn^{2+}), which means that the $5s^2$ electrons will also be lost. The electron configuration for Sn^{4+} in the ground state is

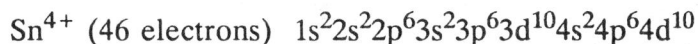

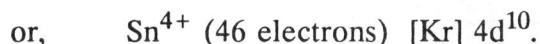

$$Sn^{4+} \text{ (46 electrons)} \quad 1s^2 2s^2 2p^6 3s^2 3p^6 3d^{10} 4s^2 4p^6 4d^{10}$$

or, Sn^{4+} (46 electrons) [Kr] $4d^{10}$.

(b)   The abbreviated orbital diagrams for the Sn atom, Sn^{2+} ion, and Sn^{4+} ion are

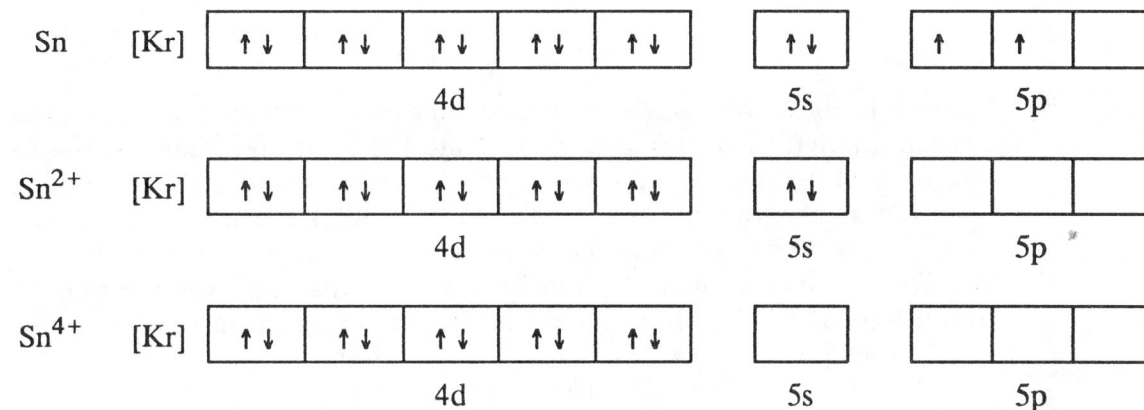

	4d	5s	5p
Sn [Kr]	↑↓ ↑↓ ↑↓ ↑↓ ↑↓	↑↓	↑ ↑ ☐
Sn^{2+} [Kr]	↑↓ ↑↓ ↑↓ ↑↓ ↑↓	↑↓	☐ ☐ ☐
Sn^{4+} [Kr]	↑↓ ↑↓ ↑↓ ↑↓ ↑↓	☐	☐ ☐ ☐

Note that the tin atom is paramagnetic with two unpaired electrons, and that the Sn^{2+} and Sn^{4+} ions are both diamagnetic with no unpaired electrons.

SECTION 8.4   ATOMIC AND IONIC RADII

After completing this section, you will be able to do the following:

■ Predict relative atomic and ionic radii by learning the trends in atomic and ionic radii in the periodic table.

REVIEW

For **main group elements** (Groups IA-VIIA,0), **atomic radii decrease from left to right**, and **increase down a group**. For **transition elements**, changes in atomic radii are small, first decreasing and then increasing across a series, and slightly increasing down a group.

For **ionic radii, cations** (positive ions) are **always smaller** than the **atoms** from which they are formed. **Anions** (negative ions) are **always larger** than the atoms from which they are formed. Species (ions or atoms) that have the same number of electrons are called **isoelectronic** (iso- meaning same). For any **isoelectronic series**, the species **decreases** in size as the nuclear charge (number of protons) **increases**.

SAMPLE PROBLEM 8.4 A

▶ ▶ Using only a periodic table, arrange the atoms of the following elements in order of decreasing radius,

S, Na, O, Rb.

Solution:

The trends for atomic radii given above indicate that atomic radii decrease as you proceed up a group and left to right in a period. Note the location of each of the four elements in the periodic table. Since you want the order of **decreasing** radii, start with the largest radius in the lower left-hand corner of the periodic table and work towards the smallest radius in the upper right-hand corner of the periodic table. In Group IA, Na is above Rb and therefore smaller than Rb. S is to the right of Na in Period 3 and therefore smaller than Na. O is above S in Group VIA and therefore smaller than S. The order of decreasing radii is

$$Rb > Na > S > O$$

SAMPLE PROBLEM 8.4 B

▶ ▶ Which is larger, (a) Ca^{2+} or K^+? (b) N^{3-} or O^{2-}? (c) Mg^{2+} or O^{2-}?

Solution:

(a) The ions Ca^{2+} and K^+ are isoelectronic with 18 electrons each, $1s^2 2s^2 2p^6 3s^2 3p^6$. But Ca has 20 protons in the nucleus (Z=20) and K has only 19 protons in the nucleus (Z=19). The higher number of protons (20) will pull in the same number of electrons more closely. Therefore, K^+ is larger than Ca^{2+}.

(b) The ions N^{3-} and O^{2-} are isoelectronic with 10 electrons each, $1s^2 2s^2 2p^6$. But N has

7 protons in the nucleus and O has 8 protons. The higher number of protons (8) will pull the same number of electrons closer to the nucleus. Therefore, N^{3-} is larger than O^{2-}.

(c) The ions Mg^{2+} and O^{2-} are isoelectronic with 10 electrons each, $1s^2 2s^2 2p^6$. Mg has 12 protons and O has 8 protons. The ion with the highest number of protons will have the smallest ionic radius. Therefore, O^{2-} is larger than Mg^{2+}.

SECTION 8.5   IONIZATION ENERGY

After completing this section, you will be able to do the following:

- Predict relative ionization energies of the elements by learning the trends in first ionization energies in the periodic table;

- Predict relative ionization energies for atoms of an element by learning the trend for successive ionization energies.

REVIEW

The **first ionization energy** of an element is the amount of energy needed to remove the least tightly bound electron (in the highest energy orbital) from a mole of gaseous atoms and thus to form a mole of gaseous **ions**. This process of removing a negatively charged electron from a positively charged nucleus is **always** an **endothermic** change,

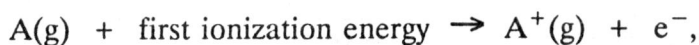

$$A(g) \; + \; \text{first ionization energy} \; \rightarrow \; A^+(g) \; + \; e^-,$$

where A is an atom of an element. For main group elements, **first ionization energies** generally increase going from left to right across a row in the periodic table, and as you move up a group. This is exactly opposite the trend observed for atomic radii (Section 8.4). As atomic radii **decrease**, first ionization energies will **increase** (up a group and left to right in a period), since more energy will be required to remove an electron from a smaller atom (where the outermost electron is closer to the nucleus).

As successive electrons are removed from an atom, the amount of energy required **always increases**, because you continue to remove negative electrons from ions with an increasing positive charge (after the first electron is removed). The energy required to remove the second most loosely bound electron from a mole of gaseous atoms is called the **second ionization potential**, and similarly for the removal of the third, fourth, fifth, etc., electrons. After all the outermost electrons have been removed from an atom and a noble gas electron configuration reached, the ionization energy required to remove additional electrons increases dramatically. The electron configuration of the noble gases is very stable, which is why there are relatively few compounds known of the Group 0 elements.

SAMPLE PROBLEM 8.5 A

▶ ▶ Indicate which species in each pair will require more energy to remove one electron: (a) Mg, Mg^{2+} (b) O, F (c) Na, K (d) Cl^-, Br^-.

Solution:

(a) Mg^{2+}. Both Mg and Mg^{2+} have 12 protons in the nucleus, but Mg^{2+} has only 10 electrons whereas Mg has 12. It will take considerably more energy to remove an electron from the smaller Mg^{2+} ion with its 2+ charge, than the larger, neutral Mg atom. Removal of an electron from the neutral atom is the **first ionization energy**. Removal of an electron from the 2+ ion is the **third ionization energy**. The ionization energies **always** become **higher** (more energy needed) as electrons are successively removed from an atom or ion of the same element.

(b) F. First ionization energies **increase** going from left to right in a row of the periodic table because atomic radii **decrease** in the same direction.

(c) Na. First ionization energies **increase** going up a group in the periodic table because atomic radii **decrease** in the same direction.

(d) Cl^-. The energy required to remove an electron from the same-charge monatomic ion will **increase** going up a group because both atomic and ionic radii **decrease** in the same direction.

SECTION 8.6   ELECTRON AFFINITY

After completing this section, you will be able to do the following:

■ From a table of electron affinities, calculate ΔH for the addition of an electron to an atom of an element;

■ Predict whether electron affinities will be positive or negative.

REVIEW

Electron affinity (i.e., attraction for an electron) is the energy **released** when an electron is added to a mole of gaseous atoms to form a mole of gaseous ions with a 1− charge,

$$A(g) \ + \ e^- \ \rightarrow \ A^-(g) \ + \ \text{thermal energy,}$$

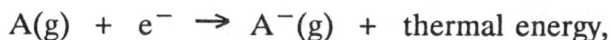

where A is an atom of an element. Because of the way **electron affinity** is defined (i.e., energy **released**), the higher the attraction of an atom for an electron (i.e., electron affinity), the more thermal energy released (have a more positive value; see Figure 8.14 of the textbook). Since the energy change is always **exothermic**, the enthalpy change, ΔH, is negative. In other words, ΔH for this change is equal in magnitude but opposite in sign to electron affinity ($\Delta H = -$electron affinity).

In general, **electron affinity increases** going from left to right across a row of the period table, and **decreases** going down a group. Some elements have a negative electron affinity, meaning that the addition of an electron to atoms of those elements is an **endothermic** process. This usually occurs when an electron is added to an atom whose electrons completely fill or half-fill subshells (N and Groups IA, IIB, and 0).

For ions, **electron affinities** are negative (endothermic) for **anions** (since a negatively charged ion will repel the addition of a negatively charged electron) and positive (exothermic) for **cations** (since a positively charged ion will attract a negatively charged electron).

SAMPLE PROBLEM 8.6 A

▶ ▶ What is ΔH for the following changes? (a) $K(g) + e^- \rightarrow K^-(g)$ (b) $Zn(g) + e^- \rightarrow Zn^-(g)$ (c) $F(g) + e^- \rightarrow F^-(g)$?

Solution:

HINT: Look up the electron affinity values in Figure 8.14 of the textbook and use the equation $\Delta H = -$(electron affinity).

(a)   The electron affinity of K is 48 kJ/mol. This means that 48 kJ are **released** when 1 mol of gaseous K atoms gains 1 mol of electrons to form 1 mol of gaseous ions, K^-. Therefore, ΔH for this **exothermic** change is

$$\Delta H \ = \ -\text{(electron affinity)} \ = \ -48 \text{ kJ/mol}$$

(b)   The electron affinity of Zn is -47 kJ/mol. This means that 47 kJ are **gained or absorbed** when 1 mol of gaseous Zn atoms gains 1 mol of electrons to form 1 mol of gaseous ions, Zn^-. Therefore, ΔH for this **endothermic** change is

$$\Delta H \ = \ -\text{(electron affinity)} \ = \ -(-47 \text{ kJ/mol}) \ = \ 47 \text{ kJ/mol}$$

(c)   The electron affinity of F is 328 kJ/mol. This means that 328 kJ are **released** when 1

Therefore, ΔH for this **exothermic** change is

$$\Delta H \;=\; -(\text{electron affinity}) \;=\; -328 \text{ kJ/mol}$$

SAMPLE PROBLEM 8.6 B

▶▶ Using a periodic table only (NOT Figure 8.14), predict whether each of the following changes will be exothermic or endothermic and explain your reasoning. (a) $Ca(g) + e^- \rightarrow Ca^-(g)$ (b) $Xe(g) + e^- \rightarrow Xe^-(g)$ (c) $Al^{3+}(g) + e^- \rightarrow Al^{2+}(g)$ (d) $Cl^-(g) + e^- \rightarrow Cl^{2-}(g)$.

Solution:

(a) Endothermic. Ca (Z=20) has all its electrons completely filling subshells, $1s^2 2s^2 2p^6 3s^2 3p^6 4s^2$. The next electron added would have to occupy a higher energy 3d orbital.

(b) Endothermic. Xe (Z=54) is a Group 0 noble gas with all its electrons completely filling subshells, $1s^2 2s^2 2p^6 3s^2 3p^6 4s^2 3d^{10} 4p^6 5s^2 4d^{10} 5p^6$. The next electron added would have to occupy the considerably higher energy 6s orbital.

(c) Exothermic. A positively charged ion will attract a negatively charged electron.

(d) Endothermic. A negatively charged ion will repel a negatively charged electron. Also, the Cl^- ion is isoelectronic with the very stable noble gas Ar, both of which would require the next electron to occupy the considerably higher energy 4s orbital.

SELF-TEST QUESTIONS

Test your understanding of the concepts and skills in this chapter by working through the multiple choice questions **BEFORE** checking the answers. A periodic table will always be available. The ANSWERS TO SELF-TEST QUESTIONS follow immediately after this section.

1. All of the following electron configurations describe different excited states of a boron atom <u>except</u>

   A. $1s^2 2s^1 2p^2$ B. $1s^2 2p^3$ C. $1s^2 2s^2 2p^1$ D. $2s^2 2p^3$ E. $1s^2 2s^2 3s^1$

2. The electron configuration of a sodium atom in the ground state (always in order of increasing energy)

is

A. $1s^2 2s^2 3s^1 2p^6$ B. $1s^2 2s^2 2p^6 3s^1$ C. $1s2s2p3s$ D. $1s^2 2s^2 2p^7$ E. $1s^2 2s^2 2p^3 3s^1$

3. The atoms of which element have the following electron configuration in the ground state,

$$1s^2 2s^2 2p^5$$

A. N   B. Cl   C. C   D. O   E. F

4. The orbital diagram for a neon atom in the ground state is

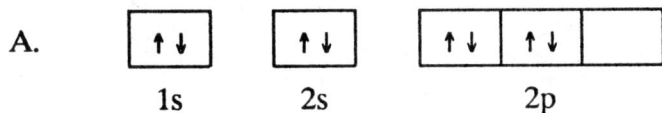

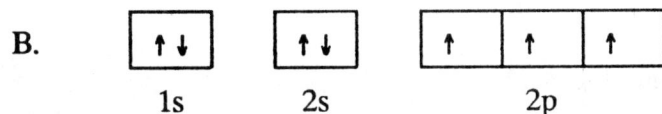

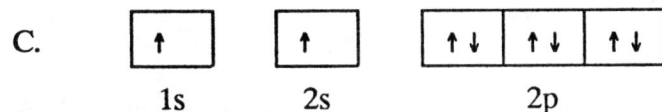

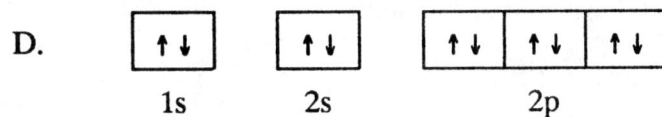

A.
| ↑↓ | ↑↓ | ↑↓ | ↑↓ | |
1s   2s   2p

B.
| ↑↓ | ↑↓ | ↑ | ↑ | ↑ |
1s   2s   2p

C.
| ↑ | ↑ | ↑↓ | ↑↓ | ↑↓ |
1s   2s   2p

D.
| ↑↓ | ↑↓ | ↑↓ | ↑↓ | ↑↓ |
1s   2s   2p

E. not given.

5. The number of unpaired electrons in the neon atom in the previous problem (#4) is

A. 0   B. 1   C. 2   D. 3   E. 4

6. The correct distribution of electrons in order of increasing energy for an atom of Mg in the ground state is

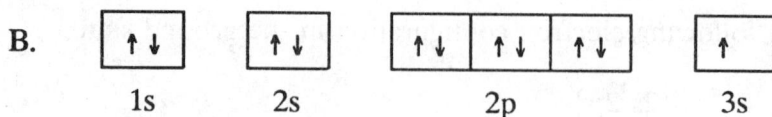

   A. $1s^2 2s^2 3s^2$

   B.

        1s   2s   2p   3s

   C. [He] $2s^2$

   D. [Ne] $3s^2$

   E. $1s^2 2s^2 2p^6 4s^2$

7. The ground-state electron configuration of Bi (in order of increasing energy) is

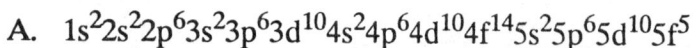

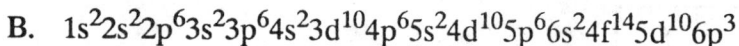

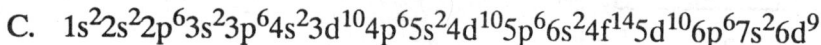

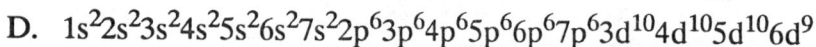

   A. $1s^2 2s^2 2p^6 3s^2 3p^6 3d^{10} 4s^2 4p^6 4d^{10} 4f^{14} 5s^2 5p^6 5d^{10} 5f^5$
   B. $1s^2 2s^2 2p^6 3s^2 3p^6 4s^2 3d^{10} 4p^6 5s^2 4d^{10} 5p^6 6s^2 4f^{14} 5d^{10} 6p^3$
   C. $1s^2 2s^2 2p^6 3s^2 3p^6 4s^2 3d^{10} 4p^6 5s^2 4d^{10} 5p^6 6s^2 4f^{14} 5d^{10} 6p^6 7s^2 6d^9$
   D. $1s^2 2s^2 3s^2 4s^2 5s^2 6s^2 7s^2 2p^6 3p^6 4p^6 5p^6 6p^6 7p^6 3d^{10} 4d^{10} 5d^{10} 6d^9$
   E. not given.

8. The following ground-state electron configuration correctly represents

$$1s^2 2s^2 2p^6 3s^2 3p^6 4s^2 3d^{10} 4p^4$$

   A. As^+ B. Br^- C. As   D. Se   E. Te

9. The correct ground-state electron configuration for Ni^{2+} (in order of increasing energy) is

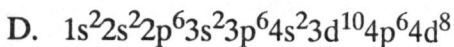

   A. $1s^2 2s^2 2p^6 3s^2 3p^6 3d^8$
   B. $1s^2 2s^2 2p^6 3s^2 3p^6 4s^2 3d^8$
   C. $1s^2 2s^2 3s^2 2p^6 3p^6 3d^8$
   D. $1s^2 2s^2 2p^6 3s^2 3p^6 4s^2 3d^{10} 4p^6 4d^8$
   E. not given.

10. The only element with the correct ground-state electron configuration (in order of increasing energy) is

A. Ba $1s^2 2s^2 2p^6 3s^2 3p^6 3d^{10} 4s^2 4p^6 4d^{10} 5s^2 5p^6 5d^{10} 6s^2$

B. Zn $1s^2 2s^2 2p^6 3s^2 3d^{10}$

C. I $1s^2 2s^2 2p^6 3s^2 3p^6 4s^2 3d^{10} 4p^6 5s^2 4d^{10} 5p^5$

D. Nb $1s^2 2s^2 2p^6 3s^2 3p^6 3d^{10} 4s^2 4p^6 5s^2 4d^3$

E. Co $1s^2 2s^2 2p^6 3s^2 3p^6 3d^7 4s^2$

11. The number of unpaired electrons in the ground state of the Fe^{2+} ion is

A. 0   B. 1   C. 2   D. 3   E. 4

12. The correct ground-state electron configuration for Mn^{3+} (in order of increasing energy) is

A. $1s^2 2s^2 2p^6 3s^2 3p^6 4s^2 3d^5$

B. $1s^2 2s^2 2p^6 3s^2 3p^6 3d^4$

C. $1s^2 2s^2 3s^2 4s^2 2p^6 3p^6 3d^5$

D. $1s^2 2s^2 2p^6 3s^2 3p^6 3d^5$

E. $1s^2 2s^2 2p^6 3s^2$

13. The species with the largest radius is

A. Cl^- B. Se^{2-} C. Br^- D. S^{2-} E. F^-

14. The ion with the same electron configuration as Se is

A. Kr^{2+} B. Rb^+ C. Br^- D. As^{3+} E. Sr^{2+}

15. The gaseous atom with the largest first ionization energy is

A. Cs   B. Mg   C. N   D. P   E. Al

16. Which of the following gaseous atoms would you predict to have the highest (most exothermic) electron affinity?

A. K   B. Br   C. P   D. S   E. Cl

17. ΔH for the following reaction is (use data from Figure 8.14 of textbook)

$$Ar(g) + e^- \rightarrow Ar^-(g)$$

A. -39 kJ/mol   B. 35 kJ/mol   C. 314 kJ/mol   D. -35 kJ/mol   E. 39 kJ/mol

ANSWERS TO SELF-TEST QUESTIONS

1. c	7. b	13. b
2. b	8. d	14. a
3. e	9. a	15. c
4. d	10. c	16. e
5. a	11. e	17. b
6. d	12. b	

PRACTICE PROBLEMS

Test your problem-solving skills in this chapter by working through the problems in the space provided **BEFORE** checking the answers. A periodic table will always be available. The ANSWERS TO PRACTICE PROBLEMS follow immediately after this section.

1.   For the P atom in the ground state, (a) sketch the energy level diagram, (b) write the electron configuration, (c) write the abbreviated electron configuration, and (d) write the orbital diagram.

2. For the bromine (Br) atom in its ground state, (a) write the electron configuration, (b) write the abbreviated electron configuration, (c) write the orbital diagram, and (d) indicate how many unpaired electrons are present.

3. For the europium (Eu) atom in its ground state, (a) write the electron configuration, (b) write the abbreviated electron configuration, (c) write the orbital diagram, and (d) indicate how many unpaired electrons are present.

4. For the strontium ion, Sr^{2+}, in the ground state, (a) write the electron configuration, and (b) give the abbreviated orbital diagram.

5. For the vanadium ion, V^{3+}, in the ground state, (a) write the electron configuration, and (b) give the abbreviated orbital diagram.

6. Arrange each series in order of increasing radius, (a) Si, Al, P   (b) Ar, S^{2-}, Cl^- (c) Na, Na^+

7. Indicate which species in each pair will require **less** energy to remove one electron, (a) Cl, Cl^- (b) Mg, Ca (c) O, S   (d) Na^+, K^+.

8. Using ONLY a periodic table (NOT Figure 8.14), predict whether each of the following changes will be exothermic or endothermic,

    (a) $O(g) + e^- \rightarrow O^-(g)$

    (b) $I(g) + e^- \rightarrow I^-(g)$

    (c) $Se^{1-}(g) + e^- \rightarrow Se^{2-}(g)$

    (d) $Ba^{2+}(g) + e^- \rightarrow Ba^{1+}(g)$.

ANSWERS TO PRACTICE PROBLEMS

1. (a)

$$\uparrow \quad \uparrow \quad \uparrow$$

$$\underline{\uparrow\downarrow}$$
3s

$$\underline{\uparrow\downarrow} \quad \underline{\uparrow\downarrow} \; \underline{\uparrow\downarrow} \; \underline{\uparrow\downarrow}$$
2s   2p

↑
Energy

$$\underline{\uparrow\downarrow}$$
1s

Ground state of P

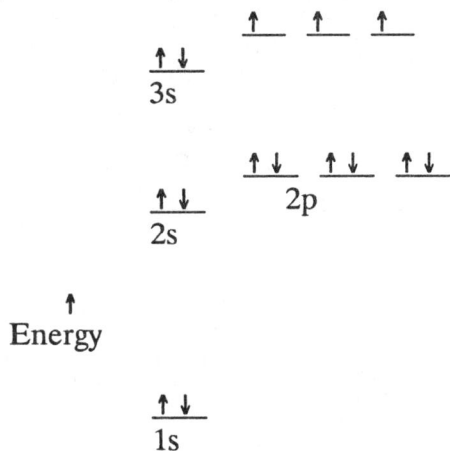

(b) P (Z=15) $1s^2 2s^2 2p^6 3s^2 3p^3$

(c) P (Z=15) [Ne] $3s^2 3p^3$

(d)

P [↑↓] [↑↓] [↑↓|↑↓|↑↓] [↑↓] [↑|↑|↑]
   1s   2s   2p   3s   3p

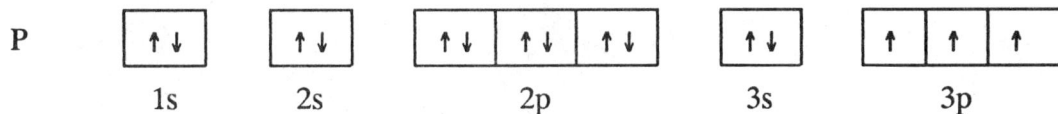

2. (a)   Br (Z=35) $1s^2 2s^2 2p^6 3s^2 3p^6 4s^2 3d^{10} 4p^5$

   or,   Br (Z=35) $1s^2 2s^2 2p^6 3s^2 3p^6 3d^{10} 4s^2 4p^5$

   (b)   Br (Z=35) [Ar] $4s^2 3d^{10} 4p^5$

   or,   Br (Z=35) [Ar] $3d^{10} 4s^2 4p^5$

   (c)

   Br [Ar] [↑↓|↑↓|↑↓|↑↓|↑↓] [↑↓] [↑↓|↑↓|↑]
                    3d   4s   4p

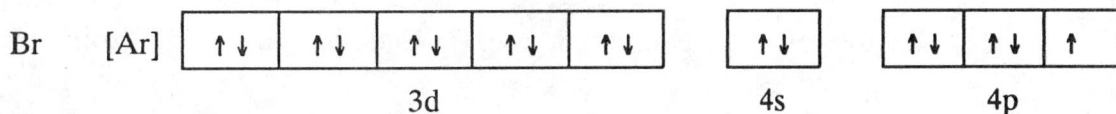

   (d)   1 unpaired electron; paramagnetic.

3.   (a)   Eu (Z=63) $1s^2 2s^2 2p^6 3s^2 3p^6 4s^2 3d^{10} 4p^6 5s^2 4d^{10} 5p^6 6s^2 4f^7$

       or,   Eu (Z=63) $1s^2 2s^2 2p^6 3s^2 3p^6 3d^{10} 4s^2 4p^6 4d^{10} 4f^7 5s^2 5p^6 6s^2$

   (b)   Eu (Z=63)   [Xe] $6s^2 4f^7$

       or,   Eu (Z=63)   [Xe] $4f^7 6s^2$

   (c)

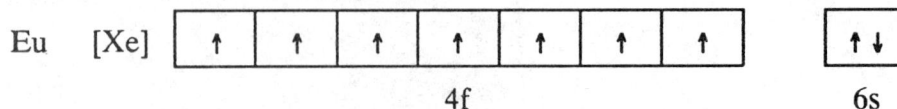

                                    4f   6s

   (d)   7 unpaired electrons; paramagnetic.

4.   (a) Sr^{2+} (36 e$^-$) $1s^2 2s^2 2p^6 3s^2 3p^6 3d^{10} 4s^2 4p^6$

       or, Sr^{2+} (36 e$^-$) [Kr]

   (b) Sr^{2+} (36 e$^-$) [Kr]   (NOTE: All higher energy orbitals are empty.)

         Diamagnetic; no unpaired electrons.

5.   (a) V^{3+} (20 e$^-$) $1s^2 2s^2 2p^6 3s^2 3p^6 3d^2$

       or, V^{3+} (20 e$^-$) [Ar] $3d^2$

   (b)

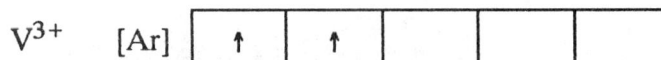

                                    3d

       Paramagnetic; 2 unpaired electrons.

6.   (a)   P < Si < Al   (b) Ar < Cl$^-$ < S^{2-} (c) Na$^+$ < Na

7.   (a)   Cl$^-$ (b) Ca   (c) S   (d) K$^+$

8.   (a)   Exothermic   (b) Exothermic   (c) Endothermic   (d) Exothermic

CHAPTER 9

CHEMICAL BONDS

SECTION 9.1   VALENCE ELECTRONS
SECTION 9.2   IONIC BONDS

After completing this section, you will be able to do the following:

■ Use electron configurations to show how ionic compounds are formed;

■ Given the formulas of a pair of ionic compounds, determine which compound has the higher lattice energy.

REVIEW

Chemical bonds are the forces that hold atoms together in molecules of elements, in compounds, and in metals. When chemical changes (reactions) occur, chemical bonds break in the reactants and form in the products. The two major types of chemical bonds are **ionic bonds** and **covalent bonds**, which are formed by the transfer and sharing of **valence electrons**, respectively. **Valence electrons** are the electrons in the **outermost, or valence, shell** -- that is, the electrons in the subshell(s) with the **highest** principal quantum number. For **main group elements** (IA-VIIA,0), the number of **valence electrons** is the **group number**. For example, all Group IA elements have **one** valence electron (or one electron in their outermost shell; see Figure 8.4 in the textbook). Since only the **valence electrons** are transferred or shared when ionic or covalent bonds are formed, the **chemical properties** of elements in a group are similar because these elements have the same number of **valence** electrons.

An **ionic bond** is the attraction between oppositely charged ions which are formed by the **transfer** of

valence electrons. Metals (with relatively low ionization energies) transfer electrons **to** non-metals (with high electron affinities) to form positive and negative ions, respectively. There are **no molecules** in ionic compounds, only alternating positive and negative ions in a three-dimensional array. The formula of an ionic compound is the lowest numerical ratio of positive and negative ions. The ionic compound potassium chloride has the formula KCl because there is one K^+ ion for every Cl^- ion. The ionic compound calcium chloride has the formula $CaCl_2$ because there are two Cl^- ions for every Ca^{2+} ion. The attraction between alternating positive and negative ions in ionic compounds is very strong, making them very hard with high melting points.

For main group elements, enough electrons are **transferred** so that each ion formed is **isoelectronic with the nearest noble gas**. This means that the metals in Groups IA and IIA **give up** (transfer) their valence electrons to form 1+ and 2+ ions, respectively, and the non-metals in Groups VIA and VIIA **gain** valence electrons (from the metals) to form 2– and 1– ions, respectively. For transition metals, the cations formed are usually not isoelectronic with the nearest noble gas owing to the varying number of d electrons in the next-to-outermost shell. As a result, transition metals typically form more than one cation.

The energy required to separate the ions of an ionic solid to an infinite distance is called the **lattice energy** of the solid. That is, the lattice energy of an ionic solid is a measure of the force of attraction between the ions in the solid. The stronger the ionic bonds, the higher the lattice energy and the harder (and higher melting) the solid. In general, **lattice energy** increases with increasing charge and decreasing size of the ion. Compounds of 2+ and 2– ions have very high lattice energies and are very stable.

SAMPLE PROBLEM 9.2 A

▶ ▶ (a) Use abbreviated electron configurations to show the formation of (i) lithium fluoride from lithium atoms and fluorine atoms, and (ii) strontium chloride from strontium atoms and chlorine atoms. (b) What are the formulas of lithium fluoride and strontium chloride?

Solution:

(a) (i) Write out the electron configurations for a Li atom and a F atom

$$[He]\, 2s^1 \; + \; [He]\, 2s^2 2p^5 \; \rightarrow$$

$$Li \; + \; F \; \rightarrow$$

For each atom to form an ion isoelectronic with the nearest very stable noble gas, Li (and all other Group IA atoms) will have to lose (transfer) its one valence electron to be isoelectronic with [He], and F (and all other Group VIIA atoms) will have to gain one electron to be isoelectronic with the noble gas [Ne]. In other words, a Li atom transfers its one valence electron to a F atom. Since a Li atom loses one electron (with no change in protons in the nucleus), a Li^+ ion is formed. And a F atom gains one electron to become a F^- ion. One Li^+ ion is needed for each F^- ion. Overall, all ionic compounds are always neutral.

$$[He]\, 2s^1 \;+\; [He]\, 2s^2 2p^5 \;\rightarrow\; [He] \;+\; [He]\, 2s^2 2p^6 \text{ or } [Ne]$$

$$Li \;+\; F \;\rightarrow\; Li^+ \;+\; F^-$$

It is the attraction between the oppositely charged ions, Li^+ and F^-, that is an ionic bond.

(ii)   Similarly, for strontium chloride,

$$[Kr]\, 5s^2 \;+\; [Ne]\, 3s^2 3p^5 \;\rightarrow$$

$$Sr \;+\; Cl \;\rightarrow$$

For each atom to form an ion isoelectronic with the nearest very stable noble gas, Sr (and all other Group IIA atoms) will have to lose (transfer) its two valence electrons to be isoelectronic with [Kr], but Cl (and all other Group VIIA atoms) only needs to gain **one** electron to be isoelectronic with the noble gas [Ar]. Since the total number of electrons lost by the metal atom(s) **must equal** the total number of electrons gained by the non-metal atom(s), **two** Cl atoms (each gaining one electron) must react with **one** Sr atom (which gives up **two** electrons) to form the neutral ionic compound strontium chloride.

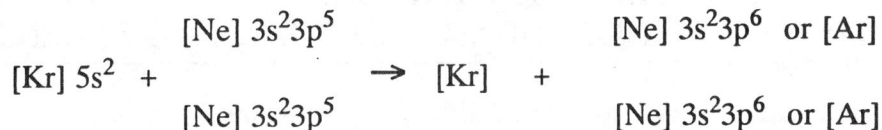

$$[Kr]\, 5s^2 \;+\; \begin{matrix}[Ne]\, 3s^2 3p^5 \\ [Ne]\, 3s^2 3p^5\end{matrix} \;\rightarrow\; [Kr] \;+\; \begin{matrix}[Ne]\, 3s^2 3p^6 \text{ or } [Ar] \\ [Ne]\, 3s^2 3p^6 \text{ or } [Ar]\end{matrix}$$

$$Sr \;+\; 2Cl \;\rightarrow\; Sr^{2+} \;+\; 2Cl^-$$

(b)   The formulas for lithium fluoride and strontium chloride are LiF and $SrCl_2$, respectively.

SAMPLE PROBLEM 9.2 B

▶▶ In each of the following pairs of solid ionic compounds, which compound has the higher lattice energy? Explain! (a) NaCl, KCl (b) Na_2O, NaF (c) BaS, CsCl.

(a) The **lattice energy** of a solid is the energy required to separate the ions of an ionic solid to an infinite distance. The stronger the ionic bonds, the higher the lattice energy. In general, **lattice energy** increases with increasing charge and decreasing size of the ion. Compounds of 2+ and 2− ions have very high lattice energies.

In NaCl and KCl, the cations have the same charge, and the anions are identical. So the only difference between the two compounds is the size of the Group IA cations. Ionic radii **decrease** as you go up a group. The Na^+ ion is smaller than the K^+ ion. Therefore, the lattice energy is higher for NaCl than KCl.

(b) In Na_2O and NaF, the cations are identical. The F^- lies further to the right in the second period and is slightly smaller than O^{2-}, favoring a higher lattice energy for NaF. But the O^{2-} ion, with its higher charge, predominates. Therefore, the lattice energy is higher for Na_2O.

(c) In BaS and CsCl, the cations and anions have different charges and sizes. The important point to remember is that 2+ and 2− ions have **very high lattice energies**. Since BaS has 2+ and 2− ions and CsCl only has 1+ and 1− ions, the lattice energy will be higher for BaS.

SECTION 9.3	SHOWING MOLECULAR STRUCTURE WITH LEWIS FORMULAS
SECTION 9.4	NONPOLAR AND POLAR COVALENT BONDS

After completing this section, you will be able to do the following:

- Define and identify nonpolar and polar covalent bonds;

- Write Lewis (electron-dot) formulas for covalently bonded compounds.

REVIEW

The second major type of **chemical bond** is the **covalent bond**, where an element achieves a stable noble gas electron configuration by **sharing** (NOT transferring) a pair of valence electrons. A **covalent bond** most often forms when two atoms approach each other, each with a half-filled valence orbital but neither atom able to transfer (or cause a transfer) of an electron from one atom to the other (which would be an ionic bond). The half-filled orbitals overlap sufficiently until a **pair of valence electrons (one from each atom) is now shared between the nuclei of the two atoms.** A **covalently bonded molecule** is then formed.

Molecular compounds consist of covalently bonded molecules. **Ionic compounds consist of ions.** The shared pair of electrons in a molecule is counted with each atom forming the bond for purposes of becoming isoelectronic with the nearest noble gas. The electron pair shared between two atoms in a molecule is called a **bonding pair** of electrons. Valence electrons around each atom in a molecule that are **not** shared are called **unshared pairs**, or **lone pairs**, or **nonbonding pairs** of electrons. If an electron pair is **shared equally** between two atoms (such as two atoms of the same element), this is a **nonpolar bond**. If the electron pair is **shared unequally** between two atoms (such as atoms of two different elements), the bond is called a **polar covalent** bond.

All the **valence** electrons are shown in a covalently bonded molecule (whether the bonds are polar or nonpolar) by using a shorthand notation called a **Lewis (or electron-dot) formula**. With this formula, the core of the atom (nucleus and inner electrons) are represented by the symbol for the element, with a dot used to represent each **valence electron**. More conveniently, a **dash** is used to represent a **pair** of valence electrons, **both bonding and nonbonding**. FROM THIS POINT ON IN THE STUDY GUIDE, all Lewis formulas will be shown with a **dash representing a pair of valence electrons**; that is,

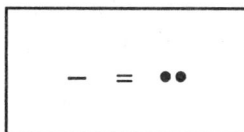

$$-\ =\ \bullet\bullet$$

Normally, **Lewis formulas** are used only for **main group** elements, in which case there will usually be eight **(an octet)** of outer s and p subshell electrons (corresponding to the outer s and p subshells of the noble gases) around each atom in a molecule, except for H, which can only have two electrons around it (corresponding to the filled 1s subshell of H). You **must** learn the following procedures to write correct **Lewis or electron-dot formulas** to account for all the valence electrons in a molecule **and** to use in predicting the shapes of molecules (Section 9.11). Shape in turn is a factor in determining physical and chemical properties.

WRITING LEWIS (ELECTRON-DOT) FORMULAS

1. Write a structural formula for the molecule or ion given. [Usually, the **central atom** (center of the molecule) will be obvious or identified, around which all the **outer atoms** are placed. Hydrogen is never a central atom.]

2. Calculate the total number of valence electrons (the group numbers for main group elements). For anions, add one electron for each negative charge. For cations, subtract one electron for each positive charge.

3. Place one pair of electrons between each pair of bonded atoms.

4. Use the remaining valence electrons in pairs to fill the octets (except H) of the **outer atoms first,** then around the **central atom.** If there are just enough valence electrons so that every atom has an octet of electrons (except H), then the Lewis formula is complete. If **not,** then follow step 5 OR 6, not both. Ions must show brackets around them with the overall charge.

5. If there are extra valence electrons after filling the octets of all the atoms, place them around the **central atom** only. [Remember, central atoms of the third period elements and beyond can have more than 8 electrons around them due to the availability of empty d and f subshells.]

6. If not enough electrons are available to give all the atoms except hydrogen an octet, move **unshared (nonbonding) pairs** of the outer atoms to form double or triple bonds (that is, two or three pairs of shared electrons, respectively). **NOTE:** As central atoms, Be has a maximum of 4 electrons, B has a maximum of 6 electrons in molecules, and other Group IIIA elements may have less than 8 electrons also.

SAMPLE PROBLEM 9.3 A

▸▸ For each of the following species, (a) write the Lewis (electron-dot) formula, (b) indicate the number of covalent bonds present and whether they are polar or nonpolar, and (c) indicate the number of nonbonding (unshared) pairs of electrons.

$$H_2O \qquad SiCl_4 \qquad Cl_2$$

Solution:

HINT: Follow the procedures given in the REVIEW for writing Lewis structures.

$\underline{\underline{H_2O}}$

(a) Step 1: H is never a central atom. The Lewis formula does not indicate the shape of the molecule (that will be discussed in Section 9.11). The structural formula is

H O H

Step 2: Since the group numbers of the main group elements give the number of valence electrons for elements in that group,

$$2\,H \times 1 \text{ valence } e^- = 2\,e^-$$
$$1\,O \times 6 \text{ valence } e^- = \underline{6\,e^-}$$
$$8\,e^-, \text{ or 4 pairs of electrons}$$

Step 3: Place one pair of e^- between each pair of bonded atoms.

H–O–H

Step 4: Place the two remaining pairs around the central atom O, since H can never have more than 1 pair of electrons around it.

H–$\overline{\underline{O}}$–H

CHECK: All valence electrons are accounted for; O has its octet and each H has 2 electrons. The Lewis structure is complete.

(b) There are two covalent bonds (O–H), and they are both polar covalent (an electron pair shared between atoms of different elements).

(c) There are two nonbonding pairs of valence electrons (around the O atom).

SiCl$_4$

(a) Step 1: In a formula with two elements, the central atom is normally the element with only one atom present. The structural formula is

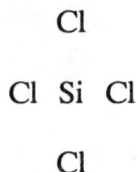

$$Cl$$

$$Cl \ \ Si \ \ Cl$$

$$Cl$$

Step 2: The total number of valence electrons is

$$4 \ Cl \ x \ 7 \ valence \ e^- \ = \ 28 \ e^-$$
$$1 \ Si \ x \ 4 \ valence \ e^- \ = \ \underline{4 \ e^-}$$
$$\overline{32 \ e^-, \ or \ 16 \ pairs \ of \ electrons}$$

Step 3: Place one pair of e$^-$ between each pair of bonded atoms.

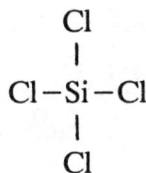

$$\begin{array}{c} Cl \\ | \\ Cl-Si-Cl \\ | \\ Cl \end{array}$$

Step 4: Place the 12 remaining pairs around the outer Cl atoms, so that each Cl has an octet of electrons around it.

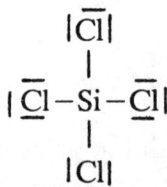

$$\begin{array}{c} |\overline{Cl}| \\ | \\ |\overline{Cl}-Si-\overline{Cl}| \\ | \\ |\underline{Cl}| \end{array}$$

CHECK: All 16 pairs of valence electrons are accounted for; all five atoms each have an octet of electrons. The Lewis structure is complete.

(b) There are four covalent bonds (Si$-$Cl), and all are polar covalent (an electron pair shared between atoms of different elements).

(c) There are 12 nonbonding pairs of valence electrons (around the 4 Cl atoms).

$\underline{\underline{Cl_2}}$

(a) Step 1: In a formula with only one element, there is no central atom. The structural formula is

Cl Cl

Step 2: The total number of valence electrons is

2 Cl x 7 valence e⁻ = 14 e⁻, or 7 pairs of electrons

Step 3: Place one pair of e⁻ between the Cl atoms.

Cl−Cl

Step 4: Place the 6 remaining pairs around the two Cl atoms.

$|\overline{\underline{Cl}}-\overline{\underline{Cl}}|$

CHECK: All 7 pairs of valence electrons are accounted for; both Cl atoms each have an octet of electrons. The Lewis structure is complete.

(b) There is one covalent bond (Cl−Cl), and it is nonpolar covalent (two atoms of the same element).

(c) There are 6 nonbonding pairs of valence electrons (around the 2 Cl atoms).

SECTION 9.5   ELECTRONEGATIVITY
SECTION 9.6   MORE ABOUT WRITING LEWIS FORMULAS

After completing this section, you will be able to do the following:

■ Predict the increasing polarity of bonds by knowing the general trends for electronegativity in the periodic table.

■ Write Lewis formulas for covalently bonded species which are ions and/or contain multiple bonds.

282

REVIEW

Electronegativity is the ability of an atom to attract to itself an electron pair that is shared with another atom. The general trends in the periodic table are that electronegativity increases both as you go across a period from left to right and as you go up a group. Overall, electronegativity increases from the lower-left corner of the periodic table to the upper-right corner (see Figure 9.3 in the textbook). The **difference in electronegativity** of the atoms joined by a covalent bond is what's important. If the difference in electronegativity between two atoms forming a covalent bond is zero, then both atoms have an equal attraction for the shared pair, which is the definition of a **nonpolar covalent bond**. The greater the difference in electronegativity (always > 0) between bonded atoms, the more **polar** the bond becomes. The **negative end** of that bond is the more electronegative element. If the difference in electronegativity becomes large enough, electrons are transferred from the less electronegative to the more electronegative atom and the bond is **ionic**. As a guide to predicting whether bonds will be mainly ionic or polar,

- Most **metal fluorides and oxides** are mainly **ionic**.

- Compounds of metals in Group **IA**, in Group **IIA** (except Be and Mg), and in the **lanthanide series** with nonmetals are mainly **ionic**.

- Compounds consisting solely of **nonmetals** are mainly **covalent**.

You will begin to write Lewis formulas for more complex molecules and ions, which may include multiple bonds and/or polyatomic ions. **Polyatomic ions** contain **two or more atoms** covalently bonded together. There are many ionic compounds which contain covalently bonded **polyatomic ions**, such as Na_2CO_3, $Ca(NO_3)_2$, and $CuSO_4$, where the cations are **monatomic** and the anions are **polyatomic**.

SAMPLE PROBLEM 9.6 A

▶▶ Of the following bonds, (a) which is nonpolar? (b) Only one bond is ionic. Which one is it? (c) Arrange the bonds in order of increasing polarity. (d) Which end of the bond is negative in each case?

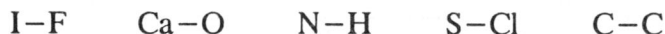

$$I-F \qquad Ca-O \qquad N-H \qquad S-Cl \qquad C-C$$

Solution:

(a)   C—C is nonpolar, since two identical atoms (same element) will always share an electron pair equally.

(b)   You can use the generalizations given in the REVIEW above, or choose the bond with the **greatest difference** in electronegativity by using the following data from Figure 9.3 in the textbook,

Bond	Difference in electronegativity
I−F	4.1 − 2.2 = 1.9
Ca−O	3.5 − 1.0 = 2.5
N−H	3.1 − 2.2 = 0.9
S−Cl	2.8 − 2.4 = 0.4
C−C	2.5 − 2.5 = 0.0

By either method, the one ionic bond must be Ca−O.

NOTE: The difference in electronegativity is always a **positive** number. Simply subtract the smaller electronegativity value from the larger value (regardless of which element appears first in the bond or its location in the periodic table).

(c) Polarity increases as the difference in electronegativity increases. From the data in part (b), the order of increasing polarity is

C−C < S−Cl < N−H < I−F < Ca−O

(d) The negative end of the bond is the end with the electronegative atom. Using either the general trends in electronegativity in the periodic table (or, alternatively, the actual electronegativity values from Figure 9.3 of the textbook),

I−F	F, which is higher in Group VIIA than I.
Ca−O	O, which is closer to the upper-right corner than Ca.
N−H	N, which is closer to the upper-right corner than H.
S−Cl	Cl, which is further to the right in the third period.
C−C	There is no negative end, since this bond is nonpolar.

SAMPLE PROBLEM 9.6 B

▶▶ Write the Lewis (electron-dot) formula for (a) SO_4^{2-}, (b) NO_2^+, and (c) Na_3PO_4.

Solution:

HINT: See the REVIEW in Section 9.3/9.4 and Sample Problem 9.3 A for procedure.

(a) Step 1: In a formula with two elements (molecule or ion), the central atom is normally the element with only one atom present. The structural formula is

284

$$O$$

$$O \quad S \quad O$$

$$O$$

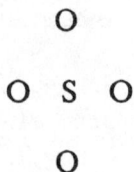

Step 2: The total number of valence electrons (including 1 electron for each negative charge on the ion) is

$$4 \, O \times 6 \text{ valence } e^- = 24 \, e^-$$
$$1 \, S \times 6 \text{ valence } e^- = 6 \, e^-$$
$$-2 \text{ charge on ion} = \underline{2 \, e^-}$$
$$32 \, e^-, \text{ or 16 pairs of electrons}$$

Step 3: Place one pair of e^- between each pair of bonded atoms.

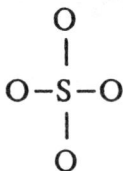

$$\begin{array}{c} O \\ | \\ O-S-O \\ | \\ O \end{array}$$

Step 4: Place the 12 remaining pairs around the outer O atoms, so that each O has an octet of electrons around it.

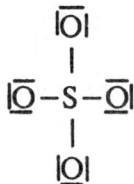

$$\begin{array}{c} |\overline{O}| \\ | \\ |\overline{O}-S-\overline{O}| \\ | \\ |\underline{O}| \end{array}$$

NOTE: As stated in the Section 9.3/9.4 REVIEW, a **dash** represents a pair of electrons, **bonding and nonbonding**.

CHECK: All 16 pairs of valence electrons are accounted for; all five atoms each have an octet of electrons. But the Lewis structure is **NOT** complete yet. This is an ion; there must be brackets around the entire ion with the overall charge shown before it is complete. The correct Lewis structure for SO_4^{2-} is

$$\left[\begin{array}{c} |\overline{O}| \\ | \\ |\overline{O}-S-\overline{O}| \\ | \\ |\underline{O}| \end{array}\right]^{2-}$$

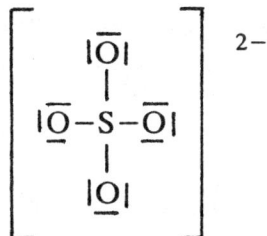

(b) Step 1: In a formula with two elements (molecule or ion), the central atom is normally the element with only one atom present. The structural formula is

$$O \quad N \quad O$$

Step 2: The total number of valence electrons (including **subtracting** one electron for each plus charge on the ion) is

$$
\begin{aligned}
2 \text{ O x 6 valence } e^- &= 12 \text{ } e^- \\
1 \text{ N x 5 valence } e^- &= 5 \text{ } e^- \\
+1 \text{ charge on ion} &= -1 \text{ } e^- \\
\hline
&16 \text{ } e^-, \text{ or 8 pairs of electrons}
\end{aligned}
$$

Step 3: Place one pair of e^- between each pair of bonded atoms.

$$O-N-O$$

Step 4: Place the 6 remaining pairs around the outer O atoms, so that each O has an octet of electrons around it.

$$|\overline{\underline{O}}-N-\overline{\underline{O}}|$$

At this point, all 8 pairs of valence electrons are accounted for, but N does NOT have an octet. To complete the Lewis formula, you will need to move unshared pairs of the outer atoms to form double or triple bonds (which is Step 6 in the procedure given in the REVIEW in Section 9.3/9.4).

Step 6: One option is to move an unshared pair from each O atom to form a second bond (now a double bond) between each O atom and the N atom.

$$\overline{\underline{O}}=N=\overline{\underline{O}}$$

No electrons have been added; **unshared** pairs around the O atoms have been converted into shared pairs around the O atoms. By so doing, the N **and** the two O atoms have an octet around them. But the formula is still not complete, because this is an ion. For ions, there **must** be brackets and an overall charge shown. The correct Lewis formula for NO_2^+ is

$$[\bar{\underline{O}}=N=\bar{\underline{O}}]^+$$

(c)   Step 1:   From earlier discussions, Na_3PO_4 is an ionic compound consisting of three monatomic sodium ions, Na^+, and one polyatomic phosphate ion, PO_4^{3-}. The Lewis formula for this compound **must** show the separate ions, since Na^+ is ionically bonded to PO_4^{3-}. For any monatomic ion, the Lewis formula is just the formula of the ion with its charge. For Na^+, the Lewis formula is

$$Na^+$$

In polyatomic ions, the atoms are covalently bonded. For PO_4^{3-}, the Lewis formula would be obtained just like SO_4^{2-} in part (a). The structural formula for PO_4^{3-} is

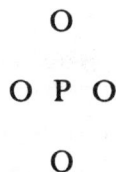

$$O$$

$$O \; P \; O$$

$$O$$

Step 2:   The total number of valence electrons (including 1 electron for each negative charge on the ion) is

$$
\begin{array}{lll}
4\ O \times 6\ \text{valence } e^- & = & 24\ e^- \\
1\ P \times 5\ \text{valence } e^- & = & 5\ e^- \\
-3\ \text{charge on ion} & = & \underline{3\ e^-} \\
& & 32\ e^-, \ \text{or 16 pairs of electrons}
\end{array}
$$

Step 3:   Place one pair of e^- between each pair of bonded atoms.

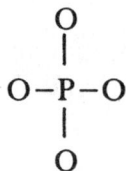

$$
\begin{array}{c}
O \\
| \\
O-P-O \\
| \\
O
\end{array}
$$

Step 4:   Place the 12 remaining pairs around the outer O atoms, so that each O has an octet of electrons around it.

$$
\begin{array}{c}
|\overline{O}| \\
| \\
|\overline{O}-P-\overline{O}| \\
| \\
|\underline{O}|
\end{array}
$$

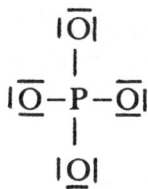

CHECK: All 16 pairs of valence electrons are accounted for; all five atoms each have an octet of electrons. Since this is an ion, there must be brackets around the entire ion with the overall charge shown before it is complete. The Lewis formula for PO_4^{3-} is

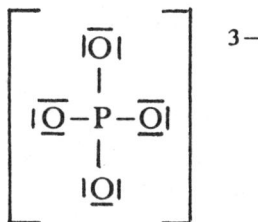

$$
\left[
\begin{array}{c}
|\overline{O}| \\
| \\
|\overline{O}-P-\overline{O}| \\
| \\
|\underline{O}|
\end{array}
\right]^{3-}
$$

The complete Lewis formula for Na_3PO_4 is

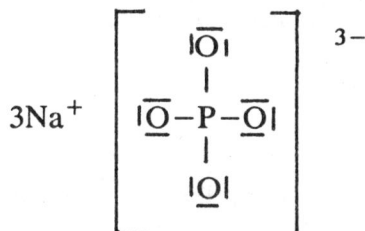

$$
3Na^+ \left[
\begin{array}{c}
|\overline{O}| \\
| \\
|\overline{O}-P-\overline{O}| \\
| \\
|\underline{O}|
\end{array}
\right]^{3-}
$$

SECTION 9.7   BOND LENGTH, BOND ENERGY, AND BOND ORDER

After completing this section, you will be able to do the following:

■ Given Lewis formulas of different compounds containing various covalent bonds between atoms of the same element, determine bond order, relative bond length, and relative bond energies.

REVIEW

Bond order is the number of bonds between two atoms in a molecule or ion. For a **single bond** (one shared pair of electrons) the bond order is **one**, for a **double bond** (two shared pairs of electrons) the bond order is **two,** and for a **triple bond** (three shared pairs of electrons) the bond order is **three. The bond length** is the distance between the centers of two bonded atoms. The **bond energy** is the average energy needed to break a given type of bond. As the **bond order increases,** the **bond length decreases,** and the **bond energy increases.**

SAMPLE PROBLEM 9.7 A

▸▸ Of the following carbon-oxygen bonds,

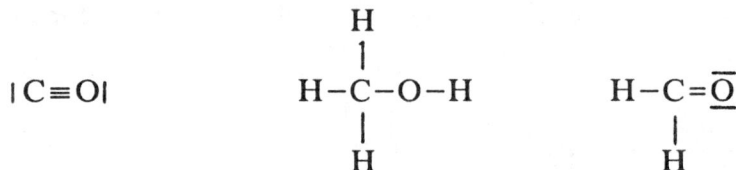

$$|C\equiv O| \qquad\qquad \begin{matrix} H \\ | \\ H-C-O-H \\ | \\ H \end{matrix} \qquad\qquad \begin{matrix} H-C=\overline{\underline{O}} \\ | \\ H \end{matrix}$$

(a) Which is the weakest? (b) Which is the shortest? (c) Which has a bond order of one?

Solution:

(a)   The carbon-oxygen bond in CH_3OH is the weakest, because the bond energy is the lowest for a carbon-oxygen single bond.

(b)   The carbon-oxygen bond in CO is the shortest, because it has the highest bond order, three.

(c)   The carbon-oxygen bond in CH_3OH is a single bond, which is always first order.

SECTION 9.8   FORMAL CHARGES

After completing this section, you will be able to do the following:

■ When several Lewis formulas are possible for a molecule or ion, **calculate formal charges on each atom to decide which formula is best.**

REVIEW

To calculate the **formal charge** (an electron bookkeeping technique not to be confused with the **actual** charge of an ion) on an atom, assume that all shared pairs of electrons are shared **equally** between the two atoms joined by the bond. Then calculate the formal charge,

Formal charge = no. valence e^- in free atom $-$ [1/2(no. valence e^- in shared pairs) + no. valence e^- in unshared pairs].

Write the formal charges over the atom to which they apply and circle them. To decide which of several possible Lewis formulas is best,

- A Lewis formula with formal charges of **zero is preferable**. Small formal charges are preferable to large formal charges.
- Lewis formulas with **negative formal charges on the more electronegative atom** are more likely than Lewis formulas with negative formal charges on the less electronegative atom.
- Lewis formulas with **unlike charges close together are more likely** than Lewis formulas with opposite charges widely separated.
- Lewis formulas with **like charges on adjacent atoms are very unlikely**.

SAMPLE PROBLEM 9.8 A

▶▶ Which of the following structures for NO_2^+ is better?

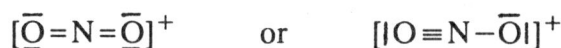

$$[\overline{O}=N=\overline{O}]^+ \qquad \text{or} \qquad [|O\equiv N-\overline{O}|]^+$$

Solution:

Using the procedure for calculating formal charges given above, the formal charges on the atoms of the **first** structure (left to right) are:

$$O = 6\ e^- \text{ (Grp VIA)} - [1/2(4 \text{ shared } e^-) + 4 \text{ unshared } e^-] = 0$$

$$N = 5\ e^- \text{ (Grp VA)} - [1/2(8 \text{ shared } e^-) + 0 \text{ unshared } e^-] = +1$$

$$O = 6\ e^- \text{ (Grp VIA)} - [1/2(4 \text{ shared } e^-) + 4 \text{ unshared } e^-] = 0$$

The formal charges on the atoms in the **second** structure (left to right) are:

$$O = 6\ e^- \text{ (Grp VIA)} - [1/2(6 \text{ shared } e^-) + 2 \text{ unshared } e^-] = +1$$

$$N = 5\ e^- \text{ (Grp VA)} - [1/2(8 \text{ shared } e^-) + 0 \text{ unshared } e^-] = +1$$

$$O = 6\ e^- \text{ (Grp VIA)} - [1/2(2 \text{ shared } e^-) + 6 \text{ unshared } e^-] = -1$$

The two Lewis structures with formal charges for NO_2^+ are

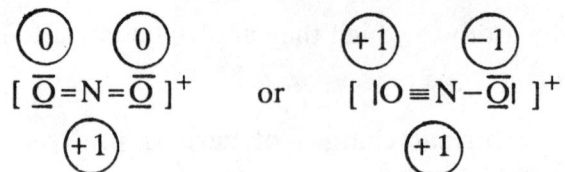

$$[\overline{O}=N=\overline{O}]^+ \quad \text{or} \quad [IO \equiv N - \overline{O}I]^+$$

The structure on the left is the best because it has the most formal charges of zero. In addition, the structure on the right has unlikely adjacent like charges.

NOTE: The formal charges must add up to zero for molecules and to the charge for ions. Formal charges of zero are often not shown. If a nonzero formal charge is shown on one or more atoms in a molecule or ion, then those atoms with no formal charge shown are assumed to be zero.

SAMPLE PROBLEM 9.8 B

▶▶ Use formal charges to decide between two possible arrangements of atoms in the NO_2^- ion.

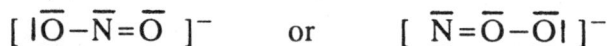

$$[I\overline{O}-\overline{N}=\overline{O}]^- \quad \text{or} \quad [\overline{N}=\overline{O}-\overline{O}I]^-$$

Solution:

See Sample Problem 9.8 A for procedure to calculate formal charges. The two structures with formal charges are

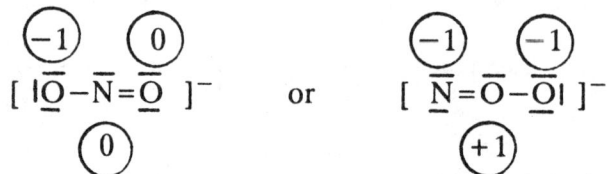

$$[I\overline{O}-\overline{N}=\overline{O}]^- \quad \text{or} \quad [\overline{N}=\overline{O}-\overline{O}I]^-$$

The left structure is the best because it has the most zero formal charges.

SECTION 9.9   RESONANCE STRUCTURES

After completing this section, you will be able to do the following:

■ Write resonance structures when they exist for molecules or ions.

REVIEW

Sometimes a single Lewis structure does not accurately reflect the actual bonding present in the molecule or ion. This is the case for the carbonate ion, CO_3^{2-},

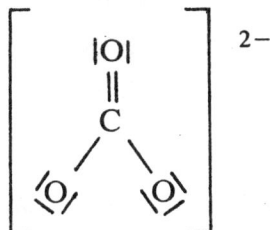

There is no reason why the double bond cannot be located between the carbon atom and each of the other two oxygen atoms. These other two Lewis structures for CO_3^{2-} along with the one drawn above are called **resonance structures** -- that is, all the atoms are shown in the **same position**, and only the electron arrangement changes. A double-headed arrow is used between resonance structures.

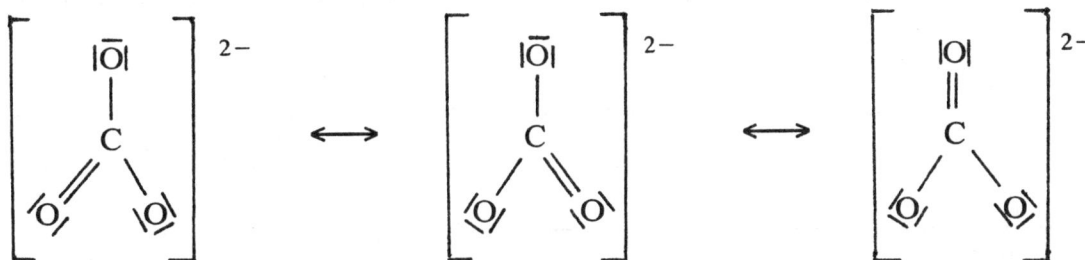

There are no **single or double carbon-oxygen bonds in CO_3^{2-}**. All three carbon-oxygen bonds are identical and have a length between that of a single and double bond. That is, all three bonds have an identical bond order somewhere between 1 and 2. The CO_3^{2-} ion is a **hybrid** of the three structures; no one structure accurately reflects the bonding in this ion. **Resonance structures** are needed when there is **more than one way to place unshared pairs to form multiple bonds.**

SAMPLE PROBLEM 9.9 A

▶▶ (a) Write the resonance structures for H_2O and SO_2, including formal charges on each atom.
(b) What are the bond orders in each case?

Solution:

(a) For H_2O, there is only **one** Lewis structure. There are no resonance structures because there is no other arrangement of electrons possible (no multiple bond present). For H_2O, the Lewis structure is

$$H - \overline{\underline{O}} - H$$

For SO_2, there are **three** ways to arrange the electrons to form double bonds, so there are **three** resonance structures,

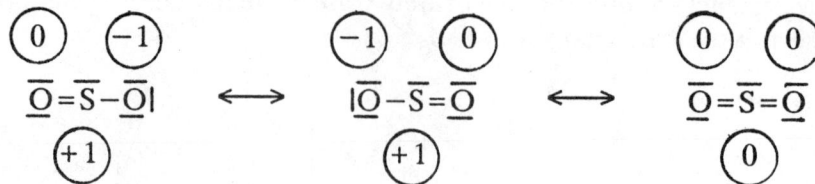

$$\overset{\text{(0)}\quad\text{(−1)}}{\overline{O}=\overline{S}-\overline{\underline{O}}|}\underset{\text{(+1)}}{} \quad \longleftrightarrow \quad \overset{\text{(−1)}\quad\text{(0)}}{|\overline{\underline{O}}-\overline{S}=\overline{O}}\underset{\text{(+1)}}{} \quad \longleftrightarrow \quad \overset{\text{(0)}\quad\text{(0)}}{\overline{O}=\overline{S}=\overline{O}}\underset{\text{(0)}}{}$$

> **NOTE:** The first two resonance structures obey the octet rule. Since sulfur is in the third period of the periodic table and can have more than eight electrons around it, a third resonance structure is possible (which with its zero formal charges turns out to best represent the structure of SO_2). [See Section 9.10 for other exceptions to the octet rule.]

(b)   The bond order is 1 for both $H-O$ bonds in H_2O. The bond order is the same for both $S-O$ bonds in the first two structures, between 1 and 2; the bond order for both $S-O$ bonds is 2 in the third structure, which best represents the structure of SO_2 because of zero formal charges.

SECTION 9.10   EXCEPTIONS TO THE OCTET RULE

After completing this section, you will be able to do the following:

■   Write Lewis formulas for covalently bonded species with either more or less than eight electrons around the central atom.

REVIEW

Central atoms beyond the second period elements can have **more than eight electrons** around them (bonding or nonbonding) due to the availability of empty d and f subshells. Some central atoms have fewer than eight electrons around them; in particular, Be has a maximum of 4 electrons, B has a maximum of 6 electrons in molecules, and other Group IIIA elements may have less than 8 electrons also.

Carefully review the **procedures for writing Lewis structures** given in the REVIEW in Section 9.3/9.4 of this study guide and at the end of Section 9.10 in the textbook. You should now be able to write the Lewis structure for many molecules and ions, by learning and applying those procedures.

SAMPLE PROBLEM 9.10 A

▸▸ Write the Lewis structures for (a) SF_4 (b) XeF_4.

HINT: Use the procedures given in the REVIEW in Section of 9.3/9.4 of this study guide and at the end of Section 9.10 in the textbook. Also see Sample Problems 9.3 A and 9.6 B.

(a)   Step 1:   In a formula with two elements (molecule or ion), the central atom is normally the element with only one atom present. The structural formula for SF_4 is

F

F S F

F

Step 2:   The total number of valence electrons (including 1 electron for each negative charge on the ion) is

$$4 \text{ F x 7 valence e}^- = 28 \text{ e}^-$$
$$1 \text{ S x 6 valence e}^- = \underline{6 \text{ e}^-}$$
$$34 \text{ e}^-, \text{ or 17 pairs of electrons}$$

Step 3:   Place one pair of e^- between each pair of bonded atoms.

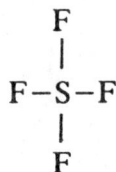

F
|
F−S−F
|
F

Step 4: Place three more pairs of electrons around each F atom to reach an octet.

$$
\begin{array}{c}
\text{|}\overline{\text{F}}\text{|} \\
\text{|} \\
\text{|}\overline{\text{F}}-\text{S}-\overline{\text{F}}\text{|} \\
\text{|} \\
\text{|}\underline{\text{F}}\text{|}
\end{array}
$$

BE CAREFUL! Even though all five atoms have an octet of electrons around them, there is still one pair of valence electrons left over. Excess electrons **always** go around the central atom (see Step 5 of the Procedures for Writing Lewis Formulas in Section 9.3/9.4).

Step 5: The correct Lewis structure for SF_4, with all valence electrons accounted for, is

$$
\begin{array}{c}
\text{|}\overline{\text{F}}\text{|} \\
\text{|} \\
\text{|}\overline{\text{F}}-\text{S}\overset{\cdot\cdot}{-}\overline{\text{F}}\text{|} \\
\text{|} \\
\text{|}\underline{\text{F}}\text{|}
\end{array}
$$

NOTE: The central atom has 10 electrons around it, 4 bonding pairs and 1 nonbonding pair of electrons.

(b) Step 1: Again, in a formula with two elements (molecule or ion), the central atom is normally the element with only one atom present. The structural formula for XeF_4 is

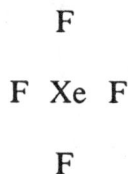

$$
\begin{array}{c}
\text{F} \\
\\
\text{F Xe F} \\
\\
\text{F}
\end{array}
$$

Step 2: The total number of valence electrons (including 1 electron for each negative charge on the ion) is

$$4 \text{ F x 7 valence } e^- = 28 \text{ } e^-$$
$$1 \text{ Xe x 8 valence } e^- = 8 \text{ } e^-$$
$$\overline{}$$
$$36 \text{ } e^-, \text{ or 18 pairs of electrons}$$

Step 3:   Place one pair of e^- between each pair of bonded atoms.

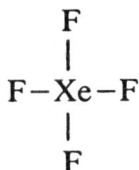

```
      F
      |
  F—Xe—F
      |
      F
```

Step 4:   Place three more pairs of electrons around each F atom to reach an octet.

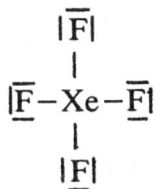

```
       |F̄|
        |
  |F̱—Xe—F̄|
        |
       |F̱|
```

BE CAREFUL! Even though all five atoms have an octet of electrons around them, there are still **two** pairs of valence electrons left over. Excess electrons **always** go around the central atom (see Step 5 of the Procedures for Writing Lewis Formulas in Section 9.3/9.4).

Step 5:   The correct Lewis structure for XeF_4, with all valence electrons accounted for, is

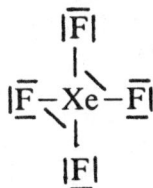

```
       |F̄|
        |  \
  |F̱ ~Xe—F̄|
        |
       |F̱|
```

NOTE: The central atom has 12 electrons around it, 4 bonding pairs and 2 nonbonding pair of electrons.

SECTION 9.11 SHAPES OF MOLECULES AND POLYATOMIC IONS: THE VALENCE SHELL ELECTRON PAIR REPULSION (VSEPR) MODEL

After completing this section, you will be able to do the following:

■ Use the valence shell electron pair repulsion (VSEPR) theory to predict the shapes of molecules and ions.

REVIEW

The **valence shell electron pair repulsion** theory, or **VSEPR**, predicts the shape of species with main group elements as central atoms. According to this theory, the most stable shape for a species is when the electron pairs **around the central atom** are as far apart as possible in three-dimensional space, or in other words, the electron repulsions are **minimized**. Only the valence electrons around the central atom determine the shape; the valence electrons around the outer atoms are too far apart to significantly affect the shape.

The most efficient arrangement (minimized repulsions) of 2, 3, 4, 5, and 6 pairs of electrons around a central atom produces the shapes of linear, trigonal planar, tetrahedral, trigonal bipyramidal, and octahedral, respectively, **only if all the pairs of electrons are bonding pairs** (that is, bonded to outer atoms). If one or more pairs of electrons around the central atom are **nonbonding**, the shape changes somewhat (see Table 9.4 in the textbook for structures, names, and examples of **total coordination numbers** 2 through 6). The **total coordination number** of the central atom is the total number of bonding pairs (outer atoms) and nonbonding pairs of electrons around the central atom. The shape of a molecule or ion only reflects the position of the **bonding pairs**, NOT the nonbonding pairs. The nonbonding pairs of electrons around the central atom **influence** the shape, but their location is not described by the shape. **Nonbonding pairs** of electrons occupy more space than bonding pairs, so the angle between nonbonding and bonding pairs is greater than the angle between adjacent bonding pairs. This is why ammonia, NH_3, with 3 bonding pairs and 1 nonbonding pair of electrons around the N atom has $H-N-H$ bond angles of $107.3°$ instead of $109.5°$, as is found in CH_4 with all 4 pairs of electrons bonding.

PREDICTING THE SHAPES OF MOLECULES AND IONS

1. Write the Lewis structure (see REVIEW in Section 9.3/9.4).

2. From the Lewis structure, find the total coordination number, i.e., count the bonding pairs (same as the number of outer atoms) and nonbonding pairs of electrons **around the central atom**. For purposes of predicting shapes, double and triple bonds **count as a single bond only**. That is, a double bond is counted as a coordination number of 1; the same is true for a triple bond.

3. Locate the coordination number **with** corresponding number of bonding and nonbonding pairs in Table 9.4 of the textbook to find the stereochemical formula and shape.

You should know why BCl_3 and NH_3 have different shapes even though both molecules have three atoms bonded to the central atom.

SAMPLE PROBLEM 9.11 A

▶ ▶ Predict the shape of $COCl_2$ (C is the central atom).

Solution:

HINT: Use the procedure given in the REVIEW for predicting the shape.

Step 1: The Lewis structure with its 24 valence electrons is

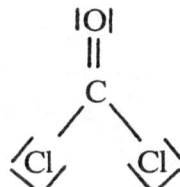

298

Remember, O more readily forms a double bond than Group VIIA elements.

Step 2: The **total coordination number** is 3. The double bond only counts as a single bond for predicting shapes.

Step 3: From Table 9.4, a molecule with a total coordination number of 3 and no nonbonding pairs is trigonal planar, just as shown for the Lewis structure.

SAMPLE PROBLEM 9.11 B

▶▶ Predict the shape of the phosphite ion, PO_3^{3-}.

Solution:

See Sample Problem 9.11 A.

Step 1: The Lewis structure with its 26 valence electrons is

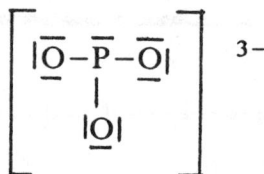

$$\left[|\overline{O}-\overline{P}-\overline{O}| \atop |\underline{O}| \right]^{3-}$$

Remember to add 3 electrons for the 3− charge on the ion.

Step 2: The **total coordination number** is 4 (3 bonding pairs or atoms, and 1 nonbonding pair).

Step 3: From Table 9.4, an ion with a total coordination number of 4, including 1 unshared pair, is trigonal pyramidal,

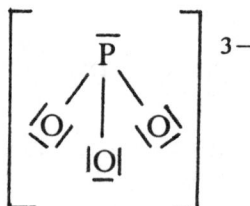

$$\left[\begin{array}{c} \overline{P} \\ \diagup | \diagdown \\ \langle O \rangle \ |\underline{O}| \ \langle O \rangle \end{array} \right]^{3-}$$

The structure should reflect the shape as much as possible. Note that the shape does NOT reflect the position of the nonbonding pair on the P atom, only the position of the four atoms.

SAMPLE PROBLEM 9.11 C

▶ ▶ Predict the shape of the chlorite ion, ClO_2^-.

See Sample Problems 9.11 A & B.

Step 1: The Lewis structure with its 20 valence electrons is

$$[\ |\overline{O} - \overline{Cl} - \overline{O}| \ ]^-$$

Remember to add 1 electron for the negative charge on the ion.

Step 2: The **total coordination number** is 4 (2 bonding pairs or atoms, and 2 nonbonding pairs).

Step 3: From Table 9.4, an ion with a total coordination number of 4, including 2 nonbonding pairs of electrons, is bent or angular.

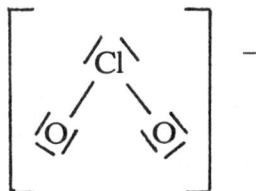

The shape reflects the position of the two bonding pairs (or two bonded atoms), not the position of the nonbonding pairs of electrons. However, this ion is bent or angular because of the repulsion of the nonbonding pairs towards each other and towards the bonding pairs.

SELF-TEST QUESTIONS

Test your understanding of the concepts and skills in this chapter by working through the multiple choice questions **BEFORE** checking the answers. A periodic table will always be available. The **ANSWERS TO SELF-TEST QUESTIONS** follow immediately after this section.

1. In order for atoms of Rb to form ionic bonds with atoms of Br,

    A. a Rb atom will give up 2 electrons.
    B. a Br atom must transfer one 4p electron to a Rb atom.
    C. a Rb atom must transfer a 5s electron to fill the 5p sublevel of Br.
    D. ions of Br and Rb must be formed, both of which have electron configurations identical to Kr.
    E. a Rb atom must transfer a 4s electron to the half-filled 4p orbital of Br.

2. In the ionic compound, K_2O,

    A. there are K^{2+} and O^{1-} ions present.
    B. each potassium ion has transferred one 4s valence electron to a half-filled 2p orbital of oxygen.
    C. the electron configurations of both the positive and negative ions are identical.
    D. valence electrons have been transferred from oxygen to potassium.
    E. the potassium ion is isoelectronic with the noble gas Kr.

3. The only compound expected to have mainly ionic bonds is

    A. SO_2 B. PH_3 C. CO_2 D. $AsCl_3$ E. SrS

4. The type of bond predicted to be formed between the following pairs of elements is identified correctly <u>except</u> for

    A. P and Cl; ionic
    B. H and S; covalent
    C. C and F; covalent
    D. Ba and O; ionic
    E. C and H; covalent

5. The bonds in the following substances are identified correctly <u>except</u> for

    A. S_8; nonpolar covalent
    B. $CaBr_2$; ionic
    C. BaO; polar covalent
    D. CO; polar covalent
    E. K_2S; ionic

6. A nonpolar covalent bond

    A. results from an equal sharing of the bonding electrons.
    B. is formed between two atoms of different electronegativities.
    C. results from an unequal sharing of the bonding electrons.
    D. has one end of the bond negative and the other end of the bond positive.
    E. is correctly described by more than one of the above answers.

7. An example of a polyatomic ion is

    A. H^+ B. O^{2-} C. K^+ D. Al^{3+} E. not given.

8. For two identical atoms to form a covalent bond,

    A. each atom must have an electronegativity of zero.
    B. one of the two atoms must supply both bonding electrons.
    C. there must be an unequal sharing of the bonding electrons.
    D. half-filled valence orbitals of each atom must overlap.
    E. more than one of the above answers is correct.

9. The only bond described correctly is

    A. $C-Cl$ ionic
    B. $H-Br$ polar covalent
    C. $H-O$ nonpolar covalent
    D. $N-N$ polar covalent
    E. $N-I$ ionic

10. When comparing a single, double, and triple bond between the same two atoms in a molecule,

    A. the single bond will be the strongest.
    B. the double bond has a bond order of three.
    C. the triple bond will be the weakest.
    D. the double bond will be the shortest.
    E. the triple bond will have the highest bond energy.

11. The formal charges on the H, C, and N, respectively, in the following molecule are

$$H-C\equiv N|$$

    A. +1, +4, +5
    B. −1, −4, −3
    C. 0, 0, 0
    D. 0, −1, +1
    E. 0, +1, −1

12. The species with the correct Lewis structures are

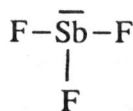

$$[\ |\overline{Cl}\text{-}\overline{I}\text{-}\overline{Cl}|\ ]^{-}$$

$$F-\overline{Sb}-F$$
$$|$$
$$F$$

$$|\overline{I}-\overline{Sn}-\overline{I}|$$

$$|\overline{F}-\overline{N}=\overline{O}$$

    ICl_2^- SbF_3 SnI_2 NOF
                                                           (N is central atom)

    A. ICl_2^- and SnI_2
    B. SbF_3 and NOF
    C. ICl_2^-, SnI_2, and NOF
    D. NOF, SbF_3, and ICl_2^-
    E. ICl_2^- and NOF

13. Both 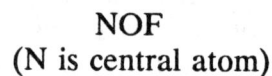 $|C\equiv O|$ and $|N\equiv N|$

    A. are examples of compounds.
    B. contain three nonbonding pairs of valence electrons.
    C. are molecules.
    D. contain two pairs of bonding electrons.
    E. are described by none of the above answers.

14. The correct Lewis structure for BrF_3 is

A. $|\overline{F}-Br-\overline{F}|$
    $|\underline{F}|$

B. $F-\overset{\frown}{Br}-F$
    F

C. $|\overline{F}-\overline{Br}-\overline{F}|$
    $\|$
    $|\overline{F}|$

D. $|\overline{F}-\overline{Br}-\overline{F}|$
    $|\underline{F}|$

E. not given

15. The shape of the SCl_2 molecule is

A. bent   B. trigonal planar   C. linear   D. trigonal bipyramidal   E. seesaw

16. Although S is bonded to 3 atoms in both SO_3 and SOF_2, SO_3 is trigonal planar and SOF_2 is trigonal pyramidal in shape. The reason for the different shapes is that

A. SO_3 has 13 pairs of valence electrons and SOF_2 has only 12 pairs of valence electrons.

B. both SOF_2 and SO_3 have four pairs of valence electrons around the central S atom, but only in SOF_2 are they all bonding.

C. the S in SO_3 has four pairs of valence electrons around it, pointing towards the four corners of a tetrahedron.

D. the S in SOF_2 has one nonbonding pair of valence electrons around it.

E. None of the above answers is correct.

17. Each of the molecules given has the shape listed except for

A. H_2S bent
B. BF_3 trigonal pyramidal
C. SF_6 octahedral
D. CI_4 tetrahedral
E. H_2CO trigonal planar (C is central atom)

18. The shape of a PH_3 molecule is

A. trigonal pyramidal
B. tetrahedral
C. bent
D. trigonal planar
E. not given

304

19. What are expected shapes of molecules I and II?

$$OF_2 \qquad H_2Te$$

$$I \qquad\qquad II$$

A. Both are linear.
B. Both are bent or angular.
C. I is linear and II is bent or angular.
D. I is bent or angular and II is linear.
E. None of the answers is correct.

ANSWERS TO SELF-TEST QUESTIONS

1. d	7. e	13. c
2. b	8. d	14. e
3. e	9. b	15. a
4. a	10. e	16. d
5. c	11. c	17. b
6. a	12. e	18. a
		19. b

PRACTICE PROBLEMS

Test your problem-solving skills in this chapter by working through the problems in the space provided **BEFORE** checking the answers. A periodic table will always be available. The **ANSWERS TO PRACTICE PROBLEMS** follow immediately after this section.

1. For NI_3, (a) write the Lewis (electron-dot) formula, (b) indicate the number of covalent bonds present and whether they are polar or nonpolar, and (c) indicate the number of nonbonding (unshared) pairs of electrons.

2.   Write the resonance structures for SO_3, including formal charges on each atom.

3.   Write the Lewis structures for (a) $AsCl_3$ (b) $BaSO_4$ (c) BI_3 (d) SeF_4 (e) BrF_5.

4.   Predict the shape of the following species: (a) $SbF_6{}^-$ (b) BCl_3 (c) H_3O^+ (d) Cl_2O (e) $NO_2{}^+$

ANSWERS TO PRACTICE PROBLEMS

1.   (a) $|\overline{\underline{I}}-\overline{N}-\overline{\underline{I}}|$ (b) 3 covalent bonds (N–I); polar
                   $|\underline{I}|$ (c) 10 unshared pairs

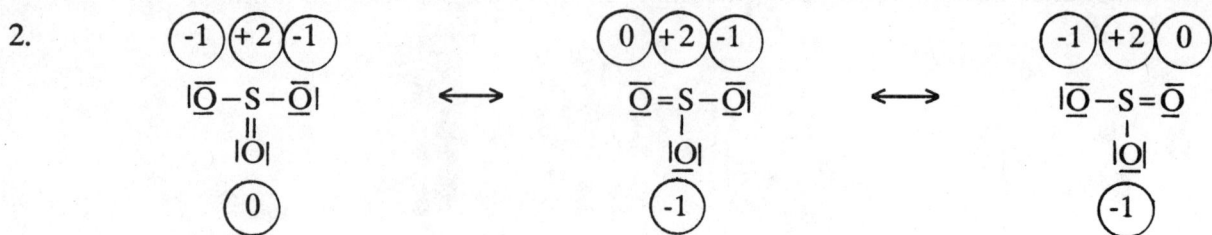

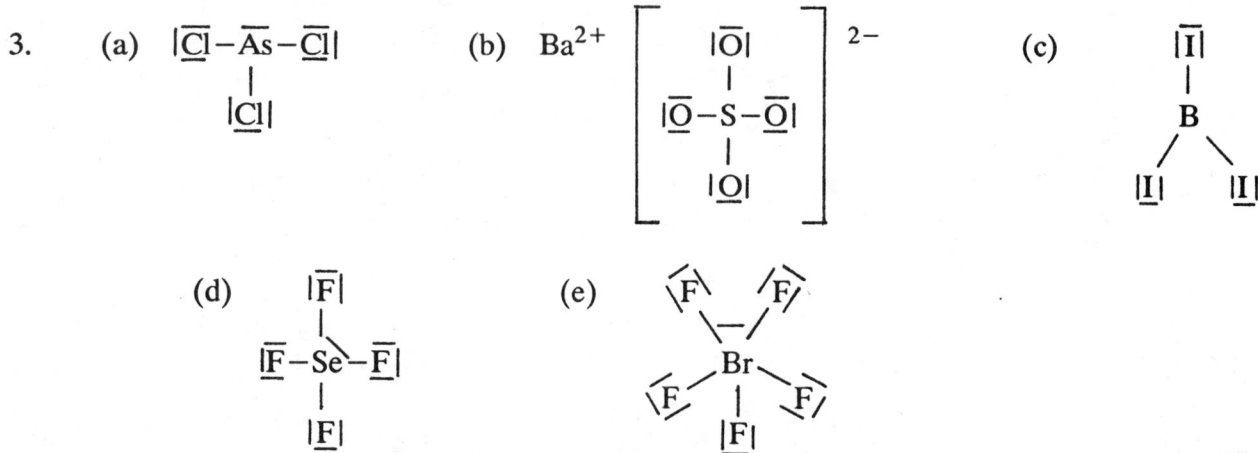

2.

(-1)(+2)(-1)	(0)(+2)(-1)	(-1)(+2)(0)
$\overline{|\underline{O}}-S-\underline{\overline{O}}|$	$\overline{O}=S-\underline{\overline{O}}|$	$\overline{|\underline{O}}-S=\overline{O}$
(with double bond to $|\underline{\overline{O}}|$ below)		
(0)	(-1)	(-1)

$\longleftrightarrow$ $\longleftrightarrow$

3. (a) $|\underline{\overline{Cl}}-\overline{As}-\underline{\overline{Cl}}|$
$|\underline{\overline{Cl}}|$

(b) Ba^{2+} $\left[\begin{array}{c} |\overline{\underline{O}}| \\ | \\ |\underline{\overline{O}}-S-\underline{\overline{O}}| \\ | \\ |\underline{\overline{O}}| \end{array} \right]^{2-}$

(c) $|\overline{\underline{I}}|$
B
$|\underline{I}|$ $|\underline{I}|$

(d) $|\overline{F}|$
$|\overline{F}-Se-\overline{F}|$
$|\underline{F}|$

(e) $\langle F \rangle \langle F \rangle$
Br
$\langle F \rangle \langle F \rangle$
$|\underline{F}|$

4. (a) SbF_6^- octahedral
    (b) BCl_3 trigonal planar
    (c) H_3O^+ trigonal pyramidal
    (d) Cl_2O bent
    (e) NO_2^+ linear

CHAPTER 10

THEORY OF CHEMICAL BONDING

SECTION 10.4   HYBRID ORBITALS

After completing this section, you will be able to do the following:

- From a Lewis formula for a molecule or ion, determine the total number of covalent bonds, the number of sigma bonds, the number of pi bonds, and the kinds of hybridized orbitals formed to explain the shape.

REVIEW

One theory to explain the formation of covalent bonds is called the **valence bond method**. According to this method, covalent bonds are formed by the partial overlap of two half-filled valence orbitals, one from each atom forming the bond. The shared pair of electrons is then attracted to the nuclei of both atoms forming the bond. However, to explain the shapes of molecules of main group elements discussed in Section 9.11, with bond angles varying from 90° to 180°, requires the formation and use of what are called **hybridized orbitals** by the central atom, particularly if it is C.

F or methane, CH_4, to have four identical $C-H$ bonds directed to the four corners of a tetrahedron with bond angles of 109.5° means that the one 2s and three 2p valence orbitals of C have to form **four identical sp^3 hybridized orbitals**; these orbitals are intermediate in size, shape, and energy between the atomic 2s and 2p orbitals. The sp^3 designation indicates the **source** of the four hybridized orbitals --- one spherical 2s orbital and three figure-eight-shaped 2p orbitals oriented 90° apart. The half-filled 1s orbital of each H atom overlaps with one half-filled sp^3 hybridized orbital of C to form four single covalent bonds (see Figure 10.7(a) in the textbook).

To account for the trigonal planar shape of molecules with one double bond like $H_2C=O$ requires that the C atom form **three identical sp^2 hybridized orbitals** pointed towards the three corners of a planar triangle, with orbitals 120° apart. These three sp^2 hybridized orbitals overlap with the half-filled 1s orbitals of the two H's and **endwise** with one of the half-filled 2p orbitals of O. The second bond between C and O is formed by the **sidewise** overlap of the one **unhybridized 2p orbital** of C and the remaining half-filled 2p orbital of O, both of which are perpendicular to the triangular plane of the four atoms (see sketch in Figure 10.11 (c) in textbook). Bonds formed by the **endwise (head-on)** overlap of orbitals are called **sigma (σ) bonds**. Bonds formed by the **sidewise** overlap of orbitals are called **pi (π) bonds**. **Single bonds are always sigma bonds. Double bonds** consist of **one sigma bond and one pi bond. Triple bonds** consist of **one sigma bond and two pi bonds**. Because of the way the bonds are formed, it is the **sigma bonds only** that **determine the shape** of a molecule or ion.

To explain the linear shapes of molecules with a triple bond like $H-C\equiv C-H$ requires that each C atom use one 2s and one 2p orbital to form **two identical sp hybridized orbitals** with an angle of 180°. The two half-filled sp hybridized orbitals of each C overlap with a half-filled 1s orbital of H and **endwise** with a half-filled sp orbital of the other C. The second and third bond between the two carbons is formed by the **sidewise** overlap of the perpendicular **unhybridized** orbitals of each C atom. A triple bond always consists of one sigma bond and two pi bonds. You cannot have a pi bond formed between two atoms unless a sigma bond is also present between those same two atoms. Again, the shape of a molecule is determined by the orientation of the **sigma bonds alone**.

In a similar manner, five identical sp^{3}d hybridized orbitals are formed when the coordination number of the central atom is 5, and six identical sp^{3}d^2 hybridized orbitals are formed when the coordination number of the central atom is 6. Remember, hybridized orbitals are formed by the central atom to account for all the electrons around the central atom that influence or determine the shape of a species -- that is, **sigma bonds and unshared electron pairs**. Table 10.1 of the textbook gives the relationship between types of hybridized orbitals necessary to explain shapes of molecules or ions with total coordination numbers 2−6.

SAMPLE PROBLEM 10.4 A

▶ ▶ For each of the following molecules, (a) what is the total number of covalent bonds? (b) How many sigma and pi bonds are there? (c) What kinds of orbitals overlap to form each of the bonds?

$$NF_3 \qquad TeF_6 \qquad SO_3 \qquad CH_3C\equiv CH$$

Solution:

HINT: Write the Lewis structure first.

NF₃

(a) The Lewis structure with 26 valence electrons is

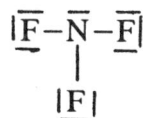

$$|\overline{\underline{F}}-\overline{N}-\overline{\underline{F}}|$$
$$|$$
$$|\underline{F}|$$

There are 3 covalent bonds, the 3 bonding pairs of electrons.

(b) 3 sigma and 0 pi bonds. Single bonds are always sigma bonds.

(c) The central atom forms enough hybridized orbitals to account for all sigma bonds and unshared pairs, which is the total coordination number 4, explained by forming 4 sp^3 hybridized orbitals (see Table 10.1 in the textbook). The three N−F bonds are formed by a sp^3 orbital of N overlapping with the half-filled 2p orbital of F. The fourth sp^3 orbital contains the unshared pair of electrons.

TeF₆

(a) The Lewis structure for TeF₆ with its 48 valence electrons is

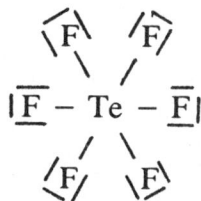

$$\begin{array}{ccc} \overline{F} & & \overline{F} \\ \backslash & & / \\ |\overline{F} - & Te & - \overline{F}| \\ / & & \backslash \\ F & & F \end{array}$$

There are six covalent bonds, since there are six bonding pairs of electrons.

(b) There are six sigma bonds and 0 pi bonds. Single bonds are always sigma bonds.

(c) The total coordination number of the central atom is 6, which means 6 sp^3d^2 orbitals are formed by Te. Each of the six sp^3d^2 orbitals of Te overlaps with the half-filled 2p orbital of F.

SO₃

(a) The Lewis structure for SO_3 with its 24 valence electrons is

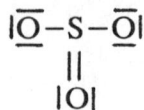

$$|\overline{O}-S-\overline{O}|$$
$$\|$$
$$|O|$$

There are four covalent bonds, since there are four bonding pairs of electrons.

(b) There are three sigma bonds and one pi bond. Single bonds are always sigma bonds, and double bonds always consist of one sigma and one pi bond.

(c) The total coordination number of the central atom is 3 (sigma bonds and unshared pairs only), which means that S forms three sp^2 hybridized orbitals. The two single S−O bonds are formed by the overlap of a 2p orbital of O with an sp^2 orbital of S. The S=O bond is formed by the endwise overlap of an sp^2 orbital of S with a 2p orbital of O and sidewise overlap of p orbitals, one from each atom.

CH₃C≡CH

(a) The Lewis structure for $CH_3C\equiv CH$ with its 16 valence electrons is

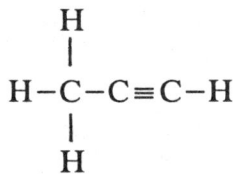

$$
\begin{array}{c}
H \\
| \\
H-C-C\equiv C-H \\
| \\
H
\end{array}
$$

There are 8 covalent bonds, since there are four bonding pairs of electrons.

(b) There are 6 sigma bonds and 2 pi bonds. A triple bond always consists of one sigma bond and one pi bond.

(c) Each C has to be considered separately. The left-most C has a total coordination number of 4 (sigma bonds), which means that four sp^3 orbitals are formed. The middle and right-most C each have a total coordination number of 2 (sigma bonds), which means that each forms two sp orbitals. The 1s orbitals of the three H's on the left end each overlaps an sp^3 orbital of the left-most C. One sp^3 orbital of the left-most C overlaps with one sp orbital of the middle C. One sp orbital of the right-most C overlaps with a 1s orbital of the right end H. The triple bond is formed by the endwise overlap of an sp orbital of the middle C with an sp orbital of the right-most C (forming a sigma bond) and the sidewise overlap of two perpendicular unhybridized 2p orbitals from each of the same C's.

| SECTION 10.7 | HOMONUCLEAR DIATOMIC MOLECULES OF OTHER ELEMENTS |
| SECTION 10.8 | BOND ORDER FROM MOLECULAR ORBITAL ENERGY LEVEL DIAGRAMS |

After completing this section, you will be able to do the following:

■ From a molecular orbital energy level diagram for a homonuclear diatomic molecule, write the electron configuration and determine the number of sigma and pi bonds as well as bond order.

REVIEW

The second method for explaining the formation of covalent bonds is called the **molecular orbital method.** According to this method, when atoms combine to form molecules, the atomic orbitals combine to form new orbitals (called **molecular orbitals**) which may extend over the entire molecule. Electrons are assigned to the molecular orbitals in a manner similar to that used for assigning electrons to atomic orbitals (i.e., by using Hund's rule and the Pauli exclusion principle). When atomic orbitals combine to form molecular orbitals, the **number** of molecular orbitals formed **must be the same** as the number of atomic orbitals combined. For example, when the 1s orbitals of **two** H atoms are combined, **two** molecular orbitals are formed, one of **lower** energy called the **bonding molecular orbital,** σ_{1s}, and one of **higher** energy called the **antibonding molecular orbital,** σ^*_{1s}. Electrons occupying bonding orbitals favor bond formation (and stability), whereas electrons occupying antibonding orbitals have the opposite effect -- that is, they make the bond (and molecule) want to fly apart. **Antibonding orbitals** always have a superscript asterisk. The resulting **molecular orbital energy level diagram** for the formation of the hydrogen molecule is shown in Figure 10.21 of the textbook. In this and all other **molecular orbital energy level diagrams,** broken lines connect each molecular orbital to the two atomic orbitals that were combined to form the molecular orbital. Electrons are added to the molecular orbitals in order of increasing energy; the capacity of a molecular orbital (bonding or antibonding) is the same as for atomic orbitals --- two. For H$_2$ in the ground state, the **molecular orbital energy level diagram** is shown with both electrons in the lowest energy σ_{1s} orbital. Electron configurations are written for molecular orbitals similar to the procedure used for atomic orbitals. For H$_2$ in the ground state, the electron configuration is

$$(\sigma_{1s})^2,$$

which says there are two electrons in the σ_{1s} molecular orbital. Parentheses are used around each molecular orbital for clarity.

In addition to H_2, **molecular orbital energy level diagrams** have been worked out for the second period **homonuclear diatomic molecules** -- that is, diatomic molecules in which both atoms are from the same element. These molecular orbitals include not only those formed from 1s atomic orbitals (like H_2), but also those molecular orbitals formed by combining 2s and 2p atomic orbitals as well. The **molecular orbital energy level diagrams** for the second period homonuclear diatomic molecules are shown in Figure 10.26 of the textbook. **Molecular orbital energy level diagrams** for second period **heteronuclear diatomic molecules and ions** are similar to those discussed for homonuclear diatomic molecules (see Figure 10.28 of the textbook).

The **bond order** for a molecule can be readily determined from **molecular orbital energy level diagrams**. Since a bond consists of two electrons, the bond order is

$$\text{Bond order} = \frac{1}{2} \text{ (number of bonding electrons } - \text{ number of antibonding electrons)}$$

A bond order of 1 is a single bond; a bond order of 2 is a double bond, etc. A bond order of 0 means that there is no bond at all and that the molecule is not stable (does not exist).

SAMPLE PROBLEM 10.7 A

▶▶ (a) Draw the molecular orbital energy level diagram for the ground state of the beryllium molecule, Be_2, and write the electron configuration. (b) How many sigma and pi bonds are there in a Be_2 molecule? (c) What is the bond order of the beryllium-beryllium bond? (d) Is the molecule paramagnetic or diamagnetic?

Solution:

HINT: Use Figure 10.26 (a) from the textbook, since that molecular orbital energy level diagram applies to homonuclear diatomic molecules between Li_2 and N_2.

(a) In a Be_2 molecule, there are **8 total electrons** (not just valence electrons). Remember, when writing electron configurations, you include **all** electrons. The electron configuration of the Be_2 molecule in its ground state is

$$\sigma^*_{2p_z} \quad -$$

$$\pi^*_{2p_x} \quad - \quad - \quad \pi^*_{2p_y}$$

$$\sigma_{2p_z} \quad -$$

$$\pi_{2p_x} \quad - \quad - \quad \pi_{2p_y}$$

$$\uparrow\downarrow \quad \sigma^*_{2s}$$

$$\uparrow\downarrow \quad \sigma_{2s}$$

$$\uparrow\downarrow \quad \sigma^*_{1s}$$

$$\uparrow\downarrow \quad \sigma_{1s}$$

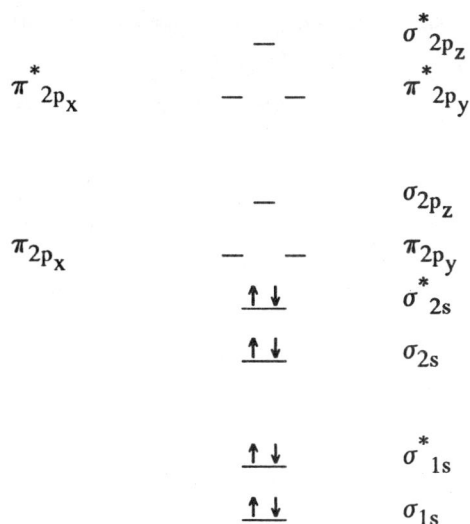

or $(\sigma_{1s})^2(\sigma^*_{1s})^2(\sigma_{2s})^2(\sigma^*_{2s})^2$.

(b) There are no electrons in any π bonding (or antibonding) orbitals, so there can be no π bonds. The bonding effect of the pair of electrons in the σ_{1s} orbital is canceled by the antibonding effect of the pair of electrons in the σ^*_{1s} orbital. Similarly, the bonding effect of the pair of electrons in the σ_{2s} orbital is canceled by the antibonding effect of the pair of electrons in the σ^*_{2s} orbital. Therefore, there are no sigma bonds.

(c) There are 2 electrons each in the σ_{1s} and σ_{2s} bonding orbitals, or 4 **bonding** electrons. There are 2 electrons each in the σ^*_{1s} and σ^*_{2s} antibonding orbitals, or 4 **antibonding** electrons. Therefore, using the formula given in the REVIEW,

$$\text{bond order} = 1/2 \ (4 - 4) = 0.$$

That is, the Be_2 molecule does not exist. Only if the bond order is at least 1 is the molecule or ion reasonably stable.

(d) Since there are no unpaired electrons, the Be_2 molecule is diamagnetic.

SELF-TEST QUESTIONS

Test your understanding of the concepts and skills in this chapter by working through the multiple choice questions **BEFORE** checking the answers. A periodic table will always be available. The ANSWERS TO SELF-TEST QUESTIONS follow immediately after this section.

1. In the molecule $SOCl_2$ (S is the central atom),

    A. there are 26 valence electrons.
    B. the total coordination number of S is 3.
    C. there are four covalent bonds.
    D. there is one pi bond.
    E. More than one of the above statements is correct.

2. In the molecule OF_2,

    A. there are only 8 valence electrons.
    B. there are four sigma bonds.
    C. the total coordination number of O is 2.
    D. there are two nonbonding pairs of electrons around O.
    E. there are four covalent bonds.

3. In the molecule $POCl_3$ (P is the central atom),

    A. there are five sigma bonds.
    B. the P atom forms four sp^3 hybridized orbitals.
    C. a half-filled 3p orbital of Cl overlaps with a sp^2 orbital of P.
    D. the shape is square planar.
    E. there are 26 valence electrons.

4. In the ClO_3^- ion,

    A. the Cl atom forms sp^2 hybridized orbitals.
    B. a p orbital of O overlaps with a p orbital of Cl.
    C. the total coordination number of Cl is 4.
    D. there is one pi bond.
    E. there are four covalent bonds.

5. In the CO_3^{2-} ion,

    A. there are 22 valence electrons.
    B. there are a total of three covalent bonds.
    C. there are two sigma bonds and two pi bonds.
    D. the coordination number of C is 4.
    E. the bond order of the carbon-oxygen bond is between 1 and 2.

6. In the ClF_3 molecule,

    A. there are 10 valence electrons.
    B. an sp^3d orbital of Cl overlaps with a 2p orbital of F.
    C. there are five covalent bonds.
    D. the shape is trigonal pyramidal.
    E. there are three sigma bonds and one pi bond.

7. In the previous six problems, the molecule or ion containing the central atom with the highest total coordination number is

    A. ClF_3 B. $POCl_3$ C. $SOCl_2$ D. ClO_3^- E. CO_3^{2-}

ANSWERS TO SELF-TEST QUESTIONS

1. a   4. c   6. b
2. d   5. e   7. a
3. b

PRACTICE PROBLEMS

Test your problem-solving skills in this chapter by working through the problems in the space provided **BEFORE** checking the answers. A periodic table will always be available. The ANSWERS TO PRACTICE PROBLEMS follow immediately after this section.

1. In the following molecules, indicate (a) the Lewis structure, (b) the total number of covalent bonds, (c) the number of sigma and pi bonds, and (d) the kinds of orbitals that overlap to form each of the bonds:

$$HC \equiv N \qquad CH_3COH$$
$$\overset{\|}{O}$$

2. (a) Draw the molecular orbital energy level diagram for the ground state of the carbon molecule, C_2, and write the electron configuration. (b) How many sigma and pi bonds are there in a C_2 molecule? (c) What is the bond order of the carbon-carbon bond? (d) Is the C_2 molecule paramagnetic or diamagnetic?

ANSWERS TO PRACTICE PROBLEMS

1.   HC≡N
      _____

   (a)   H−C≡N| (b) 4 covalent bonds   (c) 2 sigma and 2 pi bonds

   (d)   the 1s orbital of H overlaps with an sp orbital of C;
          the triple bond is formed by an sp orbital of C overlapping endwise with an sp orbital of N, and by two perpendicular 2p orbitals of C overlapping sidewise with two perpendicular 2p orbitals of N.

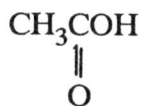

   CH_3COH
        ‖
        O
   _____

   (a)   H   (b) 8 covalent bonds   (c) 7 sigma and 1 pi bonds
              |
          H−C−C−Q̄−H
              | ‖
              H |O|

   (d)   left C: three sp^3 orbitals of C each overlap with a 1s orbital of H, and an sp^3 orbital overlaps with an sp^2 orbital of right C;

          right C: an sp^2 orbital of C overlaps with an sp^3 orbital of right O; the double bond is formed by endwise overlap of an sp^2 orbital from C with a p orbital from O and sidewise overlap of perpendicular p orbitals, one from each atom;

          right O: an sp^3 orbital of O overlaps with a 1s orbital of H.

2. See Sample Problem 10.7 A, since the same molecular orbital energy level diagram will be used.

(a) In a C_2 molecule, there are **12 total electrons**.

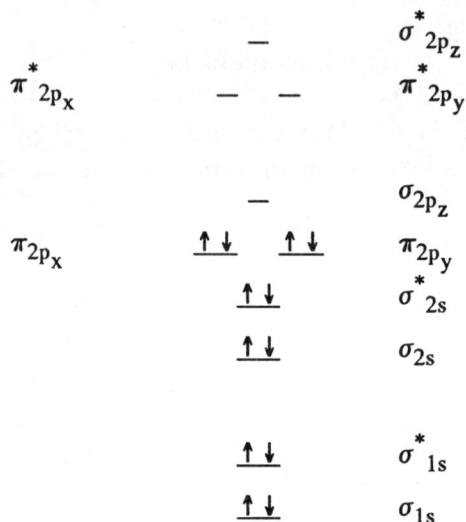

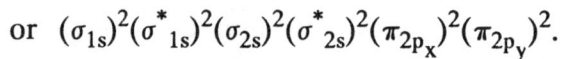

$$\sigma^*_{2p_z} \quad —$$

$$\pi^*_{2p_x} \quad — \quad — \quad \pi^*_{2p_y}$$

$$\sigma_{2p_z} \quad —$$

$$\pi_{2p_x} \quad \uparrow\downarrow \quad \uparrow\downarrow \quad \pi_{2p_y}$$

$$\uparrow\downarrow \quad \sigma^*_{2s}$$

$$\uparrow\downarrow \quad \sigma_{2s}$$

$$\uparrow\downarrow \quad \sigma^*_{1s}$$

$$\uparrow\downarrow \quad \sigma_{1s}$$

or $(\sigma_{1s})^2(\sigma^*_{1s})^2(\sigma_{2s})^2(\sigma^*_{2s})^2(\pi_{2p_x})^2(\pi_{2p_y})^2$.

(b) The bonding effect of the pair of electrons in the σ_{1s} orbital is canceled by the antibonding effect of the pair of electrons in the σ^*_{1s} orbital. Similarly, the bonding effect of the pair of electrons in the σ_{2s} orbital is canceled by the antibonding effect of the pair of electrons in the σ^*_{2s} orbital. There are two pairs of electrons in the π_{2p_x} and π_{2p_y} bonding orbitals and no electrons in the $\pi^*_{2p_x}$ and $\pi^*_{2p_y}$ antibonding orbitals. Therefore, there are no sigma bonds and 2 pi bonds.

(c) There are 8 bonding electrons and 4 antibonding electrons. The bond order is

$$\text{bond order} = 1/2 \, (8 - 4) = 2.$$

(d) Since there are no unpaired electrons, the C_2 molecule is diamagnetic.

CHAPTER 11

OXIDATION-REDUCTION REACTIONS

SECTION 11.1 OXIDATION NUMBERS

After completing this section, you will be able to do the following:

■ Given the formula of a molecule or ion, determine the oxidation number of each atom by knowing and applying the Rules for Assigning Oxidation Numbers.

REVIEW

A method is needed to (a) correctly and rapidly write and interpret formulas of compounds, whether they contain ionic and/or covalent bonds, (b) classify reactions as oxidation-reduction or not, and (c) balance oxidation-reduction reactions. The method uses what is called the **oxidation number** or **oxidation state** as a bookkeeping tool to keep track of electrons (different than formal charge described in Section 9.8). The **oxidation number** or **oxidation state** of a covalently bound element is the charge the element would have IF all the shared pairs of electrons in the Lewis formula for the species were transferred to the more electronegative atom. Oxidation numbers of covalently bound elements **are not real charges like the charges on ions**. Oxidation numbers may be positive, zero, or negative whole numbers or fractions (because it is a bookkeeping tool that works). The **oxidation number** assigned to an element **always** means **per atom** of that element. Fortunately, you will not have to draw out a Lewis structure and look up electronegativity values every time you need to assign oxidation numbers to elements in a formula, but you should **memorize** and carefully apply the following rules.

319

RULES FOR ASSIGNING OXIDATION NUMBERS

Rules must be applied in the order written:

1. The sum of the oxidation numbers of all the atoms in a species **must** equal the net charge on the species (which is zero for molecules). This means that the oxidation number of atoms in all **free elements** is always **zero**.

2. In compounds, the oxidation numbers of the elements in Groups IA, IIA, and IIIA (B and Al only) are $+1$, $+2$, and $+3$, respectively. The oxidation number of F is -1.

3. In compounds, the oxidation number of H is $+1$.

4. In compounds, the oxidation number of O is -2.

SAMPLE PROBLEM 11.1 A

▶▶ Assign oxidation numbers to each element in the following species: (a) P_4 (b) H_2SO_4 (c) $Au(NO_3)_3$ (d) $C_{12}H_{22}O_{11}$ (e) $Na_2S_4O_6$.

Solution:

HINT: Always apply the Rules for Assigning Oxidation Numbers **in the order given**.

(a) For P_4 or any other free element (meaning not bonded to any other element), the oxidation number of each atom of that element is zero. The oxidation number of phosphorus is zero.

(b) Follow the rules for assigning oxidation numbers and write the oxidation number **over** the symbol for each element and the **total** oxidation number for **all the atoms** of each element **under** the symbol for the element. For H_2SO_4, the sum of the oxidation numbers of all the atoms of all three elements must be zero (rule #1) and you can assign oxidation numbers to H (rule #3) and O (rule #4), in that order.

$$
\begin{array}{ccc}
+1 & ? & -2 \\
H_2 & S & O_4 \\
+2 & ? & -8 = 0
\end{array}
$$

Remember, the oxidation number of hydrogen is +1 (always per atom). The **total** oxidation number (of all the hydrogen atoms) is +2. Similarly for oxygen. The oxidation number of oxygen is −2. The total oxidation number of all the oxygen atoms is −8 (oxidation number above the symbol, **total** oxidation number below the symbol). In order for the sum of the oxidation numbers of all the atoms (i.e., sum of the total oxidation numbers) to be zero, the total oxidation number of S must be +6. Since there is only one S atom in the formula, +6 is the oxidation number of S. The complete assignment of oxidation numbers to H_2SO_4 is

$$\begin{array}{ccc} +1 & +6 & -2 \\ H_2 & S & O_4 \\ +2 & +6 & -8 = 0 \end{array}$$

(c)   See part (b) for details. For $Au(NO_3)_3$, the sum of the oxidation numbers of all the atoms of all three elements must be zero (rule #1) and you can assign oxidation numbers to only O (rule #4). Neither Au nor N are covered by the rules. When that happens, look for a polyatomic ion in the formula, whose charge you should know (see the table preceding Sample Problem 1.8 G). In this case the polyatomic ion is nitrate, NO_3^-, with a charge of −1. Since there are three −1 nitrates in the formula and the sum of the oxidation numbers of all the atoms must be zero, Au must be +3. Now we can look at just one of the nitrate ions to determine the oxidation number of nitrogen. For the nitrate ion,

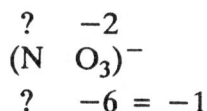

$$\begin{array}{cc} ? & -2 \\ (N & O_3)^- \\ ? & -6 = -1 \end{array}$$

Note that the sum of the oxidation numbers of all the atoms in an **ion** must equal the **charge on the ion**, in this case −1. Algebraically, the total oxidation number of N must be +5. Since there is only one N atom in the ion, the oxidation number of N is +5. The complete assignment of oxidation numbers to the nitrate ion is

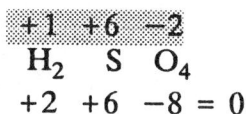

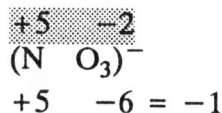

$$\begin{array}{cc} +5 & -2 \\ (N & O_3)^- \\ +5 & -6 = -1 \end{array}$$

It should be obvious that N **always has an oxidation number of +5 in the nitrate ion**, no matter what other elements are present in the compound. In $Co(NO_3)_3$, KNO_3, $Ti(NO_3)_3$, $Ba(NO_3)_2$, etc., all contain the nitrate ion in which N will always have an oxidation number of +5.

(d)   For $C_{12}H_{22}O_{11}$, the sum of the oxidation numbers of all the atoms of all three elements must be zero (rule #1) and you can assign oxidation numbers to H (rule #3) and O rule #4).

$$\begin{array}{ccc} ? & +1 & -2 \\ C_{12} & H_{22} & O_{11} \\ ? & +22 & -22 = 0 \end{array}$$

In order for the sum of the oxidation numbers of all the atoms to be zero, the **total** oxidation number of C must be zero. The oxidation number per C atom, 0/12 C atoms, is also zero. Don't be surprised at this result, since oxidation numbers can be positive, zero, or negative whole numbers or fractions. Oxidation numbers are a bookkeeping technique that works. The complete assignment of oxidation numbers to $C_{12}H_{22}O_{11}$ is

$$\begin{array}{ccc} 0 & +1 & -2 \\ C_{12} & H_{22} & O_{11} \\ 0 & +22 & -22 = 0 \end{array}$$

(e) For $Na_2S_4O_6$, the sum of the oxidation numbers of all the atoms of all three elements must be zero (rule #1) and you can assign oxidation numbers to Na (rule #2) and O rule #4).

$$\begin{array}{ccc} +1 & ? & -2 \\ Na_2 & S_4 & O_6 \\ +2 & ? & -12 = 0 \end{array}$$

In order for the sum of the oxidation numbers of all the atoms (total oxidation number) to be zero, the **total** oxidation number of sulfur must be $+10$. That is, the oxidation numbers of all 4 S atoms must add up to $+10$. The oxidation number for S in this compound is $+10/4$ S atoms, or $+2.5$. The complete assignment of oxidation numbers in $Na_2S_4O_6$ is

$$\begin{array}{ccc} +1 & +2.5 & -2 \\ Na_2 & S_4 & O_6 \\ +2 & +10 & -12 = 0 \end{array}$$

SECTION 11.3   OXIDATION NUMBERS AND NOMENCLATURE

After completing this section, you will be able to do the following:

■ Given the formula of an inorganic compound, write the correct name, and vice versa.

REVIEW

To rapidly and correctly name inorganic compounds from their formula and vice versa, **excluding acids**, you need to know (a) the procedure given in the REVIEW of Section 1.9, (b) the name, charge, and formula of the polyatomic ions (see the table preceding Sample Problem 1.8 G), and (c) the rules for assigning oxidation numbers in the REVIEW of Section 11.1. The actual charge on monatomic metal ions is the oxidation number for that ion. In iron(II) chloride, the roman numeral II is the oxidation number of iron, which is also the charge of the Fe^{2+} ion in this compound. In addition to the polyatomic ions given before Sample Problem 1.8 G, you will need to know the polyatomic ions for Group VIIA elements, in which the Group VIIA element has an odd-numbered oxidation number between +1 and +7. Using chlorine as an example (with Br and I similar),

Polyatomic Ion	Oxidation No. of Cl	Name
ClO_4^-	+7	**per**chlorate ion
ClO_3^-	+5	chlorate ion
ClO_2^-	+3	chlorite ion
ClO^-	+1	**hypo**chlorite ion

Note how the **highest** and **lowest** of the four oxidation states have prefixes meaning "above" and "below," respectively, the common middle oxidation states.

For naming inorganic acids containing oxygen, often called **oxo acids**, the first word of the name begins with the name of the **polyatomic ion**, with the last syllable replaced by **-ic** or **-ous**; the second word of the name is "acid." [In the case of polyatomic ions containing S or P, the letters "ur" or "or," respectively, will precede the "ic" or "ous" endings in order to make more sense.] The **key** to correctly naming inorganic **oxo acids** is that **-ate** polyatomic ions form **-ic** acids, and **-ite** polyatomic ions form **-ous** acids. So if you are given the formula of an acid, you need to recognize the polyatomic ion and remember its name. Similarly for writing the formula from the name.

SAMPLE PROBLEM 11.3 A

▶▶ Name each of the following: (a) V_2O_5 (b) N_2O_3 (c) K_2S (d) KIO (e) HIO_4

324

Solution:

HINT: • Carefully review Naming Inorganic Compounds (Section 1.9).

• REVIEW the name, charge, and formula of polyatomic ions in the table preceding Sample Problem 1.8 G and in the table in the REVIEW of this section.

(a) Vanadium is a metal NOT from Groups IA, IIA, or Al, so the name will require a roman numeral to indicate its oxidation number (Section 1.9, procedure #2). To find the oxidation number of V (Section 11.1),

$$\begin{array}{cc} ? & -2 \\ V_2 & O_5 \\ ? & -10 = 0 \end{array}$$

For the sum of the oxidation numbers of all the atoms to be zero, the total oxidation number of vanadium is $+10$. The oxidation number of V is $+10/2$ V atoms $= +5$. The correct name for V_2O_5 is vanadium(V) oxide. Remember, the name of binary compounds (two elements) always end with **-ide**.

(b) Since the formula, N_2O_3, contains two **non-metals**, prefixes are used to indicate the number of atoms of each element (Section 1.9, procedure #3). The correct name for N_2O_3 is dinitrogen trioxide.

(c) Since K_2S is a binary compound with a metal from "Groups IA, IIA, or Al," no roman numerals for oxidation numbers nor prefixes are needed to name it. The correct name for K_2S is potassium sulfide.

(d) Whenever a formula contains three or four elements, look for the presence of a polyatomic ion, particularly when one of the elements is O. In the case of KIO, IO^- is the **hypoiodite** polyatomic ion, comparable to the hypochlorite ion, ClO^-, shown in the table of Group VIIA polyatomic ions in the REVIEW of this section. Since K is from "Groups IA, IIA, or Al," the correct name for KIO is potassium hypoiodite.

(e) Since HIO_4 begins with hydrogen, this compound is an acid which contains the polyatomic ion, IO_4^-, called **periodate**. Remember, "ate" polyatomic ions form "ic" acids. The name begins with the name of the polyatomic ion with the "ate" replaced by the "ic." The correct name for HIO_4 is periodic acid.

SAMPLE PROBLEM 11.3 B

▶▶ Write the formula of each of the following: (a) barium chlorate (b) tin(IV) bromide (c) magnesium hypobromite (d) copper(I) oxide (e) platinum(IV) chloride.

Solution:

HINT: See HINT of Sample Problem 11.3 A.

(a)   You **must** know the symbol for barium and the formula for the polyatomic ion chlorate, as well as the oxidation number or charge for each ion, before you can write the formula from the name for this or any other compound. The charge on the barium ion of +2 (Group IIA) is the oxidation number of barium. The charge on the ClO_3^- ion is -1. Writing the oxidation number or charge above the formulas, and remembering that the sum of the charges of all the ions **must be zero** for compounds, we have

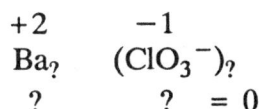

$$
\begin{array}{cc}
+2 & -1 \\
Ba_? & (ClO_3^-)_? \\
? & ? \quad = 0
\end{array}
$$

What is the **lowest number** of barium ions with a charge of +2 that will combine with the **lowest number** of chlorate ions with a charge of -1 so that the sum of the charges will be zero? These "numbers" will be the **subscripts** in the formula of the compound. It should be apparent that it will take **two** -1 chlorate ions to combine with **one** +2 barium ion so that the sum of the charges on the ions is zero. Although the sum of the charges on the ions will be zero if four -1 chlorate ions combine with two +2 barium ions, this is **incorrect** because they are **not the lowest number of ions**. [Frequently, the number of ions required of one species is the charge of the other species. That is, the number of Ba^{2+} ions in the formula is **one** (the charge on chlorate, ignoring the sign), and the number of ClO_3^- ions in the formula is **two** (the charge on barium).] The total charges on the ions and subscripts are

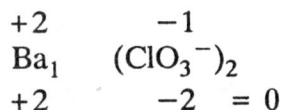

$$
\begin{array}{cc}
+2 & -1 \\
Ba_1 & (ClO_3^-)_2 \\
+2 & -2 \quad = 0
\end{array}
$$

The correct formula for barium chlorate is $Ba(ClO_3)_2$. Remember, parentheses must be used around a polyatomic ion if there are two or more of that polyatomic ion in the formula.

(b)   To start with, you know the formula contains only **two** elements, tin and bromine, because the name of binary compounds always ends with **-ide**. Second, you know the oxidation number of tin, IV = +4, and of bromine, -1 (from the table in Section 1.9). The symbols for tin and bromine you should know or look up in the periodic table. Having all the necessary information, you can determine the formula for tin(IV) bromide using the same procedure as in part (a).

$$
\begin{array}{cc}
+4 & -1 \\
\text{Sn}_? & \text{Br}_? \\
? & ? = 0
\end{array}
$$

It will take **four** -1 bromide ions to combine with **one** $+4$ tin ion in order for the oxidation numbers to add up to zero. [Again, the number of bromide ions is the oxidation number of tin; the number of tin ions is the oxidation number of bromine, ignoring the sign.] The total oxidation numbers and subscripts are

$$
\begin{array}{cc}
+4 & -1 \\
\text{Sn}_1 & \text{Br}_4 \\
+4 & -4 = 0
\end{array}
$$

The correct formula for tin(IV) bromide is SnBr_4.

(c) You should recognize magnesium hypobromite as a compound containing a metal from Group IIA and a polyatomic ion. You should know the symbol and charge (oxidation number) for magnesium, Mg^{2+}, and the formula and charge for hypobromite, BrO^-. You can now determine the formula of magnesium hypobromite using the same procedure as in parts (a) and (b). The total charges on the ions and subscripts are

$$
\begin{array}{cc}
+2 & -1 \\
\text{Mg}_? & (\text{BrO})_? \\
? & ? = 0
\end{array}
$$

It will take **two** -1 hypobromite ions to combine with **one** magnesium ion for the sum of the oxidation numbers of all the ions to add up to zero. The total oxidation numbers and subscripts are

$$
\begin{array}{cc}
+2 & -1 \\
\text{Mg}_1 & (\text{BrO})_2 \\
+2 & -2 = 0
\end{array}
$$

The correct formula for magnesium hypobromite is Mg(BrO)_2.

(d) Copper(I) oxide is a binary compound containing only the elements copper and oxygen. Knowing the symbols and oxidation numbers of each, you can determine the formula of the compound using the same procedure as in parts (a), (b), and (c).

$$
\begin{array}{cc}
+1 & -2 \\
\text{Cu}_? & \text{O}_? \\
? & ? = 0
\end{array}
$$

It will take **two** $+1$ copper ions to combine with **one** -2 oxide ion so the sum of the oxidation numbers of all the ions adds up to zero. The total oxidation numbers and subscripts are

$$+1 \quad -2$$
$$Cu_2 \quad O_1$$
$$+2 \quad -2 = 0$$

The correct formula for copper(I) oxide is Cu_2O.

(e) Platinum(IV) chloride is a binary compound containing the two elements platinum and chlorine. Knowing the symbols and oxidation numbers for each, you can determine the formula of the compound using the same procedure as in parts (a) through (d).

$$+4 \quad -1$$
$$Pt_? \quad Cl_?$$
$$? \quad ? = 0$$

It will take **four** -1 chloride ions to combine with **one** platinum ion so the sum of the oxidation numbers of the ions adds up to zero. The total oxidation numbers and subscripts are

$$+4 \quad -1$$
$$Pt_1 \quad Cl_4$$
$$+4 \quad -4 = 0$$

The correct formula for platinum(IV) chloride is $PtCl_4$.

SECTION 11.4	USING OXIDATION NUMBERS TO IDENTIFY OXIDATION-REDUCTION REACTIONS
SECTION 11.5	WRITING EQUATIONS FOR OXIDATION-REDUCTION REACTIONS

After completing this section, you will be able to do the following:

■ Identify a chemical reaction as oxidation-reduction or not;

■ In oxidation-reduction reactions, identify the species oxidized and the species reduced, as well as the oxidizing agent and the reducing agent;

■ Balance oxidation-reduction reactions by inspection.

REVIEW

In a chemical reaction, **oxidation** occurs if the oxidation number of an element **increases** during the course of the reaction. **Reduction** occurs if the oxidation number of an element **decreases** (is reduced) during the course of the reaction. **Both oxidation and reduction must occur** in a reaction; you cannot have oxidation occurring without reduction also occurring simultaneously, and vice versa. Reactions in which oxidation and reduction are occurring are called **oxidation-reduction,** or **redox,** reactions. If there is no change in oxidation number occurring for any element in a reaction, then the reaction is a **non-redox** reaction. The only way you know for sure whether a reaction is redox or not is by assigning oxidation numbers to each element in the formulas and observing whether a change in oxidation number occurs for any of the elements. However, if an element that is **in a compound** on one side of a reaction appears as a **free element** on the other side of the reaction, you can usually immediately identify it as a redox reaction.

During a redox reaction, the element **oxidized** is the one whose oxidation number **increases**; the reactant containing the element undergoing oxidation is called the **reducing agent** (because it brings about the reduction of some other element in the reaction). Similarly, the element **reduced** is the one whose oxidation number **decreases;** the reactant containing the element undergoing reduction is called the **oxidizing agent.** Remember, the element (or species containing the element) **oxidized** is the **reducing agent.** And the element (or species containing the element) **reduced** is the **oxidizing agent.** You do **NOT** have to balance a redox reaction before identifying **the element oxidized, the element reduced, the oxidizing agent,** and **the reducing agent.**

The simpler redox reactions can be balanced by inspection, just like the non-redox reactions (see REVIEW in Section 1.10). Methods for balancing more complicated redox reactions will be discussed in Sections 11.6 and 11.7.

SAMPLE PROBLEM 11.4 A

▶▶ Classify each of the following reactions as either redox or non-redox. For each reaction that is redox, indicate which element is oxidized and which element is reduced, and identify the oxidizing agent and the reducing agent.

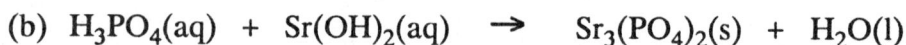

(a) $H_2S(g)$ + $HNO_3(l)$ → $S(s)$ + $NO(g)$ + $H_2O(l)$

(b) $H_3PO_4(aq)$ + $Sr(OH)_2(aq)$ → $Sr_3(PO_4)_2(s)$ + $H_2O(l)$

(c) $MnO(s)$ + $PbO_2(s)$ + $HNO_3(aq)$ → $HMnO_4(aq)$ + $Pb(NO_3)_2(aq)$ + $H_2O(l)$

HINT: Assign oxidation numbers to all the elements in all the formulas. You do NOT need to balance the reactions in order to do this problem.

(a) This reaction is a redox reaction because there is the free element S(s) on the right side with an oxidation number of zero, and S is in a compound on the left side, in which it will have an oxidation number different than zero. To verify that this is a redox reaction and to determine which element is oxidized, reduced, etc., assign oxidation numbers to all the elements following the rules in Section 11.1,

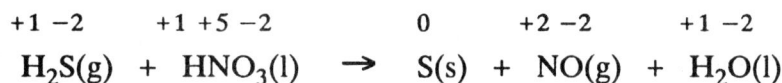

$$\overset{+1\ -2}{H_2S(g)} + \overset{+1\ +5\ -2}{HNO_3(l)} \rightarrow \overset{0}{S(s)} + \overset{+2\ -2}{NO(g)} + \overset{+1\ -2}{H_2O(l)}$$

Which elements are changing oxidation number in going from the left to the right side of the reaction? S **increases** from -2 to 0, and N **decreases** from $+5$ to $+2$. This is a redox reaction.

Element oxidized (increases oxidation number): S
Reducing agent (species that causes reduction): H_2S

Element reduced (reduces oxidation number): N
Oxidizing agent (species that causes oxidation): HNO_3

When identifying the reducing agent, you need to give the formula of the species (molecule or ion) that contains the element that is oxidized. That is, H_2S is the reducing agent because it contains the element S which is oxidized. Similarly, the oxidizing agent is HNO_3, because it contains the element N which is reduced. Remember, the element **oxidized** is in the **reducing agent**, and the element **reduced** is in the **oxidizing agent**.

(b) To determine whether this reaction is redox or not, assign oxidation numbers to all the elements in all the formulas,

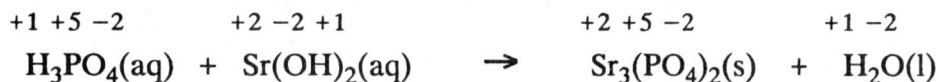

$$\overset{+1\ +5\ -2}{H_3PO_4(aq)} + \overset{+2\ -2\ +1}{Sr(OH)_2(aq)} \rightarrow \overset{+2\ +5\ -2}{Sr_3(PO_4)_2(s)} + \overset{+1\ -2}{H_2O(l)}$$

For compounds with more than one polyatomic ion, such as $Sr(OH)_2$ and $Sr_3(PO_4)_2$, you need to use only one of the polyatomic ions to assign oxidation numbers. A very common mistake is to forget that the sum of the oxidation numbers of all the atoms in a polyatomic ion must add up to the **charge on the ion**, not zero. For OH^-, the sum of the oxidation numbers is -1. For PO_4^{3-}, the sum of the oxidation numbers is -3.

Which elements are changing oxidation number in going from the left to the right side of the reaction? **None!** Neither H, P, O, nor Sr change oxidation number from the reactant to the product side. This is a non-redox reaction. For a non-redox reaction, it is impossible to have an element oxidized and reduced, and it is equally impossible for there to be an oxidizing agent and reducing agent.

(c) To determine whether this reaction is redox or not, assign oxidation numbers to all the elements in all the formulas.

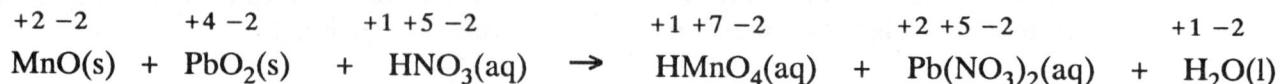

$$\overset{+2\ -2}{\text{MnO(s)}} + \overset{+4\ -2}{\text{PbO}_2\text{(s)}} + \overset{+1\ +5\ -2}{\text{HNO}_3\text{(aq)}} \rightarrow \overset{+1\ +7\ -2}{\text{HMnO}_4\text{(aq)}} + \overset{+2\ +5\ -2}{\text{Pb(NO}_3)_2\text{(aq)}} + \overset{+1\ -2}{\text{H}_2\text{O(l)}}$$

Which elements are changing oxidation number in going from the left to the right side of the reaction? Mn **increases** from +2 to +7, and Pb **decreases** from +4 to +2. This is a redox reaction.

Element oxidized (increases oxidation number): Mn
Reducing agent (species that causes reduction): MnO

Element reduced (reduces oxidation number): Pb
Oxidizing agent (species that causes oxidation): PbO$_2$

SAMPLE PROBLEM 11.4 B

▸▸ Balance the following redox reactions; indicate the element oxidized, the element reduced, the oxidizing agent, and the reducing agent.

(a) $\text{HgO(s)} \rightarrow \text{Hg(l)} + \text{O}_2\text{(g)}$

(b) $\text{KClO}_3\text{(s)} \rightarrow \text{KCl(s)} + \text{O}_2\text{(g)}$

(c) $\text{Pb(s)} + \text{AgNO}_3\text{(aq)} \rightarrow \text{Pb(NO}_3)_2\text{(aq)} + \text{Ag(s)}$

Solution:

HINT: • If a simple reaction (two or fewer reactants or products), balance by inspection (see Section 1.10 for procedure).

• If a more complex-looking reaction (more than two reactants or products), quickly determine whether the reaction is redox or non-redox by assigning oxidation numbers (mentally or by writing above the formulas).

- If complex reaction is non-redox, you have no choice but to balance by inspection.

- If complex reaction is redox, balance by methods discussed in Sections 11.6 and 11.7.

(a) For any chemical reaction, the overall and immediate goal is to balance the reaction (whether it is redox or not) by the quickest method possible. Using the HINT above, this reaction should be able to be balanced by inspection. Using the four-step procedure given in Section 1.10 for balancing by inspection, the balanced reaction is

$$2HgO(s) \rightarrow 2Hg(l) + O_2(g)$$

The assigned oxidation numbers to identify the elements oxidized and reduced, as well as the oxidizing and reducing agents, are

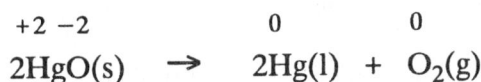

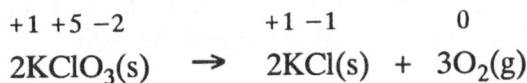

$$\overset{+2\ -2}{2HgO(s)} \rightarrow \overset{0}{2Hg(l)} + \overset{0}{O_2(g)}$$

Element oxidized (oxidation number increases): O
Reducing agent (species that causes reduction): HgO

Element reduced (oxidation number reduced): Hg
Oxidizing agent (species that causes oxidation): HgO

(b) This simple reaction can be balanced by inspection (see Section 1.10). The balanced reaction is

$$2KClO_3(s) \rightarrow 2KCl(s) + 3O_2(g)$$

The assigned oxidation numbers to identify the elements oxidized and reduced, as well as the oxidizing and reducing agents, are

$$\overset{+1\ +5\ -2}{2KClO_3(s)} \rightarrow \overset{+1\ -1}{2KCl(s)} + \overset{0}{3O_2(g)}$$

Element oxidized (oxidation number increases): O
Reducing agent (species that causes reduction): KClO_3

Element reduced (oxidation number reduced): Cl
Oxidizing agent (species that causes oxidation): KClO_3

(c) This simple reaction can be balanced by inspection (see Section 1.10). The balanced

reaction is

$$Pb(s) \; + \; 2AgNO_3(aq) \; \rightarrow \; Pb(NO_3)_2(aq) \; + \; 2Ag(s)$$

The assigned oxidation numbers to identify the elements oxidized and reduced, as well as the oxidizing and reducing agents, are

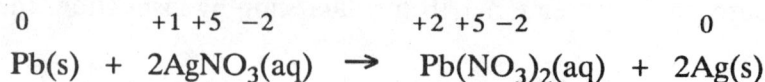

$$\overset{0}{Pb(s)} \; + \; 2\overset{+1 \; +5 \; -2}{AgNO_3(aq)} \; \rightarrow \; \overset{+2 \; +5 \; -2}{Pb(NO_3)_2(aq)} \; + \; 2\overset{0}{Ag(s)}$$

Element oxidized (oxidation number increases):   Pb
Reducing agent (species that causes reduction):   Pb(s)

Element reduced (oxidation number reduced):   Ag
Oxidizing agent (species that causes oxidation): $AgNO_3$

SECTION 11.6   CHANGE-IN-OXIDATION-NUMBER METHOD

After completing this section, you will be able to do the following:

■   Balance a redox reaction by the change-in-oxidation-number method.

REVIEW

For more complex redox reactions (usually two or more reactants or products), balancing by inspection takes too long and sometimes yields inaccurate results. The options are to use one of two specific methods: change-in-oxidation-number or half-reaction. This section discusses the change-in-oxidation-number method, and Section 11.7 covers the half-reaction method.

The procedures for balancing a redox reaction by the change-in-oxidation-number method are as follows:

BALANCING EQUATIONS BY THE CHANGE-IN-OXIDATION-NUMBER METHOD

1. Write down the reaction carefully and assign oxidation numbers to all the atoms.

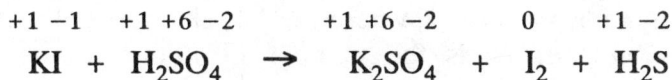

2. Identify the element oxidized and the element reduced **and** balance them **only**.

3. Balance the **total increase** and **total decrease** in oxidation number so that they are the same numerically.

4. Complete balancing by inspection, leaving H and O to last. Charge can be balanced by adding (a) H^+ to either side of the reaction if the solution is **acidic** or (b) OH^- if the solution is **basic**. Water is present in all aqueous solutions and can be used to balance H and O. **Do not change the coefficients obtained from balancing in step 3.**

5. The correctly balanced reaction will be balanced by **mass** (number of atoms of each element) **and total charge**, with the lowest whole-numbered coefficients. Add symbols to designate s, l, g, or aq.

SAMPLE PROBLEM 11.6 A

▶ ▶ Write the molecular equation for the following reaction,

$$KI(aq) + H_2SO_4(aq) \rightarrow K_2SO_4(aq) + I_2(aq) + H_2S(aq)$$

Solution:

Use the procedures given in the REVIEW.

Step 1: *Write down the reaction carefully and assign oxidation numbers to all the atoms.*

$$\overset{+1\ -1}{KI} + \overset{+1\ +6\ -2}{H_2SO_4} \rightarrow \overset{+1\ +6\ -2}{K_2SO_4} + \overset{0}{I_2} + \overset{+1\ -2}{H_2S}$$

Step 2:   *Identify the element oxidized and the element reduced and balance them only.*

Iodine is oxidized because its oxidation number **increases** from -1 to 0. Sulfur is reduced because its oxidation number **decreases** from $+6$ to -2. However, some of the sulfur (in the form of H_2SO_4) is not reduced. When that happens in any redox reaction, write down the formula twice so that one formula can be used for oxidation (or reduction) and the other formula for no change in oxidation number. In this case,

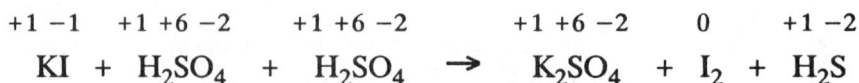

$$\overset{+1\ -1}{KI} + \overset{+1\ +6\ -2}{H_2SO_4} + \overset{+1\ +6\ -2}{H_2SO_4} \rightarrow \overset{+1\ +6\ -2}{K_2SO_4} + \overset{0}{I_2} + \overset{+1\ -2}{H_2S}$$

Don't forget the last part of Step 2, **balance just the atoms of the elements undergoing oxidation and reduction** (because these are the only atoms changing oxidation number, which is the basis for balancing the reaction in Step 3). **Forgetting this part is the most common mistake made when balancing redox reactions by this method.** There are two iodine atoms on the right, so you must start with two iodine atoms on the left. There is one sulfur in one of the H_2SO_4 formulas (either one) and one S in H_2S, so the sulfur **undergoing reduction** is balanced. [The other sulfur atoms not undergoing reduction will be balanced in Step 4.] With just the elements undergoing oxidation and reduction balanced,

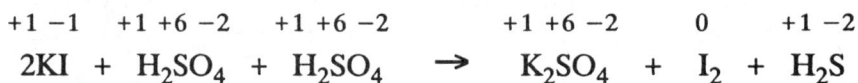

$$\overset{+1\ -1}{2KI} + \overset{+1\ +6\ -2}{H_2SO_4} + \overset{+1\ +6\ -2}{H_2SO_4} \rightarrow \overset{+1\ +6\ -2}{K_2SO_4} + \overset{0}{I_2} + \overset{+1\ -2}{H_2S}$$

Step 3:   *Balance the **total increase** and **total decrease** in oxidation number so that they are the same numerically.*

The easiest way to keep track of this step is to draw a bracket connecting the atoms of the element undergoing oxidation, and another bracket connecting the atoms of the element undergoing reduction. For simplicity, one bracket should be above the reaction (either one) and the other bracket below the reaction.

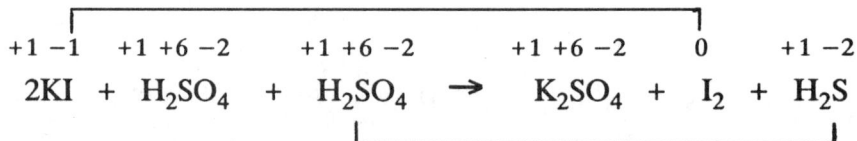

$$\overset{+1\ -1}{2KI} + \overset{+1\ +6\ -2}{H_2SO_4} + \overset{+1\ +6\ -2}{H_2SO_4} \rightarrow \overset{+1\ +6\ -2}{K_2SO_4} + \overset{0}{I_2} + \overset{+1\ -2}{H_2S}$$

Above the top line, write the **total change** in oxidation number for **all** the iodine atoms, Δox no $= 2(1) = 2$, where the Greek letter delta, Δ, means "total change in," the first number, 2, is the number of iodine atoms, and the number in the parentheses is the change in oxidation number per iodine atom. That is, the "total change in oxidation number $= 2$ I atoms x oxidation number change of $1/I$ atom $=$

2." Under the bottom bracket, write the **total change** in oxidation number for **all** the sulfur atoms undergoing reduction, Δox no = 1(8) = 8.

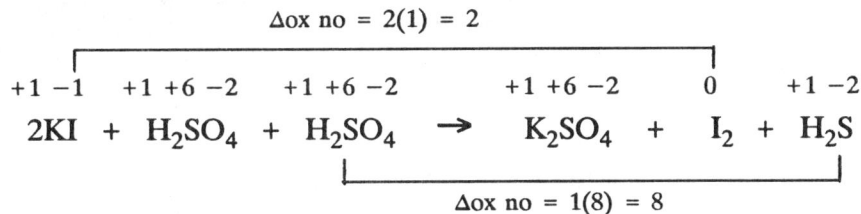

$$\Delta\text{ox no} = 2(1) = 2$$

$$\overset{+1\ -1}{2KI} + \overset{+1\ +6\ -2}{H_2SO_4} + \overset{+1\ +6\ -2}{H_2SO_4} \rightarrow \overset{+1\ +6\ -2}{K_2SO_4} + \overset{0}{I_2} + \overset{+1\ -2}{H_2S}$$

$$\Delta\text{ox no} = 1(8) = 8$$

The total oxidation number increase **must always equal** the total oxidation number decrease. Look for the **smallest whole number** that the oxidation number increase and the oxidation number decrease divide into evenly. In this case, what is the smallest whole number that 2 and 8 divide into evenly? The answer is 8. This means that the total oxidation number increase and the total oxidation number decrease must each be 8. For S, there already is an oxidation number change of 8. For iodine, the oxidation number change of 2 divides into 8 four times; therefore four times as many iodine atoms will be needed to undergo an oxidation number change of 8. Show this information above the bracket for iodine **and multiply the coefficients in front of the formulas containing iodine at both ends of the bracket by four.** You have now balanced the two elements undergoing oxidation and reduction, iodine and sulfur (also see Sample Problem 11.2 in the textbook for the interpretation of the information above and below the brackets).

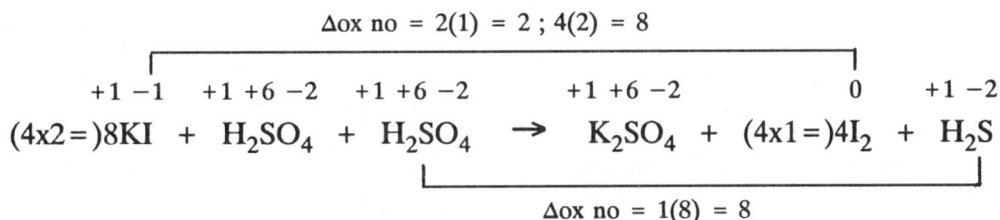

$$\Delta\text{ox no} = 2(1) = 2 \ ; \ 4(2) = 8$$

$$\overset{+1\ -1}{(4\times2=)8KI} + \overset{+1\ +6\ -2}{H_2SO_4} + \overset{+1\ +6\ -2}{H_2SO_4} \rightarrow \overset{+1\ +6\ -2}{K_2SO_4} + \overset{0}{(4\times1=)4I_2} + \overset{+1\ -2}{H_2S}$$

$$\Delta\text{ox no} = 1(8) = 8$$

Step 4: *Complete balancing by inspection, leaving H and O to last. Charge can be balanced by adding (a) H^+ to either side of the reaction if the solution is **acidic** or (b) OH^- if the solution is **basic**. Water is present in all aqueous solutions and can be used to balance H and O. **Do not change the coefficients obtained from balancing in step 3.***

Balancing the potassium at 8 on each side and the sulfur NOT undergoing reduction at 4 on each side, we have

$$8KI + 4H_2SO_4 + H_2SO_4 \rightarrow 4K_2SO_4 + 4I_2 + H_2S$$

All elements are balanced except H and O. There are 10 H and 20 O on the left and 2 H and 16 O on the right. You must **add** 8 H and 4 O to the **right side**, which is $4H_2O$. When you are writing a **molecular** equation as stated in this problem, you balance the H and O with H_2O only. H^+ and OH^- are used to balance when you are told the reaction is taking place in acid or base solution, respectively.

Step 5:   *The correctly balanced reaction will be balanced by **mass** (number of atoms of each element) **and total charge**, with the lowest whole-numbered coefficients. Add symbols to designate s, l, g, or aq.*

Combining the H_2SO_4 and adding $4H_2O$ to the right side, the completely balanced reaction with s, l, g, aq symbols is

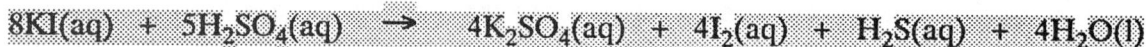

$$8KI(aq) + 5H_2SO_4(aq) \rightarrow 4K_2SO_4(aq) + 4I_2(aq) + H_2S(aq) + 4H_2O(l)$$

CHECK: Make a final check to see that there are the same number of atoms of each element on both sides of the reaction.

SAMPLE PROBLEM 11.6 B

▶▶ Write the net ionic equation for the following reaction, which takes place in acidic solution:

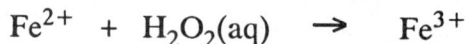

$$Fe^{2+} + H_2O_2(aq) \rightarrow Fe^{3+}$$

Solution:

See Sample Problem 11.6 A for detailed explanations of each step.

Step 1:   The assignment of oxidation numbers is

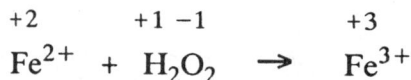

$$\overset{+2}{Fe^{2+}} + \overset{+1 \ -1}{H_2O_2} \rightarrow \overset{+3}{Fe^{3+}}$$

NOTE: You will **correctly** assign an oxidation number of -1 to oxygen **only if** you use the oxidation rules **in the order they are given.**

Step 2:   Iron is oxidized because its oxidation number **increases** from $+2$ to $+3$. But what is reduced? There can **never** be oxidation without reduction. If hydrogen were reduced to H_2, that formula would have appeared in the reaction to begin with. The only other possibility is that oxygen is **reduced** from -1 to its common -2 oxidation state. Since this reaction takes place in acidic solution, you can add H^+ or H_2O to either side of the reaction. Add one H_2O to the right side,

$$\overset{+2}{Fe^{2+}} + \overset{+1 \ -1}{H_2O_2} \quad \rightarrow \quad \overset{+3}{Fe^{3+}} + \overset{+1 \ -2}{H_2O}$$

Don't forget the last part of step 2 (the most frequent error), which is to **balance just the elements undergoing oxidation and reduction**. The iron atoms are balanced, but the 2 O on the left will require a 2 in front of H_2O on the right,

$$\overset{+2}{Fe^{2+}} + \overset{+1 \ -1}{H_2O_2} \quad \rightarrow \quad \overset{+3}{Fe^{3+}} + \overset{+1 \ -2}{2H_2O}$$

Step 3: The total increase in oxidation number by iron is 1, and the total decrease in oxidation number by the two oxygen atoms is 2. The smallest whole number that both 1 and 2 divide into evenly is obviously 2. The oxidation number increase and decrease will be balanced at 2 each. Oxygen is already 2, as shown in the information below the bracket. The change in oxidation number of iron must be multiplied by 2 to equal the oxidation number change of oxygen, which means that the coefficients in front of the iron ions at both ends of the bracket must be multiplied by two.

$$\Delta ox \ no = 1(1) = 1; \ 2(1) = 2$$
$$\overset{+2}{2Fe^{2+}} + \overset{+1 \ -1}{H_2O_2} \quad \rightarrow \quad \overset{+3}{2Fe^{3+}} + \overset{+1 \ -2}{2H_2O}$$
$$\Delta ox \ no = 2(1) = 2$$

Step 4: Complete the balancing by inspection, leaving O and H to last. Before balancing H and O (which is already balanced because it was reduced) **in acidic** (or basic) solution, **balance the total charges on each side of the reaction first**. There is a total charge of $4+$ on the left and $6+$ on the right. In **acidic solution**, you can only add H^+ or H_2O as needed to complete the balancing. To balance the total charge at $+6$ on each side, add $2H^+$ to the left side. Now make the final H and O check. There are 4 H and 2 O on both the left and the right. Thus, they are correctly balanced.

$$2H^+ + 2Fe^{2+} + H_2O_2 \quad \rightarrow \quad 2Fe^{3+} + 2H_2O$$

Step 5: The reaction, correctly balanced **by charge and mass**, is

$$2H^+ + 2Fe^{2+} + H_2O_2(aq) \quad \rightarrow \quad 2Fe^{3+} + 2H_2O(l)$$

CHECK: Make a final check to see that there are the same number of atoms of each element on both sides of the reaction **and** that the total charge is balanced on each side of the reaction. **A chemical reaction is correctly** balanced only when both mass (number of atoms) and charge are balanced!

SAMPLE PROBLEM 11.6 C

▶ ▶ Write the net ionic equation for the following reaction, which takes place in basic solution:

$$Bi_2O_3(s) + OCl^- \rightarrow BiO_3^- + Cl^-$$

Solution:

See Sample Problem 11.6 A for detailed procedure, as well as Sample Problem 11.6 B.

Step 1:   The assignment of oxidation numbers is

$$\overset{+3\ -2}{Bi_2O_3} + \overset{-2\ +1}{OCl^-} \rightarrow \overset{+5\ -2}{BiO_3^-} + \overset{-1}{Cl^-}$$

BE CAREFUL!   In OCl^-, follow the rules for assigning oxidation numbers in the order that they are written.   Oxygen is assigned -2 first, and then chlorine is assigned $+1$ (not its more common -1 as a monatomic anion), so that the oxidation numbers add up to the charge on the ion, $1-$.

Step 2:   Bismuth is oxidized because its oxidation number **increases** from $+3$ to $+5$. Chlorine is reduced because its oxidation number is **reduced** from $+1$ to -1.   Are the atoms of bismuth and chlorine balanced?   No!   You **must** balance the atoms of the elements changing oxidation number (undergoing oxidation and reduction) at this point, **before** proceeding to the next step.   Chlorine is balanced at one each, but you will need 2 bismuth on the right to balance the two on the left.

$$\overset{+3\ -2}{Bi_2O_3} + \overset{-2\ +1}{OCl^-} \rightarrow \overset{+5\ -2}{2BiO_3^-} + \overset{-1}{Cl^-}$$

Step 3:   The total increase in oxidation number by the two bismuth atoms is 4, and the total decrease in oxidation number by the chlorine atom is 2.   The oxidation number increase and decrease will be equal at the smallest whole number that 2 and 4 divide into evenly, namely 4.   Bismuth is already 4.   The number of chlorine atoms will have to be multiplied by 2 to have a total decrease in oxidation number of 4.

$$\Delta ox\ no = 2(2) = 4$$

$$\overset{+3\ -2}{Bi_2O_3} + \overset{-2\ +1}{2OCl^-} \rightarrow \overset{+5\ -2}{2BiO_3^-} + \overset{-1}{2Cl^-}$$

$$\Delta ox\ no = 1(2) = 2\ ;\ 2(2) = 4$$

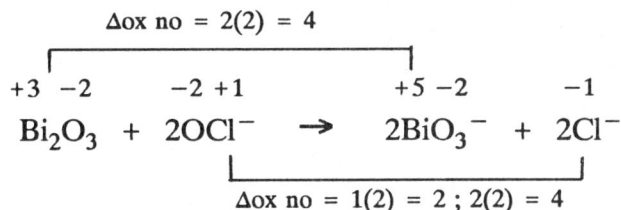

Step 4: Complete the balancing by inspection, leaving O and H to last. All the elements are balanced at this point except O and H. **In basic (or acidic) solution, balance the total charge on each side of the reaction first**, before balancing H and O. There is a total charge of 2– on the left and 4– on the right. In **basic solution**, you can only add OH^- or H_2O as needed to complete the balancing. To balance the total charge at 4– on each side, add $2OH^-$ to the left side. Now make the final H and O check. There are 2 H and 7 O on the left and 0 H and 6 O on the right. You must add 2 H and 1 O, or one H_2O, to the right side.

$$2OH^- + Bi_2O_3 + 2OCl^- \rightarrow 2BiO_3^- + 2Cl^- + H_2O$$

Step 5: The reaction, correctly balanced **by charge and mass**, is

$$2OH^- + Bi_2O_3(s) + 2OCl^- \rightarrow 2BiO_3^- + 2Cl^- + H_2O(l)$$

CHECK: Make a final check to see that there are the same number of atoms of each element on both sides of the reaction **and** that the total charge is balanced on each side of the reaction. **A chemical reaction is correctly balanced only when both mass (number of atoms) and charge are balanced!**

SECTION 11.7 HALF-REACTION METHOD

After completing this section, you will be able to do the following:

■ Balance a redox reaction by the half-reaction method.

REVIEW

For balancing more complex redox reactions (containing two or more reactants or products), the options are to use the change-in-oxidation-number method discussed in Section 11.6 or the half-reaction method. The former is often quicker if you are able to assign oxidation numbers to all the atoms. If that is not the case, then the half-reaction method can be used because oxidation numbers are not required for all the atoms to balance reactions by this method. The half-reaction method becomes particularly useful when discussing electrochemistry in Chapter 19.

The half-reaction method is based on dividing the overall reaction into two **half-reactions**: oxidation and reduction. In the **oxidation half-reaction**, an element **increases** its oxidation number because it **loses electrons**. In the **reduction half-reaction**, an element **reduces** its oxidation number because it **gains electrons**. Each half-reaction is balanced separately, and then the equation for the overall reaction is obtained by combining the half-reactions so that the **total electron loss in oxidation equals the total electron gain in reduction**. This principle was the basis for the change-in-oxidation-number method discussed in Section 11.6. In that method, the total oxidation number increase had to equal the total oxidation number decrease. In effect, the change-in-oxidation-number method says that the total electron loss in oxidation (which **causes the oxidation number to increase**) has to equal the total electron gain (which **causes the oxidation number to decrease**).

Frequently, **oxidation** is defined as the **loss of electrons** which causes the oxidation number to increase (because the element is losing negatively charged electrons), and **reduction** is defined as the **gain of electrons** which causes the oxidation number to decrease or be reduced (because the element is gaining negatively charged electrons). The electrons lost by one element undergoing oxidation are the very **same electrons** gained by (usually) another element undergoing reduction. Since matter (including electrons) can neither be created nor destroyed, the number of electrons lost by oxidation **must always equal** the number of electrons gained in reduction.

The procedures for balancing a redox reaction by the half-reaction method are as follows:

BALANCING EQUATIONS
BY THE HALF-REACTION METHOD

1. Divide the reaction into an oxidation half-reaction and a reduction half-reaction and balance each half-reaction by following steps 2-6, **in the order shown.**
2. Balance elements other than O and H (if any) first.
3. Use H_2O to balance O.
4. Use H^+ to balance H.
5. Use e^- to balance charge.
6. If the reaction takes place in **basic** solution, add enough OH^- ions to each side to neutralize any H^+ ions shown from step 4. Combine H^+ and OH^- to form H_2O.
7. Combine the two half-reactions so that electrons cancel, and simplify.

Note that steps 1-5 are the same for reactions taking place in **both acidic and basic** solution. Step 6 is used **only** if the reaction is taking place in basic solution.

SAMPLE PROBLEM 11.7 A

▶ ▶ Use the half-reaction method to balance the following reaction in acidic solution. Identify the element oxidized, the element reduced, the species acting as the oxidizing agent, and the species acting as the reducing agent.

$$Ag^+ + AsH_3(g) \rightarrow H_3AsO_4(aq) + Ag(s)$$

Solution:

HINT: Use the procedure given in the REVIEW of this section.

Step 1:   Assign oxidation numbers to identify the oxidation and reduction half-reactions,

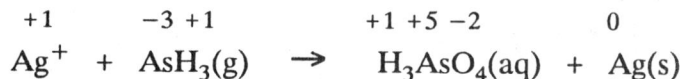

$$\overset{+1}{Ag^+} + \overset{-3\ +1}{AsH_3(g)} \rightarrow \overset{+1\ +5\ -2}{H_3AsO_4(aq)} + \overset{0}{Ag(s)}$$

Since the oxidation number of silver is reduced from +1 to 0, the reduction half-reaction is

$$Ag^+ \rightarrow Ag$$

Since the oxidation number of arsenic is increased from −3 to +5, the oxidation half-reaction is

$$AsH_3 \rightarrow H_3AsO_4$$

Either half-reaction can be balanced first. Let's go through steps 2-5 for the oxidation half-reaction first, then the reduction half-reaction. Step 6 doesn't apply, since this reaction is in acidic solution.

Step 2:   The only element other than O and H is As, and it is already balanced.

$$AsH_3 \rightarrow H_3AsO_4$$

Step 3:   Balance the oxygen with H_2O by placing $4H_2O$ on the left side,

$$4H_2O + AsH_3 \rightarrow H_3AsO_4$$

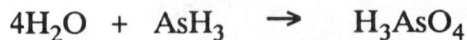

Step 4:   Balance the hydrogen with H^+ by placing $8H^+$ on the right side,

$$4H_2O + AsH_3 \rightarrow H_3AsO_4 + 8H^+$$

Step 5:   Balance the charge with e^- by placing $8e^-$ on the right side,

$$4H_2O + AsH_3 \rightarrow H_3AsO_4 + 8H^+ + 8e^-$$

The electrons must appear on the right side for two reasons: (a) to balance the charge, and (b) to show that this is an oxidation half-reaction (electrons are always lost or given up in oxidation). **This half-reaction is now balanced by mass (atoms) and charge.** [There can only be H^+ ions (NOT OH^- ions) appearing in a final balanced reaction taking place in acidic solution.] The **same procedure** must be used for the **reduction half-reaction** taking place in acidic solution.

Step 2:   The reduction half-reaction has all the atoms already balanced. Thus, steps 3 and 4 are unnecessary.

$$Ag^+ \rightarrow Ag$$

Step 5:   Balance the charge with e^- by placing $1e^-$ on the left side,

$$Ag^+ + e^- \rightarrow Ag$$

The electron must appear on the left side for two reasons: (a) to balance the charge, and (b) to show that this is a reduction half-reaction (electrons are always gained in reduction). **This half-reaction is now balanced by mass and charge.**

Step 6:   Applies to basic solutions only.

Step 7:   To combine the reactions so that the electrons cancel, multiply the reduction half-reaction by 8 and add the two half-reactions together,

$$4H_2O + AsH_3 \rightarrow H_3AsO_4 + 8H^+ + 8e^-$$

$$+ \quad 8 \times [\, Ag^+ + e^- \rightarrow Ag\,]$$

$$4H_2O + AsH_3 + 8Ag^+ + \cancel{8e^-} \rightarrow H_3AsO_4 + 8H^+ + 8Ag + \cancel{8e^-}$$

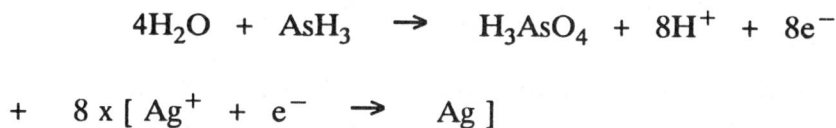

Then simplify and add s, l, g, and aq for non-ionic species,

$$4H_2O(l) + AsH_3(g) + 8Ag^+ \rightarrow H_3AsO_4(aq) + 8H^+ + 8Ag(s)$$

CHECK: Make a final check to see that there are the same number of atoms of each element on both sides of the reaction **and** that the total charge is balanced on each side of the reaction. **A chemical reaction is correctly balanced only when both mass (number of atoms) and charge are balanced!**

Element oxidized: As
Element reduced: Ag
Oxidizing agent: Ag^+
Reducing agent: AsH_3

SAMPLE PROBLEM 11.7 B

▶▶ Use the half-reaction method to balance the following reaction in basic solution. Identify the element oxidized, the element reduced, the species acting as the oxidizing agent, and the species acting as the reducing agent.

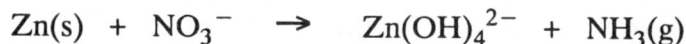

$$Zn(s) + NO_3^- \rightarrow Zn(OH)_4^{2-} + NH_3(g)$$

Solution:

Step 1: Assign oxidation numbers to identify the oxidation and reduction half-reactions,

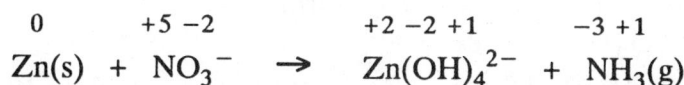

$$\overset{0}{Zn(s)} + \overset{+5\ -2}{NO_3^-} \rightarrow \overset{+2\ -2\ +1}{Zn(OH)_4^{2-}} + \overset{-3\ +1}{NH_3(g)}$$

Since the oxidation number of zinc is increased from 0 to +2, the oxidation half-reaction is

$$Zn \rightarrow Zn(OH)_4^{2-}$$

Since the oxidation number of nitrogen is reduced from $+5$ to -3, the reduction half-reaction is

$$NO_3^- \rightarrow NH_3$$

Either half-reaction can be balanced first. Let's go through steps 2-6 for the oxidation half-reaction first, then the reduction half-reaction.

Step 2: The only element other than O and H is Zn, and it is already balanced.

$$Zn \rightarrow Zn(OH)_4^{2-}$$

Step 3: Balance the oxygen with H_2O by placing $4H_2O$ on the left side,

$$4H_2O + Zn \rightarrow Zn(OH)_4^{2-}$$

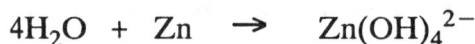

Step 4: Balance the hydrogen with H^+ by placing $4H^+$ on the right side,

$$4H_2O + Zn \rightarrow Zn(OH)_4^{2-} + 4H^+$$

Step 5: Balance the charge with e^- by placing $2e^-$ on the right side,

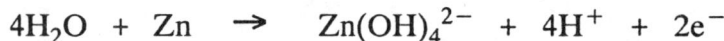

$$4H_2O + Zn \rightarrow Zn(OH)_4^{2-} + 4H^+ + 2e^-$$

The electrons must appear on the right side for two reasons: (a) to balance the charge, and (b) to show that this is an oxidation half-reaction (electrons are always lost or given up in oxidation). This half-reaction is now balanced by mass (atoms) and charge in **acidic** solution (for simplicity). However, this reaction is taking place in **basic** solution, so we must do step 6.

Step 6: Add enough OH^- ions to neutralize H^+ by adding $4OH^-$ to **each side**,

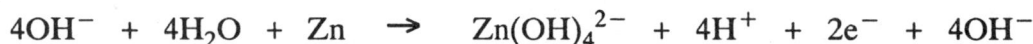

$$4OH^- + 4H_2O + Zn \rightarrow Zn(OH)_4^{2-} + 4H^+ + 2e^- + 4OH^-$$

The $4H^+$ and $4OH^-$ on the right side of the reaction will immediately combine to form $4H_2O$ (1 H^+ and 1 OH^- form 1 H_2O).

$$4OH^- + 4H_2O + Zn \rightarrow Zn(OH)_4^{2-} + 4H_2O + 2e^-$$

This oxidation half-reaction is now balanced by mass (atoms) and charge in basic

solution. [There can only be OH^- ions (NOT H^+ ions) appearing in a final balanced reaction taking place in basic solution.] The **same procedure** must be used for the **reduction half-reaction** in basic solution.

Step 2: The only element other than O and H is N and it is already balanced.

$$NO_3^- \rightarrow NH_3$$

Step 3: Balance the oxygen with H_2O by placing $3H_2O$ on the right side,

$$NO_3^- \rightarrow NH_3 + 3H_2O$$

Step 4: Balance the hydrogen with H^+ by placing $9H^+$ on the left side,

$$9H^+ + NO_3^- \rightarrow NH_3 + 3H_2O$$

Step 5: Balance the charge with e^- by placing $8e^-$ on the left side,

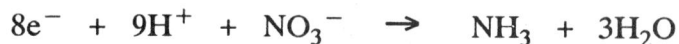

$$8e^- + 9H^+ + NO_3^- \rightarrow NH_3 + 3H_2O$$

The electrons must appear on the left side for two reasons: (a) to balance the charge, and (b) to show that this is a reduction half-reaction (electrons are always gained in reduction). This half-reaction is now balanced by mass (atoms) and charge in **acidic** solution (for simplicity). However, this reaction is taking place in **basic** solution, so we must do step 6.

Step 6: Add enough OH^- ions to neutralize H^+ by adding $9OH^-$ to **each side**,

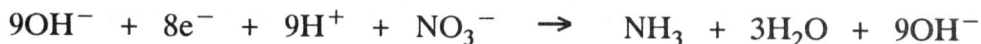

$$9OH^- + 8e^- + 9H^+ + NO_3^- \rightarrow NH_3 + 3H_2O + 9OH^-$$

The $9H^+$ and $9OH^-$ on the left side of the reaction will immediately combine to form $9H_2O$ (1 H^+ and 1 OH^- form 1 H_2O).

$$9H_2O + 8e^- + NO_3^- \rightarrow NH_3 + 3H_2O + 9OH^-$$

This reduction half-reaction is now balanced by mass (atoms) and charge in basic solution. [There can only be OH^- ions (NOT H^+ ions) appearing in a final balanced reaction taking place in basic solution.]

Step 7: To combine the reactions so that the electrons cancel, multiply the oxidation half-reaction by 4 and add the two half-reactions together,

$$4 \times [\, 4OH^- + 4H_2O + Zn \rightarrow Zn(OH)_4^{2-} + 4H_2O + 2e^- \,]$$

$$+ \quad 9H_2O \ + \ 8e^- \ + \ NO_3^- \quad \rightarrow \quad NH_3 \ + \ 3H_2O \ + \ 9OH^-$$

$$16OH^- \ + \ 25H_2O \ + \ 4Zn \ + \ NO_3^- \ + \ 8e^- \quad \rightarrow \quad 4Zn(OH)_4^{2-} \ + \ NH_3 \ + \ 19H_2O \ + \ 9OH^- \ + \ 8e^-$$

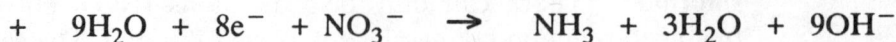

Then simplify and add s, l, g, and aq for non-ionic species,

$$7OH^- \ + \ 6H_2O(l) \ + \ 4Zn(s) \ + \ NO_3^- \quad \rightarrow \quad 4Zn(OH)_4^{2-} \ + \ NH_3(g)$$

CHECK: Make a final check to see that there are the same number of atoms of each element on both sides of the reaction **and** that the total charge is balanced on each side of the reaction. **A chemical reaction is correctly balanced only when both mass (number of atoms) and charge is balanced!**

Element oxidized: Zn
Element reduced: N
Oxidizing agent: NO_3^-
Reducing agent: Zn(s)

SELF-TEST QUESTIONS

Test your understanding of the concepts and skills in this chapter by working through the multiple choice questions **BEFORE** checking the answers. A periodic table will always be available. The ANSWERS TO SELF-TEST QUESTIONS follow immediately after this section.

1. The oxidation number of Cl in $Ni(ClO_4)_2$ is

    A. +3   B. +5   C. +7   D. +8   E. not given

2. The oxidation numbers of Ir, Cr, and O, respectively, in $Ir_2(Cr_2O_7)_3$ are

   A. +3, +6, −2   B. +2, +3, −2   C. +6, +36, −42   D. +2, +6, −2   E. not given

3. The oxidation number of phosphorus is +5 in all of the following species <u>except</u>

   A. $AlPO_4$ B. $Ca_3(PO_3)_2$ C. PF_5 D. PCl_6^- E. H_3PO_4

4. The oxidation numbers of La and C, respectively, in the compound $La_2(CO_3)_3$ are

   A. +2, +6   B. +5, +3   C. +2, +4   D. +4, +4   E. +3, +4

5. The formula for each of the following compounds is written correctly <u>except</u> for

   A. KBr   B. LiO   C. H_2S D. $MgCl_2$ E. Li_2S

6. Mn_2O_3 is correctly named

   A. trimanganese dioxide.
   B. manganese oxide.
   C. manganese(VI) oxide.
   D. manganese(III) oxide.
   E. dimanganese oxide.

7. The formula for cobalt(III) bromide is

   A. CoBr   B. Co_3Br_2 C. $CoBr_3$ D. Co_2Br_3 E. Co_3Br

8. The correct formula for chromium(III) sulfate is

   A. $Cr_2(SO_4)_3$ B. $CrSO_4$ C. Cr_3SO_4 D. $Cr_3(SO_4)_2$ E. not given

9. The correct name for $Ni(IO_2)_2$ is

   A. nickel iodite.
   B. nickel iodide.
   C. nickel(II) hypoiodite.
   D. nickel diiodite.
   E. nickel(II) iodite.

10. The correct formula for tantalum(V) sulfide is

    A. Ta_2S_5 B. Ta_5S C. TaS   D. Ta_5S_2 E. Ta_3S_5

11. The correct name for $Bi_2(CO_3)_3$ is

    A. bismuth(II) carbonate.
    B. dibismuth carbonate.
    C. bismuth(III) carbonate.
    D. bismuth(III) bicarbonate.
    E. not given.

12. The correct formula for sulfurous acid is

    A. SO_3 B. HSO_2 C. H_2SO_4 D. H_2SO_3 E. H_2S

13. In the following reaction,

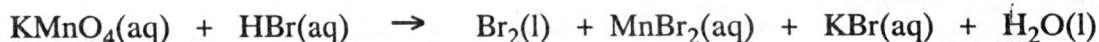

$$KMnO_4(aq) \ + \ HBr(aq) \ \rightarrow \ Br_2(l) \ + \ MnBr_2(aq) \ + \ KBr(aq) \ + \ H_2O(l)$$

    A. Mn is oxidized.
    B. Br is oxidized.
    C. HBr is the oxidizing agent.
    D. $KMnO_4$ is the reducing agent.
    E. K is reduced.

14. When the following reaction is correctly balanced with the lowest whole-numbered coefficients, there will be

$$Al_2(SO_4)_3(aq) \ + \ BaCl_2(aq) \ \rightarrow \ BaSO_4(s) \ + \ AlCl_3(aq)$$

    A. 2 $Al_2(SO_4)_3$ B. 2 $BaSO_4$ C. 1 $AlCl_3$ D. 3 $Al_2(SO_4)_3$ E. 3 $BaCl_2$

15. When the following reaction is balanced with the lowest whole-numbered coefficients, there will be

$$Na(s) \ + \ H_2O(l) \ \rightarrow \ NaOH(aq) \ + \ H_2(g)$$

    A. 1 Na   B. 2 H_2 C. 2 H_2O D. 1 NaOII   E. 3 H_2

16. In the previous reaction (#15),

   A. H_2O is the oxidizing agent.
   B. the oxidation number of Na decreases.
   C. Na is reduced.
   D. O is oxidized.
   E. O has an oxidation number of -1 on NaOH.

17. When the following reaction is balanced with the lowest whole-numbered coefficients, there will be

$$Ni(NO_3)_2(aq) + BiH_3(g) \rightarrow HNO_3(aq) + H_3BiO_3(aq) + Ni(s)$$

   A. $5\ HNO_3$ B. $3\ H_2O$ C. $6\ Ni(NO_3)_2$ D. $2\ BiH_3$ E. $3\ H_3BiO_3$

18. When the following reaction in basic solution is balanced with the lowest whole-numbered coefficients, the coefficient of OH^- will be

$$Cl_2(g) + Fe(OH)_3(s) \rightarrow FeO_4^{2-} + Cl^-$$

   A. 2   B. 3   C. 8   D. 10   E. 6

ANSWERS TO SELF-TEST QUESTIONS

1. c	7. c	13. b
2. a	8. a	14. e
3. b	9. e	15. c
4. e	10. a	16. a
5. b	11. c	17. b
6. d	12. d	18. d

PRACTICE PROBLEMS

Test your problem-solving skills in this chapter by working through the problems in the space provided **BEFORE** checking the answers. A periodic table will always be available. The ANSWERS TO PRACTICE PROBLEMS follow immediately after this section.

1. Write the molecular equation for the following reaction:

$$Cr_2O_3(s) \ + \ Na_2CO_3(s) \ + \ KNO_3(aq) \ \longrightarrow \ Na_2CrO_4(aq) \ + \ CO_2(g) \ + \ KNO_2(aq)$$

2. Write the net ionic equation for the following reaction in basic solution:

$$PbO_2(s) \ + \ Sb(s) \ \longrightarrow \ PbO(s) \ + \ SbO_2^-$$

3. Write the net ionic equation for the following reaction in acidic solution:

$$Cl^- \ + \ Cr_2O_7^{2-} \ \longrightarrow \ CrCl_3(s) \ + \ Cl_2(g)$$

4. Use the half-reaction method to balance the following reaction in acidic solution:

$$Zn(s) \ + \ NO_3^- \ \longrightarrow \ Zn^{2+} \ + \ NH_4^+$$

5.   Use the half-reaction method to balance the following reaction in basic solution:

$$MnO_4^- + C_2O_4^{2-} \rightarrow MnO_2(s) + CO_3^{2-}$$

6.   Use the half-reaction method to balance the following reaction in acidic solution:

$$Mn^{2+} + BiO_3^- \rightarrow MnO_4^- + Bi^{3+}$$

7.   Use the half-reaction method to balance the following reaction in basic solution:

$$ClO_3^- + N_2H_4(l) \rightarrow NO_3^- + Cl^-$$

ANSWERS TO PRACTICE PROBLEMS

1. $Cr_2O_3(s) + 2Na_2CO_3(s) + 3KNO_3(aq) \rightarrow 2Na_2CrO_4(aq) + 2CO_2(g) + 3KNO_2(aq)$

2. $2OH^- + 3PbO_2(s) + 2Sb(s) \rightarrow 3PbO(s) + 2SbO_2^- + H_2O(l)$

3. $14H^+ + 12Cl^- + Cr_2O_7^{2-} \rightarrow 2CrCl_3(s) + 3Cl_2(g) + 7H_2O(l)$

4. $10H^+ + 4Zn(s) + NO_3^- \rightarrow 4Zn^{2+} + NH_4^+ + 3H_2O(l)$

352

5. $4OH^- + 2MnO_4^- + 3C_2O_4^{2-} \rightarrow 2MnO_2(s) + 6CO_3^{2-} + 2H_2O(l)$

6. $14H^+ + 2Mn^{2+} + 5BiO_3^- \rightarrow 2MnO_4^- + 5Bi^{3+} + 7H_2O(l)$

7. $6OH^- + 7ClO_3^- + 3N_2H_4(l) \rightarrow 6NO_3^- + 7Cl^- + 9H_2O(l)$

CHAPTER 12

LIQUIDS, SOLIDS, AND CHANGES IN STATE

SECTION 12.2   INTERMOLECULAR ATTRACTIONS

After completing this section, you will be able to do the following:

- Define and differentiate the three kinds of van der Waals forces --- dipole-dipole attractions, Hydrogen bonding, and London forces;

- Predict which molecules will hydrogen bond; determine the relative strengths of London forces.

REVIEW

The two covalent bonds binding oxygen to hydrogen in a water molecule are **intramolecular attractions**, that is, forces within or inside the molecule. Whether a substance is a solid, liquid, or gas at a specific temperature depends on the strength of the attractive forces **between** molecules, which are called **intermolecular attractions** or **van der Waals forces**. There are three kinds of **intermolecular attractions** or **van der Waals forces** --- dipole-dipole interactions, hydrogen bonds, and London forces.

Dipole-dipole attractions are attractions between the negative end of the dipole of one molecule and the positive end of the dipole of a neighboring molecule. Because dipoles result from **partial** positive and negative charges, the dipole-dipole attractions are not nearly as strong as the attractions between ions with full positive and negative charges. The attractions between Na^+ and Cl^- ions in NaCl are much stronger than the attraction between the polar molecules (dipoles) of covalently bonded HCl.

Hydrogen bonds are unusually strong dipole-dipole attractions present between molecules that have hydrogen bonded to nitrogen, oxygen, or fluorine. Because nitrogen, oxygen, and fluorine are the most electronegative elements and also the smallest atoms of any element except hydrogen, covalent bonds between these elements and hydrogen are **more polar** than bonds between hydrogen and any other elements. As a result of the large partial charges and small sizes of these atoms, dipole-dipole attractions **between** molecules with hydrogen-fluorine, hydrogen-oxygen, and hydrogen-nitrogen bonds are unusually strong. For a H atom to **form a hydrogen bond**, the H atom **must** be covalently bonded **to N, O, or F** and be close to a non-bonding pair of electrons of a **N, O, or F** atom on a neighboring molecule. For this reason, H_2O molecules form hydrogen bonds with each other; the same is true of hydrogen fluoride, HF. Molecules that contain N, O, or F **not** bonded to H (CO_2 or $H_2C=O$) can **accept** a H bond with one of their unshared pairs of electrons, but they cannot provide a hydrogen atom to form a hydrogen bond with neighboring molecules.

London forces are the attractions between molecules that result from temporary dipoles (or induced dipoles) caused by the movement of electrons. At any given instant, the charge density distribution of electrons around the nucleus of an atom is **unsymmetrical**. This condition sets up a temporary or induced dipole, which in turn induces similar temporary dipoles in neighboring molecules. London forces **increase** with (a) the polarizability (ease of inducing a temporary dipole) of the outer electrons of the atoms in the molecule, (b) the sizes of molecules of similar shapes, and (c) the more non-spherical shape of the molecule. All molecules are attracted to other molecules by London forces. Polar molecules have **both** London forces and dipole-dipole attractions. For most molecules, London forces are more important than dipole-dipole attractions. That is, the **polarizability** of the larger atom (London forces) is more important and produces stronger forces than the dipole-dipole attractions of the more electronegative smaller atom.

SAMPLE PROBLEM 12.2 A

▶ ▶ Of the following molecules, (a) which can hydrogen bond to other molecules of the same kind? (b) Which can only accept hydrogen bonds (from H_2O, HF, etc.)?

$$H_2NOH \qquad O_2 \qquad HOOH \qquad N_2O_4 \qquad H_2$$

Solution:

(a) Any molecule which has H **bonded to** N, O, or F can form hydrogen bonds. The only molecules meeting this definition are H_2NOH (2 H's bonded to N and 1 H bonded to O) and HOOH. O_2 and N_2O_4 don't contain H. H_2 does not have H bonded to N, O, or F.

(b) Any molecule which contains O, N, or F with no H bonded to it can **only accept** H bonds. That is, the molecule does not have a H bonded to N, O, or F to form a H bond with another molecule. The molecules O_2 and N_2O_4 can **only accept** H bonds.

SAMPLE PROBLEM 12.2 B

▶▶ Which substance has stronger London forces: (a) argon gas or radon gas? (b) $CH_3CH_2CH_2Br$ or CH_3CH_2Br? (c) $CH_3(CH_2)_2CH_3$ or $(CH_3)_2CHCH_3$?

Solution:

HINT: The factors to take into consideration are polarizability of atoms, sizes of molecules, and shapes of molecules.

(a) You are comparing atoms only, so polarizability is the important factor. The larger the atom, the more polarizable the atom is and the stronger the London forces. Radon gas will have the stronger London forces.

(b) These molecules have similar shapes (essentially linear). The longer the molecule, the greater the contact between molecules and the stronger the London forces. $CH_3CH_2CH_2Br$ is longer and will have stronger London forces.

(c) These molecules have identical numbers of atoms but different shapes. $CH_3(CH_2)_2CH_3$ is essentially linear and $(CH_3)_2CHCH_3$ is T-shaped. The less spherical the shape (the more linear), the more contact between molecules and the higher the London forces. $CH_3(CH_2)_2CH_3$ is less spherical and will have the stronger London forces.

SECTION 12.4   VAPOR PRESSURE AND BOILING POINT

After completing this section, you will be able to do the following:

■ Define boiling point and normal boiling point;
■ Interpret vapor pressure-temperature curves for various substances.

REVIEW

The **equilibrium vapor pressure** is the pressure of the vapor in equilibrium with a liquid. The equilibrium vapor pressure of a specific liquid depends on the intermolecular forces in the liquid and on temperature. It does **not** depend on the volume of the vapor or on the volume or surface area of the liquid. The **boiling point** of a liquid is the temperature at which the liquid's vapor pressure equals the pressure on the sample. The **normal boiling point** of a liquid is the boiling point at which the pressure on the sample is standard atmospheric pressure (1 atm, or 760 mmHg). Liquids that have **high vapor pressures** and **low normal boiling points** are called **volatile** liquids. Liquids that have **low vapor pressures** and **high normal boiling** points are called **non-volatile** liquids.

SAMPLE PROBLEM 12.4 A

▸ ▸ From Figure 12.24 in the textbook, (a) what is the normal boiling point of liquid bromine? (b) What is the boiling point of bromine at 200 mmHg? (c) What must the pressure be if bromine boils at 45 °C? (d) Which has the weaker intermolecular attractions, Br_2 or H_2O?

Solution:

(a) The normal boiling point is the temperature at which the vapor pressure of the liquid equals 1 atm, or 760 mmHg. In Figure 12.24, the intersection of the horizontal dashed line (representing 760 mmHg) and the vapor pressure curve of Br_2 is the normal boiling point of Br_2. The normal boiling point of Br_2 is ~58 °C.

(b) Where the horizontal dashed line (representing 200 mmHg) intersects with the vapor pressure curve of Br_2 is the boiling point of Br_2 at 200 mmHg. The boiling point of Br_2 at 200 mmHg is ~24 °C. Only at 760 mmHg is the boiling point of a liquid called the **normal** boiling point.

(c) To find the pressure at which a liquid will boil at a specific temperature, locate the temperature on the horizontal scale and draw a vertical line up to the vapor pressure curve. Where the vertical temperature line intersects with the vapor pressure curve is the pressure at that specific boiling point. Liquid Br_2 will boil at 45 °C if the pressure is ~475 mmHg.

(d) At a particular pressure, the liquid which boils at the **lower** temperature has **weaker** intermolecular attractions. That is, the weaker the intermolecular attractive forces, the less energy (heat) will be required to pull the molecules apart and form a vapor. Arbitrarily pick 200 mmHg. At this pressure Br_2 boils at ~24 °C and H_2O boils at ~67 °C. Br_2 boils at the lower temperature, so Br_2 has the weaker intermolecular attractions. The same conclusion would be reached for any other pressure that intersects the vapor pressure curves of both Br_2 and H_2O.

SECTION 12.7   PHASE DIAGRAMS

After completing this section, you will be able to do the following:

■ Interpret a phase diagram of a substance.

REVIEW

A **phase diagram** summarizes information about solid, liquid, and vapor phase changes. Phase diagrams are drawn by plotting the pressures and temperatures at which solid and liquid (s-l), liquid and vapor (l-v), and solid and vapor (s-v) are in equilibrium (see Figures 12.29 and 12.32 in the textbook for phase diagrams of water and carbon dioxide, respectively). Lines in phase diagrams show equilibria between two phases. At the **triple point**, all three phases are in equilibrium at a fixed temperature and pressure. The **critical point** is the **end** of the vapor pressure curve. The temperature at the critical point is called the **critical temperature, t_c,** and the pressure at the critical point is called the **critical pressure, P_c.** At temperatures **above** the critical temperature, a substance cannot be liquified no matter how great the pressure. Above the **critical point**, a substance is neither a gas nor a liquid but a **supercritical fluid**. Supercritical fluids have low viscosities like gases but are as dense as liquids. (Hence, the word fluid is used, which includes liquids and gases.) When a substance changes directly from a solid to a gas without melting, the process is called **sublimation**.

A phase diagram for a substance (a) tells the state in which the substance exists under different combinations of temperature and pressure, and (b) can be used to predict what will happen if the temperature of a substance is changed at constant pressure or if the pressure on a substance is changed at constant temperature.

SAMPLE PROBLEM 12.7 A

▸▸ Refer to the phase diagram for carbon dioxide in the textbook (Figure 12.32). (a) Will carbon dioxide be a solid, liquid, gas, or supercritical fluid at (i) -25 °C and 25 atm? (ii) -65 °C and 15 atm? (iii) 40 °C and 80 atm? (b) At what temperature and pressure will all three phases (s, l, and g) exist? (c) Does carbon dioxide sublime? If so, under what conditions? (d) Describe what will happen if a sample of carbon dioxide at -75 °C and 6 atm is warmed to 25 °C at a constant pressure of 6 atm.

Solution:

(a) (i) Find the point on the phase diagram (-25 °C and 25 atm) and note the phase describing that area. Carbon dioxide is a liquid at -25 °C and 25 atm.

(ii) Carbon dioxide is a solid at -65 °C and 15 atm.

(iii) Carbon dioxide is a supercritical fluid at 40 °C and 80 atm. From Figure 12.32, carbon dioxide will be a supercritical fluid as long as the temperature is above 31.0 °C **and** the pressure is greater than 73 atm.

(b) All three phases of carbon dioxide will be present (are at equilibrium) at the triple point, -56.6 °C and 5.11 atm.

(c)   **Yes**, carbon dioxide does change directly from solid to gas without melting (that is, it sublimes) at the pressures and temperatures of points along the curve from the triple point to the intersection with the horizontal supercooled liquid line.

(d)   At 6 atm, draw a horizontal line across the phase diagram. As the solid sample at -75 °C and 6 atm warms at a constant pressure of 6 atm, the temperature will reach ~ -57 °C at which point the temperature will stay constant until all the solid has melted. Then liquid CO_2 will warm to ~ -51 °C at which point the temperature will remain constant until all the liquid has turned to vapor. Above -51 °C, the CO_2 vapor will simply warm until 75 °C is reached.

SECTION 12.8   TYPES OF CRYSTALS

After completing this section, you will be able to do the following:

■   Classify crystals as one of four types: molecular, ionic, covalent network, and metallic.

REVIEW

The physical properties of crystalline solids depend on (a) the nature of the particles that make up the solid, and (b) the attractive forces between them. Using these two criteria as the basis for classification, crystals can be divided into four types: molecular, ionic, covalent network, and metallic.

In a **molecular crystal**, the unit particles are **molecules**. **Molecular crystals** are characterized as being soft substances with low melting points and no electrical conduction as a solid or melt (liquid near its freezing point) owing to **relatively weak forces of attraction** ---- London forces and (in polar molecules) dipole-dipole attractions or hydrogen bonding. In an **ionic crystal**, the unit particles are ions. **Ionic crystals** are characterized as being hard and brittle with high melting points and good electrical conductors as a melt owing to a **strong attraction between oppositely charged ions**. In **covalent network crystals**, the unit particles are atoms. **Covalent network crystals** are characterized as being very hard and very high melting substances that do not conduct an electrical current owing to a **three-dimensional network of very strong covalent bonds**. The covalent bonds are the attractive forces between the atoms in a covalent network solid, such as in diamond. In **metallic crystals**, the "unit particles" are metal ions at fixed positions in a sea of mobile valence electrons. **Metallic crystals** are characterized by being shiny, ductile, and malleable substances that may be soft or hard, low or high melting, and good conductors as solids and as a melt owing to the **intermediate attractive forces between metal ions and mobile valence electrons**.

SAMPLE PROBLEM 12.8 A

▶ ▶ For each of the following crystals, (a) classify them as one of the four types. (b) Describe the attractive forces holding the unit particles in place.

Ice SiO_2 (sand)   Ag   KCl

Solution:

Ice

(a) Ice (H_2O) is a compound of two non-metals that is relatively low melting and a non-conductor. Ice is a molecular solid.

(b) The attractive forces are London forces and hydrogen bonds.

SiO_2 (sand)

(a) Individual crystals of SiO_2 are hard, very high melting, and do not conduct an electrical current (which make beaches much more enjoyable). SiO_2 (sand) is a covalent network solid. In spite of its formula, SiO_2 is a network of SiO_4 tetrahedra with Si covalently bonded to four oxygen atoms.

(b) The attractive forces are covalent bonds between Si and O atoms.

Ag

(a) Silver is shiny, ductile, malleable, and a good conductor of electricity (wiring, electrodes, etc.). Ag is a metallic solid.

(b) The attractive forces are the attraction between Ag^+ ions and its mobile valence electrons.

KCl

(a) KCl is a compound of a metal from Group IA (with relatively low electronegativity) and a non-metal from Group VIIA (with relatively high electronegativity) that is hard and has a high melting point. KCl is an ionic solid.

(b) The attractive forces are the attraction between K^+ ions and Cl^- ions.

SECTION 12.11   CALCULATION OF ATOMIC AND IONIC RADII AND AVOGADRO'S NUMBER

After completing this section, you will be able to do the following:

■ Calculate atomic radii from x-ray diffraction data.

REVIEW

Crystals of metals contain atoms in **simple cubic** or, more commonly, **face-centered cubic** or **body-centered cubic** arrangements. Each of these three crystal arrangements has a different **unit cell**. A **unit cell** is the smallest unit of the structure that, if repeated in three dimensions, would produce the crystal. For all three types of arrangements, the unit cell is cubic in shape.

For a **simple cubic** unit cell, there is one metal atom at each of the eight corners of the cube (see Figure 12.37 of the textbook). The number of metal atoms **per unit cell** is **NOT** eight, but **one**, since each corner atom in a unit cell is shared with the corners of 7 other adjacent unit cells. That is, in three dimensions, 1/8 of a corner atom is counted with each unit cell. For a **face-centered cubic** unit cell, there are metal atoms at each of the eight corners and a metal atom at the center of the six faces (see Figure 12.43 (d) in the textbook). The number of atoms per unit cell for the **face-centered cubic** arrangement is **four** (1/8 x 8 corner atoms + 1/2 x 6 face atoms). Each face atom is shared by two adjacent unit cells. For a **body-centered cubic** unit cell, there are atoms at each of the eight corners and one atom at the **center** of the cube (see Figure 12.44 in the textbook). The number of atoms per unit cell for the **body-centered** arrangement is **two** (1/8 x 8 corner atoms + 1 atom in the center of the cube). The atom in the center of the cube is shared with no other unit cells.

When the crystal arrangement and unit cell data are known for a substance, the atomic or ionic radius can be calculated using the Pythagorean theorem, $A^2 = B^2 + C^2$, where A is the hypotenuse and B and C are the two sides of a right triangle.

SAMPLE PROBLEM 12.11 A

▶ ▶ Iridium crystallizes in a face-centered cubic arrangement. If the edge of a unit cell of iridium is 383 pm long, what is the radius of the iridium atom?

Solution:

See Figure 12.43 (d) in the textbook for a drawing of a face-centered cubic unit cell. The two adjacent edges and diagonal of the same face make a right triangle. Since the atoms of iridium are touching along the diagonal of a face, the length of the diagonal is four times the radius of the iridium atom, r, or 4r (the diameter of the center atom of the face is 2r). The lengths of the two edges forming the other two sides of the triangle are known, 383 pm. According to the Pythagorean theorem,

$$A^2 = B^2 + C^2$$

where A is the hypotenuse of the triangle (diagonal of the face, 4r) and B and C are the two sides of the triangle (two adjacent cell edges, each 383 pm). Substituting in the information,

$$(4r)^2 = (383 \text{ pm})^2 + (383 \text{ pm})^2 = 2.93 \times 10^5 \text{ pm}^2$$

Solving for the radius of the iridium atom,

$$r = 135 \text{ pm}$$

SELF-TEST QUESTIONS

Test your understanding of the concepts and skills in this chapter by working through the multiple choice questions **BEFORE** checking the answers. A periodic table will always be available. The ANSWERS TO SELF-TEST QUESTIONS follow immediately after this section.

1. Which of the following molecules will hydrogen bond to other molecules of the same kind?

   A. C_2H_6 B. NH_2NH_2 C. CH_3F D. H_3COCH_3 E. more than one of those given

2. Argon has a higher melting point (condensation temperature) than neon because of

   A. stronger dipole-dipole attractions.
   B. stronger dipole-induced dipole attractions.
   C. stronger London forces.
   D. hydrogen bonding.
   E. ion-dipole forces.

3. The molecules in a sample of solid sulfur dioxide, SO_2, are attracted to each other by

   A. hydrogen bonding and London forces.
   B. hydrogen bonding.
   C. dipole-dipole attractions, hydrogen bonding, and London forces.
   D. London forces only.
   E. dipole-dipole attractions and London forces.

4. Given the following phase diagram, what is the process indicated by going from point B to F?

   A. sublimation
   B. melting
   C. boiling
   D. condensation
   E. freezing

   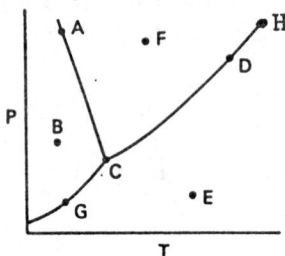

5. Using the same phase diagram (#4), a liquid can be present at conditions corresponding to points

   A. A, C, G, and D.
   B. A, B, and G.
   C. C, D, E, and G.
   D. A, C, D, F, and H.
   E. A, C, D, and H only.

6. A substance is a non-conducting liquid at room temperature. The type of crystal this substance will most likely form below its melting point is

   A. covalent network.
   B   ionic.
   C. molecular.
   D. metallic.
   E. a salt.

7. Copper crystallizes in a face-centered cubic arrangement. The number of copper atoms per unit cell is

   A. 4   B. 2   C. 1   D. 3   E. 14

8. What is the simplest formula of a solid containing A, B, and C atoms in a cubic arrangement in which the A atoms occupy the corner positions, the B atoms occupy the body-center positions, and the C atoms occupy the faces of the unit cell?

A. ABC_6 B. ABC C. A_8BC_6 D. A_4BC_3 E. ABC_3

ANSWERS TO SELF-TEST QUESTIONS

1. b   4. b   7. a
2. c   5. d   8. e
3. e   6. c

PRACTICE PROBLEMS

Test your problem-solving skills in this chapter by working through the problems in the space provided **BEFORE** checking the answers. A periodic table will always be available. The ANSWERS TO PRACTICE PROBLEMS follow immediately after this section.

1. Using Figure 12.24 in the textbook, (a) what is the normal boiling point of normal octane, $CH_3(CH_2)_6CH_3$? (b) What is the boiling point of ethyl alcohol, CH_3CH_2OH, at 600 mmHg? (c) What must the pressure be if water boils at 50 °C? (d) Arrange the four substances in order of increasing intermolecular attractions.

2.   A metallic solid crystallizes in a face-centered cubic arrangement, with a unit cell edge of 375 pm. What is the radius of the metal ion?

ANSWERS TO PRACTICE PROBLEMS

1.   (a) ~125 °C (b) ~72 °C   (c) ~100 mmHg   (d) $Br_2 < CH_3CH_2OH < H_2O < CH_3(CH_2)_6CH_3$

2.   radius = 133 pm

CHAPTER 13

SOLUTIONS REVISITED

SECTION 13.3   EFFECT OF TEMPERATURE ON SOLUBILITY

After completing this section, you will be able to do the following:

◼ Interpret solubility-temperature curves for typical solids.

REVIEW

The solubilities of most, but not all, solids are greater at high temperatures than at low temperatures. A **solubility curve** shows how the solubility of a substance changes with temperature. **Solubility curves** for several typical substances are shown in Figure 13.5 of the textbook. A point on a solubility curve will tell you the solubility of that substance in g/100 g water (Figure 13.5) at a specific temperature. For example, at 20 °C the solubility of solid NaOH in water is 110 g NaOH/100 g water. That is, when you have dissolved 110 g of NaOH in 110 g of water at 20 °C, you will have a **saturated** solution. Dissolving **less** than 110 g of NaOH in 100 g of water at that same temperature will produce an **unsaturated** solution. Having **more** than 110 g of NaOH dissolved in 100 g of water at that same temperature will be a **supersaturated** solution (see Section 4.1 for definitions and examples of **unsaturated, saturated, and supersaturated** solutions). Thus, **at a given temperature**, any point **above** the solubility curve is a **supersaturated** solution, any point **on the curve** is a **saturated** solution, and any point **below** the curve is an **unsaturated** solution.

SAMPLE PROBLEM 13.3 A

▶▶ Referring to Figure 13.5 in the textbook, (a) what is the solubility of $C_{12}H_{22}O_{11}$ at 20 °C? (b) Will a solution of 160 g of NaOH in 100 g of water at 52 °C be unsaturated, saturated, or supersaturated? (c) What is the minimum quantity of water needed to dissolve 50.0 g of NaOH at 40 °C?

Solution:

(a) Draw a vertical line from 20 °C up to the solubility curve of $C_{12}H_{22}O_{11}$ (top curve); then draw a horizontal line from the point of intersection to the solubility axis. The solubility of $C_{12}H_{22}O_{11}$ at 20 °C is 204 g/100 g water.

(b) Locate the point of 160 g solubility/100 g water and 52 °C. Then find the solubility curve for NaOH. The point is **above the solubility curve,** so the solution of 160 g of NaOH in 100 g of water will be supersaturated. For NaOH at 52 °C, a **saturated** solution would contain ~152 g/100 g water; an **unsaturated** solution would contain any amount less than ~152 g NaOH/100 g of water.

(c) The solubility of NaOH at 40 °C is 130 g/100 g water (see part (a) for method), which is a conversion factor. Using the general problem-solving technique, the minimum amount of water needed to dissolve 50.0 g of NaOH is

$$? \text{ g water } = 50.0 \text{ g NaOH } \times \frac{100 \text{ g water}}{130 \text{ g NaOH}} = 38 \text{ g water}$$

SECTION 13.5	TWO MORE CONCENTRATION UNITS: MOLALITY AND MOLE FRACTION

After completing this section, you will be able to do the following:

■ Use mole fractions to calculate partial pressures of gases in gaseous mixtures;

■ Interconvert concentrations in mole fraction, molality, and molarity.

REVIEW

Molality, m, is the number of moles of solute dissolved in one kilogram of solvent,

$$m \text{ (molality)} = \frac{\text{moles solute}}{\text{kg of solvent}}$$

The **mole fraction** of component A in a solution is the fraction of the total number of moles in the solution that is A,

$$X_A = \frac{n_A}{n_A + n_B + n_C + \ldots},$$

where n_A, n_B, n_C, etc., are the number of moles of components A, B, C, respectively.

Mole fractions can be used to calculate the partial pressures of the individual gases in a mixture of gases from the total pressure of the mixture. The partial pressure of a component of the mixture, p_A, is equal to the mole fraction of the component, X_A, times the total pressure, P_{total},

$$p_A = X_A P_{total}$$

Molality, mole fraction, and molarity (Section 4.8) can be interconverted. That is, given any one of the three concentration units for a solution, you can convert it to either of the other two concentration units. Whenever you are converting to or from molarity, the density of the solution must be known.

SAMPLE PROBLEM 13.5 A

▶▶ In a mixture of 30.0 g of benzene (C_6H_6) and 65.0 g of toluene (C_7H_8), (a) what is the mole fraction of each component? (b) What is the molality of the solution if toluene is the solvent?

Solution:

(a) Before you can calculate the mole fraction of each component, you need to calculate the number of moles of each component (molecular masses are calculated from atomic masses on the inside front cover).

$$n_{benzene} = 30.0 \text{ g } C_6H_6 \times \frac{1 \text{ mol } C_6H_6}{78.114 \text{ g } C_6H_6} = 0.384 \text{ mol } C_6H_6$$

$$n_{toluene} = 65.0 \text{ g } C_7H_8 \times \frac{1 \text{ mol } C_7H_8}{92.141 \text{ g } C_7H_8} = 0.705 \text{ mol } C_7H_8$$

Now you can calculate the mole fraction of each component.

$$X_{benzene} = \frac{n_{benzene}}{n_{benzene} + n_{toluene}} = \frac{0.384}{0.384 + 0.705} = 0.353$$

$$X_{toluene} = \frac{n_{toluene}}{n_{benzene} + n_{toluene}} = \frac{0.705}{0.384 + 0.705} = 0.647$$

CHECK: The sum of the mole fractions for all the components in a solution must add up to one, i.e., $0.353 + 0.647 = 1.000$. When the mole fractions are known for all but one component, the mole fraction for the last component can be found by difference. For example, once the mole fraction of benzene was found to be 0.353, the mole fraction of toluene can be calculated by difference, $1.000 - 0.353 = 0.647$.

(b)   Using the formula for molality given in the REVIEW, the number of moles of benzene from part (a), and converting the mass of the solvent toluene to kg, we have

$$m = \frac{\text{mol } C_6H_6}{\text{kg } C_7H_8} = \frac{0.384 \text{ mol } C_6H_6}{0.0650 \text{ kg } C_7H_8} = \frac{5.91 \text{ mol } C_6H_6}{\text{kg } C_7H_8} = 5.91 \text{ m}$$

SAMPLE PROBLEM 13.5 B

▶▶ What is the partial pressure of each gas in a mixture of 5.75 mol of helium gas and 2.86 mol of hydrogen gas if the total pressure is 725.4 mmHg?

Solution:

The partial pressure of a gas, A, in a mixture of gases is

$$p_A = X_A P_{total}$$

You will need to calculate the mole fraction of each gas in the mixture before calculating the partial pressure of each gas.

$$X_{He} = \frac{n_{He}}{n_{He} + n_{H_2}} = \frac{5.75 \text{ mol He}}{5.75 \text{ mol He} + 2.86 \text{ mol } H_2} = 0.668$$

$$X_{H_2} = 1.000 - 0.668 = 0.332$$

Then use the equation given above for calculating the partial pressure of each component,

$$P_{He} = X_{He}P_{total} = 0.668(725.4 \text{ mmHg}) = 485 \text{ mmHg}$$

$$P_{H_2} = X_{H_2}P_{total} = 0.332(725.4 \text{ mmHg}) = 241 \text{ mmHg}$$

CHECK: The partial pressures of all the component gases must add up to the total pressure (i.e., 485 mmHg + 241 mmHg = 726 mmHg, which with rounding off is essentially the P_{total} of 725.4 mmHg).

SAMPLE PROBLEM 13.5 C

▶▶ If aqueous H_2SO_4 solution is 4.75 m, (a) what is the mole fraction of H_2SO_4 in the solution? (b) If the density of the solution is 1.0875 g/mL, what is the molarity of the solution?

Solution:

(a) To find the mole fraction of H_2SO_4, you will need to calculate first the number of moles of each component, H_2SO_4 and H_2O. By definition, a 4.75 m H_2SO_4 solution contains 4.75 mol H_2SO_4 in 1.000 kg of H_2O. To find the number of moles of water in 1.000 kg of water,

$$? \text{ mol } H_2O = 1000 \text{ g } H_2O \times \frac{1 \text{ mol } H_2O}{18.0153 \text{ g } H_2O} = 55.51 \text{ mol } H_2O$$

Now calculate the mole fraction of H_2SO_4,

$$X_{H_2SO_4} = \frac{n_{H_2SO_4}}{n_{H_2SO_4} + n_{H_2O}} = \frac{4.75 \text{ mol } H_2SO_4}{4.75 \text{ mol } H_2SO_4 + 55.51 \text{ mol } H_2O} = 0.0788$$

(b) Remember the definition of molarity (Section 4.8),

$$M \text{ (molarity)} = \frac{\text{moles of solute}}{\text{L solution}}$$

You already know there are 4.75 mol H_2SO_4 in a 4.75 m H_2SO_4 solution. The word solution, however, means **solute + solvent**, that is, H_2SO_4 + H_2O. From the information given in the problem, you will need to determine the number of grams of solution (H_2SO_4 and H_2O) and convert this mass to volume in liters using the density of the solution.

$$? \text{ g soln.} = \text{g } H_2SO_4 + \text{g } H_2O = ? \text{ g } H_2SO_4 + 1000 \text{ g } H_2O.$$

Using the molecular mass of H_2SO_4 from the inside back cover,

$$? \text{ g } H_2SO_4 = 4.75 \text{ mol } H_2SO_4 \times \frac{98.079 \text{ g } H_2SO_4}{1 \text{ mol } H_2SO_4} = 466 \text{ g } H_2SO_4$$

The grams of solution are

$$? \text{ g soln.} = 466 \text{ g } H_2SO_4 + 1000 \text{ g } H_2O = 1466 \text{ g soln.}$$

Using the density of the solution, the volume in liters of the solution is

$$? \text{ L soln.} = 1466 \text{ g soln} \times \frac{1 \text{ mL soln}}{1.0875 \text{ g soln}} \times \frac{1 \text{ L soln}}{1000 \text{ mL soln}} = 1.348 \text{ L}$$

Now the molarity of the solution can be calculated.

$$? \text{ M} = \frac{4.75 \text{ mol } H_2SO_4}{1.348 \text{ L soln}} = 3.52 \text{ M}$$

SECTION 13.7   COLLIGATIVE PROPERTIES

After completing this section, you will be able to do the following:

■ Calculate the vapor pressure of a solution of a non-volatile substance;

■ Calculate the boiling points and freezing points of solutions;

■ Determine the molecular mass by freezing point depression;

■ Calculate molecular mass from osmotic pressure data.

REVIEW

Properties of solutions that depend only on the **number** (not identity) of solute particles are called **colligative properties**. The four **colligative properties** of solutions (which are the same for all **non-volatile** solutes) are **vapor pressure lowering, boiling point elevation, freezing point depression, and osmotic pressure**.

The vapor pressure of a solution containing a **non-volatile** component in a volatile solvent is equal to the mole fraction of solvent in the solution times the vapor pressure of the pure solvent, since the vapor pressure of a **non-volatile** component in a solution is essentially zero. That is, the vapor pressure of the volatile solvent **is** the vapor pressure of the solution. To calculate the vapor pressure of a solution containing a non-volatile component in a volatile solvent, we must use Raoult's law (Section 13.6 in the textbook),

$$VP_{A \text{ in solution}} = X_{A \text{ in solution}} VP_{\text{pure A}},$$

where A is the volatile solvent, and component B is the non-volatile component. Because the mole fraction of solvent in a solution must be less than one (due to the mole fraction of non-volatile solute present), the vapor pressure of a solution of a non-volatile solute is always **lower** than the vapor pressure of the pure solvent.

Because a non-volatile solute lowers the vapor pressure of the solvent, a higher temperature must be reached for the vapor pressure of the solvent to be equal to atmospheric pressure. That is, the **boiling point** of a solution of a **non-volatile** solute is always **higher** than the boiling point of the pure solvent. At the freezing point for the same kind of solution, the solvent freezes out first, making the remaining solution more concentrated, and the **freezing point decreases**. A **non-volatile** solute **raises the boiling point** and **lowers the freezing point** of the solvent. The magnitude of the boiling point elevation and freezing point depression are proportional to the **molality** (not molarity) of the solution,

$$\Delta bp = K_{bp}m \qquad \text{and} \qquad \Delta fp = K_{fp}m,$$

where K_{bp} and K_{fp} are constants characteristic of the **solvent**. These equations can be used to calculate boiling points, freezing points, and molalities of solutions containing a non-volatile solute in a volatile solvent. These same equations can also be used if the solute is an electrolyte, because we will always assume that the electrolyte dissociates 100%. That is, a 0.10 m NaCl solution is 0.20 m in solute particles (Na^+ and Cl^- ions) for calculating Δbp and Δfp. The freezing point depression equation, $\Delta fp = K_{fp}m$, can also be used to calculate the molecular mass (g/mol) of the solute by having sufficient data to calculate the molality of the solution first and then the molecular mass of the solute.

The spontaneous net movement of solvent molecules (typically water) through a semipermeable membrane **from** a dilute solution (higher mole fraction of solvent) **into** a more concentrated solution (lower mole fraction of solvent) is called **osmosis**. A **semipermeable membrane** is one that will only allow solvent molecules to pass through (in both directions). As long as the concentrations of the solutions on both sides of the membrane are different, **osmosis** (net flow of water molecules) will continue. The pressure that is needed to **stop** the net movement of solvent molecules through a semipermeable membrane (that is, to stop osmosis) is called the **osmotic pressure** of the solution. The **osmotic pressure** needed to stop osmosis is determined by

$$\pi = MRT,$$

where M is the molarity of the solution, R is the gas constant, and T is the Kelvin temperature. This equation is used to calculate osmotic pressure and the molecular mass of biologically important molecules.

372

SAMPLE PROBLEM 13.7 A

▶▶ What will be the vapor pressure of a solution at 25 °C in which a non-volatile solute has a mole fraction of 0.250 and benzene is the solvent? The vapor pressure of benzene at 25 °C is 93.4 torr.

Solution:

The vapor pressure of the non-volatile solute is zero. The vapor pressure of the solution is just the vapor pressure of the volatile solvent benzene. According to Raoult's law (Section 13.6 in textbook),

$$VP_{benzene} = X_{benzene}VP_{pure\ benzene}$$

If the mole fraction of the solute is 0.250, then the mole fraction of benzene must be 1.000 − 0.250, or 0.750. Substituting into the equation,

$$VP_{benzene} = (0.750)(93.4\ torr) = \boxed{70.1\ torr}$$

SAMPLE PROBLEM 13.7 B

▶▶ For a 0.150 m solution of toluene (non-volatile, non-electrolyte) in the solvent benzene, (a) what is the boiling point? (b) What is the freezing point?

Solution:

(a) The boiling point of the solution is equal to the boiling point of the solvent plus the boiling point change, or

$$bp_{soln} = bp_{solvent} + \Delta bp$$

From Table 13.3 in the textbook, the boiling point of the solvent benzene is 80.1 °C and the boiling point constant, K_{bp}, is 2.53 °C/m. To find the change in boiling point, Δbp, use the equation given in the REVIEW,

$$\Delta bp = K_{bp}m = (2.53\ °C/m)(0.150\ m) = 0.380\ °C$$

The boiling point of the solution is

$$bp_{soln} = bp_{solvent} + \Delta bp = 80.1\ °C + 0.380\ °C = \boxed{80.5\ °C}$$

(b) The freezing point of the solution is equal to the freezing point of the solvent plus the freezing point change, or

$$fp_{soln} = fp_{solvent} + \Delta fp$$

From Table 13.3 in the textbook, the freezing point of the solvent benzene is 5.5 °C and the freezing point constant, K_{fp}, is −5.12 °C/m. To find the change in freezing point, Δfp, use the equation given in the REVIEW,

$$\Delta fp = K_{fp}m = (-5.12 \text{ °C/m})(0.150 \text{ m}) = -0.768 \text{ °C}$$

The freezing point of the solution is

$$fp_{soln} = fp_{solvent} + \Delta fp = 5.5 \text{ °C} + (-0.768 \text{ °C}) = 4.7 \text{ °C}$$

SAMPLE PROBLEM 13.7 C

▶▶ What is the freezing point of a 0.65 m $Mg(NO_3)_2$ solution?

Solution:

See Sample Problem 13.7 B(b) for detailed procedure.

$$fp_{soln} = fp_{solvent} + \Delta fp$$

The freezing point of the solvent water is 0.0 °C and the freezing point constant, K_{fp}, is −1.85 °C/m. To find the change in freezing point, Δfp, the molality will have to be multiplied by three, since three ions (1 Mg^{2+} and 2 NO_3^-) are produced when one $Mg(NO_3)_2$ unit completely dissociates. (Remember, colligative properties depend on the **number** of particles in solution.) The change in freezing point for the 0.65 m $Mg(NO_3)_2$ solution is

$$\Delta fp = K_{fp}m = (-1.85 \text{ °C/m})(3)(0.65 \text{ m}) = -3.6 \text{ °C}$$

The freezing point of the solution is

$$fp_{soln} = fp_{solvent} + \Delta fp = 0.0 \text{ °C} + (-3.6 \text{ °C}) = -3.6 \text{ °C}$$

SAMPLE PROBLEM 13.7 D

▶▶ A mixture of 0.454 g of an unknown solute and 8.16 g of benzene freezes at 1.4 °C. What is the molecular mass of the unknown solute?

Solution:

Before you can find the molecular mass (g/mol) of the unknown solute, you will need to calculate the molality of the solution (which has moles of solute in it). However, before solving the freezing point change equation for molality, you must first find the freezing point change for this solution from the actual freezing point. In Sample Problem 13.7 C, we saw that

$$fp_{soln} = fp_{solvent} + \Delta fp$$

Solving this equation for the change in freezing point,

$$\Delta fp = fp_{soln} - fp_{solvent} = 1.4 \text{ °C} - 5.5 \text{ °C} = -4.1 \text{ °C}$$

Now calculate the molality of the solution using $K_{fp} = -5.12$ °C/m for benzene (from Table 13.3 in textbook),

$$m = \frac{\Delta fp}{K_{fp}} = \frac{-4.1 \text{ °C}}{-5.12 \text{ °C}} = 0.80 \text{ m} = \frac{0.80 \text{ mol unknown}}{\text{kg benzene}}$$

We now know the concentration of the solution in molality. We also know the concentration of the solution from the other information given in the statement of the problem, 0.454 g unknown/8.16 g benzene, or (converting the solvent to kg) 0.454 g unknown/0.008 16 kg benzene. (The reason for changing the solvent to kg will be evident with the last calculation in this problem.) These two concentration units are conversion factors from which we can find the molecular mass of the unknown by applying our general problem-solving technique. [This method of calculating molecular mass will also be used in the next problem, Sample Problem 13.7 E.] Remember, molecular mass is the mass in grams of one mole, or g/mol. Use the two conversion factors so that grams of unknown is in the numerator and moles of unknown is in the denominator (and there is only one way the conversion factors can be used to arrive at these units).

$$? \frac{g}{mol} \text{ (MM)} = \frac{0.454 \text{ g unknown}}{0.008 \text{ 16 kg benzene}} \times \frac{1 \text{ kg benzene}}{0.80 \text{ mol unknown}} = 7.0 \times 10^1 \text{ g/mol}$$

SAMPLE PROBLEM 13.7 E

▸▸ A solution of 0.75 g of a polypeptide in 500.0 mL of solution has an osmotic pressure of 6.50 mmHg at 23.0 °C. What is the molecular mass of the polypeptide?

Solution:

Applying the formula containing osmotic pressure discussed in the REVIEW, you can calculate the molarity (M) of the solution, using the value for R containing the pressure in mmHg.

$$M = \frac{\pi}{RT} = \frac{6.50 \text{ mmHg}}{(62.36 \text{ mmHg} \cdot L \cdot mol^{-1} \cdot K^{-1})(273.2 + 23.0 \text{ K})}$$

$$= 3.52 \times 10^{-4} \text{ M} = 3.52 \times 10^{-4} \text{ mol polypeptide/L soln.}$$

You know the concentration of the solution in units of molarity **and** g/L from the statement of the problem, 0.75 g polypeptide/500.0 mL soln, or (changing the volume to liters) 0.75 g polypeptide/0.5000 L soln. Now you can calculate the molecular mass (g/mol) of the polypeptide, using the exact same method as in Sample Problem 13.7 D.

$$? \frac{g}{mol} \text{ (MM)} = \frac{0.75 \text{ g polypeptide}}{0.5000 \text{ L soln.}} \times \frac{1 \text{ L soln.}}{3.52 \times 10^{-4} \text{ mol polypeptide}} = 4.3 \times 10^3 \text{ g/mol}$$

SELF-TEST QUESTIONS

Test your understanding of the concepts and skills in this chapter by working through the multiple choice questions **BEFORE** checking the answers. A periodic table will always be available. The ANSWERS TO SELF-TEST QUESTIONS follow immediately after this section.

1. Referring to Figure 13.5 in the textbook, the solubilities of all the following substances increase with increasing temperature except for

   A. $C_{12}H_{22}O_{11}$ B. $Ce_2(SO_4)_3$ C. NaCl   D. NaOH   E. KCl

2. Referring to Figure 13.5 in the textbook, arrange the following substances in order of decreasing solubility at 50 °C

   KCl   NaOH   NaCl $Ce_2(SO_4)_3$ KNO_3

   A. $Ce_2(SO_4)_3$ > NaCl > KCl > KNO_3 > NaOH

   B. $Ce_2(SO_4)_3$ > KCl > NaCl > KNO_3 > NaOH

   C. NaOH > KNO_3 > NaCl > KCl > $Ce_2(SO_4)_3$

   D. NaOH > KNO_3 > KCl > NaCl > $Ce_2(SO_4)_3$

   E. $Ce_2(SO_4)_3$ > NaOH > KNO_3 > KCl > NaCl

3. Referring to Figure 13.5 in the textbook, a solution containing 25 g of a solid dissolved in 100 g of water at 10 °C will be

   A. saturated if the solid is $C_{12}H_{22}O_{11}$.

   B. supersaturated if the solid is $KClO_3$.

   C. unsaturated if the solid is KNO_3.

   D. supersaturated if the solid is NaCl.

   E. unsaturated if the solid is $Ce_2(SO_4)_3$.

4. The change in boiling point for a 1.31 m solution of naphthalene in benzene is

$$(K_{bp} = 2.53 \text{ °C/m for benzene})$$

A. 1.93 °C B. 10.7 °C C. 4.31 °C D. 7.41 °C E. 3.31 °C

5. Addition of a nonvolatile solute to a solvent

    A. raises the freezing point and the boiling point.
    B. lowers the freezing point and the boiling point.
    C. raises the freezing point and lowers the boiling point.
    D. lowers the freezing point and raises the boiling point.
    E. affects only the freezing point.

6. The aqueous solution with the lowest freezing point, assuming 100% dissociation of electrolytes, is

    A. 0.010 m $FeCl_3$.

    B. 0.010 m $BaCl_2$.

    C. 0.010 m NaCl.

    D. 0.0050 m $C_{12}H_{22}O_{11}$ (sucrose).

    E. 0.010 m HNO_3.

ANSWERS TO SELF-TEST QUESTIONS

1. b	3. b	5. d
2. d	4. e	6. a

PRACTICE PROBLEMS

Test your problem-solving skills in this chapter by working through the problems in the space provided **BEFORE** checking the answers. A periodic table will always be available. The ANSWERS TO PRACTICE PROBLEMS follow immediately after this section.

1.   If the mole fraction of NaCl in an aqueous solution is 0.028 80 and the density of the solution is 1.064 g/mL, (a) what is the molality of the solution? (b) What is the molarity of the solution?

2.   What is the vapor pressure of an aqueous solution at 25 °C in which sucrose, $C_{12}H_{22}O_{11}$, has a mole fraction of 0.143? The vapor pressure of water at 25 °C is 23.76 mmHg.

3.   For a 0.650 m solution of urea in water, calculate the boiling point and the freezing point. For water, K_{bp} = 0.51 °C/m, and K_{fp} = −1.85 °C/m.

4.   What is the mole fraction of non-volatile sucrose in an aqueous solution that has a vapor pressure of 20.00 mmHg at 25 °C. The vapor pressure of pure water at 25 °C is 23.76 mmHg.

5.   A solution consisting of 0.15 g of an unknown non-volatile in 5.70 g of benzene had a freezing point of 4.90 °C. What is the molecular mass of the unknown? For benzene, mp = 5.50 °C, and K_{fp} = −5.12 °C/m.

6.   A certain water-soluble polymer has a molecular mass of 5000.0 g/mol. What is the osmotic pressure in mmHg of a 100.0-mL aqueous solution containing 0.025 g of the polymer, when the solution is at 25 °C?

ANSWERS TO PRACTICE PROBLEMS

1. (a) In a one-mole sample of the solution, there will be 0.0288 mol NaCl and $(1 - 0.0288) =$ 0.9712 mol H_2O. Convert 0.9712 mol H_2O to kg H_2O.

    $$\text{m of NaCl soln.} = \frac{0.028\ 80 \text{ mol NaCl}}{0.017\ 50 \text{ kg } H_2O} = \boxed{1.646 \text{ m}}$$

    (b) You can calculate molarity from either mole fraction or molality (which is shorter). Using the molality calculated in part (a),

    $$? \text{ g NaCl} = 1.646 \text{ mol NaCl} \times \frac{58.4425 \text{ g NaCl}}{1 \text{ mol NaCl}} = 96.20 \text{ g NaCl}$$

    $$? \text{ g soln.} = \text{g NaCl} + \text{g } H_2O = 96.20 \text{ g NaCl} + 1000 \text{ g } H_2O = 1096 \text{ g soln.}$$

    $$? \text{ L soln.} = 1096 \text{ g soln.} \times \frac{1 \text{ mL soln.}}{1.064 \text{ g soln.}} \times \frac{1 \text{ L soln.}}{1000 \text{ mL soln.}} = 1.030 \text{ L soln.}$$

    $$? \text{ M} = \frac{1.646 \text{ mol NaCl}}{1.030 \text{ L soln}} = \boxed{1.598 \text{ M}}$$

2. 20.4 mmHg

3. bp = 100.3 °C; fp = −1.2 °C

4. $$X_{H_2O} = \frac{VP_{soln}}{VP_{pure\ water}} = \frac{20.00 \text{ mmHg}}{23.76 \text{ mmHg}} = 0.8418$$

    $$X_{sucrose} = 1.0000 - 0.8418 = \boxed{0.1582}$$

5. $$m = \frac{-0.60 \text{ °C}}{-5.12 \text{ °C/m}} = 0.12 \text{ m}$$

    $$? \frac{g}{\text{mol}} \text{ (MM)} = \frac{0.15 \text{ g unknown}}{0.005\ 70 \text{ kg benzene}} \times \frac{1 \text{ kg benzene}}{0.12 \text{ mol unknown}} = \boxed{220 \text{ g/mol}}$$

6. $M = \dfrac{0.025 \text{ g polymer}}{0.1000 \text{ L soln.}} \times \dfrac{1 \text{ mol polymer}}{5000.0 \text{ g polymer}} = 5.0 \times 10^{-5} \text{ M}$

$\pi = MRT = (5.0 \times 10^{-5} \text{ mol} \cdot \text{L}^{-1})(62.36 \text{ mmHg} \cdot \text{L} \cdot \text{mol}^{-1} \cdot \text{K}^{-1})(298.2 \text{ K}) = 0.93 \text{ mmHg}$

CHAPTER 14

CHEMICAL EQUILIBRIUM

SECTION 14.2 EQUILIBRIUM CONSTANTS AND EQUILIBRIUM CONSTANT EXPRESSIONS

After completing this section, you will be able to do the following:

■ Write the equilibrium constant expression for a reaction at equilibrium;

■ Given one or more of the substances in a reaction, determine whether equilibrium can be reached.

REVIEW

Chemical equilibrium is reached when the **rates** of the forward and reverse reactions are **equal**, at which point the **concentrations** of the aqueous and gaseous reactants and products are **constant** (NOT necessarily equal). Equilibrium can be reached starting with the substances on either side of the reaction. If the majority of the starting materials for the forward reaction are used up by the time equilibrium is reached, the **point of equilibrium,** or **equilibrium position,** lies to the right (greater than 50% towards completion). If only a small amount (usually 5-15%) of the starting materials are used up by the time equilibrium is reached, the **equilibrium position** lies to the left. There are an infinite number of equilibrium positions possible. Equilibrium reactions that involve only a single phase (s, l, g, or aq) are called **homogeneous equilibria**. Equilibrium reactions that involve more than one phase (s and l, aq and g, etc.) are called **heterogeneous equilibria**.

Because the rates of the forward and reverse reactions are equal at equilibrium, an expression can be written for every reaction at equilibrium, called the **equilibrium constant expression,** which is equal to the **equilibrium constant, K_c.** The subscript c indicates that concentrations in molarity were used to calculate the value of the constant (as opposed to K_p, where pressures of gases are used). The equilibrium constant, K_c, is unique for a particular chemical equation and is determined experimentally. The **equilibrium constant** for a reaction is constant as long as the temperature stays constant. The value of the equilibrium constant gives you an idea of where the position of equilibrium lies for a reaction. Generally, if $K_c > 10^2$, then the equilibrium position lies to the right. If $K_c < 10^{-2}$, then the equilibrium position lies to the left. If the value of K_c is intermediate, $10^{-2} < K_c < 10^2$, then the point of equilibrium is near the center (forward reaction 40-60% completed) and equilibrium is easily observed experimentally.

You **must know** the following procedures for correctly writing equilibrium constant expressions:

WRITING EQUILIBRIUM CONSTANT EXPRESSIONS

1. **Balance** the equilibrium reaction (write an equation for the equilibrium).

2. Put the product (result of multiplication) of the equilibrium concentrations of the **products** in the **numerator,** and the product of the equilibrium concentrations of the **reactants** in the **denominator.**

3. **Omit** pure liquids (l), pure solids (s), and solvents in dilute solution from the equilibrium constant expression (because their concentrations are always constant and do not change when a reaction occurs).

4. Each concentration in the equilibrium expression must be raised to a power equal to the coefficient of the species in the balanced reaction, with an unwritten exponent understood to be one.

The **equilibrium constant expression** and **equilibrium constant** for a reaction can be used to determine whether a reaction is at equilibrium and to calculate equilibrium concentrations of reactants and products from starting concentrations (Section 14.4).

SAMPLE PROBLEM 14.2 A

▶▶ Write the equilibrium constant expression for the following reactions at equilibrium:

(a) $N_2(g) + H_2(g) \rightleftarrows NH_3(g)$

(b) $AgNO_3(aq) + BaCl_2(aq) \rightleftarrows AgCl(s) + Ba(NO_3)_2(aq)$

(c) $SiCl_4(l) + H_2O(g) \rightleftarrows SiO_2(s) + HCl(g)$

Solution:

HINT:
- Use the procedures given in the REVIEW.
- Brackets around a number always mean concentration units of **molarity**.
- The symbols (aq) and (g) do not appear after the concentrations in the equilibrium constant expression.

(a) You **must always** start with a **balanced** reaction to write a correct equilibrium constant expression (or perform any stoichiometric calculations). This reaction as given is **NOT** balanced; the equilibrium constant expression written for this unbalanced reaction would be incorrect. Balance the reaction,

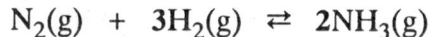

$$N_2(g) + 3H_2(g) \rightleftarrows 2NH_3(g)$$

Now write the equilibrium constant expression with the concentrations of the products multiplied together in the numerator, and the concentration of the reactants multiplied together in the denominator. Remember, **products** in **numerator**, and **reactants** in **denominator**.

$$\frac{[NH_3]^2}{[N_2][H_2]^3} = K_c$$

The order of the reactants in the denominator (or numerator for two or more products) is immaterial. That is, it could also be $[H_2]^3[N_2]$. If the value of K_c was given for this reaction, then it could be substituted into the above equation.

(b) This reaction is **NOT** balanced! Balance the reaction first, then write the equilibrium constant expression, remembering that only formulas with (aq) or (g) appear in the equilibrium constant expression because the concentrations of pure solids and liquids are constant.

$$2AgNO_3(aq) + BaCl_2(aq) \rightleftarrows 2AgCl(s) + Ba(NO_3)_2(aq)$$

$$\frac{[Ba(NO_3)_2]}{[AgNO_3]^2[BaCl_2]} = K_c$$

(c)   This reaction also is **NOT** balanced! Before doing anything further, **always** check to see if the reaction is balanced. If not, then balance it and write the equilibrium constant expression. Formulas identified as (s) and (l) do NOT appear in the equilibrium constant expression.

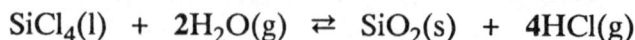

$$SiCl_4(l) + 2H_2O(g) \rightleftarrows SiO_2(s) + 4HCl(g)$$

$$\frac{[HCl]^4}{[H_2O]^2} = K_c$$

SAMPLE PROBLEM 14.2 B

▸▸ Which of the following can reach equilibrium through the reaction

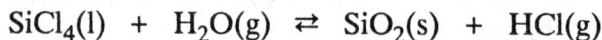

$$SiCl_4(l) + H_2O(g) \rightleftarrows SiO_2(s) + HCl(g)$$

if placed in a closed, dry container (no water vapor present) and allowed to stand for a sufficient length of time?

(a) pure $SiO_2(s)$ (b) pure $SiCl_4(l)$ (c) $SiO_2(s)$ and $HCl(g)$ (d) a mixture of $SiCl_4(l)$ and $SiO_2(s)$ (e) $SiCl_4(l)$ and $HCl(g)$ (f) $SiCl_4(l)$ and $H_2O(g)$.

Solution:

(a)   For $SiO_2(s)$ to react, there must be $HCl(g)$ present. Equilibrium cannot be reached.

(b)   For $SiCl_4(l)$ to react, there must be $H_2O(g)$ present. Equilibrium cannot be reached.

(c) $SiO_2(s)$ and $HCl(g)$ will react to produce $SiCl_4(l)$ and $H_2O(g)$ and eventually establish equilibrium.

(d)   Neither $SiCl_4(l)$ nor $SiO_2(s)$ will react, since the former needs $H_2O(g)$ and the latter needs $HCl(g)$. Since neither $H_2O(g)$ nor $HCl(g)$ are present, equilibrium cannot be established.

(e) $SiCl_4(l)$ needs $H_2O(g)$ to react, and $HCl(g)$ needs $SiO_2(s)$ to react. Since neither $H_2O(g)$ nor $SiO_2(s)$ are present, equilibrium cannot be established.

(f) $SiCl_4(l)$ and $H_2O(g)$ will react to produce $SiO_2(s)$ and $HCl(g)$ and eventually establish equilibrium.

SECTION 14.3   DETERMINATION OF VALUES OF EQUILIBRIUM CONSTANTS

After completing this section, you will be able to do the following:

■ Calculate the numerical value of the equilibrium constant, K_c, from equilibrium concentrations or initial concentrations and one equilibrium concentration;

■ Calculate the numerical value of the equilibrium constant, K_p, from equilibrium partial pressures or K_c for the same equilibrium reaction;

■ Given the value of K_c for a reaction at equilibrium, calculate K_c for any multiple of that reaction or the reverse of that reaction.

REVIEW

The **equilibrium constant expression** can be used to calculate the value of K_c for a reaction **only if the equilibrium** concentrations of all the aqueous and gaseous reactants and products are known. If the **equilibrium** concentrations of all the reactants and products are given in the statement of the problem, then substitute those values into the equilibrium constant expression and calculate K_c (Sample Problem 14.3 A). If the **initial** concentrations (before the reaction has begun) **and** one equilibrium concentration are given, then the equilibrium concentrations of all the reactants and products can be determined and then substituted into the equilibrium constant expression to solve for K_c (Sample Problem 14.3 B).

Equilibrium constants are determined by experiment and can be calculated from concentrations, K_c, or (where gases are present in the reaction) from partial pressures, K_p. The relationship between the two kinds of equilibrium constants for the same equilibrium reaction is

$$K_p = K_c(RT)^{\Delta n_g},$$

where Δn_g is the number of moles of **gaseous products** minus the number of moles of **gaseous reactants**

$$\Delta n_g = n_{gas\ products} - n_{gas\ reactants}$$

If $\Delta n_g = 0$, then $K_p = K_c$. An example of this sort of problem is given in Sample Problem 14.3 C.

An equilibrium constant, K_c or K_p, is specific for the equilibrium reaction given. The equilibrium constant for the **reverse** of this same reaction is

$$K_{new} = \frac{1}{K_{old}}$$

The equilibrium constant for any **multiple** of this reaction, K_{new}, is equal to the original equilibrium constant, K_{old}, raised to the power equal to the multiple, i.e.

$$K_{new} = K_{old}^x,$$

where x can be a whole number or a fraction. That is, if the equilibrium reaction is doubled (all the coefficients multiplied by two), then x = 2. If the equilibrium reaction is multiplied by 1/3 (all the coefficients multiplied by 1/3), then x = 1/3 (Sample Problem 14.3 D).

SAMPLE PROBLEM 14.3 A

▶ ▶ What is the numerical value of K_c for the equilibrium

$$CH_3OH(g) \rightleftharpoons CO(g) + 2H_2(g)$$

if $[CH_3OH] = 0.0209$, $[CO] = 0.0790$, and $[H_2] = 0.158$ at equilibrium?

Solution:

Check to make sure the reaction is balanced. (It is.) Then write the equilibrium constant expression for the reaction, remembering that it is **always** products over reactants.

$$\frac{[CO][H_2]^2}{[CH_3OH]} = K_c$$

The only way that you can calculate an equilibrium constant from the equilibrium constant expression is if you know **concentrations at equilibrium**. Only at equilibrium is the equilibrium constant expression equal to K_c. Since the concentrations given are equilibrium values, substitute them into the equilibrium expression and solve for K_c. Don't forget to square the $[H_2]$.

$$\frac{(0.0790)(0.158)^2}{(0.0209)} = 0.0944 = K_c$$

SAMPLE PROBLEM 14.3 B

▶▶ A mixture that was initially 0.525 M in $Br_2(g)$ and 0.950 M in $NO(g)$ was allowed to reach equilibrium. If, at equilibrium, $[Br_2] = 0.105$, what is the numerical value of K_c for the equilibrium

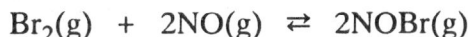

$$Br_2(g) + 2NO(g) \rightleftharpoons 2NOBr(g)$$

Solution:

See Sample Problem 14.3 A. You **must** have equilibrium concentrations of **all** the reactants and **products** in order to calculate K_c from the equilibrium expression. You can determine the equilibrium concentrations of NO and NOBr by first creating a table of known data underneath the reaction as follows:

	$Br_2(g)$	+	$2NO(g)$	$\rightleftharpoons$	$2NOBr(g)$
Initial, M	0.525 M		0.950 M		0.00 M
Change, ΔM	?		?		?
Equilibrium, M	0.105 M		?		?

The first row of data under each reactant and product shows the **initial** concentrations of each (no reaction has begun). Unless told otherwise, always assume an **initial** concentration of **zero** for any reactant or product with no specified concentration. The second row of data is the **change** in concentration that occurs between the initial and equilibrium conditions (which we don't know yet). The third row shows the equilibrium concentration data known at this time. **Initial** plus **change** concentration data yield the equilibrium concentration.

Our goal is to determine the **equilibrium concentrations** of NO and NOBr so that we can calculate K_c using the equilibrium expression for this reaction. Since we know two of the three pieces of information for Br_2 (initial and equilibrium concentrations), we can calculate the change in concentration that had to occur from initial to equilibrium values. That value is the difference between initial and equilibrium concentrations, 0.525 M $-$ 0.105 M, or 0.420 M. Adding that data to our table, we now have

	$Br_2(g)$	+	$2NO(g)$	$\rightleftarrows$	$2NOBr(g)$
Initial, M	0.525 M		0.950 M		0.00 M
Change, ΔM	−0.420 M				
Equilibrium, M	0.105 M				

The minus sign indicates that 0.420 M Br_2 has been **consumed or used up** to reach equilibrium. The next step is to complete the change in concentration data for NO and NOBr so that their equilibrium concentrations can be calculated. The data in the Change row **always** reflect the stoichiometry of the reaction. That is, looking at the balanced reaction, when 1 mol of Br_2 reacts, it requires 2 mol of NO and produces 2 mol of NOBr. Therefore, when 0.420 M Br_2 reacts to reach equilibrium, twice as much NO will react (2 x 0.420 M) and twice as much NOBr will be produced (2 x 0.420 M). Adding these data to the table, we can calculate the equilibrium concentrations of NO and NOBr by **always adding the change** in concentration to the **initial** concentration.

	$Br_2(g)$	+	$2NO(g)$	$\rightleftarrows$	$2NOBr(g)$
Initial, M	0.525 M		0.950 M		0.00 M
Change, ΔM	−0.420 M		−0.840 M		0.840 M
Equilibrium, M	0.105 M		0.110 M		0.840 M

Note that the change in concentration of the reactions is always negative (meaning consumed), whereas the change in concentration for the product(s) is positive (being formed). Now the equilibrium concentrations can be substituted into the equilibrium constant expression and K_c calculated.

$$\frac{[NOBr]^2}{[Br_2][NO]^2} = \frac{(0.840)^2}{(0.105)(0.110)^2} = 555 = K_c$$

SAMPLE PROBLEM 14.3 C

▶▶ At 400 °C, K_p = 5.88 x 10^3 atm^2 for the reaction

$$2NH_3(g) \rightleftarrows 3H_2(g) + N_2(g)$$

What is the value of K_c for this same reaction at 400 °C?

From the REVIEW, the relationship between K_p and K_c is

$$K_p = K_c(RT)^{\Delta n_g}$$

Solving the equation for K_c,

$$K_c = \frac{K_p}{(RT)^{\Delta n_g}}$$

Since the constant K_p has units of atm, $R = 0.082\ 06\ \text{L} \bullet \text{atm} \bullet \text{mol}^{-1} \bullet \text{K}^{-1}$, $T = 273 + 400 = 673\ \text{K}$, and $\Delta n_g = (4\ \text{mol gaseous products}) - (2\ \text{mol gaseous reactants}) = 2$. Substituting these values into the equation and solving for K_c,

$$K_c = \frac{5.88 \times 10^3\ \text{atm}^2}{[(0.082\ 06\ \text{L} \bullet \text{atm} \bullet \text{mol}^{-1} \bullet \text{K}^{-1})(673\ \text{K})]^2}$$

$$= \frac{5.88 \times 10^3}{3.05 \times 10^3\ \text{L}^2 \bullet \text{mol}^{-2}} = 1.93\ \text{M}^2$$

The value of K_c is 1.93.

SAMPLE PROBLEM 14.3 D

▶▶ For the equilibrium

$$2HCN(g) \rightleftharpoons H_2(g) + C_2N_2(g),$$

$K_p = 4.00 \times 10^{-4}$. What is the value of K_p for the equilibrium

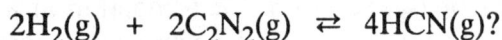

$$2H_2(g) + 2C_2N_2(g) \rightleftharpoons 4HCN(g)?$$

How is the new reaction different than the old (original) reaction? First, the new reaction is the reverse of the old reaction. From the REVIEW,

$$K_{new} = \frac{1}{K_{old}}$$

Second, the new reaction is double the old reaction. That is, all the coefficients in the new reaction are double the coefficients in the old reaction. From the REVIEW,

$$K_{new} = K_{old}{}^x,$$

where $x = 2$. We could calculate a new K_p for each of these successive steps, or alternatively we could combine them by taking the reciprocal of K_{old} and squaring it.

$$K_{new} = \left[\frac{1}{K_{old}} \right]^x$$

Substituting in the values for K_p (old) and $x = 2$,

$$K_{new} = \left[\frac{1}{4.00 \times 10^{-4}} \right]^2 = 6.25 \times 10^6$$

The value of K_p for the new reaction is 6.25 x 10⁶.

SECTION 14.4   CALCULATIONS INVOLVING EQUILIBRIUM CONSTANTS

After completing this section, you will be able to do the following:

- Determine if a reaction is at equilibrium or the direction of the reaction to reach equilibrium;

- Calculate equilibrium concentrations of all reactants and products from K_c.

REVIEW

When the equilibrium constant, K_c or K_p, is known for a reaction at a particular temperature and a set of concentrations for all reactants and products is given, you can determine whether these concentrations are at equilibrium or not by calculating the **reaction quotient, Q**. The quotient in the reaction quotient is identical to that in the equilibrium constant expression -- that is, the product of products divided by the product of reactants, with each raised to a power equal to the coefficient in the balanced reaction. For example, the equilibrium constant expression and the reaction quotient for the reaction, $aA + bB \rightleftharpoons cC + dD$, is

$$K_c = \frac{C^c D^d}{A^a B^b} \qquad \text{and} \qquad Q = \frac{C^c D^d}{A^a B^b}$$

Any concentrations can be used in the reaction quotient; only equilibrium concentrations can be used in the equilibrium constant expression. To determine whether a set of concentrations is at equilibrium, calculate Q and compare its value with the value of K_c. If

> $Q < K_c$ Net forward reaction will take place.
>
> $Q = K_c$ System is at equilibrium (no net change will take place).
>
> $Q > K_c$ Net reverse reaction will take place.

That is, if $Q < K_c$, then the reaction is not at equilibrium because the numerator (where the products are) is too small and the denominator (where the reactants are) is too large. To reach equilibrium, a net reaction will take place in the **forward direction** (to the **right**), which will increase the concentrations of the products and decrease the concentrations of the reactants until equilibrium concentrations are obtained, at which time Q will equal K_c. When the reaction has reached equilibrium, the rates of the forward and reverse reactions will be equal. A **net reaction in the forward direction** means that the forward reaction is occurring at a faster rate than the reverse reaction until equilibrium is reached and the rates become equal. Similarly, if $Q > K_c$, the reaction is not at equilibrium because the numerator (products) is too large and the denominator (reactants) is too small. To reach equilibrium, by means of lowering the concentrations in the numerator (products) and increasing the concentrations in the denominator (reactants), a net reaction will take place in the **reverse direction** (to the **left**). When equilibrium is reached, $Q = K_c$.

When the equilibrium constant, K_c or K_p, and the initial concentrations are known, the equilibrium concentrations can be found by letting x (or any other letter) equal the concentration of one of the starting materials consumed (used up) by the time equilibrium is reached. Then by completing the data table under the reaction (see Sample Problem 14.3 B), the equilibrium concentrations **in terms of x** can be determined. Finally, by substituting the equilibrium concentrations in terms of x into the equilibrium constant expression for that reaction, you will have an equation with one unknown, x. Several techniques can be used to solve for x, after which the numerical value of the equilibrium concentrations can be calculated. The following table summarizes the general procedures for calculating equilibrium concentrations, including strategies for solving the equilibrium constant expression for x.

CALCULATING EQUILIBRIUM CONCENTRATIONS AND PRESSURES

1. Write the equilibrium reaction and equilibrium constant expression.

2. Let x equal the concentration (or pressure) of one of the starting materials that is consumed by the time equilibrium is reached, and complete the data table under the reaction. (To simplify the algebra, try to select a starting material with a coefficient of one.)

3. Substitute the equilibrium concentrations or pressures (in terms of x) into the equilibrium constant expression and solve for x. To solve for x, first look for a perfect square (Sample Problem 14.4 C). If a perfect square is not present, then assume that x is negligible when subtracted from or added to any concentration, and solve the equilibrium constant expression. **CHECK** to see if x is negligible (usually it is if $10^{-3} < K < 10^3$). If x is not negligible, then the method of successive approximations must be used (Sample Problem 14.4 D).

4. Use the value found for x to calculate equilibrium concentrations of all reactants and products.

5. You can check your work by substituting the calculated equilibrium concentrations into the equilibrium constant expression and comparing the value obtained with K_c or K_p.

SAMPLE PROBLEM 14.4 A

▶▶ Determine whether each of the following reaction mixtures is at equilibrium for the reaction in which $K_c = 5.67$

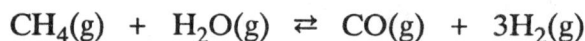

$$CH_4(g) + H_2O(g) \rightleftharpoons CO(g) + 3H_2(g)$$

If the mixture is not at equilibrium, indicate in which direction the reaction will take place in order to reach equilibrium.

(a)   The concentration of all reactants and products is 0.100 M.

(b) $[CO] = 0.150$, $[H_2] = 0.250$, $[CH_4] = 1.20 \times 10^{-2}$, and $[H_2O] = 3.50 \times 10^{-3}$.

Solution:

HINT: Calculate the reaction quotient, Q, and compare it to K_c.

(a) $Q = \dfrac{[CO][H_2]^3}{[CH_4][H_2O]} = \dfrac{[0.100]^4}{[0.100]^2} = 0.0100 < K_c \ (= 5.67)$

Since $Q < K_c$, there will be a net reaction in the forward direction, or to the right, in order to reach equilibrium.

(b) $Q = \dfrac{[CO][H_2]^3}{[CH_4][H_2O]} = \dfrac{[0.150][0.250]^3}{[1.20 \times 10^{-2}][3.50 \times 10^{-3}]} = 55.8 > K_c \ (= 5.67)$

Since $Q > K_c$, there will be a net reaction in the reverse direction, or to the left, in order to reach equilibrium.

SAMPLE PROBLEM 14.4 B

▶▶ After equilibrium has been reached for the following reaction, it is observed that $[O_2] = 1.25$ and $[NO] = 2.25$. If $K_c = 1.20$, what is the equilibrium concentration of NO_2?

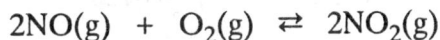

$$2NO(g) + O_2(g) \rightleftarrows 2NO_2(g)$$

Solution:

You know the value of the equilibrium constant and all the equilibrium concentrations except one species. Write the equilibrium constant expression and substitute in all the equilibrium concentrations given in order to find the equilibrium concentration of NO_2.

$$\dfrac{[NO_2]^2}{[NO]^2[O_2]} = \dfrac{[NO_2]^2}{(2.25)^2(1.25)} = K_c = 1.20$$

$[NO_2]^2 = 7.59$ and $[NO_2] = 2.76$.

CHECK: Substitute all the equilibrium concentrations into the equilibrium expression and the numerical result (if the calculations are correct) should be equal to K_c.

SAMPLE PROBLEM 14.4 C

▶▶ If the initial concentrations of O_2 and N_2 are each 1.25×10^{-2} M in the following reaction for which $K_c = 1.72$, what are the concentrations of all species at equilibrium?

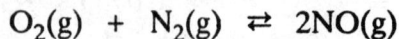

$$O_2(g) + N_2(g) \rightleftarrows 2NO(g)$$

Solution:

Whenever you only know the equilibrium constant and **initial** concentrations for a reaction (such as this one), follow the procedure for calculating equilibrium concentrations given in the REVIEW. First, write the equilibrium constant expression for this reaction,

$$\frac{[NO]^2}{[O_2][N_2]} = K_c = 1.72$$

Second, write down the data table (see Sample Problem 14.3 B) with initial concentrations shown under the reaction, always assuming **zero initial concentration** for any species whose initial concentration is not given,

	$O_2(g)$	+	$N_2(g)$	$\rightleftarrows$	$2NO(g)$
Initial, M	0.0125		0.0125		0.00
Change, ΔM	?		?		?
Equilibrium, M	?		?		?

How much O_2 and N_2 are consumed by the time equilibrium is reached, 0.001 M or 0.002 M, etc.? We don't know at this point. So let x equal the concentration of one of the starting species (preferably with a coefficient of one) that is used up (consumed) by the time equilibrium is reached. Since both of the starting materials, O_2 and N_2, have coefficients of one, we could specify either one. Arbitrarily, let x = $[O_2]$ be the concentration of a species that is consumed by the time equilibrium is reached, and indicate in the data table the **changes in concentration** for N_2 and NO. These **must ALWAYS reflect the stoichiometry of the reaction.** When **one** mole of O_2 reacts, it requires **one** mole of N_2 and forms **two** moles of NO.

$$O_2(g) \quad + \quad N_2(g) \quad \rightleftarrows \quad 2NO(g)$$

	$O_2(g)$	$N_2(g)$	$2NO(g)$
Initial, M	0.0125	0.0125	0.00
Change, ΔM	$-x$	$-x$	$2x$
Equilibrium, M	?	?	?

The equilibrium concentrations are obtained in terms of x by always adding the change in concentration to the initial concentration.

	$O_2(g)$	$N_2(g)$	$2NO(g)$
Initial, M	0.0125	0.0125	0.00
Change, ΔM	$-x$	$-x$	$2x$
Equilibrium, M	$(0.0125-x)$	$(0.0125-x)$	$2x$

Third, substitute the **equilibrium** concentrations into the equilibrium constant expression and solve for x,

$$\frac{[NO]^2}{[O_2][N_2]} = \frac{(2x)^2}{(0.0125-x)^2} = K_c = 1.72$$

Remember, to check step #3 of the procedure in the REVIEW. Look for a perfect square first before using approximation methods. In this case there is a perfect square on the left-hand side (and you can always find the square root of the numerical value of K_c on the right side). Taking the square root of both sides of the equation,

$$\frac{2x}{(0.0125-x)} = 1.31$$

Multiplying both sides of the equation by $(0.0125-x)$ and solving for x,

$$x = 4.95 \times 10^{-3}$$

This is an exact value of x no matter whether the equilibrium constant is large, small, or in-between. No approximation methods are needed. In this problem, the percentage of O_2 (and N_2) consumed by the time equilibrium was reached is

$$\% \text{ reacted} = \frac{x}{\text{Initial Conc.}} \times 100 = \frac{4.95 \times 10^{-3}}{0.0125} \times 100 = 39.6\%$$

Fourth, use the value of x to calculate the numerical concentrations at equilibrium.

	$O_2(g)$	+	$N_2(g)$	$\rightleftharpoons$	$2NO(g)$
Initial, M	0.0125		0.0125		0.00
Change, ΔM	$-x$		$-x$		$2x$
Equilibrium, M	$(0.0125-x)$		$(0.0125-x)$		$2x$
	7.6×10^{-3}		7.6×10^{-3}		9.90×10^{-3}

Finally, you can check your results by substituting the (calculated) equilibrium concentrations into the equilibrium constant expression. If correct, the quotient will be equal to the value of K_c. For this problem,

$$\frac{[NO]^2}{[O_2][N_2]} = \frac{(9.90 \times 10^{-3})^2}{(7.6 \times 10^{-3})^2} = 1.7 = K_c \text{ (to two s.f.)}$$

SAMPLE PROBLEM 14.4 D

▶▶ If the initial concentration of CH_3OH is 2.50 M in the following reaction for which $K_c = 0.0943$, calculate the concentrations of all species at equilibrium.

$$CH_3OH(g) \rightleftharpoons CO(g) + 2H_2(g)$$

Solution:

Whenever you only know the equilibrium constant and **initial** concentrations for a reaction (such as this one), follow the procedure for calculating equilibrium concentrations given in the REVIEW (also see Sample Problem 14.4 C for a detailed application of procedure). First, write the equilibrium constant expression for this reaction,

$$\frac{[CO][H_2]^2}{[CH_3OH]} = K_c = 0.0943$$

Second, write down the data table (see Sample Problem 14.3 B) with initial concentrations shown under the reaction, always assuming **zero initial concentration** for any species whose initial concentration is not given,

	CH$_3$OH(g)	$\rightleftarrows$	CO(g)	+	2H$_2$(g)
Initial, M	2.50		0.00		0.00
Change, ΔM	?		?		?
Equilibrium, M	?		?		?

How much CH$_3$OH is consumed by the time equilibrium is reached, 0.100 M or 0.500 M, etc.? We don't know at this point. So let x = [CH$_3$OH] be the concentration of a species that is consumed by the time equilibrium is reached, and indicate in the data table the **change in concentration** for CO and H$_2$. These must **ALWAYS reflect the stoichiometry of the reaction.** When **one** mole of CH$_3$OH reacts, it forms **one** mole of CO and **two** moles of H$_2$. So when x moles/L of CH$_3$OH react, x moles/L of CO and 2x moles/L of H$_2$ will be formed. **The coefficients of x must be in the same ratio as the coefficients in the balanced reaction!**

	CH$_3$OH(g)	$\rightleftarrows$	CO(g)	+	2H$_2$(g)
Initial, M	2.50		0.00		0.00
Change, ΔM	−x		x		2x
Equilibrium, M	?		?		?

The equilibrium concentrations are obtained in terms of x by always adding the change in concentration to the initial concentration.

	CH$_3$OH(g)	$\rightleftarrows$	CO(g)	+	2H$_2$(g)
Initial, M	2.50		0.00		0.00
Change, ΔM	−x		x		2x
Equilibrium, M	(2.50−x)		x		2x

Third, substitute the **equilibrium** concentrations into the equilibrium constant expression and solve for x,

$$\frac{[CO][H_2]^2}{[CH_3OH]} = \frac{(x)(2x)^2}{(2.50-x)} = K_c = 0.0943$$

Remember, to check step #3 of the procedure in the REVIEW. Look for a perfect square first before using approximation methods. In this case there is no perfect square, so an approximation method normally is the shortest route to the solution. The method of **successive**

approximations uses successively better approximations of the starting substance (CH_3OH) until there is no longer any **change in the value of x**. If the value of the equilibrium constant is $10^{-3} < K_c < 10^3$, then often the first approximation is sufficient. The closer the value of the equilibrium constant is to the value of the concentrations used, the greater the need for more approximations. Generally, two to three approximations are sufficient for most applications. The **first approximation** is always to assume that x is zero with respect to the initial concentration of CH_3OH -- that is, x is so small compared with the initial 2.50 M CH_3OH that it is mathematically insignificant. Making this assumption, you can solve the equilibrium constant expression for x.

$$\frac{[CO][H_2]^2}{[CH_3OH]} = \frac{(x)(2x)^2}{2.50} = \frac{4x^3}{2.50} = K_c = 0.0943$$

Solving for x (cube roots can be readily done with a scientific calculator), we obtain x = 0.389. Is x significant with respect to the [CH_3OH] at equilibrium? At equilibrium, [CH_3OH] = 2.50 − x = 2.50 − 0.389 = 2.11 M. The value 2.11 M is different than the initial 2.50, so x **is** significant and **cannot** be ignored for CH_3OH. [By the time equilibrium is reached, 15.6% of the CH_3OH has reacted, (0.389/2.50) x 100.] So our first approximation for the equilibrium concentration of CH_3OH (2.50 M) was not good enough. A **second** approximation is needed for CH_3OH at equilibrium, (2.50 − x), using the value of x obtained from the first approximation. Substitute the second approximation into the equilibrium expression and solve for x,

$$\frac{[CO][H_2]^2}{[CH_3OH]} = \frac{(x)(2x)^2}{(2.50 - 0.389)} = \frac{4x^3}{2.11} = K_c = 0.0943$$

Solving for x, we obtain x = 0.368. Is the second approximation good enough? Continue the successive-approximations process until there is no longer any change in the equilibrium concentration of CH_3OH. A summary of the **successive-approximations** process for this problem is

Approximation	[CH_3OH]$_{equil}$	x
1st	2.50 − 0.00 = 2.50	0.389
2nd	2.50 − 0.389 = 2.11	0.368
3rd	2.50 − 0.368 = 2.13	0.369
4th	2.50 − 0.369 = 2.13	

There is very little change in the approximation of the concentration of CH_3OH at equilibrium between the second and third approximation. The fourth approximation confirms that x =

0.369.

Fourth, use the value of x to calculate the numerical concentrations at equilibrium.

	$CH_3OH(g)$	$\rightleftharpoons$	$CO(g)$	$+$	$2H_2(g)$
Initial, M	2.50		0.00		0.00
Change, ΔM	$-x$		x		2x
Equilibrium, M	$(2.50-x)$		x		2x
	2.13 M		0.369 M		0.738 M

Finally, you can check your results by substituting the (calculated) equilibrium concentrations into the equilibrium constant expression. If correct, the quotient will be equal (allowing for rounding errors) to the value of K. For this problem,

$$\frac{[CO][H_2]^2}{[CH_3OH]} = \frac{(0.369)(0.738)^2}{(2.13)} = 0.0944 = K_c$$

SECTION 14.5 USING LE CHATELIER'S PRINCIPLE TO PREDICT SHIFTS IN CHEMICAL EQUILIBRIA

After completing this section, you will be able to do the following:

■ Use Le Chatelier's principle to predict the direction in which chemical equilibria will shift as a result of changes in concentration, pressure, and temperature.

REVIEW

Le Chatelier's principle states that if a stress (change in concentration, pressure, or temperature) is imposed on a system at equilibrium, the system will relieve the stress by shifting the position of equilibrium to the right or left.

Changes in **concentration** do not change the value of the equilibrium constant as long as the temperature remains constant. The point or position of equilibrium will shift as the concentration of reactants and products changes. For any reaction **at equilibrium,**

● If the concentration of any gaseous or aqueous reactant is increased or product decreased, the equilibrium will shift to the right to relieve the stress.

- If the concentration of any gaseous or aqueous reactant is decreased or product increased, the equilibrium will shift to the left to relieve the stress.

Changing the **pressure** does not change the value of the equilibrium constant, as long as the temperature remains constant. For any reaction **at equilibrium**,

- Adding a gaseous reactant or product increases the concentration of that reactant or product, and removing a gaseous reactant or product lowers the concentration of that reactant or product. The effect on a reaction at equilibrium of adding or removing a gaseous reactant or product is identical to that indicated for changes in concentration above.

- If the pressure of a gaseous reaction mixture (the entire reaction) is **increased** (by decreasing the volume), the equilibrium will shift in the direction of the **smaller number of moles of gases**. (The stress of too much pressure will be lowered by shifting in the direction of less gas molecules, which restores equilibrium.)

- If the pressure of a gaseous reaction mixture (the entire reaction) is **decreased** (by increasing the volume), the equilibrium will shift in the direction of the **larger number of moles of gases**. (The stress of too little pressure will be increased by shifting in the direction of more gas molecules, which restores equilibrium.)

- Increasing the total pressure by adding an **inert gas** (one that does not react with reactants or products) at constant volume does **not** shift gaseous equilibria.

Changes in **temperature** usually **do change** the value of the equilibrium constant. (Changes in concentration and pressure do not change the value of the equilibrium constant, but a change in temperature does.) If the enthalpy change, $\Delta H°$, is given for a reaction, write in the thermal energy change as a reactant or product. (Exothermic means thermal energy is a product, and endothermic means thermal energy is a reactant.) For a reaction **at equilibrium**,

- If the temperature is **increased**, the stress (of too high a temperature) will be relieved by the equilibrium shifting to the **left** for an **exothermic** reaction, and by the equilibrium shifting to the **right** for an **endothermic** reaction.

- If the temperature is **decreased**, the stress (of too low a temperature) will be relieved by the equilibrium shifting to the **right** for an **exothermic** reaction, and by the equilibrium shifting to the **left** for an **endothermic** reaction.

A **catalyst** is a substance that speeds up a reaction but is not used up in the reaction. Addition of a catalyst increases the rate or speed of both the forward and reverse reactions equally, but it does **not** shift the position of equilibrium.

SAMPLE PROBLEM 14.5 A

▶ ▶ Predict the effect of each of the following changes on the equilibrium reaction:

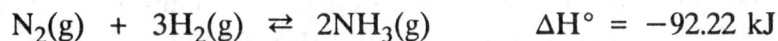

$$N_2(g) + 3H_2(g) \rightleftarrows 2NH_3(g) \qquad \Delta H° = -92.22 \text{ kJ}$$

(a)   decreasing the concentration of $H_2(g)$
(b)   increasing the concentration of $NH_3(g)$
(c)   adding a catalyst
(d)   decreasing the pressure by increasing the volume at constant temperature
(e)   increasing the pressure by adding inert He gas at constant temperature and volume
(f)   increasing the pressure by decreasing the volume at constant temperature.

Solution:

HINT: Le Chatelier's principle only applies to a concentration, pressure, or temperature change of a reaction already at equilibrium.

(a)   The equilibrium will shift to the left to replace the $[H_2]$.
(b)   The equilibrium will shift to the left to lower the excess $[NH_3]$.
(c)   Adding a catalyst does not shift the equilibrium, only the speed with which equilibrium is reached.
(d)   To relieve the stress of too little pressure, the equilibrium will shift in the direction of the larger number of moles of gases, which for this reaction is to the left (4 mol of gases on the left and 2 mol of gases on the right).
(e)   Increasing the pressure with an inert gas (at constant T and V) does not change the pressure of any other gas; there is no shift in the equilibrium.
(f)   To relieve the stress of too much pressure, the equilibrium will shift in the direction of the smaller number of moles of gases, which for this reaction is to the right (4 mol of gases on the left and 2 mol of gases on the right).

SELF-TEST QUESTIONS

Test your understanding of the concepts and skills in this chapter by working through the multiple choice questions **BEFORE** checking the answers. A periodic table will always be available. The ANSWERS TO SELF-TEST QUESTIONS follow immediately after this section.

1. Before the reaction $A + B \rightarrow C + D$ begins, both A and B are 1.00 M and there are no products C and D. By the time equilibrium is reached,

    A. the concentrations of A and B will be zero.
    B. the forward reaction will stop.
    C. the rate of the reverse reaction will be zero.
    D. the concentrations of A, B, C, and D must be equal.
    E. None of the above statements is correct.

2. The correct equilibrium constant expression for the following reaction is

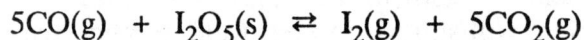

    $$5CO(g) + I_2O_5(s) \rightleftarrows I_2(g) + 5CO_2(g)$$

    A. $K_c = [I_2][CO_2] / [CO]$

    B. $K_c = [I_2][CO_2]^5 / [CO]^5$

    C. $K_c = [I_2][CO_2]^5 / [CO]^5[I_2O_5]$

    D. $K_c = [I_2][CO_2] / [CO][I_2O_5]$

    E. $K_c = [CO]^5 / [I_2][CO_2]^5$

3. If $K_c = 0.27$ at 90 °C for the equilibrium reaction given, the value of K_p is

    $$N_2O_4(g) \rightleftarrows 2NO_2(g)$$

    A. 810   B. 0.27   C. 8.0   D. 2.0   E. 9.1×10^{-3}

4. After the following reaction reached equilibrium, the gas pressure of NH_3 was 0.242 atm and of CO_2 was 0.121 atm. The value of K_p is

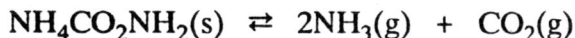

$$NH_4CO_2NH_2(s) \rightleftarrows 2NH_3(g) + CO_2(g)$$

A. 7.09×10^{-3}　　　B. 5.86×10^{-2}　　　C. 1.77×10^{-3}　　　D. 2.93×10^{-2}　　　E. 0.363

5. What is the value of K_c for the following reaction if $[NO_2] = 0.0235$, $[O_2] = 0.000\ 750$, and $[NO] = 0.001\ 50$ at equilibrium?

$$2NO_2(g) \rightleftarrows 2NO(g) + O_2(g)$$

A. 3.06×10^{-7}　　　B. 4.79×10^{-5}　　　C. 3.06×10^{-6}　　　D. 6.11×10^{-6}　　　E. 3.27×10^5

6. If $K_c = 2.5 \times 10^{-9}$ at 27 °C for the equilibrium

$$N_2(g) + 3H_2(g) \rightleftarrows 2NH_3(g)$$

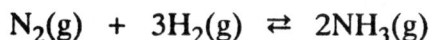

the value of K_c for the following reaction at 27 °C is

$$NH_3(g) \rightleftarrows 1/2N_2(g) + 3/2H_2(g)$$

A. 1.6×10^{17}　　　B. 2.0×10^8　　　C. 8.0×10^8　　　D. 4.0×10^6　　　E. 2.0×10^4

7. After the following reaction reached equilibrium at 2000 °C, $[NO] = 6.66 \times 10^{-3}$, $[N_2] = 0.100$, and $[O_2] = 1.08$. The value of K_c is

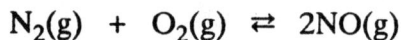

$$N_2(g) + O_2(g) \rightleftarrows 2NO(g)$$

A. 9.26×10^{-3}　　　B. 1.03×10^{-4}　　　C. 4.11×10^{-4}　　　D. 6.18×10^{-3}　　　E. 2.43×10^3

8. If $K_c = 0.255\ M^2$ at 1.200×10^3 K for the reaction given, the value of K_p for this reaction at the same temperature is

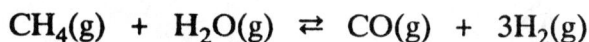

$$CH_4(g) + H_2O(g) \rightleftarrows CO(g) + 3H_2(g)$$

A. $2.54 \times 10^7\ atm^2$.
B. $25.1\ atm^2$.
C. $2.63 \times 10^{-5}\ atm^2$.
D. $2.47 \times 10^3\ atm^2$.
E. $0.255\ atm^2$.

9. For the following reaction, $K_c = 62.5$ and the equilibrium concentrations of HI and H_2 are 0.20 M and 0.10 M, respectively. The equilibrium concentration of I_2 is

$$H_2(g) + I_2(g) \rightleftarrows 2HI(g)$$

A. 156 M   B. 0.040 M   C. 0.032 M   D. 6.4 x 10^{-3} M   E. 8.0 x 10^{-4} M

10. If the initial concentration of HBr(g) is 0.250 M in the following reaction for which $K_c = 4.59$ x 10^{-7}, the equilibrium concentration of H_2 is

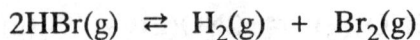

$$2HBr(g) \rightleftarrows H_2(g) + Br_2(g)$$

A. 0.125 M.
B. 1.70 x 10^{-4} M.
C. 3.39 x 10^{-4} M.
D. 2.70 x 10^{-3} M.
E. 1.35 x 10^{-3} M.

USE THE FOLLOWING DATA FOR THE NEXT THREE QUESTIONS

At 100 °C, $K_c = 2.19 \times 10^{-10}$ for $COCl_2(g) \rightleftarrows CO(g) + Cl_2(g)$

11. Which set of concentrations represents the system at equilibrium?

	$[COCl_2]$	$[CO]$	$[Cl_2]$
A.	2.19×10^{-1}	1.00×10^{-3}	1.00×10^{-3}
B.	5.00×10^{-2}	3.31×10^{-6}	3.31×10^{-6}
C.	8.57×10^{-2}	4.50×10^{-7}	5.73×10^{-6}
D.	2.00×10^{-4}	1.48×10^{-5}	1.48×10^{-5}
E.	5.48×10^{-14}	1.58×10^{-2}	1.58×10^{-2}

12. For which set of concentrations in the previous question (#11) must the net reaction shift to the right (forward direction) to reach equilibrium?

A. A   B. B   C. C   D. D   E. E

13. If the initial concentration of $COCl_2(g)$ was 0.250 M at 100 °C, what would be the concentrations at equilibrium of $CO(g)$ and $Cl_2(g)$, respectively?

	$[CO]$	$[Cl_2]$
A.	0.250	0.250
B.	2.96×10^{-5}	2.96×10^{-5}
C.	0.125	0.125
D.	1.48×10^{-5}	1.48×10^{-5}
E.	7.40×10^{-6}	7.40×10^{-6}

14. Given the following reaction at equilibrium at a high temperature, which of the following changes will not cause the equilibrium to shift to the right?

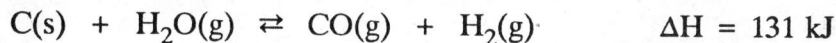

$$C(s) + H_2O(g) \rightleftarrows CO(g) + H_2(g) \qquad \Delta H = 131 \text{ kJ}$$

A. Remove some $H_2(g)$.
B. Add a catalyst.
C. Add a little more $C(s)$ (no significant change in volume).
D. Increase the volume at constant temperature.
E. More than one of the above is correct.

15. Consider the following equilibrium reaction,

$$4NH_3(g) + 3O_2(g) \rightleftarrows 2N_2(g) + 6H_2O(l) \qquad \Delta H = -1530 \text{ kJ}$$

If you could change the temperature and pressure, what combination would result in a shift of the equilibrium position to the right?

A. higher T, higher P
B. higher T, lower P
C. lower T, lower P
D. lower T, higher P
E. The equilibrium is not sensitive to either T or P.

16. For which of the following systems at equilibrium will an increase in the container volume at constant temperature cause the equilibrium to shift to the right?

A. $2H_2(g) + O_2(g) \rightleftarrows 2H_2O(g)$

B. $CS_2(g) + 4H_2(g) \rightleftarrows CH_4(g) + 2H_2S(g)$

C. $2NO_2(g) \rightleftarrows N_2O_4(g)$

D. $2NaHCO_3(s) \rightleftarrows Na_2CO_3(s) + CO_2(g) + H_2O(g)$

E. $CO_2(g) + H_2(g) \rightleftarrows CO(g) + H_2O(g)$

17. For the following reaction at equilibrium, which of the following is true?

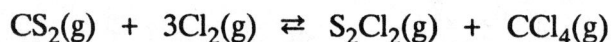

$$CS_2(g) + 3Cl_2(g) \rightleftarrows S_2Cl_2(g) + CCl_4(g)$$

A. If the concentration of $S_2Cl_2(g)$ is increased, the equilibrium will shift to the right.
B. If the pressure is decreased by increasing the volume at constant temperature, the equilibrium will shift to the left.
C. If the concentrations of $CS_2(g)$ and $Cl_2(g)$ are decreased, the equilibrium will not be affected.
D. If the pressure is increased by decreasing the volume at constant temperature, the equilibrium will shift to the left.
E. More than one of the above answers is correct.

18. For which of the reactions at equilibrium and at constant temperature will decreasing the volume cause the equilibrium to shift to the right?

A. $N_2(g) + 3H_2(g) \rightleftarrows 2NH_3(g)$

B. $SO_2Cl_2(g) \rightleftarrows SO_2(g) + Cl_2(g)$

C. $NH_4Cl(s) \rightleftarrows NH_3(g) + HCl(g)$

D. $H_2(g) + Cl_2(g) \rightleftarrows 2HCl(g)$

E. $I_2(g) \rightleftarrows 2I(g)$

ANSWERS TO SELF-TEST QUESTIONS

1. e	7. c	13. e
2. b	8. d	14. e*
3. c	9. d	15. d
4. a	10. b	16. d
5. c	11. b	17. b
6. e	12. c	18. a

* both b and c correct

PRACTICE PROBLEMS

Test your problem-solving skills in this chapter by working through the problems in the space provided **BEFORE** checking the answers. A periodic table will always be available. The ANSWERS TO PRACTICE PROBLEMS follow immediately after this section.

1.   Write the equilibrium constant expression for the following reactions,

(a) $H_2(g) + O_2(g) \rightleftarrows H_2O(g)$

(b) $N_2H_6CO_2(s) \rightleftarrows NH_3(g) + CO_2(g)$

(c) $2PbS(s) + 3O_2(g) \rightleftarrows 2PbO(s) + 2SO_2(g)$

(d) $TiCl_4(g) + H_2O(g) \rightleftarrows TiO_2(s) + HCl(g)$

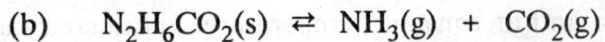

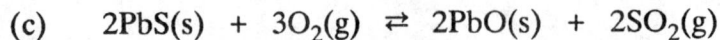

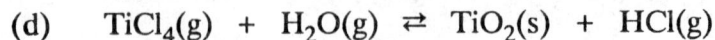

2.   If $K_c = 55$ at 425 °C for the first equilibrium reaction given below, what is the value of K_c for the second equilibrium reaction at the same temperature?

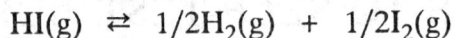

$$H_2(g) + I_2(g) \rightleftarrows 2HI(g)$$

$$HI(g) \rightleftarrows 1/2H_2(g) + 1/2I_2(g)$$

3.   After the following reaction reached equilibrium in a 12.0-L flask at 700 °C, there were 2.50 mol of H_2, 1.35×10^{-5} mol of S_2, and 8.70 mol of H_2S. The value of K_c for this reaction at 700 °C is

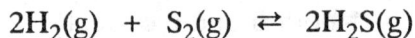

$$2H_2(g) + S_2(g) \rightleftarrows 2H_2S(g)$$

4.   If $K_c = 3.4 \times 10^{81}$ L/mol for the following reaction at 25 °C

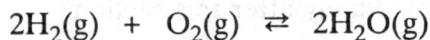

$$2H_2(g) + O_2(g) \rightleftarrows 2H_2O(g)$$

(a) what is the value of K_p under these conditions?

(b) What is the value of K_c for the following reaction at 25 °C?

$$2H_2O(g) \rightleftharpoons 2H_2(g) + O_2(g)$$

5.   If there are 11.2 atm each of $H_2O(g)$ and $CO(g)$ and 8.8 atm each of $H_2(g)$ and $CO_2(g)$ at equilibrium, what is the value of K_p for the following reaction?

$$H_2(g) + CO_2(g) \rightleftharpoons H_2O(g) + CO(g)$$

6.   If the initial pressure of $SO_3(g)$ is 1.50 atm at 700 °C in the following reaction for which $K_p = 1.6$ x 10^{-5} at 700 °C, what are the equilibrium pressures of all the reactants and products at the same temperature?   [HINT: Successive approximations will be needed.]

$$2SO_3(g) \rightleftharpoons 2SO_2(g) + O_2(g)$$

7. $K_c = 2.264$ x 10^{-4} at 1132 °C for the following reaction

$$2H_2S(g) \rightleftharpoons 2H_2(g) + S_2(g)$$

(a)   What is the equilibrium concentration of S_2 if the equilibrium concentrations of H_2 and H_2S are 0.100 M and 0.250 M, respectively?

(b)   Is the reaction at equilibrium if the concentrations of all species are 0.0200 M? If not, indicate in which direction a net reaction will take place.

(c)   What is the value of K_c for the following reaction at the same temperature?

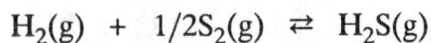

$$H_2(g) + 1/2 S_2(g) \rightleftarrows H_2S(g)$$

8.   At a certain temperature, $K_c = 4.7 \times 10^{-2}$ for the reaction

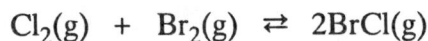

$$Cl_2(g) + Br_2(g) \rightleftarrows 2BrCl(g)$$

(a)   Would this reaction be at equilibrium at the same temperature if $[Cl_2] = 0.040$, $[Br_2] = 0.060$, and $[BrCl] = 0.0040$? If not, in which direction will the reaction shift to reach equilibrium?

(b)   If $[Cl_2] = 0.0050$ and $[BrCl] = 0.0030$ at equilibrium, what is the equilibrium concentration of Br_2?

(c)   If $[Cl_2] = 1.3 \times 10^{-2}$ and $[Br_2] = 1.3 \times 10^{-2}$ initially, what will be the equilibrium concentration of all species under the same conditions?

(d) If, initially, $[BrCl] = 1.3 \times 10^{-2}$ and no Cl_2 or Br_2 is present, what will be the equilibrium concentration of all species under the same conditions?

ANSWERS TO PRACTICE PROBLEMS

1. Reactions (a), (c), and (d) must be balanced before the equilibrium constant expression can be written.

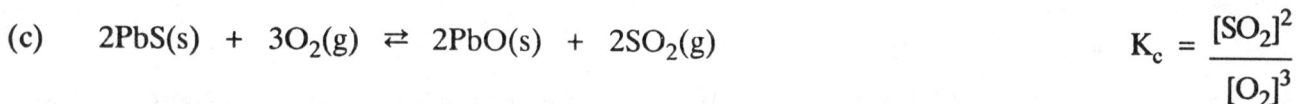

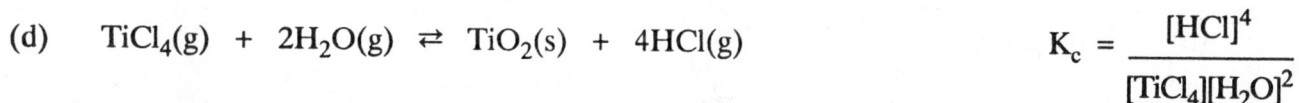

(a) $2H_2(g) + O_2(g) \rightleftarrows 2H_2O(g)$

$$K_c = \frac{[H_2O]^2}{[H_2]^2[O_2]}$$

(b) $N_2H_6CO_2(s) \rightleftarrows 2NH_3(g) + CO_2(g)$

$$K_c = [NH_3]^2[CO_2]$$

(c) $2PbS(s) + 3O_2(g) \rightleftarrows 2PbO(s) + 2SO_2(g)$

$$K_c = \frac{[SO_2]^2}{[O_2]^3}$$

(d) $TiCl_4(g) + 2H_2O(g) \rightleftarrows TiO_2(s) + 4HCl(g)$

$$K_c = \frac{[HCl]^4}{[TiCl_4][H_2O]^2}$$

2. $K_{new} = \left[\dfrac{1}{K_{old}}\right]^{1/2} = \left[\dfrac{1}{55}\right]^{1/2} = 0.13$

3. $K_c = \dfrac{[H_2S]^2}{[H_2]^2[S_2]} = \dfrac{(8.70 \text{ mol}/12.0 \text{ L})^2}{(2.50 \text{ mol}/12.0 \text{ L})^2(1.35 \times 10^{-5} \text{ mol}/12.0 \text{ L})} = 1.08 \times 10^7$

4. (a) $K_p = K_c(RT)^{\Delta n_g} = (3.4 \times 10^{81} \ L \cdot mol^{-1})[(0.082 \ L \cdot atm \cdot mol^{-1} \cdot K^{-1})(298 \ K)]^{-1}$

$= 1.4 \times 10^{80} \ atm^{-1}$

(b) $K_{new} = \left[\dfrac{1}{K_{old}}\right] = \left[\dfrac{1}{3.4 \times 10^{81}}\right] = 2.9 \times 10^{-82}$

5. $K_p = \dfrac{p_{H_2O} \ p_{CO}}{p_{H_2} \ p_{CO_2}} = \dfrac{(11.2 \ atm)(11.2 \ atm)}{(8.8 \ atm)(8.8 \ atm)} = 1.6$

6. Using successive approximations: $p_{SO_3} = 1.46 \ atm$, $p_{SO_2} = 0.040 \ atm$, and $p_{O_2} = 0.020 \ atm$.

7. (a) $[S_2]_{equil} = 1.42 \times 10^{-3}$

(b) $Q = 0.0200 > K = 2.264 \times 10^{-4}$; net reverse reaction (to the left) will occur.

(c) $K_{new} = \left[\dfrac{1}{K_{old}}\right]^{1/2} = \left[\dfrac{1}{2.264 \times 10^{-4}}\right]^{1/2} = 66.46$

8. (a) $Q = 6.7 \times 10^{-3} < K = 4.7 \times 10^{-2}$; net forward reaction (to the right) will occur.

(b) $[Br_2]_{equil} = 3.8 \times 10^{-2}$

(c) Perfect square, $x = 1.3 \times 10^{-3}$. Equilibrium concentrations are: $[Cl_2] = 0.012$, $[Br_2] = 0.012$, and $[BrCl] = 2.6 \times 10^{-3}$.

(d) Perfect square, $x = 5.9 \times 10^{-3}$. Equilibrium concentrations are: $[Cl_2] = 5.9 \times 10^{-3}$, $[Br_2] = 5.9 \times 10^{-3}$, and $[BrCl] = 1 \times 10^{-3}$.

CHAPTER 15

ACIDS AND BASES

SECTION 15.1   THE BRONSTED-LOWRY DEFINITIONS

After completing this section, you will be able to do the following:

■ Identify and write formulas of Bronsted-Lowry acids and bases.

REVIEW

The **classical** or **Arrhenius** definitions of acids and bases are limited to aqueous solutions; **acids** increase the concentration of hydrogen ions, H^+, and bases increase the concentration of hydroxide ions, OH^-, when dissolved in water. Since a number of reactions that take place in the gas phase or in solvents other than water resemble acid-base reactions, a broader definition of acids and bases is used, called the **Bronsted-Lowry** definition. According to the **Bronsted-Lowry** definitions, an **acid** is any species that can **donate a proton** (H^+ ion). A **base** is any species that can **accept a proton**. In other words, a Bronsted-Lowry **acid** is a **proton donor** and a **base** is a **proton acceptor**. All of the acids and bases classified under the Arrhenius definition are included under the Bronsted-Lowry definition plus many more species as well. [Arrhenius acids increase the H^+ concentration, which makes them proton donors; Arrhenius bases increase the OH^- concentration, and OH^- is a proton acceptor.]

An **acid-base** reaction under the Bronsted-Lowry definition consists of the **transfer of a proton** from an acid to a base. Every acid-base reaction has an acid and a base for reactants **and** an acid and a base for products. For example, water acts as a Bronsted-Lowry (B-L) base when it reacts with the weak acid, HF, in the forward reaction,

$$HF(aq) \quad + \quad H_2O(l) \quad \rightleftarrows \quad H_3O^+(aq) \quad + \quad F^-(aq)$$

B-L acid   B-L base   B-L acid   B-L base

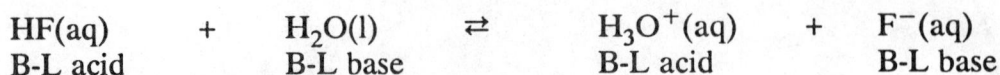

The F^- ion is a B-L base because it accepts a proton from the acid H_3O^+ in the **reverse** reaction. Acids and bases that are related by loss or gain of H^+ are called **conjugate acid-base pairs**. In this reaction, the acid HF and the base F^- are a conjugate acid-base pair. Similarly, the base H_2O and the acid it forms, H_3O^+, are a conjugate acid-base pair. Members of conjugate acid-base pairs are often marked with the same subscript, as shown below.

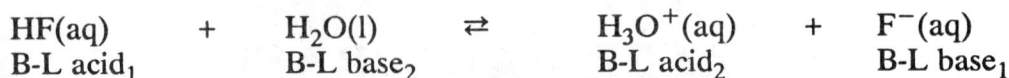

$$HF(aq) \quad + \quad H_2O(l) \quad \rightleftarrows \quad H_3O^+(aq) \quad + \quad F^-(aq)$$

B-L acid$_1$ B-L base$_2$ B-L acid$_2$ B-L base$_1$

Note that the formulas of a conjugate acid-base pair differ only by a H^+ (that is, a H atom with a 1+ charge). You should be able to identify conjugate acid-base pairs in an acid-base reaction; given the formula of a Bronsted-Lowry acid or base, you should be able to write the formula of the other member of the pair.

Species that can react as either an acid or a base (in different reactions), like water and the bicarbonate ion (HCO_3^-), are said to be **amphoteric**.

SAMPLE PROBLEM 15.1 A

▶▶ For the following Bronsted-Lowry acid-base reactions, identify the conjugate acid base pairs using the same system as in the REVIEW (the same subscript for both members of the conjugate acid-base pair).

(a) $H_2O(l) + ClO^-(aq) \rightleftarrows OH^-(aq) + HClO(aq)$

(b) $HCO_3^-(aq) + OH^-(aq) \rightleftarrows CO_3^{2-}(aq) + H_2O(l)$

(c) $HCO_3^-(aq) + HCl(aq) \rightleftarrows H_2CO_3(aq) + Cl^-(aq)$

(d) $NH_2OH(aq) + H_2O(l) \rightleftarrows NH_3OH^+(aq) + OH^-(aq)$

Solution:

HINT:
- To identify a conjugate acid-base pair, start with either reactant and look for a formula on the product side that differs by only a H^+.

- One of the reactants is an acid and the other reactant is a base, but the reactants are never a conjugate acid-base pair. Similarly for the product side of the reaction.

(a) $H_2O(l)$ + $ClO^-(aq)$ $\rightleftarrows$ $OH^-(aq)$ + $HClO(aq)$

        $Acid_1$ $Base_2$ $Base_1$ $Acid_2$

(b) $HCO_3^-(aq)$ + $OH^-(aq)$ $\rightleftarrows$ $CO_3^{2-}(aq)$ + $H_2O(l)$

        $Acid_1$ $Base_2$ $Base_1$ $Acid_2$

(c) $HCO_3^-(aq)$ + $HCl(aq)$ $\rightleftarrows$ $H_2CO_3(aq)$ + $Cl^-(aq)$

        $Base_1$ $Acid_2$ $Acid_1$ $Base_2$

(d) $NH_2OH(aq)$ + $H_2O(l)$ $\rightleftarrows$ $NH_3OH^+(aq)$ + $OH^-(aq)$

        $Base_1$ $Acid_2$ $Acid_1$ $Base_2$

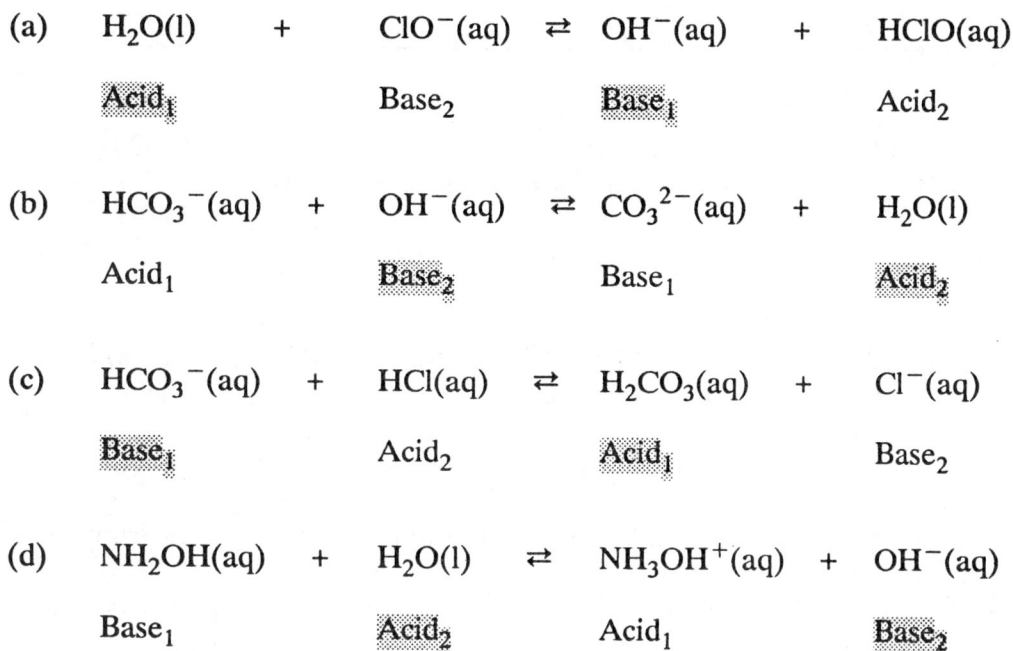

NOTE: One of the conjugate acid-base pairs is shaded for each reaction.

CHECK: Make sure the formulas of the conjugate acid-pairs differ by only a H^+.

SAMPLE PROBLEM 15.1 B

▶▶ For each of the following Bronsted-Lowry bases, write the formula of the conjugate acid: (a) ClO_2^- (b) NH_3 (c) N_2H_4 (d) HPO_4^{2-} (e) OH^-

Solution:

HINT: The conjugate acid will have one **more** H^+ than the base; that is, the formula of the acid has one more H atom and a charge one unit more positive than the base.

    (a) $HClO_2$ The formula has one more H and the charge went from 1− to zero.

    (b) NH_4^+ The formula has one more H and the charge went from zero to 1+.

    (c) $N_2H_5^+$ The formula has one more H and the charge went from zero to 1+.

    (d) $H_2PO_4^-$ The formula has one more H and the charge went from 2− to 1−.

    (e) H_2O The formula has one more H and the charge went from 1− to zero.

SAMPLE PROBLEM 15.1 C

▶ ▶ For each of the following Bronsted-Lowry acids, write the formula of the conjugate base: (a) NH_4^+ (b) HPO_4^{2-} (c) H_3PO_3 (d) H_2SeO_4 (e) $H_3C_6H_5O_7$

Solution:

HINT: The conjugate base has one **less** H^+ than the acid; that is, the formula of the base has one less H atom and a charge one unit less positive than the acid.

(a) NH_3 The formula has one less H and the charge went from 1+ to zero.

(b) PO_4^{3-} The formula has one less H and the charge went from 2− to 3−.

(c) $H_2PO_3^-$ The formula has one less H and the charge went from zero to 1−.

(d) $HSeO_4^-$ The formula has one less H and the charge went from zero to 1−.

(e) $H_2C_6H_5O_7^-$ The formula has one less H and the charge went from zero to 1−.

              NOTE: The ionizable hydrogens are those written first in the formula.

SECTION 15.2   THE ION PRODUCT FOR WATER
SECTION 15.3   THE pH AND OTHER "p" SCALES

After completing this section, you will be able to do the following:

■ For any solution, given one of the four quantities $[H^+]$, $[OH^-]$, pH, or pOH, be able to calculate the other three quantities.

REVIEW

The equilibrium reaction and equilibrium constant expression for the ionization of water is

$$2H_2O(l) \rightleftharpoons H_3O^+(aq) + OH^-(aq) \qquad\qquad [H^+][OH^-] = 1.0 \times 10^{-14} = K_w,$$

where the H_3O^+ is replaced by the more common H^+ in the equilibrium expression, and the subscript indicates the specific equilibrium reaction for water. When pure water ionizes, equal amounts of H^+ and OH^- are produced. From the equilibrium expression for water, the concentrations of each are

$$[H^+] = [OH^-] = 1.0 \times 10^{-7}$$

Any solution (no matter how many solutes are present) that has the same concentrations of H^+ and OH^- as pure water is called a **neutral** solution. If a solution at equilibrium has a $[H^+] > 1.0 \times 10^{-7}$, it is called an **acidic** solution. If a solution at equilibrium has a $[H^+] < 1.0 \times 10^{-7}$, it is called a **basic** solution. No matter whether a solution is acidic, basic, or neutral, the equilibrium expression for water (sometimes called the ion product for water) **must be satisfied,** that is,

$$[H^+][OH^-] = 1.0 \times 10^{-14} = K_w$$

Knowing either the $[H^+]$ or $[OH^-]$ for a solution, you can calculate the other concentration.

Because the concentrations of hydrogen ions commonly range from 1.0 M to 10^{-14} M, a logarithmic scale is used called the **pH scale** The **pH** of a solution is the negative (base ten) logarithm of the molar hydrogen ion concentration,

$$pH = -\log [H^+],$$

where a **logarithm** (base ten) is the power to which 10 has to be raised to equal the number. The number $1000 = 10^3$; the log of 1000 = 3. For numbers which are not multiples of ten, you will need a scientific calculator to find the logarithm (see Appendix A.3 of the textbook for a discussion and examples of logarithms). You must be able to calculate the pH from the $[H^+]$ and vice versa. The log of one is zero, log 1 = 0, and the log of a product is the sum of the individual logs,

$$\log 1.0 \times 10^4 = \log 1.0 + \log 10^4 = 0 + 4 = 4.$$

The logarithmic term for $[OH^-]$ is pOH.

$$pOH = -\log [OH^-]$$

A very convenient equation to also remember and use (which is derived from the equilibrium expression for water) is

$$pH + pOH = 14.00$$

Every solution (whether acidic, basic, or neutral) **has H^+ and OH^- ions present** and therefore a corresponding pH and pOH. Given the $[H^+]$, $[OH^-]$, pH or pOH of a solution, you **must** be able to calculate the other three quantities. The following "square" conceptual diagram summarizes the equations needed to calculate the four quantities.

$$pH = -\log [H^+]$$

$$pH \quad \Leftrightarrow \quad [H^+]$$

$$pH + pOH = 14.00 \qquad \updownarrow \qquad \updownarrow \qquad [H^+][OH^-] = 1.0 \times 10^{-14}$$

$$pOH \quad \Leftrightarrow \quad [OH^-]$$

$$pOH = -\log [OH^-]$$

If you are given the $[H^+]$ and want to calculate pH or vice versa, the equation that contains those two quantities is the definition of pH. If you are given the pH of a solution and want to calculate the $[OH^-]$ present (that is, go "diagonally" across the square), there are two equally correct paths, both of which have two steps. You can calculate pOH from pH by using the left-hand equation and then convert pOH to $[OH^-]$ using the bottom equation (definition of pOH). Or, you can calculate $[H^+]$ from pH by the top equation and then convert $[H^+]$ to $[OH^-]$ with the right-hand equation. The path you choose depends on which mathematical procedures are easier for you to do.

Now we can also summarize the definition of an acidic, basic or neutral solution in terms of $[H^+]$ **and** pH,

Solution	$[H^+]$	pH
Acidic	$> 1.0 \times 10^{-7}$	< 7.00
Neutral	$= 1.0 \times 10^{-7}$	$= 7.00$
Basic	$< 1.0 \times 10^{-7}$	> 7.00

Although we could have added a third and fourth column to this table containing $[OH^-]$ and pOH information, you will find it most useful to learn the definition of an acidic, basic, and neutral solution by the criteria for $[H^+]$ and pH.

SAMPLE PROBLEM 15.3 A

▶▶ In a solution that has $[H^+] = 8.0 \times 10^{-10}$, (a) what is the $[OH^-]$, (b) What is the pH? (c) What is the pOH? (d) Is the solution acidic, basic, or neutral?

Solution:

HINT: See the "square" conceptual diagram in the REVIEW for equations and applications.

(a) The equation that contains both $[H^+]$ (given) and $[OH^-]$ (sought) is the equilibrium expression for water. Solving that equation for $[OH^-]$,

$$[H^+][OH^-] \;=\; 1.0 \times 10^{-14}$$

$$[OH^-] \;=\; \frac{1.0 \times 10^{-14}}{8.0 \times 10^{-10}} \;=\; 1.3 \times 10^{-5}$$

(b) The only equation that contains $[H^+]$ and pH is the definition of pH. Solving that equation for pH,

$$pH \;=\; -\log[H^+] \;=\; -\log[8.0 \times 10^{-10}] \;=\; 9.10$$

That is, calculate the log first and change the sign.

NOTE: • pH (and pOH) use base 10 logs, which is the "log" on most calculators. The "ln" key is used for calculating base e logs.
• The number of digits to the **right** of the decimal place in a pH should be the same as the number of significant figures in the hydrogen ion concentration. In this problem, there are **two** digits to the right of the decimal place in the pH because $[H^+]$ is given to **two** significant figures.

(c) Although $[H^+]$, $[OH^-]$, and pH are now known for this solution and you could calculate the pOH from any of the three quantities (see "square" conceptual diagram in REVIEW), by far the easiest procedure is to calculate the pOH from pH by using the equation

$$pH \;+\; pOH \;=\; 14.00$$

$$pOH \;=\; 14.00 \;-\; pH \;=\; 14.00 \;-\; 9.10 \;=\; 4.90$$

(d) The solution is basic, since $[H^+] < 1.0 \times 10^{-7}$ and pH > 7.00 (and $[OH^-] > 1.0 \times 10^{-7}$ and pOH < 7.00).

SAMPLE PROBLEM 15.3 B

▸▸ In a solution in which the pH = 1.70, (a) what is the $[H^+]$? (b) What is the $[OH^-]$? (c) What is the pOH? (d) Is the solution acidic, basic, or neutral?

Solution:

(a) The equation that has pH and $[H^+]$ in it is the definition of pH. Solve that equation for $\log[H^+]$ first,

$$\log[H^+] \;=\; -pH,$$

and then take the antilog of both sides of the equation to find the $[H^+]$,

$$[H^+] = \text{antilog}(-pH) = \text{antilog}(-1.70) = 2.0 \times 10^{-2}$$

NOTE:
- Enter -1.7 in your calculator and then (generally) press the "INV" and "log" keys to calculate the antilog.
- If unsure of your result, you can always check your answer by taking the $-\log$ of the $[H^+]$ and (if correct) you will come up with the pH you started with.
- The answer has **two** significant figures because there are two digits to the right of the decimal place in the pH.

(b)   The equation with both $[H^+]$ (calculated in part (a)) and $[OH^-]$ in it is the equilibrium expression for water. Solving that equation for $[OH^-]$,

$$[H^+][OH^-] = 1.0 \times 10^{-14}$$

$$[OH^-] = \frac{1.0 \times 10^{-14}}{[H^+]} = \frac{1.0 \times 10^{-14}}{2.0 \times 10^{-2}} = 5.0 \times 10^{-13}$$

(c)   When the pH is known, it is easiest to find the pOH from

$$pH + pOH = 14.00$$

$$pOH = 14.00 - pH = 14.00 - 1.70 = 12.30$$

(d)   The solution is acidic, since $[H^+] > 1.0 \times 10^{-7}$ and the pH < 7.00 (and $[OH^-] < 1.0 \times 10^{-7}$ and pOH > 7.00).

SECTION 15.4   CONCENTRATIONS OF HYDROGEN IONS IN AQUEOUS SOLUTIONS OF ACIDS

After completing this section, you will be able to do the following:

■   For a weak acid, calculate K_a from the percent ionization;

■   For any acid, determine whether it is strong or weak and calculate the $[H^+]$, pH, $[OH^-]$, and pOH.

REVIEW

To calculate the $[H^+]$, pH, $[OH^-]$, or pOH of an acid, you **must** know whether the acid is strong or weak. Strong acids are 100% ionized; there is no equilibrium established. The stoichiometry of the reaction (coefficients) will tell you the $[H^+]$ directly. On the other hand, weak acids are **NOT** 100% ionized and equilibrium **IS** established. You must use the equilibrium constant expression (see Sample Problems 14.4 C & D) to calculate the $[H^+]$, etc., for a weak acid. The equilibrium constant for a weak acid, K_a, has the subscript "a" to identify it as the equilibrium constant for an acid ionization or acid dissociation reaction. [The subscripts of equilibrium constants are used to identify specific types of equilibrium reactions to which they apply, e.g., K_w, K_a, K_b, etc.]

For weak acids, the value of K_a must be given, either in the statement of the problem or, more commonly, in a table such as Table 15.3 in the textbook. There are no equilibrium constants to be given for strong acids.

You can calculate the equilibrium constant for any reaction only if you know the concentrations of all the aqueous and gaseous reactants and products **at equilibrium**. You can be given the equilibrium concentrations of all the reactants and products directly (Sample Problem 14.3 A), or you can be given initial concentrations and one equilibrium concentration (Sample Problem 14.3 B), or for weak acids you can be given the initial concentration and **percent ionization** to reach equilibrium (Sample Problem 15.4 A). The **percent ionization** tells you exactly how much of the initial concentration of the weak acid has ionized by the time equilibrium has been reached, which is the "Change in concentration" information in the data table for the reaction. Then the equilibrium concentrations and equilibrium constant, K_a, can be calculated.

SAMPLE PROBLEM 15.4 A

▶ ▶ If a 0.020 M HN_3 solution is 3.0% ionized by the time equilibrium is reached, what is the value of the equilibrium constant, K_a, for HN_3?

Solution:

You will need to calculate the equilibrium concentrations of all the reactants and products before substituting them into the equilibrium expression and calculating K_a. Start by writing the equilibrium reaction and set up the data table.

	HN$_3$(aq)	$\rightleftharpoons$	H$^+$(aq)	+	N$_3{}^-$(aq)
Initial, M	0.020		0.0		0.0
Change, ΔM	?		?		?
Equilibrium, M	?		?		?

How much of the weak acid, HN$_3$, is used up by the time equilibrium is reached? The answer is 3.0%, or $(0.030)(0.020\ M) = 6.0 \times 10^{-4}\ M$. Completing the "Change in concentration" data (which must always reflect the stoichiometry of the reaction) and adding them to the initial concentration data, the equilibrium concentrations can be calculated.

	HN$_3$(aq)	$\rightleftharpoons$	H$^+$(aq)	+	N$_3{}^-$(aq)
Initial, M	0.020		0.0		0.0
Change, ΔM	-6.0×10^{-4}		6.0×10^{-4}		6.0×10^{-4}
Equilibrium, M	0.019		6.0×10^{-4}		6.0×10^{-4}

Now substitute the equilibrium concentrations into the equilibrium expression and calculate K$_a$.

$$\frac{[\text{H}^+][\text{N}_3{}^-]}{[\text{HN}_3]} = \frac{(6.0 \times 10^{-4})^2}{0.019} = 1.9 \times 10^{-5} = K_a$$

SAMPLE PROBLEM 15.4 B

▶▶ What is the [H$^+$] and pH of a 0.080 M HNO$_3$ solution?

Solution:

HINT: Determine whether the acid is strong or weak before proceeding further.

HNO$_3$ is a strong acid (no K$_a$ given in the problem or in Table 15.3 in the textbook). All strong acids are always 100% ionized (no equilibrium established). Since the ionization reaction for HNO$_3$ shows all coefficients to be one,

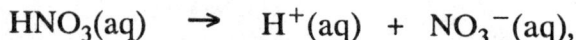

$$\text{HNO}_3(\text{aq}) \rightarrow \text{H}^+(\text{aq}) + \text{NO}_3{}^-(\text{aq}),$$

0.080 M HNO$_3$ will ionize completely into 0.080 M H$^+$ and 0.080 M NO$_3{}^-$. Therefore, [H$^+$] = [NO$_3{}^-$] = 0.080 M.

The pH can be found from the definition of pH.

$$pH = -\log[H^+] = -\log(8.0 \times 10^{-2}) = 1.10$$

SAMPLE PROBLEM 15.4 C

▶▶ For a 0.30 M hypochlorous acid solution, HOCl, calculate the (a) $[H^+]$ (b) pH, and (c) percent ionization.

Solution:

(a) Is the acid strong or weak? From Table 15.3 in the textbook, $K_a = 3.2 \times 10^{-8}$ for HOCl; HOCl is a weak acid, and an equilibrium data table and equilibrium constant expression are needed to calculate the $[H^+]$ (similar to Sample Problems 14.4 C and D). Letting x equal the concentration of HOCl that reacts to reach equilibrium, the ionization reaction and data table are

	HOCl(aq)	$\rightleftharpoons$	H^+(aq)	+	OCl^-(aq)
Initial, M	0.30		0.0		0.0
Change, ΔM	$-x$		x		x
Equilibrium, M	$(0.30-x)$		x		x

Substitute the equilibrium concentrations in terms of x into the equilibrium constant expression,

$$\frac{[H^+][OCl^-]}{[HOCl]} = \frac{x^2}{(0.30-x)} = K_a = 3.2 \times 10^{-8}$$

Since there is no perfect square (and there won't be for weak acids and bases), assume x is zero with respect to the initial concentration of 0.30 M HOCl and solve the expression for x,

$$\frac{x^2}{0.30} = 3.2 \times 10^{-8} \qquad x = 9.8 \times 10^{-5}$$

Is x negligible compared with 0.30 M? Yes, since

$$[HOCl]_{equil} = 0.30 - x = 0.30 - (9.8 \times 10^{-5}) = 0.30 \text{ M}$$

No further approximations for HOCl at equilibrium are needed. Thus,

$$[H^+] = 9.8 \times 10^{-5}.$$

(b) The pH can be found by

$$pH = -\log[H^+] = -\log(9.8 \times 10^{-5}) = 4.01$$

(c) The percent ionization is the "Change in concentration" of HOCl (x) compared to the initial concentration of HOCl times 100

$$\% \text{ ionization} = \frac{[HOCl]_{change}}{[HOCl]_{initial}} \times 100 = \frac{(9.8 \times 10^{-5})}{0.30} \times 100 = 0.033\%$$

Less than 1.00% of the HOCl has ionized by the time equilibrium has been reached.

SECTION 15.5   CONCENTRATIONS OF HYDROXIDE IONS IN AQUEOUS SOLUTIONS OF BASES
SECTION 15.6   HYDROLYSIS

After completing this section, you will be able to do the following:

■ For a weak base, calculate K_b from the percent ionization or from K_a of the conjugate acid;

■ For any base, determine whether it is strong or weak and calculate the $[OH^-]$, pH, $[H^+]$, and pOH.

REVIEW

To calculate the $[H^+]$, pH, $[OH^-]$, or pOH of a base, you **must** know whether the base is strong or weak. Strong bases are 100% ionized; there is no equilibrium established. The stoichiometry of the reaction (coefficients) will tell you the $[OH^-]$ directly. On the other hand, weak bases are **NOT** 100% ionized and equilibrium **IS** established. You must use the equilibrium constant expression (like Sample Problems 15.4 C) to calculate the $[OH^-]$, etc., for a weak base (Sample Problem 15.5 C). The equilibrium constant for a weak base, K_b, has the subscript "b" to identify it as the equilibrium constant for a base ionization or base dissociation reaction. Since the formulas of most weak bases do not contain the hydroxide ion, OH^-, you will need to write the reaction between the base and water to show the formation of OH^- ion. For example, the conjugate base of the weak acid HNO_2 is NO_2^-. The salt, $NaNO_2$, is an ionic substance which ionizes completely (remember salts are 100% ionized) into Na^+ and NO_2^- ions. An ion will only react with water if a weak electrolyte (including weak acids and weak bases) is formed. From the salt $NaNO_2$, only the base NO_2^- will react with water to form the weak acid, HNO_2,

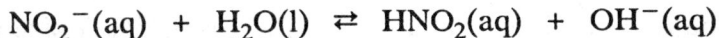

$$NO_2^-(aq) \ + \ H_2O(l) \ \rightleftarrows \ HNO_2(aq) \ + \ OH^-(aq)$$

This solution and solutions of all bases are **ALWAYS** basic ($[H^+] < [OH^-]$; pH > 7.00). The equilibrium constant for this reaction, K_b, is for the base NO_2^- reacting with water to form a weak acid **AND OH$^-$** ions. The OH^- ion **MUST** appear on the product side for every reactant acting as a base, just as the H^+ ion **MUST** appear on the product side for every reactant acting as an acid. The base NO_2^- and the acid HNO_2 are a conjugate acid-base pair, whose equilibrium constants, K_a and K_b are related to the equilibrium constant for water, K_w,

$$K_a \ x \ K_b \ = \ K_w$$

Knowing the value of K_a for HNO_2 from Table 15.3 in the textbook will allow you to calculate K_b for NO_2^-. For weak bases, the value of K_b will be given in the statement of the problem only if it is an organic base not listed in Table 15.4 of the textbook or the K_a for the conjugate acid is not listed in Table 15.3 of the textbook. There are no equilibrium constants to be given for strong bases.

The reaction just discussed between NO_2^- ion and water is a Bronsted-Lowry acid-base reaction. However, when one of the reactants is water, the more specific name given to this acid-base reaction is **hydrolysis**. **Hydrolysis** is the reaction of an anion (i.e., NO_2^-) with water to produce the conjugate acid of the anion (i.e., HNO_2) and hydroxide ion or the reaction of a cation with water to produce the conjugate base of the cation and the hydrogen ion. Salts contain anions of acids and cations of bases. You need to be able to predict whether a solution of a salt will be acidic, basic, or neutral. Salts are always 100% ionized. A cation or anion of a salt will react with water (that is, hydrolysis occurs) **only if a weak electrolyte is formed** (i.e., weak acid, weak base, H_2O, etc.). If no weak acid or weak base is formed, then no reaction with water occurs and the solution is neutral (Sample Problem 15.5 D).

You can calculate the equilibrium constant for any reaction only if you know the concentrations of all the aqueous and gaseous reactants and products **at equilibrium**. You can be given the equilibrium concentrations of all the reactants and products directly (Sample Problem 14.3 A), or you can be given initial concentrations and one equilibrium concentration (Sample Problem 14.3 B), or for weak bases you can be given the initial concentration and **percent ionization** to reach equilibrium (Sample Problem 15.4 A and Sample Problem 15.5 A). The **percent ionization** tells you exactly how much of the initial concentration of the weak base has ionized by the time equilibrium has been reached, which is the "Change in concentration" information in the data table for the reaction. Then the equilibrium concentrations and equilibrium constant, K_b, can be calculated.

SAMPLE PROBLEM 15.5 A

▸▸ If a 0.050 M solution of trimethylamine, $(CH_3)_3N$, is 3.55% ionized by the time equilibrium is reached, calculate the value of the equilibrium constant, K_b.

Solution:

HINT: Like inorganic NH_3, organic compounds with a N atom bonded to one or more carbon-containing groups are also bases. It is the N in NH_3 and the organic bases which accepts the proton from H_2O. To show the OH^- ion formed as a product for organic bases and all inorganic bases which do not contain the hydroxyl ion, OH^-, in their formula, **water must be shown as a reactant.**

You will need to calculate the equilibrium concentrations of all the reactants and products before substituting them into the equilibrium expression and calculating K_b. Start by writing the equilibrium reaction and set up the data table (the concentration of liquid water is constant).

	$(CH_3)_3N(aq)$	+	$H_2O(l)$	⇌	$(CH_3)_3NH^+(aq)$	+	$OH^-(aq)$
Initial, M	0.050				0.0		0.0
Change, ΔM	?				?		?
Equilibrium, M	?				?		?

How much of the weak base, $(CH_3)_3N$, is used up by the time equilibrium is reached? The answer is 3.55%, or $(0.0355)(0.050$ M$) = 1.78 \times 10^{-3}$ M. Completing the "Change in concentration" data (which must always reflect the stoichiometry of the reaction) and adding them to the initial concentration data, the equilibrium concentrations can be calculated

	$(CH_3)_3N(aq)$	+	$H_2O(l)$	⇌	$(CH_3)_3NH^+(aq)$	+	$OH^-(aq)$
Initial, M	0.050				0.0		0.0
Change, ΔM	-1.78×10^{-3}				1.78×10^{-3}		1.78×10^{-3}
Equilibrium, M	0.0482				1.78×10^{-3}		1.78×10^{-3}

Now substitute the equilibrium concentrations into the equilibrium expression and calculate K_b,

$$\frac{[(CH_3)_3NH^+][OH^-]}{[(CH_3)_3N]} = \frac{(1.78 \times 10^{-3})^2}{0.0482} = 6.6 \times 10^{-5} = K_b$$

SAMPLE PROBLEM 15.5 B

▶▶ What is the value of K_b for the base carbonate ion, CO_3^{2-}?

Solution:

If CO_3^{2-} is a base, then write the reaction with water to form the conjugate acid and OH^-,

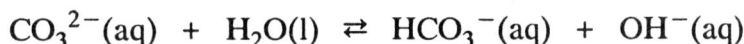

$$CO_3^{2-}(aq) + H_2O(l) \rightleftarrows HCO_3^-(aq) + OH^-(aq)$$

Since $K_a \times K_b = K_w$ for any conjugate acid-base pair, and from Table 15.3 $K_a = 5.6 \times 10^{-11}$ for HCO_3^-, you can easily find K_b for the base CO_3^{2-}.

$$K_b\,(CO_3^{2-}) = \frac{K_w}{K_a\,(HCO_3^-)} = \frac{1.0 \times 10^{-14}}{5.6 \times 10^{-11}} = 1.8 \times 10^{-4}$$

SAMPLE PROBLEM 15.5 C

▶▶ What is the pH of a 0.10 M KF solution?

Solution:

KF is a salt, which is always 100% ionized. Which ion, if any, forms a weak acid or a weak base? Only the F^- ion will react with water because it forms the weak acid, HF.

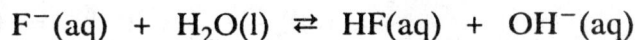

$$F^-(aq) + H_2O(l) \rightleftarrows HF(aq) + OH^-(aq)$$

K_b for F^- is calculated from the value of K_a for HF given in Table 15.3

$$K_b\,(F^-) = \frac{K_w}{K_a\,(HF)} = \frac{1.0 \times 10^{-14}}{3.5 \times 10^{-4}} = 2.9 \times 10^{-11}$$

The most important thing to remember when working with bases is that one of the products is OH^- ion, **NOT** H^+ ion. After calculating the $[OH^-]$, you can calculate the pH, $[H^+]$, or pOH. Start by indicating the initial concentrations of all reactants and products. (The procedure from this point on is virtually identical to that of Sample Problem 15.4 C).

$$F^-(aq) \quad + \quad H_2O(l) \quad \rightleftarrows \quad HF(aq) \quad + \quad OH^-(aq)$$

Initial, M	0.10	0.0	0.0
Change, ΔM	?	?	?
Equilibrium, M	?	?	?

How much of the weak base, F^-, is used up by the time equilibrium is reached? We don't know, so we let x equal the concentration of F^- that reacts to reach equilibrium. Completing the data table,

$$F^-(aq) \quad + \quad H_2O(l) \quad \rightleftarrows \quad HF(aq) \quad + \quad OH^-(aq)$$

Initial, M	0.10	0.0	0.0
Change, ΔM	$-x$	x	x
Equilibrium, M	$(0.10-x)$	x	x

Substitute the equilibrium concentrations in terms of x into the equilibrium constant expression

$$\frac{[HF][OH^-]}{[F^-]} = \frac{x^2}{(0.10-x)} = K_b = 2.9 \times 10^{-11}$$

Since there is no perfect square (and there won't be for weak acids and weak bases), assume x is zero with respect to the initial concentration of 0.10 M F^- and solve the expression for x,

$$\frac{x^2}{0.10} = 2.9 \times 10^{-11} \qquad x = 1.7 \times 10^{-6}$$

Is x negligible compared with 0.10 M F^-? Yes, since

$$[F^-]_{equil} = 0.10 - x = 0.10 - (1.7 \times 10^{-6}) = 0.10$$

No further approximations for HOCl at equilibrium are needed. Remember, for bases, you are calculating OH^- ion, not H^+ ion

$$x = [OH^-] = 1.7 \times 10^{-6}$$

To find pH from $[OH^-]$, it is probably easiest to calculate the pOH and then the pH. (However, you are equally correct if you calculate $[H^+]$ first and then pH.)

$$pOH = -\log[OH^-] = -\log(1.7 \times 10^{-6}) = 5.77$$

and the pH is found by

$$pH = 14.00 - pOH = 14.00 - 5.77 = 8.23$$

CHECK: The pH of all bases **MUST** be > 7.00!

SAMPLE PROBLEM 15.6 A

►► Predict whether a water solution of each of the following salts will be acidic, basic, or neutral. If acidic or basic, justify with a balanced net ionic reaction.

(a) KCN (b) NH_4Cl (c) $NaC_2H_3O_2$ (d) $Ca(NO_3)_2$ (e) $ZnCl_2 \cdot 6H_2O$

Solution:

HINT: Salts are always 100% ionized. Does either ion of the salt react with water to form a weak acid or a weak base? If no, the solution is neutral. If yes, the solution is acidic or basic (unless **both** ions react equally with water and produce a neutral solution).

(a) $K^+(aq) + H_2O(l) \rightarrow$ No reaction   KOH is a strong base.

$CN^-(aq) + H_2O(l) \rightleftharpoons HCN(aq) + OH^-(aq)$

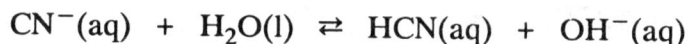

HCN is a weak acid and OH^- is formed.

The solution will be basic.

(b) $NH_4^+(aq) + H_2O(l) \rightleftharpoons NH_3(aq) + H_3O^+(aq)$

NH_3 is a weak base and H_3O^+ is formed.

$Cl^-(aq) + H_2O(l) \rightarrow$ No reaction   HCl is a strong acid.

The solution will be acidic.

(c) $Na^+(aq) + H_2O(l) \rightarrow$ No reaction   NaOH is a strong base.

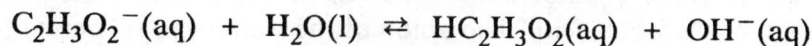

$C_2H_3O_2^-(aq) + H_2O(l) \rightleftharpoons HC_2H_3O_2(aq) + OH^-(aq)$

$HC_2H_3O_2$ is a weak acid and OH^- is formed.

The solution will be basic.

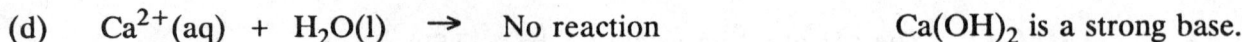

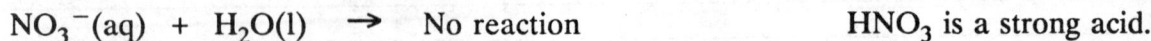

(d) $Ca^{2+}(aq) + H_2O(l) \rightarrow$ No reaction $\qquad$ $Ca(OH)_2$ is a strong base.

$NO_3^-(aq) + H_2O(l) \rightarrow$ No reaction $\qquad$ HNO_3 is a strong acid.

The solution will be neutral.

(e) $Zn(H_2O)_6^{2+}(aq) + H_2O(l) \rightleftharpoons Zn(H_2O)_5(OH)^+(aq) + H_3O^+(aq)$

$\qquad\qquad$ $Zn(H_2O)_6^{2+}$ is a weak acid (see Table 15.3 in textbook).

$Cl^-(aq) + H_2O(l) \rightarrow$ No reaction $\qquad$ HCl is a strong acid.

The solution will be acidic.

| SECTION 15.7 | THE COMMON ION EFFECT |
| SECTION 15.8 | BUFFER SOLUTIONS |

After completing this section, you will be able to do the following:

- Calculate the $[H^+]$ in a solution that contains both a weak acid and a salt of the weak acid;

- Identify buffer solutions;

- Determine (a) the $[H^+]$ and pH of a buffer and (b) the $[H^+]$ and pH after the addition of a strong acid or strong base to the buffer.

REVIEW

So far, you have learned how to calculate the hydrogen ion concentration in a solution of a weak acid (Sample Problem 15.4 C) and how to calculate the hydrogen ion concentration in a solution of a salt of a weak acid (Sample Problem 15.5 C). In this section, we will combine the two so that you can calculate the hydrogen ion concentration in a solution of a weak acid and a salt of the weak acid. The salt is 100% ionized. The reaction of concern is the equilibrium reaction of the weak acid. The salt of the weak acid is a **second source** of the anion of the weak acid. The anion of the weak acid is **common** to both the weak acid and salt of the weak acid. The presence of the anion of the weak acid from the salt shifts the weak acid equilibrium to the left and represses the ionization of the weak acid. That is, in the presence of salt furnishing the anion of a weak acid initially, not as much of the weak acid will ionize to reach equilibrium.

Solutions that contain approximately equal concentrations of a weak acid and a salt of the weak acid (Sample Problem 15.7 A) or a weak base and a salt of the weak base resist significant changes in pH when moderate amounts of a strong acid or strong base are added. These solutions are called **buffer solutions** because they buffer the solution against (resist) significant changes in pH. For example, if the buffer solution formed by HNO_2 and $NaNO_2$ contains 0.10 M of each at equilibrium, we have both an acid and a base present to resist changes in pH as a strong acid or base is added,

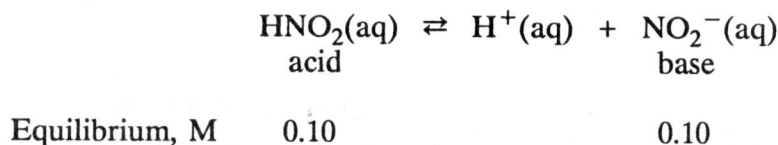

$$HNO_2(aq) \rightleftarrows H^+(aq) + NO_2^-(aq)$$
$$\text{acid} \qquad\qquad\qquad \text{base}$$

Equilibrium, M   0.10   0.10

If a strong base is added (NaOH), equilibrium is temporarily destroyed and the acid (HNO_2) will immediately react with the NaOH to form more NO_2^- and restore equilibrium (respond to the right). On the other hand, if a strong acid (HCl) is added to this same buffer, equilibrium will again be temporarily destroyed until the base present in the buffer (NO_2^-) can react with the HCl to form more HNO_2 and restore equilibrium (respond to the left). To summarize, when a strong base is added to a buffer consisting of a weak acid and one of its salts, the reaction will respond quantitatively in the forward direction; when a strong acid is added, the reaction will respond quantitatively in the reverse direction.

SAMPLE PROBLEM 15.7 A

▶▶ What is the $[H^+]$ and pH of a solution formed by the addition of 0.20 mol of $NaHCO_3$ to 1.00 L of 0.10 M carbonic acid, H_2CO_3? Assume no change in volume. K_a for H_2CO_3 = 4.3 x 10^{-7}; K_a for HCO_3^- is negligible.

Solution:

The $NaHCO_3$ is 0.20 M and is 100% ionized into 0.20 M Na^+ and 0.20 M HCO_3^-. The reaction of interest is the equilibrium reaction for the release of the first hydrogen by H_2CO_3,

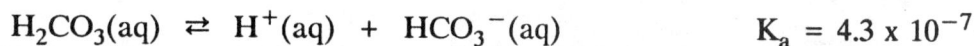

$$H_2CO_3(aq) \rightleftarrows H^+(aq) + HCO_3^-(aq) \qquad\qquad K_a = 4.3 \times 10^{-7}$$

Set up the data table and note that H_2CO_3 and HCO_3^- are present **initially.**

	$H_2CO_3(aq)$	$\rightleftarrows$	$H^+(aq)$ +	$HCO_3^-(aq)$
Initial, M	0.10		0.00	0.20 (from 0.20 M $NaHCO_3$)
Change, ΔM	?		?	?
Equilibrium, M	?		?	?

As usual, let x equal the concentration of H_2CO_3 that ionizes to reach equilibrium.

$$H_2CO_3(aq) \quad \rightleftarrows \quad H^+(aq) \quad + \quad HCO_3^-(aq)$$

Initial, M	0.10	0.00	0.20 (from 0.20 M $NaHCO_3$)
Change, ΔM	$-x$	x	x
Equilibrium, M	$(0.10-x)$	x	$(0.20 + x)$

By the time equilibrium is reached, there are two sources of the **common ion**, HCO_3^-: 0.20 M from $NaHCO_3$ and x M from the ionization of carbonic acid. Substitute the equilibrium concentrations in terms of x into the equilibrium expression, assume x is negligible compared to 0.10 M and 0.20 M, and solve the expression for x.

$$\frac{[H^+][HCO_3^-]}{[H_2CO_3]} = \frac{x(0.20)}{0.10} = K_a = 4.3 \times 10^{-7}$$

$$x = [H^+] = 2.2 \times 10^{-7} \quad \text{and} \quad pH = -\log[H^+] = 6.70$$

It should be apparent that x is negligible compared to 0.10 M H_2CO_3 and 0.20 M HCO_3^-.

SAMPLE PROBLEM 15.8 A

▶▶ Which of the following combinations of solutions could be buffers?

(a) HNO_3 and $NaNO_3$ (b) NH_4Cl and NH_3 (c) H_2SO_4 and H_2SO_3 (d) $HC_2H_3O_2$ and $Cu(C_2H_3O_2)_2$ (e) $HOCl$ and KBr

Solution

HINT: A buffer must contain a weak acid and a salt of the weak acid or a weak base and a salt of the weak base.

(a)   This cannot be a buffer solution. HNO_3 is a strong acid.

(b)   This is a buffer solution. NH_3 is a weak base

$$NH_3(aq) + H_2O(l) \rightleftarrows NH_4^+(aq) + OH^-(aq)$$

A salt of NH_3 must contain the common ion, NH_4^+. NH_4Cl (or NH_4Br, NH_4NO_3, $(NH_4)_2CO_3$, etc.) is a salt of the weak base.

(c) This cannot be a buffer solution. H_2SO_3 is a weak acid, and a salt of H_2SO_3 must contain the common ion, SO_3^{2-}. H_2SO_4 is a strong acid, not a salt of H_2SO_3.

(d) This is a buffer solution. $HC_2H_3O_2$ is a weak acid, and a salt of $HC_2H_3O_2$ must contain the common ion, $C_2H_3O_2^{-}$. $Cu(C_2H_3O_2)_2$ (or $Ni(C_2H_3O_2)_2$, $KC_2H_3O_2$, etc.) is a salt of $HC_2H_3O_2$.

(e) This cannot be a buffer solution. $HOCl$ is a weak base, and a salt of $HOCl$ must contain the common ion, OCl^{-}. KBr is a salt but does not contain the common ion, OCl^{-}.

SAMPLE PROBLEM 15.8 B

▶▶ (a) What is the $[H^{+}]$ and pH in a solution that is 0.20 M in $NaHCO_3$ and 0.10 M in H_2CO_3? $K_a = 4.3 \times 10^{-7}$ for H_2CO_3; K_a for HCO_3^{-} is negligible. (b) What will the $[H^{+}]$ and pH be after 0.005 mol of NaOH has been added to 1.0 L of the buffer solution? (Assume no change in volume.)

Solution:

(a) This solution contains a weak acid, H_2CO_3, and a salt of the weak acid, $NaHCO_3$, which is a buffer solution. This problem was worked for you as a common ion problem (Sample Problem 15.7 A) with the equilibrium concentrations and pH of

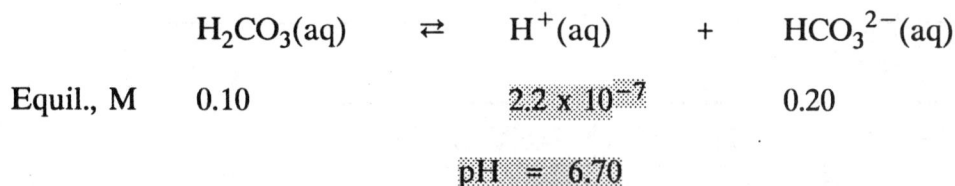

$$H_2CO_3(aq) \quad \rightleftarrows \quad H^{+}(aq) \quad + \quad HCO_3^{2-}(aq)$$

Equil., M	0.10	2.2×10^{-7}	0.20

$$pH = 6.70$$

(b) Since you are adding a strong base to the buffer (always assume to be at equilibrium), the buffer can counteract it by the acid in the buffer (H_2CO_3) reacting quantitatively with the NaOH (forward direction) to form more HCO_3^{-} and to establish a **new set of initial concentrations that must re-establish equilibrium**. The net ionic reaction and quantitative changes for the neutralization of the 0.005 M NaOH by H_2CO_3 are

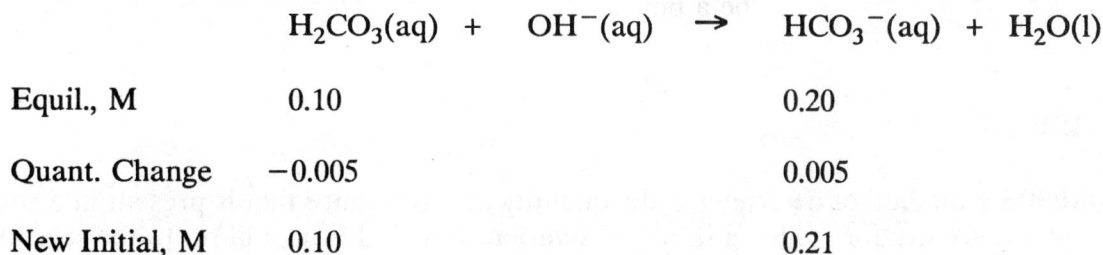

$$H_2CO_3(aq) \quad + \quad OH^{-}(aq) \quad \rightarrow \quad HCO_3^{-}(aq) \quad + \quad H_2O(l)$$

Equil., M	0.10		0.20
Quant. Change	−0.005		0.005
New Initial, M	0.10		0.21

(Recall that numbers ending in a "5" are always rounded up when rounding for the correct number of significant figures.) Now equilibrium is re-established following the normal procedures for a weak acid. Letting x equal the concentration of H_2CO_3 that ionizes to reach equilibrium

	$H_2CO_3(aq)$	$\rightleftarrows$	$H^+(aq)$	$+$	$HCO_3^{2-}(aq)$
Initial, M	0.10		~0.0		0.21
Change, ΔM	$-x$		x		x
Equil., M	$(0.10-x)$		x		$(0.21+x)$

Assuming x is small compared to (0.10) and (0.21), solve the expression for x, which is the $[H^+]$, and calculate the pH.

$$\frac{[H^+][HCO_3^-]}{[H_2CO_3]} = \frac{x(0.21)}{0.10} = K_a = 4.3 \times 10^{-7}$$

$$x = [H^+] = 2.0 \times 10^{-7} \quad \text{and} \quad pH = 6.70$$

The $[H^+]$ and pH of the buffer solution is the same before (part (a)) and after the addition of the strong base. The solution is "buffered" against large changes in pH. Depending upon the quantity of strong base added, there may be a small change in pH (usually less than one pH unit).

SECTION 15.10   TITRATION REVISITED

After completing this section, you will be able to do the following:

- For the titration of a strong acid with a strong base, calculate the pH before, at, and after the equivalence point;

- For the titration of a weak acid with a strong base, calculate the pH before, at, and after the equivalence point.

REVIEW

Titration is a method of determining the quantity of a substance that is present in a solution. A solution of known concentration, called a **standard solution,** is added from a buret to the solution being analyzed.

Addition of standard solution is stopped when the **endpoint** of the titration is reached, which is usually signaled by a color change of an appropriate indicator (see Section 15.9 in the textbook) or by a change in conductivity. If the means of measuring the endpoint has been chosen correctly, the **endpoint** occurs at the **equivalence point,** which is the point in the titration where the quantity of reactant (standard solution) called for by the equation for the reaction has been added.

To calculate the pH at any point of a titration, you need to take into account whether the acid and base are strong or weak. For the titration of a **strong acid with a strong base**, you are adding a strong base to the strong acid. As strong base is added, the solution will remain acidic (pH < 7.00) up until the equivalence point (see Sample Problem 15.15 in the textbook). At the equivalence point, the reaction is complete (quantitative) with no strong base or strong acid, only the salt of a strong acid and a strong base and water. Neither ion in the salt will hydrolyze, so the solution is neutral. Beyond the equivalence point, only excess OH^- (from the strong base), the salt of the strong acid and strong base, and water are present. The pH is determined by the amount of excess OH^- (see Sample Problem 15.16 in the textbook).

For the titration of a **weak acid with a strong base**, you are adding a strong base to a weak acid. As strong base is added up to the equivalence point, there will be excess weak acid and the salt of the weak acid present, which is a **buffer solution**. Up to the equivalence point, the pH is calculated for a buffer just like in Sample Problem 15.7 A, which is a common ion and buffer problem. **At the equivalence point,** all of the weak acid has reacted to form salt. Because the salt is a salt of a weak acid and a strong base, the anion of the salt hydrolyzes to form the weak acid and OH^-. The pH is **not 7.00** at the equivalence point of a titration of a weak acid with a strong base; the solution is basic or the pH is > 7.00, just like in Sample Problem 15.5 C. **After the equivalence point,** strong base is being added to a solution containing the salt of the weak acid. Although the salt hydrolyzes to form some OH^-, it is ordinarily negligible compared to the excess OH^- from the strong base; the pH will be determined primarily by the concentration of excess base. Sample Problem 15.10 A illustrates the calculation of pH before, at, and after the equivalence point of a titration of a weak acid with a strong base.

SAMPLE PROBLEM 15.10 A

▶▶ If 25.00 mL of 0.1000 M CH_3COOH is titrated with 0.1000 M NaOH, what is the pH after adding the following amounts of NaOH? (a) 5.00 mL (b) 25.00 mL (c) 35.00 mL

Solution:

HINT: Think of the calculation as a two-step process. First, the acid and base react quantitatively to the right (that is, the maximum amount of product formed by the limiting reactant). Second, look at what is present to determine whether an equilibrium reaction will occur or not.

(a) The weak acetic acid CH_3COOH reacts quantitatively with the strong base NaOH. It is best to work with moles rather than molarity for the quantitative reaction. The number of moles of acetic acid present initially in 25.00 mL of 0.1000 M CH_3COOH

and the number of moles of base added in 5.00 mL of 0.1000 M NaOH are

$$? \text{ mol CH}_3\text{COOH} = 0.02500 \text{ L soln.} \times \frac{0.1000 \text{ mol CH}_3\text{COOH}}{1 \text{ L soln.}} = 0.002\ 500 \text{ mol CH}_3\text{COOH}$$

$$? \text{ mol NaOH} = 0.00500 \text{ L soln.} \times \frac{0.1000 \text{ mol NaOH}}{1 \text{ L soln.}} = 0.000\ 500 \text{ mol NaOH}$$

By now you should be able to convert volume in mL to L by mentally moving the decimal point three places to the left. **NOTE** that the number of moles of any substance in solution is the volume in liters multiplied by the molarity of the solution.

The acid-base reaction taking place quantitatively (with NaOH as the limiting reactant) and the "modified data table" are

$$CH_3COOH(aq) + NaOH(aq) \rightarrow NaC_2H_3O_2(aq) + H_2O(l)$$

	CH₃COOH(aq)	NaOH(aq)	NaC₂H₃O₂(aq)
Initial, mol	0.002 500	0.000 500	0.000 000
Change, Δmol	−0.000 500	−0.000 500	0.000 500
Reaction complete, mol	0.002 000	0.000 000	0.000 500

The "modified data table" is helpful to keep track of how the number of moles of reactants and products change for the quantitative reaction. What is present after the quantitative reaction occurs (that is, the "Reaction complete" row of information)? There are 0.002 000 mol of a weak acid and 0.000 500 mol of a salt of the weak acid. By definition, this is a **buffer solution**. Convert moles to molarity for both the weak acid and salt by dividing the number of moles by the **total volume**, 0.030 00 L,

$$? \text{ M CH}_3\text{COOH} = \frac{0.002\ 000 \text{ mol CH}_3\text{COOH}}{0.030\ 00 \text{ L soln}} = 0.06667 \text{ CH}_3\text{COOH}$$

$$? \text{ M CH}_3\text{COO}^- = \frac{0.000\ 500 \text{ mol CH}_3\text{COO}^-}{0.030\ 00 \text{ L soln}} = 0.0167 \text{ M CH}_3\text{COO}^-$$

Now write the equilibrium reaction for the weak acid, CH_3COOH, set up the data table for this buffer solution, and calculate the $[H^+]$ and pH just like in Sample Problem 15.7 A.

$$CH_3COOH(aq) \rightleftharpoons H^+(aq) + CH_3COO^-(aq)$$

Initial, M	0.06667	0.00	0.0167
Change, ΔM	$-x$	x	x
Equilibrium, M	$(0.06667-x)$	x	$(0.0167 + x)$

Assuming x is negligible compared to 0.06667 and 0.0167, we obtain

$$\frac{[H^+][CH_3COO^-]}{[CH_3COOH]} = \frac{x(0.0167)}{0.06667} = K_a = 1.8 \times 10^{-5}$$

$$x = [H^+] = 7.2 \times 10^{-5} \quad \text{and} \quad pH = -\log[H^+] = 4.14$$

It should be apparent that x is negligible compared to 0.06667 M CH_3COOH and 0.0167 M CH_3COO^-.

NOTE: For any volume of NaOH added up to the equivalence point, you have a buffer solution whose pH is calculated identical to part (a).

(b)   The number of moles of NaOH in 25.00 mL of a 0.1000 M NaOH solution is

$$? \text{ mol NaOH} = 0.02500 \text{ L soln.} \times \frac{0.1000 \text{ mol NaOH}}{1 \text{ L soln.}} = 0.002\ 500 \text{ mol NaOH}$$

Start again with the quantitative reaction of adding 0.002 500 mol NaOH to 0.002 500 mol CH_3COOH,

$$CH_3COOH(aq) + NaOH(aq) \rightarrow NaC_2H_3O_2(aq) + H_2O(l)$$

Initial, mol	0.002 500	0.002 500	0.000 000
Change, mol	$-0.002\ 500$	$-0.002\ 500$	0.002 500
Reaction complete, mol	0.000 000	0.000 000	0.002 500

What is present after the quantitative reaction is complete? There is no excess acid or base, which is the definition of the **equivalence point**. Only the salt of the strong base and weak acid is present, $NaC_2H_3O_2$. The acetate ion from the salt undergoes hydrolysis to form the weak acid CH_3COOH. Converting moles of CH_3COO^- to molarity

$$? \text{ M CH}_3\text{COO}^- = \frac{0.002\ 500 \text{ mol CH}_3\text{COO}^-}{0.050\ 00 \text{ L soln.}} = 0.050\ 00 \text{ M CH}_3\text{COO}^-$$

Now write the equilibrium reaction for the hydrolysis of the weak base acetate ion to form the weak acid CH_3COOH, set up the data table, and calculate the $[H^+]$ and pH just like Sample Problem 15.5 C.

$$CH_3COO^-(aq) + H_2O(l) \rightleftarrows CH_3COOH(aq) + OH^-(aq)$$

Initial, M	0.050 00		0.0	0.0
Change, ΔM	$-x$		x	x
Equil., M	$(0.050\ 000-x)$		x	x

Assuming x is negligible compared to 0.050 000,

$$\frac{[CH_3COOH][OH^-]}{[CH_3COO^-]} = \frac{x^2}{(0.050\ 00)} = K_b = 5.6 \times 10^{-10}$$

where K_b for acetate ion is calculated from K_a in Table 15.3 in the textbook and K_w,

$$K_b\ (CH_3COO^-) = \frac{K_w}{K_a\ (CH_3COOH)} = \frac{1.0 \times 10^{-14}}{1.8 \times 10^{-5}} = 5.6 \times 10^{-10}$$

$$x = [OH^-] = 5.3 \times 10^{-6}; \quad pOH = 5.28; \quad pH = 8.72$$

It should be apparent that x is negligible compared to 0.050 000.

(c)   The number of moles of NaOH in 35.00 mL of a 0.1000 M NaOH solution is

$$? \text{ mol NaOH} = 0.03500 \text{ L soln. } \times \frac{0.1000 \text{ mol NaOH}}{1 \text{ L soln.}} = 0.003\ 500 \text{ mol NaOH}$$

Start again with the quantitative reaction of adding 0.003 500 mol NaOH to 0.002 500 mol CH_3COOH, with CH_3COOH the limiting reactant.

$$CH_3COOH(aq) + NaOH(aq) \rightarrow NaC_2H_3O_2(aq) + H_2O(l)$$

Initial, mol	0.002 500	0.003 500	0.000 000
Change, Δmol	$-0.002\ 500$	$-0.002\ 500$	0.002 500
Reaction complete, mol	0.000 000	0.001 000	0.002 500

What is present after the quantitative reaction is complete? There is excess base and the salt $NaC_2H_3O_2$. We are **beyond the equivalence point**. We will work through the hydrolysis equilibrium just like part (b) to show you that the $[OH^-]$ (and pH) is determined essentially by the excess strong base alone; the concentration of OH^- contributed by the hydrolysis reaction is negligible. Converting moles of NaOH and CH_3COO^- to molarity, we obtain

$$? \text{ M NaOH} = \frac{0.001\ 000 \text{ mol NaOH}}{0.060\ 00 \text{ L soln.}} = 0.016\ 67 \text{ M NaOH}$$

$$? \text{ M } CH_3COO^- = \frac{0.002\ 500 \text{ mol } CH_3COO^-}{0.060\ 00 \text{ L soln}} = 0.041\ 67 \text{ M } CH_3COO^-$$

Now write the equilibrium reaction for the hydrolysis of the weak base acetate ion to form the weak acid CH_3COOH, set up the data table noting the initial concentration of OH^- from the excess NaOH, and calculate the $[H^+]$ and pH just like in part (b),

$$CH_3COO^-(aq) + H_2O(l) \rightleftarrows CH_3COOH(aq) + OH^-(aq)$$

Initial, M	0.041 67	0.0	0.016 67
Change, ΔM	$-x$	x	x
Equil., M	$(0.041\ 67 - x)$	x	$(0.016\ 67 + x)$

Assuming x is negligible compared to 0.041 67 and 0.016 67,

$$\frac{[CH_3COOH][OH^-]}{[CH_3COO^-]} = \frac{x(0.016\ 67)}{(0.041\ 67)} = K_b = 5.6 \times 10^{-10}$$

$$x = 1.4 \times 10^{-9} \quad \text{and} \quad [OH^-] = (0.016\ 67 + x) = 0.016\ 67$$

The [OH$^-$] at equilibrium is determined by the excess strong base only. The pH is

$$pOH = 1.78; \quad pH = 12.22$$

It should be apparent that x is also negligible compared to 0.041 67.

NOTE: For virtually any volume of NaOH added **beyond** the equivalence point, you have salt of a weak acid and strong base and excess strong base. The [OH$^-$] is essentially determined by the excess strong base left after the quantitative reaction occurs. As we just saw, you ordinarily will not need to go through an equilibrium calculation to calculate the [OH$^-$]; use the [OH$^-$] left after the quantitative reaction.

SECTION 15.11 POLYPROTIC ACIDS

After completing this section, you will be able to do the following:

- Calculate the equilibrium concentrations of all species in a polyprotic acid.

REVIEW

Monoprotic acids have **only one** ionizable hydrogen (HCl, HNO$_3$, HClO$_4$, etc.). **Polyprotic acids** have **more than one** ionizable hydrogen, which you can ordinarily recognize if the formula of the acid **begins** with more than one hydrogen (H$_2$SO$_4$, H$_2$CO$_3$, H$_3$PO$_4$, etc.). **Polyprotic acids** ionize stepwise with an equilibrium constant for each step, numbered to indicate which ionizable hydrogen is being released in that reaction. For example, oxalic acid (H$_2$C$_2$O$_4$) is polyprotic with ionization reactions and equilibrium constants (from Table 15.7 in the textbook) of

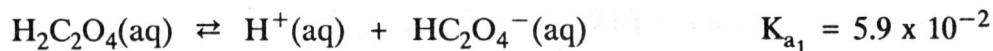

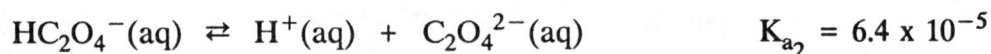

$$H_2C_2O_4(aq) \rightleftarrows H^+(aq) + HC_2O_4^-(aq) \qquad K_{a_1} = 5.9 \times 10^{-2}$$

$$HC_2O_4^-(aq) \rightleftarrows H^+(aq) + C_2O_4^{2-}(aq) \qquad K_{a_2} = 6.4 \times 10^{-5}$$

Because it takes more energy to remove successive hydrogen ions from a negatively charged anion, the values of successive K$_a$'s for a polyprotic acid decrease by a factor of 1000 or more. Note that the equilibrium constant for the ionization of the second hydrogen in H$_2$C$_2$O$_4$, K$_{a_2}$, is also the equilibrium constant for the release of the one hydrogen ion by the weak acid HC$_2$O$_4^-$. You can calculate the equilibrium concentration of all species in a polyprotic acid by working through each ionization reaction. Usually the [H$^+$] is determined by the first ionization step; that is, the H$^+$ contributed by the second step and beyond is negligible.

SAMPLE PROBLEM 15.11 A

▶▶ In a 0.10 M solution of hydrosulfuric acid, H_2S, what are $[H^+]$, $[HS^-]$, and $[S^{2-}]$?

Solution:

Two equilibria exist at the same time in a solution of hydrosulfuric acid, with K_{a_1} and K_{a_2} found in Table 15.7 of the textbook,

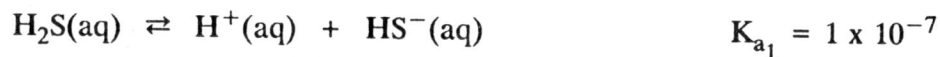

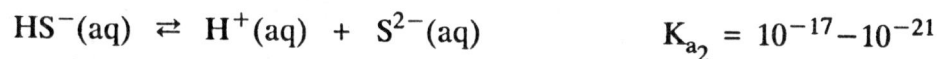

$$H_2S(aq) \rightleftarrows H^+(aq) + HS^-(aq) \qquad\qquad K_{a_1} = 1 \times 10^{-7}$$

$$HS^-(aq) \rightleftarrows H^+(aq) + S^{2-}(aq) \qquad\qquad K_{a_2} = 10^{-17} - 10^{-21}$$

Set up the data table for the **first** equilibrium reaction and let x equal the concentration of H_2S that ionizes to reach equilibrium,

	$H_2S(aq)$	$\rightleftarrows$	$H^+(aq)$	+	$HS^-(aq)$
Initial, M	0.10		0.0		0.0
Change, ΔM	$-x$		x		x
Equil., M	$(0.10-x)$		x		x

Assuming x is small compared to 0.10, substitute the equilibrium concentrations into the equilibrium expression and solve for x

$$\frac{[H^+][HS^-]}{[H_2S]} = \frac{x^2}{0.10} = K_{a_1} = 1 \times 10^{-7}$$

$$x = [H^+] = [HS^-] = 1 \times 10^{-4}$$

and

$$[H_2S] = (0.10 - x) = 0.10 - (1 \times 10^{-4}) = 0.10$$

The HS^- formed in the first step is the reactant for the second step, and the H^+ formed in the first step is the initial $[H^+]$ in the second step. [We will see shortly that the $[H^+]$ contributed by the second step is negligible.] Set up the data table for the **second** equilibrium reaction and let x equal the concentration of HS^- that ionizes to reach equilibrium

$$HS^-(aq) \quad \rightleftarrows \quad H^+(aq) \quad + \quad S^{2-}(aq)$$

	$HS^-(aq)$	$H^+(aq)$	$S^{2-}(aq)$
Initial, M	1×10^{-4}	1×10^{-4}	0.0
Change, ΔM	$-x$	x	x
Equil., M	$(1 \times 10^{-4} - x)$	$(1 \times 10^{-4} + x)$	x

Assuming x is small compared to 1×10^{-4}, substitute the equilibrium concentrations into the equilibrium expression and solve for x,

$$\frac{[H^+][S^{2-}]}{[HS^-]} = \frac{(1 \times 10^{-4})x}{1 \times 10^{-4}} = K_{a_2} = 10^{-17} \cdot 10^{-21}$$

$$x = [S^{2-}] = K_{a_2} = 10^{-17} \cdot 10^{-21}$$

and the concentrations of HS^- and H^+ are

$$[HS^-] = (1 \times 10^{-4} - x) = 1 \times 10^{-4}$$

$$[H^+] = (1 \times 10^{-4} + x) = 1 \times 10^{-4}$$

NOTE: The $[H^+]$ is determined by the first step only.

SELF-TEST QUESTIONS

Test your understanding of the concepts and skills in this chapter by working through the multiple choice questions BEFORE checking the answers. A periodic table will always be available. The ANSWERS TO SELF-TEST QUESTIONS follow immediately after this section.

1. The conjugate acid of the base $H_2PO_4^-$ is

   A. PO_4^{3-} B. H_3PO_4 C. HPO_4^{2-} D. H^+ E. not given

2. In the following reactions, a conjugate acid-base pair is shaded correctly for

   A. $HC_2H_3O_2(aq)$ + $Cl^-(aq)$ $\rightleftarrows$ $HCl(aq)$ + $C_2H_3O_2^-(aq)$.

   B. $NH_3(aq)$ + $HCO_3^-(aq)$ $\rightleftarrows$ $NH_4^+(aq)$ + $CO_3^{2-}(aq)$.

   C. $H_2O(l)$ + $HClO_4(aq)$ $\rightleftarrows$ $H_3O^+(aq)$ + $ClO_4^-(aq)$.

   D. $(CH_3)_3P(l)$ + $HNO_3(aq)$ $\rightleftarrows$ $(CH_3)_3PH^+(aq)$ + $NO_3^-(aq)$.

   E. none of the above.

3. The only correct conjugate acid-base pair is

   A. H_3PO_4/HPO_4^{2-}.

   B. H_2SO_4/H_2SO_3.

   C. HNO_3/NO_2^-.

   D. H_3O^+/H^+.

   E. H_2O/OH^-.

4. In the following acid-base reaction

$$PH_3(g) + H_2O(l) \rightleftarrows PH_4^+(aq) + OH^-(aq)$$

A. PH_4^+ is the conjugate acid of the base PH_3.

B. H_2O is a base.

C. PH_3 is the conjugate base of the acid H_2O.

D. PH_4^+ is a base.

E. more than one of the above answers is correct.

QUESTIONS 5-9 REFER TO THE FOLLOWING TABLE OF DATA				
Solution	pH	pOH	$[H^+]$	$[OH^-]$
K	10.60			
L			4.0×10^{-4}	
M		1.80		
N				8.0×10^{-12}

5. The $[II^+]$ of solution K is

   A. 4.1×10^{10}
   B. 6.2×10^{-10}
   C. 2.5×10^{-11}
   D. 4.1×10^{-10}
   E. 2.5×10^{11}

6. For solution L,

   A. the $[OH^-] = 2.5 \times 10^{-10}$.
   B. the pH $= 3.40$.
   C. the pOH $= 11.60$.
   D. both A and C are correct.
   E. A, B, and C are correct.

7. The $[H^+]$ for solution M is

    A. 1.6×10^{-2}
    B. 6.3×10^{-12}
    C. 1.6×10^{12}
    D. 6.3×10^{-13}
    E. 1.8×10^{-13}

8. The pOH of solution N is

    A. 12.99.
    B. 2.90.
    C. 1.11.
    D. 10.40.
    E. 11.10.

9. Comparing solutions K, L, M, and N,

    A. all are acidic.
    B. the pH of solutions L and N are both less than 7.
    C. only solution M is basic.
    D. solution K has the highest $[H^+]$.
    E. solution M has the lowest $[OH^-]$.

---

10. In a 0.50 M HNO_3 solution, the

    A. $[H^+] = 1.0$
    B. $[NO_3^-] = 0.50$
    C. $[OH^-] = 1.0 \times 10^{-14}$
    D. pH = 0.00
    E. pOH = 14.00

11. The pH of a 0.064 M KOH solution is

    A. 1.19.
    B. 6.40.
    C. 12.81.
    D. 2.81.
    E. 11.19.

12. In a 2.0×10^{-2} M solution of a monoprotic weak acid HA ($K_a = 2.0 \times 10^{-6}$),

    A. $[H^+] = 4.0 \times 10^{-8}$.
    B. pOH = 12.30.
    C. pH = 3.70.
    D. $[HA] = 2.0 \times 10^{-4}$.
    E. $[A^-] = 0.020$.

13. In a 0.010 M HCN solution, the

    A. $[H^+] = 0.010$.
    B. $[CN^-] = 0.010$.
    C. $[HCN] = 0.0$.
    D. $[H^+] = [CN^-]$.
    E. None of the above is correct.

14. The only solution with the correct pH given is

    A. 0.1 M H_2SO_4; 1.00

    B. 10 M HCl; -1.00

    C. 0.60 HNO_3; -0.22

    D. 0.0040 M $Ca(OH)_2$; 11.60

    E. 0.050 M $HC_2H_3O_2$; 4.70

15. The pH of a 0.80 M H_2SO_4 solution is

    A. 0.097.
    B. -0.20.
    C. 1.60.
    D. 0.80.
    E. 0.20.

16. A solution with a pH of 10.60 would

    A. be 4.0×10^{-4} M NaOH.
    B. be 2.5×10^{-2} M HCl.
    C. have a pOH of 4.40.
    D. have the $[H^+] > [OH^-]$.
    E. be described by more than one of the above.

17. After a 0.020 M NH_3 solution has reached equilibrium,

    A. $[NH_4^+] = 0.020.$
    B. $[OH^-] = 0.010.$
    C. $[OH^-] > [NH_3].$
    D. $[H^+] = 6.0 \times 10^{-4}.$
    E. pH = 10.78.

18. The HCO_3^- ion can act as a base according to the following reaction, with $K_b = 2.3 \times 10^{-8}$. By the time a 0.5 M HCO_3^- solution has reached equilibrium,

$$HCO_3^-(aq) + H_2O(l) \rightleftarrows H_2CO_3(aq) + OH^-(aq)$$

    A. $[HCO_3^-] = 0$
    B. pH < 7.00
    C. $[H_2CO_3] < [HCO_3^-]$
    D. $[OH^-] < 0.5$
    E. none of the above is correct

19. The pH of a 3.1×10^{-2} M solution of uric acid is (only one ionizable hydrogen; $K_a = 1.3 \times 10^{-4}$)

    A. 5.40   B. 1.70   C. 2.70   D. 6.40   E. 3.30

20. The correct $[H^+]$ and $[OH^-]$, respectively, in a 0.025 M $HClO_4$ solution is

    A. 0.025; 4.0×10^{-13}
    B. 0.025;   0.025
    C. 0.050; 2.0×10^{-13}
    D. 0.158; 6.3×10^{-14}
    E. 1.58×10^{-8}; 6.32×10^{-7}

21. If $K_b = 1.8 \times 10^{-9}$ for pyridine, C_5H_5N, which choice correctly identifies the behavior of its conjugate acid in pure water <u>and</u> has the correct value of K_a?

A. $C_5H_5NH^+(aq) + H_2O(l) \rightleftarrows C_5H_5N(aq) + H_3O^+(aq)$ $K_a = 5.6 \times 10^{-6}$

B. $C_5H_5N^+(aq) + H_2O(l) \rightleftarrows C_5H_5NOH(aq) + H^+(aq)$ $K_a = 1.0 \times 10^{-14}$

C. $C_5H_5N(l) + H_2O(l) \rightleftarrows C_5H_5NH^+(aq) + OH^-(aq)$ $K_a = 1.8 \times 10^{-9}$

D. $C_5H_5NH^+(aq) + OH^-(aq) \rightleftarrows H_2O(l) + C_5H_5N(l)$ $K_a = 1.8 \times 10^{-9}$

E. $C_5H_5N(l) + H_3O^+(aq) \rightleftarrows C_5H_5NH^+(aq) + OH^-(aq)$ $K_a = 5.6 \times 10^{-6}$

22. Which of the following 0.1 M solutions has the greatest percent ionization?

A. HNO_2 B. $HOCl$ C. $HC_2H_3O_2$ D. HNO_3 E. NH_3

23. Nicotine $(C_{10}H_{14}N_2)$ is a base $(K_b = 9.55 \times 10^{-9})$. What is the value of K_a for the nicotinium ion, $C_{10}H_{14}N_2H^+$?

A. 9.55×10^{-23}
B. 9.55×10^{-9}
C. 1.05×10^{-6}
D. 1.05×10^{8}
E. 9.77×10^{-5}

24. The pH of a 0.250 M benzoic acid, $HC_7H_5O_2$, solution is

A. 2.39   B. 2.09   C. 1.79   D. 4.79   E. 4.19

25. The $[OH^-]$ and the percent ionization of a 0.080 M solution of analine, $C_6H_5NH_2$, is

A. 5.9×10^{-6}; 0.0073
B. 2.1×10^{-5}; 0.026
C. 7.3×10^{-5}; 0.091
D. 4.3×10^{-10}; 0.000 000 54
E. 1.7×10^{-9}; 0.000 002 1

26. The pH of a 0.100 M HOBr solution is $(K_a = 2.5 \times 10^{-9})$

A. 5.40   B. 3.80   C. 4.30   D. 4.80   E. 3.20

27. The pH of a 0.025 M HNO_3 solution is

    A. 0.025   B. 1.60   C. 1.90   D. 0.90   E. 1.00

28. The only solution that will be basic is

    A. KCl   B. $HC_2H_3O_2$ C. $Fe(H_2O)_6^{3+}$ D. NH_4^+ E. CH_3NH_2

29. The only solution that will be acidic is

    A. KNO_3 B. KF   C. NaCN   D. NaOH   E. NH_4Cl

30. The only pair of compounds that could be a buffer solution is

    A. $HC_2H_3O_2$; KF

    B. H_2SO_4; $NaHSO_4$

    C. HOCl; KOCl

    D. HNO_3; KOH

    E. HNO_2; $NaNO_3$

31. The pH of a buffer solution consisting of 2.00 M NH_3 and 0.500 M NH_4Cl is

    A. 9.86   B. 4.14   C. 8.72   D. 4.74   E. 11.81

32. Solutions of all of the following pairs of compounds could be buffer solutions except for

    A. $KC_2H_3O_2$; $HC_2H_3O_2$

    B. KNO_3; $NaNO_3$

    C. NH_3; NH_4NO_3

    D. HCOOH; HCOONa

    E. $CuCO_3$; $NaHCO_3$

33. The pH of a solution consisting of 0.50 M CH_3NH_2 (weak base) and 0.10 M CH_3NH_3Cl (salt of weak base) is

    A. 11.36   B. 2.64   C. 12.10   D. 3.34   E. 3.72

34. When a strong acid is titrated with a strong base, the equivalence point is reached when

    A. the rate of the forward reaction equals the rate of the reverse reaction.
    B. the pH of the solution becomes constant.
    C. the quantity of reactant called for by the equation for the reaction has been added.
    D. one-half of the acid and base has reacted.
    E. the pH is 5.0

35. The $[H^+]$ in a 0.10 M H_2CO_3 solution is

    A. 4.3×10^{-7} B. 7.0×10^{-4} C. 4.8×10^{-11} D. 1.0×10^{-7} E. 2.1×10^{-4}

ANSWERS TO SELF-TEST QUESTIONS

1. b	13. d	25. a
2. d	14. b	26. d
3. e	15. b	27. b
4. a	16. a	28. e
5. c	17. e	29. e
6. b	18. c	30. c
7. d	19. c	31. a
8. e	20. a	32. b
9. b	21. a	33. a
10. b	22. d	34. c
11. c	23. c	35. e
12. c	24. a	

PRACTICE PROBLEMS

Test your problem-solving skills in this chapter by working through the problems in the space provided **BEFORE** checking the answers. A periodic table will always be available. The ANSWERS TO PRACTICE PROBLEMS follow immediately after this section.

1. If $K_a = 2.1 \times 10^{-12}$ for the acid saccharin, $HC_7H_4NO_3S$, what is the value of K_b for its conjugate base, saccharide ion $(C_7H_4NO_3S^-)$?

$$HC_7H_4NO_3S(aq) \rightleftarrows H^+(aq) + C_7H_4NO_3S^-(aq)$$

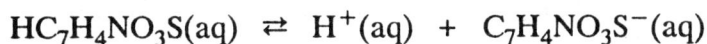

2. The pH and pOH of the following strong acid and strong base are

   (a) 0.0065 M HBr

   (b) 0.0076 M NaOH

3. The $[H^+]$ and pH of each of the following solutions are

   (a) 1.5 M HCN

   (b) 0.10 M NH_2OH (the reactants are $NH_2OH + H_2O$; $K_b = 6.6 \times 10^{-9}$)

4.   Calculate the pH and percent ionization of 0.050 M HOCl

5.   If K_a = 4.5 x 10^{-11} for the acid hydroquinone, $HC_6H_5O_2$, what is the value of K_b for the conjugate base, hydroquinonide ion $(C_6H_5O_2^-)$?

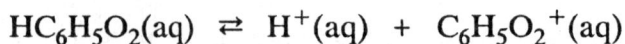

$$HC_6H_5O_2(aq) \rightleftarrows H^+(aq) + C_6H_5O_2^+(aq)$$

6.   The pH and $[OH^-]$ of solutions of the following strong acid and strong base are

(a) 0.0032 M HI

(b) 0.000 75 M KOH

7.   The $[OH^-]$ and pH of each of the following solutions are

(a) 0.50 M $(CH_3)_2NH$

(b) 0.65 M HCOOH

8. Calculate the pH and percent ionization of 0.40 M HF.

9. What is the pH of a 0.20 M $NaNO_2$ solution?

10. What is the pH of 0.15 M pyridine (C_5H_5N)? (K_b = 1.8 x 10^{-9} for pyridine)

11. Calculate the pH of the following solutions:

   (a)   0.70 M HF (K_a = 4 x 10^{-4}) and 0.60 M KF

   (b)   after adding 0.40 mol of HCl to 1.0 L of the buffer solution in (a). (Assume no change in volume.)

(c)   after adding 0.30 mol of NaOH to 1.0 L of the buffer solution in (a). (Assume no change in volume.)

12.   Calculate the pH of the following solutions

(a)   0.050 M HCOOH and 0.030 M HCOONa

(b)   after adding 0.010 mol of HCl to 1.0 L of the buffer solution in (a). (Assume no change in volume.)

(c)   after adding 0.030 mol of NaOH to 1.0 L of the buffer solution in (a). (Assume no change in volume.)

13.   In a 0.10 M $H_2C_6H_6O_6$ (vitamin C) solution, what are $[H^+]$, $[HC_6H_6O_6^-]$, $[C_6H_6O_6^{2-}]$ and $[H_2C_6H_6O_6]$?   [For $H_2C_6H_6O_6$, $K_{a_1} = 7.9 \times 10^{-5}$ and $K_{a_2} = 1.6 \times 10^{-12}$]

14. If 25.0 mL of 0.060 M HN_3 (K_a = 1.9 x 10^{-5}) is titrated with 0.030 M KOH, what will be the pH after

(a) 20.0 mL of KOH have been added?

(b) 50.0 mL of KOH have been added?

ANSWERS TO PRACTICE PROBLEMS

1. $$K_b = \frac{K_w}{K_a} = \frac{1.0 \times 10^{-14}}{2.1 \times 10^{-12}} = 4.8 \times 10^{-3}$$

2. (a) 100% ionized; [H^+] = 0.0065; pH = 2.19; pOH = 11.81

   (b) 100% ionized; [OH^-] = 0.0076; pOH = 2.12; pH = 11.88

3. (a) Set up the data table and assume x is negligible compared to 1.5 M HCN.
   $x = [H^+] = 2.7 \times 10^{-5}$; pH = 4.57.

   (b) $NH_2OH(l) + H_2O(l) \rightleftarrows NH_3OH^+(aq) + OH^-(aq)$

   Set up data table and assume x is negligible compared to 0.10 M NH_2OH.
   $x = [OH^-] = 2.6 \times 10^{-5}$; [$H^+$] = 3.8 x 10^{-10}; pH = 9.42.

458

4. Set up the data table and assume x is negligible compared to 0.05 M HOCl.
$x = [H^+] = 4.0 \times 10^{-5}$; pH = 4.40.

$$\% \text{ ionization} = \frac{[HOCl]_{change}}{[HOCl]_{initial}} \times 100 = \frac{4.0 \times 10^{-5}}{0.05} \times 100 = 0.080\%$$

5. $$K_b = \frac{K_w}{K_a} = \frac{1 \times 10^{-14}}{4.5 \times 10^{-11}} = 2.2 \times 10^{-4}$$

6. (a) $[H^+] = 0.0032$; pH = 2.49; $[OH^-] = 3.1 \times 10^{-12}$

   (b) $[OH^-] = 0.000\ 75$; pOH = 3.12; pH = 10.88.

7. (a) $(CH_3)_2NH(l) + H_2O(l) \rightleftharpoons (CH_3)_2NH_2^+(aq) + OH^-(aq)$

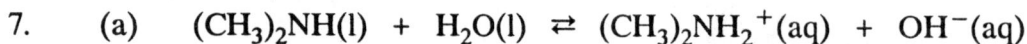

   Set up the data table and assume x is negligible compared to 0.50 M $(CH_3)_2NH$.
   $x = [OH^-] = 1.6 \times 10^{-2}$; pOH = 1.80; pH 12.20.

   (b) Set up the data table and assume x is negligible compared to 0.65 M HCOOH.
   $x = [H^+] = 1.1 \times 10^{-2}$; pH = 1.96; $[OH^-] = 9.1 \times 10^{-13}$

8. Set up the data table and assume x is negligible compared to 0.40 M HF.
$x = [H^+] = 1.2 \times 10^{-2}$; pH = 1.92.

$$\% \text{ ionization} = \frac{[HF]_{change}}{[HF]_{initial}} \times 100 = \frac{1.2 \times 10^{-2}}{0.40} \times 100 = 3.0\%$$

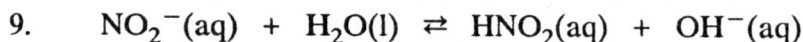

9. $NO_2^-(aq) + H_2O(l) \rightleftharpoons HNO_2(aq) + OH^-(aq)$

$$K_b (NO_2^-) = \frac{K_w}{K_a (HNO_2)} = \frac{1 \times 10^{-14}}{4.6 \times 10^{-4}} = 2.2 \times 10^{-11}$$

Set up the data table and assume x is negligible compared to 0.20 M NO_2^-.
$x = [OH^-] = 2.1 \times 10^{-6}$; pOH = 5.68; pH = 8.32.

10. $C_5H_5N(aq) + H_2O(l) \rightleftharpoons C_5H_5NH^+(aq) + OH^-(aq)$ $\qquad$ $K_b = 1.8 \times 10^{-9}$

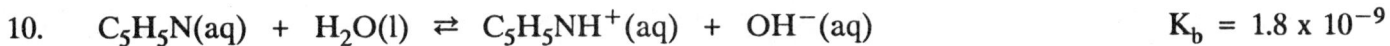

Set up the data table and assume x is negligible compared to 0.15 M C_5H_5N.
$x = [OH^-] = 1.64 \times 10^{-5}$; pOH = 4.78; pH = 9.22

11.   (a)   Buffer solution; $[H^+] = 4.1 \times 10^{-4}$; pH = 3.39.

(b)   New initial concentrations after adding HCl are 1.1 M HF and 0.20 M F^-; after re-establishing equilibrium, $[H^+] = 1.9 \times 10^{-3}$; pH = 2.72.

(c)   New initial concentrations after adding NaOH are 0.40 M HF and 0.90 M F^-; after re-establishing equilibrium, $[H^+] = 1.6 \times 10^{-4}$; pH = 3.80.

12.   (a)   Buffer solution; $[H^+] = 3.0 \times 10^{-4}$; pH = 3.52.

(b)   New initial concentrations after adding HCl are 0.06 M HCOOH and 0.02 M $HCOO^-$; after re-establishing equilibrium, $[H^+] = 5.4 \times 10^{-4}$; pH = 3.27.

(c)   New initial concentrations after adding HCl are 0.02 M HCOOH and 0.06 M $HCOO^-$; after re-establishing equilibrium, $[H^+] = 6.0 \times 10^{-5}$; pH = 4.22.

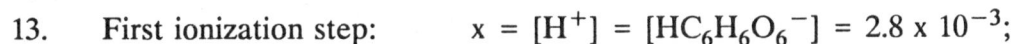

13.   First ionization step: $\qquad$ $x = [H^+] = [HC_6H_6O_6^-] = 2.8 \times 10^{-3}$;

$[H_2C_6H_6O_6] = (0.10-x) = 0.10$

Second ionization step: $\qquad$ $x = [C_6H_6O_6^{2-}] = K_{a_2} = 1.6 \times 10^{-12}$

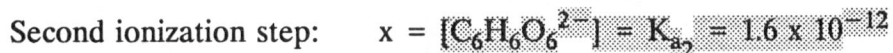

$[H^+] = (2.8 \times 10^{-3} + x) = 2.8 \times 10^{-3}$

$[HC_6H_6O_6^-] = (2.8 \times 10^{-3} - x) = 2.8 \times 10^{-3}$

14.   (a)   After quantitative reaction, 9.0×10^{-4} mol $HN_3/0.045$ L = 0.020 M and 6.0×10^{-4} mol $KN_3/0.045$ L = 0.013 M, which is a buffer solution. Write the equilibrium equation for HN_3 and solve as a buffer problem.

$[H^+] = 2.9 \times 10^{-5}$; pH = 4.54.

(b)   After quantitative reaction, only 1.5×10^{-3} mol KN_3/0.075 L = 0.020 M, which means you are at the equivalence point. Hydrolysis of the salt occurs with the following equilibrium reaction and K_b

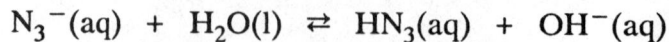

$$N_3^-(aq) \ + \ H_2O(l) \ \rightleftharpoons \ HN_3(aq) \ + \ OH^-(aq)$$

$$K_b \ (N_3^-) = \frac{K_w}{K_a \ (HN_3)} = \frac{1 \times 10^{-14}}{1.9 \times 10^{-5}} = 5.3 \times 10^{-10}$$

From the equilibrium calculation

$$x = [OH^-] = 3.3 \times 10^{-6}; \ pOH = 5.48; \ pH = 8.52.$$

CHAPTER 16

MORE ABOUT EQUILIBRIA

SECTION 16.2   ACIDITY, BASICITY, AND THE PERIODIC TABLE

After completing this section, you will be able to do the following:

■ Predict the strengths of acids.

REVIEW

The strength of an acid is related to the strength of the bond between hydrogen and the non-metal; the longer and/or more polar this bond becomes, the stronger the acid. For **binary acids** (hydrogen plus a non-metal) the strength of the acid **increases** as you go **down a group** ($HCl < HBr < HI$) and **increases** as you go across a row in the periodic table from **left to right** ($NH_3 < H_2O < HF$).

For **oxo acids** (hydrogen plus a polyatomic ion containing oxygen), the strength of the acid is related to the polarity of the $O-H$ bond. For **oxo acids with similar structures**, the acid strength **increases** as the **electronegativity of the central atom increases** ($HOI < HOBr < HOCl$). For oxo acids with differing numbers of extra oxygens (oxygens that are not part of $-OH$ groups), acid strength **increases** as the number of extra oxygens **increases** ($HOCl < HClO_2 < HClO_3 < HClO_4$). More generally, without a Lewis structure to show you the extra oxygens, you can usually predict that the acid strength of the oxo acids of the same non-metal will **increase** as the number of oxygen atoms in the formula **increases** (as shown by the $HOCl$ to $HClO_4$ series at the end of the last sentence).

461

SAMPLE PROBLEM 16.2 A

▶▶ Which member of each of the following pairs is the stronger acid? Explain!
(a) CH_3CH_2OH or CH_3CH_2SH (b) $HBrO_2$ or $HOBr$ (c) H_2S or HCl (d) $HClO_2$ or HNO_2
(e) CH_3CH_2OH or CH_3CH_2COOH

Solution:

HINT: Find the difference between the two formulas; how will the strength of the $O-H$ bond be affected?

(a)   In these two organic acids the only difference between the two formulas is that the ionizable hydrogen is bonded to O in CH_3CH_2OH and to S in CH_3CH_2SH. The $O-H$ bond is more polar than the $S-H$ bond because O is more electronegative than S. The more polar the bond, the easier it is to break the bond and furnish H^+. CH_3CH_2OH is the stronger acid.

(b)   These are two oxo acids of the same nonmetal (Br) whose formulas differ only in the number of oxygen atoms. The larger the number of oxygen atoms, the stronger the acid. $HBrO_2$ is the stronger acid.

(c)   These are two binary acids with S and Cl in the same row of the periodic table. As indicated in the REVIEW, acid strength of binary acids increases as you go from left to right in a row. HCl is the stronger acid.

(d)   Here we have two oxo acids containing the same number of oxygens and differing only in the central atom. The more electronegative the central atom, the more polar the $O-H$ bond and the stronger the acid. Cl is closer to the upper right corner of the periodic table, so Cl is the more electronegative element. $HClO_2$ is the stronger acid.

(e)   These two organic acids differ only in the number of oxygen atoms. The higher the number of oxygen atoms, the stronger the acid. CH_3CH_2COOH is the stronger acid.

SECTION 16.3   LEWIS ACIDS AND BASES

After completing this section, you will be able to do the following:

■ Define and identify Lewis acids and bases.

REVIEW

Compared to the Arrhenius definition of acids and bases, the Bronsted-Lowry definition is the same for an acid but broadens the definition of a base. To account for still other reactions that resemble acid-base reactions but that do not involve the transfer of a proton, an even more general definition of acids and bases was put forth by Lewis, which broadens the definition of an acid beyond that of Bronsted-Lowry. According to the Lewis definition, a **Lewis acid** is a species that accepts an electron pair to form a covalent bond; thus, a Lewis acid must have an **empty valence orbital** available (or create one) to accept the electron pair from the Lewis base. A **Lewis base** donates an electron pair to form a covalent bond; thus, a Lewis base must have a non-bonding pair of valence electrons to form a covalent bond with the Lewis acid. To determine whether a species is a Lewis acid or base, it will be helpful to draw the Lewis formula as done earlier.

A summary of the three definitions of acids and bases is given in the following table.

Definition	Acid	Base
Arrhenius	Increases $[H^+]$	Increases $[OH^-]$
Bronsted-Lowry	Proton donor	Proton acceptor
Lewis	Electron pair acceptor	Electron pair donor

SAMPLE PROBLEM 16.3 A

▸▸ Of the following species, which will probably act as a Lewis acid, a Lewis base, or neither?

$$BCl_3 \quad Fe^{3+} \quad CH_3CH_2CH_3 \quad Mg^{2+} \quad (CH_3)_3N$$

Solution:

HINT: A Lewis acid has to have (or create) an empty valence orbital to accept and share a pair of electrons. A Lewis base must have a non-bonding pair of valence electrons to donate and share with the acid.

BCl_3 is a Lewis acid.
If you draw the Lewis structure for BCl_3, you will see that B uses its three valence electrons to form three $B-Cl$ bonds and has one empty 2p orbital with which to accept an electron pair. Also, B does not have any non-bonding pairs of electrons to act as a base.

Fe^{3+} is a Lewis acid.
Fe^{3+} has five 3d valence electrons in separate orbitals. It pairs up electrons to create a vacant 3d orbital rather than donate and share a pair of electrons. The positively charged metal ions

464

in general attract a pair of electrons from a base.

CH₃CH₂CH₃ is neither.

If you draw out the Lewis structure, you will see that none of the carbon atoms has or can create an empty orbital (cannot form a double bond between C and H) to function as an acid, and that there are no unshared pairs of valence electrons to function as a base.

Mg²⁺ is a Lewis acid.

Mg^{2+} has empty 3s and 3p valence orbitals to accept an electron pair. Mg^{2+} has no unshared pair of valence electrons to function as a base.

(CH₃)₃N is a Lewis base.

The N has one non-bonding pair of electrons to donate to and share with a Lewis acid. N does not have any nor can it create any empty valence orbitals to function as a base.

SAMPLE PROBLEM 16.3 B

▶▶ Identify the Lewis acid and Lewis base in each of the following reactions:

(a) $Cd^{2+}(aq) + 4CN^-(aq) \rightleftarrows Cd(CN)_4{}^{2-}(aq)$

(b) $Zn(OH)_2(s) + 2OH^-(aq) \rightleftarrows Zn(OH)_4{}^{2-}(aq)$

(c) $SO_3(g) + H_2O(l) \rightarrow H_2SO_4(aq)$

Solution:

(a) $Cd^{2+}(aq) + 4CN^-(aq) \rightleftarrows Cd(CN)_4{}^{2-}(aq)$
     acid   base

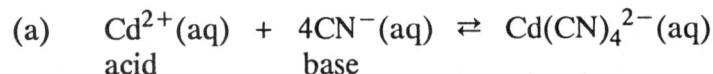

Metal ions typically have an empty valence orbital to accept an electron pair; anions typically have one or more unshared pairs to donate.

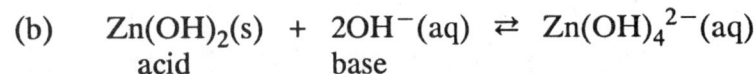

(b) $Zn(OH)_2(s) + 2OH^-(aq) \rightleftarrows Zn(OH)_4{}^{2-}(aq)$
      acid   base

Same as part (a).

(c) $SO_3(g) + H_2O(l) \rightarrow H_2SO_4(aq)$
     acid   base

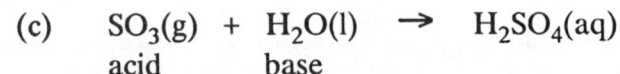

Water has two unshared pairs of valence electrons; water does not have and cannot create an empty valence orbital. SO_3 creates an empty valence orbital by converting the one double-bonded oxygen to a single bond.

SECTION 16.5 THE SOLUBILITY PRODUCT CONSTANT AND SOLUBILITY

After completing this section, you will be able to do the following:

■ Given the K_{sp} for a slightly soluble solid, calculate the solubility of the solid, and vice versa.

REVIEW

Another kind of equilibrium reaction is that between a slightly soluble solid and its ions in solution. At equilibrium, the rate of dissolution equals the rate of precipitation. By convention, the solid is always the reactant. For example, the equilibrium reaction for the slightly soluble lithium carbonate, Li_2CO_3, is

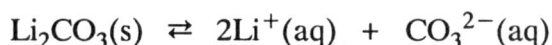

$$Li_2CO_3(s) \rightleftarrows 2Li^+(aq) + CO_3^{2-}(aq)$$

The equilibrium constant for this kind of equilibrium is called the **solubility product constant,** or K_{sp}. The equilibrium constant expression for this reaction is (remember, pure solids do not appear in the equilibrium expression)

$$[Li^+]^2[CO_3^{2-}] = K_{sp}$$

The amount of solid that dissolves to reach equilibrium is the **solubility.** The **solubility** of a solid can be calculated from the K_{sp} for the reaction (Sample Problem 16.5 A), and the K_{sp} can be calculated from the **solubility** (Sample Problem 16.5 B). Table 16.5 in the textbook gives the K_{sp} and corresponding solubilities for a number of slightly soluble solids.

SAMPLE PROBLEM 16.5 A

▶▶ What is the solubility of $NiCO_3$ in grams/100 cm³?

Solution:

HINT: Set up a data table and let x equal the number of moles of $NiCO_3$ that dissolve in a liter of solution to reach equilibrium, which is exactly the definition of solubility; that is, x = solubility.

Write out the equilibrium reaction and set up the data table, letting x equal the number of moles of $NiCO_3$ that dissolve in a liter of solution to reach equilibrium, i.e., the solubility.

$$NiCO_3(s) \rightleftarrows Ni^{2+}(aq) + CO_3^{2-}(aq)$$

Initial, M		0.0	0.0
Change, ΔM	$-x$	x	x
Equilibrium, M		x	x

Since $NiCO_3$ is a solid, there are no initial or equilibrium concentrations. Substitute into the equilibrium expression, find K_{sp} in Table 16.5 in the textbook, and solve for x,

$$[Ni^{2+}][CO_3^{2-}] = x^2 = 1.2 \times 10^{-7} = K_{sp}$$

$$x = \text{solubility} = [Ni^{2+}] = [CO_3^{2-}] = 3.5 \times 10^{-4} \text{ M}$$

The molar solubility of $NiCO_3$ is 3.5×10^{-4}; that is, 3.5×10^{-4} mol of solid $NiCO_3$ dissolves in a liter of solution and ionizes into 3.5×10^{-4} M each of Ni^{2+} and CO_3^{2-} by the time equilibrium is reached. But the units of solubility requested in this problem are grams/100 cm^3 solution. Converting moles/L to grams/100 cm^3

$$\text{solubility } NiCO_3 = \frac{3.5 \times 10^{-4} \text{ mol } NiCO_3}{1 \text{ L soln.}} \times \frac{118.70 \text{ g } NiCO_3}{1 \text{ mol } NiCO_3} \times \frac{1 \text{ L soln.}}{1000 \text{ cm}^3} \times 100 \text{ cm}^3 = 0.0042 \text{ g } NiCO_3 \text{ in } 100 \text{ cm}^3$$

SAMPLE PROBLEM 16.5 B

▶▶ What is the value of the K_{sp} for $PbBr_2$ at 20 °C if it has a solubility of 0.50 g/100 cm^3 at 20 °C?

Solution:

HINT: Before you can calculate the value of **any equilibrium constant**, you must know the equilibrium concentrations of all species.

Convert the solubility from g/100 cm^3 to mol/L.

$$\text{solubility } PbBr_2, \text{ M} = \frac{0.50 \text{ g } PbBr_2}{100 \text{ cm}^3 \text{ soln.}} \times \frac{1 \text{ mol } PbBr_2}{367.0 \text{ g } PbBr_2} \times \frac{1000 \text{ cm}^3}{1 \text{ L soln.}} = 0.014 \text{ M}$$

Write out the equilibrium reaction and set up a data table, remembering that the solubility, by definition, **IS** the concentration of solid ($PbBr_2$) that dissolves in a liter of solution to reach equilibrium

$$PbBr_2(s) \rightleftarrows \quad Pb^{2+}(aq) \; + \; 2Br^-(aq)$$

		Pb^{2+}	Br^-
Initial, M		0.000	0.000
Change, ΔM	-0.014	0.014	0.028
Equilibrium, M		0.014	0.028

As always, the data in the **Change row** reflect the stoichiometry of the reaction. When 1 mol of solid $PbBr_2$ dissolves, **1** mol of Pb^{2+} and **2** mol of Br^- are formed. Now substitute into the equilibrium expression and calculate the value of the K_{sp} for $PbBr_2$,

$$[Pb^{2+}][Br^-]^2 \; = \; (0.014)(0.028)^2 \; = \; 1.1 \times 10^{-5} = K_{sp}$$

SECTION 16.7 CALCULATIONS INVOLVING K_{sp}

After completing this section, you will be able to do the following:

- Determine whether a precipitate will form when two solutions of known volume and molarity are mixed;

- After a precipitate is formed by mixing together two solutions, determine the equilibrium concentration of one ion from the known concentration of the other ion;

- Determine the molar solubility in a solution containing a common ion.

REVIEW

The three types of calculations that can be done using K_{sp} are indicated above. To determine whether a precipitate will form when two solutions of known volume and concentration are mixed together, you need to calculate the reaction quotient, Q, just like in Section 14.4. Remember, if

$$Q < K_c \quad \text{Net forward reaction occurs until } Q = K_c$$

$$Q = K_c \quad \text{System is at equilibrium}$$

$$Q > K_c \quad \text{Net reverse reaction occurs until } Q = K_c.$$

468

For slightly soluble substances whose equilibrium constant is designated as K_{sp}, the reaction quotient Q is given the special name **ion product** and is represented by the symbol Q_{sp}. The calculation of the ion product, Q_{sp}, is essentially the same as the reaction quotient, Q. If

$$Q_{sp} < K_{sp} \quad \text{Solid dissolves (net forward reaction occurs) until } Q_{sp} = K_{sp}$$

$$Q_{sp} = K_{sp} \quad \text{Solution is saturated (system is at equilibrium)}$$

$$Q_{sp} > K_{sp} \quad \text{Solid will precipitate (net reverse reaction occurs) until } Q_{sp} = K_{sp}.$$

When you see a solid precipitating out of a solution, this means that $Q_{sp} > K_{sp}$ and the reaction is trying to reach equilibrium by the net reverse reaction occurring. When the precipitation stops (but not the forward and reverse reactions), essentially equilibrium has been reached. This application is illustrated by Sample Problem 16.7 A.

The second application involves mixing two solutions to form a precipitate. By knowing the final concentration of one ion, you can determine the concentration of the other ion (Sample Problem 16.7 B).

The third application is to calculate the solubility of a solid in a solution other than pure water. One variation is to dissolve a solid in a solution that contains an ion that will complex with one of the ions from the solid (see Sample Problem 16.7 in the textbook). Another variation is to dissolve a solid in a solution containing a common ion (Sample Problem 16.7 C).

SAMPLE PROBLEM 16.7 A

▶▶ Will a precipitate form if 35.0 mL of 2.0 x 10^{-4} M Ca(NO$_3$)$_2$ are mixed with 15.0 mL of 5.0 x 10^{-4} M NaF?

HINT: Calculate the concentrations of the solutions after dilution before calculating the ion product, Q_{sp}.

Of the four ions present in solution, Ca^{2+}, NO_3^-, Na^+, and F^-, only CaF_2 is insoluble because it has a K_{sp} given in Table 16.6 of the textbook (the solubility rules in Table 4.4 of the textbook are also useful in predicting insoluble compounds). The equilibrium involved and K_{sp} (from Table 16.6 in the textbook) are

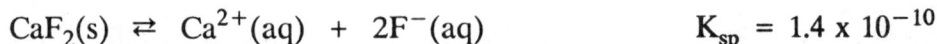

$$CaF_2(s) \rightleftharpoons Ca^{2+}(aq) + 2F^-(aq) \qquad K_{sp} = 1.4 \times 10^{-10}$$

To find out whether a precipitate will form, you will need to calculate the ion product, Q_{sp}, for the conditions given. However, whenever you mix two volumes of solutions, the concentrations of each solution are diluted. A very direct approach to calculate the concentration of a solution after dilution is to use the formula for dilution, $M_{bd} \times V_{bd} = M_{ad} \times V_{ad}$ (see Sample Problem 4.8 D). To find the concentration after dilution, M_{ad}, of the $Ca(NO_3)_2$ solution in a total volume of 50.0 mL of solution,

$$? \ M_{ad} \ = \ M_{bd} \ \times \ \frac{V_{bd}}{V_{ad}} \ = \ 2.0 \times 10^{-4} \ M \ \times \ \frac{35.0 \ mL}{50.0 \ mL} \ = \ 1.4 \times 10^{-4} \ M$$

Similarly, to find the M_{ad} of the NaF solution,

$$? \ M_{ad} \ = \ M_{bd} \ \times \ \frac{V_{bd}}{V_{ad}} \ = \ 5.0 \times 10^{-4} \ M \ \times \ \frac{15.0 \ mL}{50.0 \ mL} \ = \ 1.5 \times 10^{-4} \ M$$

Now calculate the ion product, Q_{sp}, to see if a precipitate will form or not,

$$Q_{sp} \ = \ [Ca^{2+}][F^-]^2 \ = \ (1.4 \times 10^{-4})(1.5 \times 10^{-4})^2 \ = \ 3.2 \times 10^{-12}$$

$$Q_{sp} \ (= 3.2 \times 10^{-12}) \ < \ K_{sp} \ (= 1.4 \times 10^{-10})$$

Therefore, no precipitate will occur. In fact, this is an unsaturated solution and cannot reach equilibrium because there is no solid present to dissolve and allow a net forward reaction to occur.

SAMPLE PROBLEM 16.7 B

▶▶ What $[Ca^{2+}]$ must be used to reduce $[CO_3^{2-}]$ to 2×10^{-8}?

Solution:

The $[CO_3^{2-}]$ is what must be present when the following equilibrium is reached,

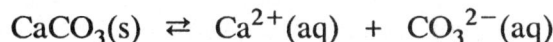

$$CaCO_3(s) \ \rightleftarrows \ Ca^{2+}(aq) \ + \ CO_3^{2-}(aq)$$

Since you know the $[CO_3^{2-}]$ at equilibrium, solve the equilibrium expression for $[Ca^{2+}]$,

$$[Ca^{2+}][CO_3^{2-}] = K_{sp} = 5 \times 10^{-9}$$

and

$$[Ca^{2+}] \ = \ \frac{K_{sp}}{[CO_3^{2-}]} \ = \ \frac{5 \times 10^{-9}}{2 \times 10^{-8}} \ = \ 3 \times 10^{-1}$$

SAMPLE PROBLEM 16.7 C

▶ ▶ What is the molar solubility of $NiCO_3$ in (a) pure water, and (b) 0.10 M K_2CO_3?

(a)   The molar solubility of $NiCO_3$ in pure water was calculated in Sample Problem 16.5 A (before converting the solubility to grams/100 cm^3). The results were

$$\text{solubility of } NiCO_3 = 3.5 \times 10^{-4} \text{ M}$$

Unless stated otherwise in a problem like 16.5 A, pure water can be assumed to be the solvent.

(b)   When $NiCO_3$ dissolves in a 0.10 M K_2CO_3 solution, the common ion CO_3^{2-} is present, which suppresses the solubility; that is, not as much $NiCO_3$ will dissolve in 0.10 M K_2CO_3 because of the presence of 0.10 M CO_3^{2-} from K_2CO_3. From this point on, the problem is worked just like in Sample Problem 15.7 A.

Write out the equilibrium reaction and set up the data table, letting x equal the number of moles of $NiCO_3$ that dissolve in a liter of solution to reach equilibrium (which is the solubility),

	$NiCO_3(s)$	$\rightleftarrows$	$Ni^{2+}(aq)$	+	$CO_3^{2-}(aq)$
Initial, M			0.0		0.10 (from K_2CO_3)
Change, ΔM	$-x$		x		x
Equilibrium, M			x		(0.10 + x)

Assuming x is small compared to 0.10, substitute the equilibrium concentrations into the equilibrium expression and solve for x,

$$[Ni^{2+}][CO_3^{2-}] = x(0.10) = 1.2 \times 10^{-7} = K_{sp}$$

$$x = \text{solubility} = 1.2 \times 10^{-6} \text{ M}$$

NOTE: In the presence of a common ion, the solubility of $NiCO_3$ is less (1.2×10^{-6}) than in pure water (3.5×10^{-4}).

SELF-TEST QUESTIONS

Test your understanding of the concepts and skills in this chapter by working through the multiple choice questions BEFORE checking the answers. A periodic table will always be available. The ANSWERS TO SELF-TEST QUESTIONS follow immediately after this section.

1. When arranged in order of increasing acid strength (from left to right), the correct choice is

    A. $HIO < HIO_3 < HIO_2$.

    B. $HIO < HIO_2 < HIO_3$.

    C. $HIO_2 < HIO < HIO_3$.

    D. $HIO_3 < HIO_2 < HIO$.

    E. $HIO_3 < HIO < HIO_2$.

2. The weakest acid among the following is

    A. $HOCl$ B. $HClO_4$ C. $HClO_2$ D. HCl E. $HClO_3$

3. The molar solubility of AgBr in pure water at 25 °C is

    A. 4×10^{-7} B. 5×10^{-13} C. 7×10^{-7} D. 3×10^{-13} E. 6×10^{-4}

4. If a saturated solution of palladium(II) thiocyanate, $Pd(SCN)_2$, in pure water is found to have $[Pd^{2+}]$ = 2.22×10^{-8}, what is the value of K_{sp} for $Pd(SCN)_2$?

    A. 9.86×10^{-16} B. 4.93×10^{-16} C. 1.09×10^{-23} D. 2.22×10^{-8} E. 4.38×10^{-23}

5. Of the following, the least soluble solid in pure water is

    A. AgI B. CuI C. $AgCl$ D. $CaCO_3$ E. $BaSO_4$

6. The correct solubility product expression for Bi_2S_3 is

    A. $[Bi^{2+}][S^{2-}]^3$ B. $[2Bi^{3+}]^2[3S^{2-}]^3$ C. $[Bi_2][S_3]$ D. $[Bi^{3+}]^2[S^{2-}]^3$ E. $[Bi^{3+}][S^{2-}]$

7. In which of the following solutions is the molar solubility of calcium carbonate the lowest (least)?

   A. 0.40 M sodium carbonate
   B. 0.50 M sodium chloride
   C. pure water
   D. 0.10 M calcium chloride
   E. 0.050 M calcium nitrate

8. The correct solubility product expression for $Fe(OH)_3$ is

   A. $[Fe^{3+}][OH^-]$

   B. $[3Fe^{3+}][OH^-]^3$

   C. $[Fe^{3+}][OH^-]^3$

   D. $[Fe^{3+}][3OH^-]^3$

   E. $[Fe^{3+}]^3[OH^-]^3$

9. At 25 °C, a maximum of 0.015 g of $Zn(CN)_2$ (117 g/mol) will dissolve in 1.00 L of pure water. The value of K_{sp} for $Zn(CN)_2$ at the same temperature is

   A. 2.2×10^{-12} B. 3.4×10^{-6} C. 1.4×10^{-5} D. 8.8×10^{-12} E. 4.4×10^{-12}

ANSWERS TO SELF-TEST QUESTIONS

1. b	4. e	7. a
2. a	5. a	8. c
3. c	6. d	9. d

PRACTICE PROBLEMS

Test your problem-solving skills in this chapter by working through the problems in the space provided **BEFORE** checking the answers. A periodic table will always be available. The ANSWERS TO PRACTICE PROBLEMS follow immediately after this section.

1. Identify the stronger of the two acids HIO_3 and HIO and explain why.

2. Identify the Lewis acid and Lewis base in each of the following reactions:

   (a) $OH^-(aq) + CO_2(g) \rightleftarrows HCO_3^-(aq)$

   (b) $Fe^{3+}(aq) + SCN^-(aq) \rightleftarrows FeSCN^{2+}(aq)$

   (c) $Cl^-(aq) + AlCl_3(aq) \rightleftarrows AlCl_4^-(aq)$

   (d) $SO_2(g) + OH^-(aq) \rightleftarrows HSO_3^-(aq)$

3. What is the molar solubility of $Ca(OH)_2$ in pure water at 25 °C?

4. What is the molar solubility of AgCl in a 0.0010 M NaCl solution at 25°C?

474

5.   What is the molar solubility of $Ni(OH)_2$ in pure water at 25 °C?

($K_{sp} = 1.6 \times 10^{-16}$)

6.   What is the molar solubility of $PbCrO_4$ in a 0.010 M Na_2CrO_4 solution at 25 °C?

($K_{sp} = 2 \times 10^{-16}$)

7.   Write the balanced equilibrium reaction and solubility product expression for each of the following substances:

(a) Ag_3PO_4

(b) $CaCO_3$

(c) Cr_2S_3

8.   What is the molar solubility of ZnS in a 0.10 M Na_2S solution at 25 °C?

($K_{sp} = 2.5 \times 10^{-22}$)

ANSWERS TO PRACTICE PROBLEMS

1. HIO_3 is the stronger of the two acids because there are more oxygen atoms bonded to the central iodine atom, which makes the $O-H$ bond weaker and easier to break.

2.

Lewis acid		Lewis base
(a)	CO_2	OH^-
(b)	Fe^{3+}	SCN^-
(c)	$AlCl_3$	Cl^-
(d)	SO_2	OH^-

3. $K_{sp} = 5 \times 10^{-6}$; solubility is 1×10^{-2} M.

4. $K_{sp} = 1.6 \times 10^{-10}$; common ion is present; solubility is 1.6×10^{-7} M

5. $K_{sp} = 1.6 \times 10^{-16}$; solubility is 3.4×10^{-6} M

6. $K_{sp} = 2 \times 10^{-16}$; common ion is present; solubility is 2×10^{-14} M

7.
   (a) $Ag_3PO_4(s) \rightleftarrows 3Ag^+(aq) + PO_4^{3-}(aq)$ $\qquad K_{sp} = [Ag^+]^3[PO_4^{3-}]$

   (b) $CaCO_3(s) \rightleftarrows Ca^{2+}(aq) + CO_3^{2-}(aq)$ $\qquad K_{sp} = [Ca^{2+}][CO_3^{2-}]$

   (c) $Cr_2S_3(s) \rightleftarrows 2Cr^{3+}(aq) + 3S^{2-}(aq)$ $\qquad K_{sp} = [Cr^{3+}]^2[S^{2-}]^3$

8. $K_{sp} = 2.5 \times 10^{-22}$; common ion is present; solubility is 2.5×10^{-21} M

CHAPTER 17

CHEMICAL THERMODYNAMICS REVISITED: A CLOSER LOOK AT ENTHALPY, ENTROPY, AND EQUILIBRIUM

SECTION 17.1   BOND ENERGIES

After completing this section, you will be able to do the following:

- Calculate ΔH_{rxn} from bond energies.

REVIEW

Earlier, we found that the standard enthalpy change for a reaction could be calculated from standard enthalpies of formation (Section 6.9)

$$\Delta H_{rxn}^{\circ} = \text{sum of standard enthalpies of formation of } \textbf{products} - \text{sum of standard enthalpies of formation of } \textbf{reactants}$$

However, in some instances, the standard enthalpies of formation are unavailable. A second method of calculating the enthalpy change for a reaction, ΔH_{rxn}, is to use **bond energies or bond dissociation energies**. **Bond dissociation energy** is the energy needed to break the bond between two atoms in a molecule. Energy is required to break a bond (endothermic), and energy is released when a bond is formed (exothermic). When a chemical reaction occurs, one or more bonds in the reactant molecules are broken and new bonds are formed between different atoms to make products. Using the bond energies in Table 17.1 in the textbook, the enthalpy change for a reaction, ΔH_{rxn}, is equal to the sum of the bond dissociation energies of all the bonds in the reactants (endothermic) **minus** the bond dissociation energies

478

of all the bonds in the products (exothermic), i.e.

ΔH_{rxn} = sum of bond dissociation − sum of bond dissociation
                    energies of **reactants**   energies of **products**

Bond dissociation energies are always **positive** (because energy is **always** required to break a bond). The negative sign is required before the bond dissociation energies of the **products** because you are forming (not breaking) bonds, which releases energy.

SAMPLE PROBLEM 17.1 A

▸▸ Use bond dissociation energies to estimate ΔH_{rxn} at 25 °C for the reaction

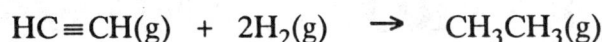

$$HC \equiv CH(g) \ + \ 2H_2(g) \ \rightarrow \ CH_3CH_3(g)$$

Solution:

Write out the Lewis formula first so you don't miss any of the bonds.

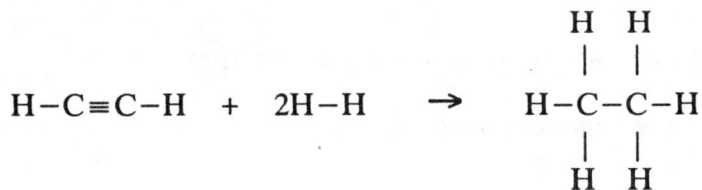

$$H-C \equiv C-H \ + \ 2H-H \ \rightarrow \ \begin{matrix} & H & H \\ & | & | \\ H- & C- & C-H \\ & | & | \\ & H & H \end{matrix}$$

Count the number of bonds of each kind (which are actually moles of bonds) for the reactants and products and substitute this information into the equation given in the REVIEW using **BD** for bond dissociation energy,

ΔH_{rxn} = [2 mol C−H x BD(C−H) + 1 mol C≡C x BD(C≡C) + 2 mol H−H x BD(H−H)]

       − [6 mol C−H x BD(C−H) + 1 mol C−C x BD(C−C)]

From Table 17.1, substitute the bond dissociation energies (in kJ/mol) into the equation and calculate ΔH_{rxn}

$$\Delta H_{rxn} = [2 \text{ mol C}-\text{H} \times (414 \text{ kJ/mol C}-\text{H}) + 1 \text{ mol C} \equiv \text{C} \times (812 \text{ kJ/mol C} \equiv \text{C}) + 2 \text{ mol H}-\text{H} \times (436 \text{ kJ/mol H}-\text{H})]$$
$$- [6 \text{ mol C}-\text{H} \times (414 \text{ kJ/mol C}-\text{H}) + 1 \text{ mol C}-\text{C} \times (347 \text{ kJ/mol C}-\text{C})]$$

$$= 2512 \text{ kJ (reactants)} - 2831 \text{ kJ (products)} = -319 \text{ kJ}$$

NOTE: It takes 2831 kJ to break all the bonds in one mole of the product, CH_3CH_3. To form all the bonds in 1 mol of CH_3CH_3 from isolated gaseous C and H atoms, 2831 kJ are released (exothermic), which is shown by -2831 kJ.

SECTION 17.3   ENTROPY

After completing this section, you will be able to do the following:

- Predict the sign of the entropy change for changes in physical state, size, and temperature;

- Calculate the standard entropy change for a reaction from standard entropies.

REVIEW

Entropy is a measure of disorder or randomness. Every spontaneous change increases the entropy of the universe according to the second law of thermodynamics (Section 17.2 in the textbook). Generally, entropy increases as a substance changes from solid to liquid to gas, the largest increase occurring from liquid to gas because of the largest increase in volume. For substances in the same physical state, entropy increases as you go down a group of the periodic table. For species with similar structures, the more atoms bonded together, the greater the number of bond orientations and the higher the entropy. And entropy increases as the temperature of the substance increases.

The **standard entropy, S°,** of a substance is the entropy in a standard state (pure liquids and pure solids under 1 atm and ideal gas at a partial pressure of 1 atm). Standard entropy is the difference between the entropy of perfect crystals of the substance at absolute zero and the entropy of the substance in a standard state. Thus **elements** as well as compounds have **non-zero** entropies. In Appendix E of the textbook, you will find the standard heat of formation, $\Delta H_f°$, the standard free energy (to be discussed in Section 17.7) of formation, $\Delta G_f°$, and the standard entropy given for selected substances. **NOTE** that for **elements,** $\Delta H_f°$ and $\Delta G_f°$ are **always zero** but the standard entropy, S°, is **not zero.** Also note in this appendix (and all other tables of thermodynamic data) that the values of $\Delta H_f°$ and $\Delta G_f°$ have units of kJ/mol, whereas the unit for S° is **J/K•mol,** since the magnitudes of entropy changes are usually small compared to the magnitudes of enthalpy and free energy changes.

Like the standard enthalpy change for a reaction, $\Delta H_{rxn}°$, the standard entropy change is the sum of the standard entropies of the **products** minus the sum of the standard entropies of the **reactants,** i.e.

$$\Delta S_{rxn}^\circ = \begin{array}{l} \text{sum of standard entropies} \\ \text{of } \textbf{products} \end{array} - \begin{array}{l} \text{sum of standard entropies} \\ \text{of } \textbf{reactants} \end{array}$$

$$= \Sigma S_{products}^\circ - \Sigma S_{reactants}^\circ,$$

where each standard entropy, S°, must be multiplied by the coefficient of the substance in the chemical equation.

SAMPLE PROBLEM 17.3 A

▶▶ For each of the following pairs of substances, predict which substance will have the higher entropy.

(a) $CO_2(s)$ at 0 K or $CO_2(g)$ at 0 °C
(b) $CaCl_2(aq)$ or $CaCl_2(s)$
(c) 2.50 mol $O_2(g)$ at 10 atm or 2.50 mol $O_2(g)$ at 5 atm
(d) Ne(g) or Kr(g)
(e) $CH_3COOH(l)$ or $HCOOH(l)$

Solution:

(a) $CO_2(g)$ at 0°C has the higher entropy. The same substance as a gas always has a more random, disordered arrangement of its atoms and molecules than in the solid phase.

(b) $CaCl_2(aq)$ has the higher entropy. The Ca^{2+} and Cl^- ions are much more randomly distributed in a liquid than a solid.

(c) 2.50 mol $O_2(g)$ at 5 atm has the higher entropy. At a lower pressure, the same number of moles of a gas will occupy a greater volume, allowing for a more random distribution of the gas molecules.

(d) Kr(g) has the higher entropy. Entropy increases as you go down a group; Kr is lower in group 0 than Ne.

(e) CH₃COOH(l) has the higher entropy. Both of these substances have the identical −COOH structure. For substances with similar structures, the more atoms bonded together, the higher the entropy.

SAMPLE PROBLEM 17.3 B

▶▶ Using standard entropies, calculate $\Delta S_{rxn}°$ at 25 °C for the reactions

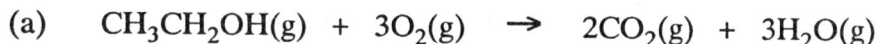

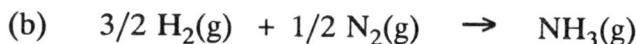

(a) $CH_3CH_2OH(g) + 3O_2(g) \rightarrow 2CO_2(g) + 3H_2O(g)$

(b) $3/2\ H_2(g) + 1/2\ N_2(g) \rightarrow NH_3(g)$

Solution:

(a) To avoid errors, look up the standard entropies for each substance in Table 17.2 or Appendix E (remember, S° for elements is not zero) and write them under the formulas in the reaction

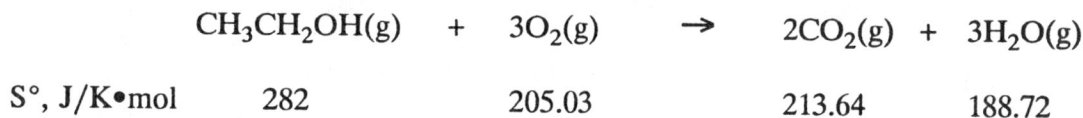

	$CH_3CH_2OH(g)$	+	$3O_2(g)$	$\rightarrow$	$2CO_2(g)$	+	$3H_2O(g)$
S°, J/K•mol	282		205.03		213.64		188.72

NOTE: Check the standard entropy data carefully! If you are using Appendix E, the standard entropy data is the **last** column. Also, you will find two values of S° each for CH_3CH_2OH and H_2O, one for the liquid state and one for the gaseous state. You must **always** select the value of S° that has the same physical state as given in the reaction. For this problem, both CH_3CH_2OH and H_2O are in the gas phase.

Now use the equation given in the REVIEW to calculate $\Delta S_{rxn}°$, remembering that each S° must be multiplied by the coefficient of the substance in the chemical equation,

$$\Delta S_{rxn}° = \Sigma S°_{products} - \Sigma S°_{reactants}$$

$$= [2\ mol\ CO_2(213.64\ J/K \cdot mol\ CO_2) + 3\ mol\ H_2O(188.72\ J/K \cdot mol\ H_2O)]$$

$$- [1\ mol\ CH_3CH_2OH(282\ J/K \cdot mol\ CH_3CH_2OH) + 3\ mol\ O_2(205.03\ J/K \cdot mol\ O_2)]$$

$$= [427.28\ J/K + 566.16\ J/K] - [282\ J/K + 615.09\ J/K]$$

$$= +96\ J/K$$

CHECK: Does the sign of ΔS° make sense? Yes. The answer just calculated says that the products in the reaction have an entropy 96 J/K **higher** than the reactants; in other words, the products are more disordered than the reactants. Qualitatively, you can predict the sign of ΔS° by looking at the change in the number of moles of gases, Δn_g, for this reaction. Remember, $\Delta n_g = n_{g\ prod} - n_{g\ react} = 5 - 4 = +1$ for this reaction. Thus, the sign of the entropy change is predicted to be positive and agrees with the quantitative calculation.

(b) Following the same procedure as in part (a), look up the standard entropies for each substance in Table 17.2 or Appendix E (remember, S° for elements is not zero) and write them under the formulas in the reaction

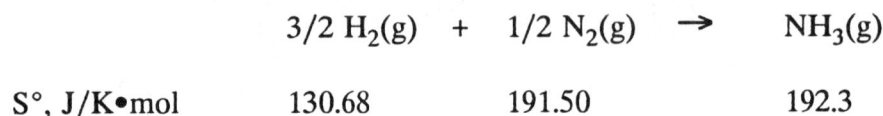

$$3/2\ H_2(g)\quad +\quad 1/2\ N_2(g)\quad \rightarrow\quad NH_3(g)$$

S°, J/K•mol   130.68   191.50   192.3

Now use the equation given in the REVIEW to calculate $\Delta S_{rxn}°$, remembering that each S° must be multiplied by the coefficient of the substance in the chemical equation,

$$\Delta S_{rxn}° \quad = \quad \Sigma S°_{products} \quad - \quad \Sigma S°_{reactants}$$

$$= \quad [1\ mol\ NH_3(192.3\ J/K\cdot mol\ NH_3)]$$

$$-[3/2\ mol\ H_2(130.68\ J/K\cdot mol\ H_2) + 1/2\ mol\ N_2(191.50\ J/K\cdot mol\ N_2)]$$

$$= \quad [192.3\ J/K]\ -\ [196.02\ J/K\ +\ 95.75\ J/K]$$

$$= \quad -99.5\ J/K$$

CHECK: Does the sign of ΔS° make sense? Yes. The answer just calculated says that the products in the reaction have an entropy 99.5 J/K **lower** than the reactants; in other words, the products are less disordered (or more ordered) than the reactants. Qualitatively, you can predict the sign of ΔS° by looking at the change in the number of moles of gases, Δn_g, for this reaction. Remember, $\Delta n_g = n_{g\ prod} - n_{g\ react} = 1 - 2 = -1$ for this reaction. Thus, the sign of the entropy change is predicted to be positive and agrees with the quantitative calculation.

After completing this section, you will be able to do the following:

■ Given a reaction and the signs of ΔH and ΔS, determine whether that reaction will be spontaneous or non-spontaneous and what effect temperature will have (if any);

■ For a given reaction, calculate $\Delta G°$ from $\Delta H°$ and $\Delta S°$.

REVIEW

The **enthalpy change, ΔH**, indicates the thermal energy change for a system under constant pressure. Many (but not all) exothermic reactions are spontaneous. The **entropy change ΔS**, indicates the change in randomness or disorder for a system. Systems that experience an increase in entropy (disorder) are spontaneous (second law of thermodynamics). In the real world, systems (physical and chemical changes) undergo simultaneous changes in both enthalpy and entropy. To combine these two thermodynamic functions, a third thermodynamic function called the **Gibbs free energy, G**, is used. The term "free energy" is used because ΔG represents the maximum amount of energy released by a physical or chemical change (occurring at constant temperature and pressure) that is "free" or "available" to do useful work. That is, it is the amount of free energy available after taking into account the enthalpy change and entropy change of the physical and chemical change. The three thermodynamic functions and the temperature dependence of the entropy change are related by the following **very important equation**

$$\Delta G = \Delta H - T\Delta S$$

at constant temperature and pressure, and where T is in Kelvin.

This equation can be used **qualitatively** to determine whether a reaction is spontaneous or not, and the effect of temperature, simply by knowing the signs of ΔH and ΔS. We already know that spontaneous changes usually are associated with a negative sign ($-$) for ΔH and a positive ($+$) sign for ΔS. For both terms (of opposite sign) to be additive and indicate a spontaneous process (T is always positive), there must be a negative sign ($-$) in front of $T\Delta S$ ($\Delta G = \Delta H + (-T\Delta S) = \Delta H - T\Delta S$). That is, for a spontaneous process, ΔH is $-$ and although ΔS is positive, the quantity "$-T\Delta S$" is negative. Thus, the free energy change, ΔG, is **negative** for **spontaneous** processes and positive for non-spontaneous processes. In a similar manner we can look at the other three combinations of signs for ΔH and ΔS to predict whether ΔG will be positive (non-spontaneous) or negative (spontaneous). If ΔH and ΔS have **opposite** signs, temperature does not determine the direction of spontaneous change. To see this more clearly, focus on the **signs** of ΔH and $-T\Delta S$ (not ΔS), which are added together to give the sign of ΔG,

ΔH	(ΔS)	−TΔS	ΔG	Result
−	(+)	−	−	Spontaneous at all temperatures
+	(−)	+	+	Non-spontaneous at all temperatures.

If ΔH and ΔS have the **same sign**, temperature **does determine** the direction of spontaneous change. That is, ΔH and −TΔS will have opposite signs, which means that the magnitude of ΔH vs. −TΔS will determine the sign of ΔG. Since temperature is the variable, the spontaneity will change depending on whether the temperature is high or low. A summary of these two combinations of signs is

ΔH	(ΔS)	−TΔS	ΔG	Result
−	(−)	+	−	Spontaneous at **low** temperatures
			+	Non-spontaneous at **high** temperatures
+	(+)	−	+	Non-spontaneous at **low** temperatures
			−	Spontaneous at **high** temperatures

That is, at **low** temperatures (small numerical values of T), the magnitude of "−TΔS" will be more than offset by the magnitude of ΔH; the **sign of ΔG will be the same as ΔH**. At **high** temperatures (larger numerical values of T), the magnitude of the "−TΔS" term will predominate; **the sign of ΔG will be the same as −TΔS**.

The **Gibbs free energy** equation can be used to calculate ΔG for a reaction from the values of ΔH and ΔS, which you have calculated earlier (Sample Problems 6.9 C & D and 17.3 B).

At this point, it will be helpful to keep in mind the meaning of the **signs** for the three thermodynamic functions, ΔH, ΔS, and ΔG. While we may sometimes use ΔH or ΔS to predict whether a reaction is spontaneous or not, only ΔG provides that information. A useful summary is provided below

Function	Sign (−)	Sign (+)
ΔH	Heat evolved	Heat absorbed
ΔS	Less disorder	More disorder
ΔG	Spontaneous	Non-spontaneous

SAMPLE PROBLEM 17.6 A

▶▶ For the following reaction at 25 °C and standard conditions,

$$H_2C=CH_2(g) \; + \; 3O_2(g) \; \rightarrow \; 2CO_2(g) \; + \; 2H_2O(g)$$

(a)   calculate $\Delta H°$ using the standard enthalpies of formation in Appendix E;
(b)   calculate $\Delta S°$ using the standard entropies in Appendix E;
(c)   use the free energy equation, $\Delta G = \Delta H - T\Delta S$, to determine whether the reaction will be (i) spontaneous at all temperatures, (ii) not spontaneous at any temperature, (iii) spontaneous at low temperatures but non-spontaneous at high temperatures, or (iv) non-spontaneous at low temperatures but spontaneous at high temperatures;
(d)   calculate $\Delta G°$ using the free energy equation.

Solution:

(a)   To avoid errors, look up the standard enthalpies of formation for each substance in Appendix E and write them under the formulas in the reaction

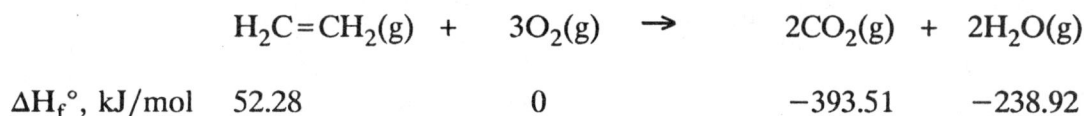

$$H_2C=CH_2(g) \; + \quad 3O_2(g) \quad \rightarrow \qquad 2CO_2(g) \; + \; 2H_2O(g)$$

$\Delta H_f°$, kJ/mol   52.28   0   −393.51   −238.92

Now use the equation given in Sample Problems 6.9 C and D to calculate $\Delta H_{rxn}°$, remembering that each $\Delta H_f°$ must be multiplied by the coefficient of the substance in the chemical equation,

$$\Delta H_{rxn}° \;\; = \;\; \Sigma \Delta H_f°{}_{products} \quad - \quad \Sigma \Delta H_f°{}_{reactants}$$

$$= \quad [2 \text{ mol } CO_2(-393.51 \text{ kJ/mol } CO_2) + 2 \text{ mol } H_2O(-238.92 \text{ kJ/mol } H_2O)]$$

$$- \; [1 \text{ mol } H_2C=CH_2(52.28 \text{ kJ/mol } H_2C=CH_2) + 3 \text{ mol } O_2(0 \text{ kJ/mol } O_2)]$$

$$= \quad [-787.02 \text{ kJ} - 477.84 \text{ kJ}] - [52.28 \text{ kJ}]$$

$$= \quad -1317.14 \text{ kJ}$$

(b)   Follow the same procedure as in part (a), except that you are now looking up standard entropies in Appendix E,

$$H_2C=CH_2(g) \;+\; 3O_2(g) \;\rightarrow\; 2CO_2(g) \;+\; 2H_2O(g)$$

| $S°$, J/K•mol | 219.4 | 205.03 | 213.64 | 188.72 |

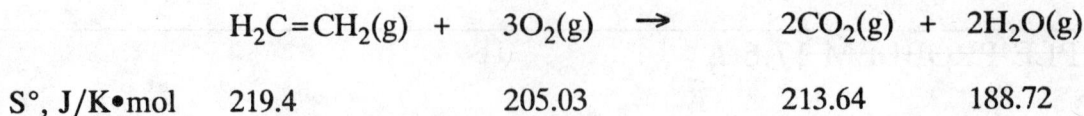

Now use the equation given in Sample Problem 17.3 B to calculate $\Delta S_{rxn}°$, remembering that each $S°$ must be multiplied by the coefficient of the substance in the chemical equation,

$$\Delta S_{rxn}° \;=\; \Sigma S°_{products} \;-\; \Sigma S°_{reactants}$$

$$=\; [2 \text{ mol } CO_2(213.64 \text{ J/K} \cdot \text{mol } CO_2) + 2 \text{ mol } H_2O(188.72 \text{ J/K} \cdot \text{mol } H_2O)]$$

$$-\; [1 \text{ mol } H_2C=CH_2(219.4 \text{ J/K} \cdot \text{mol } H_2C=CH_2) + 3 \text{ mol } O_2(205.03 \text{ J/K} \cdot \text{mol } O_2)]$$

$$=\; [427.28 \text{ J/K} + 377.44 \text{ J/K}] - [219.4 \text{ J/K} + 615.09 \text{ J/K}]$$

$$=\; -29.8 \text{ J/K}$$

(c)   To determine whether this reaction is spontaneous or not, look at the signs of $\Delta H°$ (part a) and $\Delta S°$ (part b) to see what the sign of $\Delta G°$ must be from the Gibbs free energy equation using the same procedure as discussed in the REVIEW.

$$\Delta G° \;=\; \Delta H° \;-\; T\Delta S°$$
$$(-) \;-\; (-)$$
$$=\; (-) \;+\; \qquad =\; -\,(\text{low T})$$
$$+\,(\text{high T})$$

This reaction is spontaneous ($\Delta G°$ negative) at low temperatures but non-spontaneous ($\Delta G°$ positive) at high temperatures. A temperature of 25 °C is considered low on the kelvin scale. Therefore, this reaction should be spontaneous. Now do part (d) to calculate the value and sign of $\Delta G_{rxn}°$.

(d)   When the values of $\Delta H°$ and $\Delta S°$ are known for a reaction, you can immediately calculate $\Delta G°$ from the Gibbs free energy equation (a second method of calculating $\Delta G°$ will be discussed in Section 17.7)

$$\Delta G° \;=\; \Delta H° \;-\; T\Delta S°$$

Now substitute into the equation the values of $\Delta H°$ (in kJ), $\Delta S°$ (converted to kJ/K•mol by moving the decimal point three places to the left), and T in Kelvin,

$$\Delta G° = \Delta H° - T\Delta S°$$

$$= (-1317.14 \text{ kJ}) - 298 \text{ K}(-0.0298 \text{ kJ/K})$$

$$= -1317.14 \text{ kJ} + 8.88 \text{ kJ}$$

$$= -1308.26 \text{ kJ}$$

CHECK: The reaction is spontaneous; the magnitude of "$-T\Delta S$" is significantly less than ΔH at the relatively low temperature of 25 °C.

SECTION 17.7   CALCULATION OF $\Delta G°$ FROM $\Delta G_f°$

After completing this section, you will be able to do the following:

■ For a given reaction, calculate $\Delta G_{rxn}°$ from standard free energies of formation, $\Delta G_f°$.

REVIEW

Another way to calculate the standard free energy change for a reaction, $\Delta G_{rxn}°$, is from the free energies of formation, $\Delta G_f°$, which are the standard free energy changes for the formation of **1 mol** of a substance from the **elements in their standard states**. Free energies of formation are found in Appendix E and in Table 17.5 of the textbook. Like $\Delta H_f°$, the $\Delta G_f°$ of elements in their standard state is zero. The calculation of $\Delta G_{rxn}°$ from $\Delta G_f°$, of $\Delta H_{rxn}°$ from $\Delta H_f°$, and of $\Delta S_{rxn}°$ from $S°$ are summarized for easy reference.

$$\Delta G_{rxn}° = \text{sum of standard free energies} - \text{sum of standard free energies}$$
$$\text{of formation of \textbf{products}} \quad \text{of formation of \textbf{reactants}}$$

$$= \sum \Delta G_f°_{\,products} - \sum \Delta G_f°_{\,reactants},$$

$$\Delta H_{rxn}° = \text{sum of standard enthalpies} - \text{sum of standard enthalpies}$$
$$\text{of formation of \textbf{products}} \quad \text{of formation of \textbf{reactants}}$$

$$= \sum \Delta H_f°_{\,products} - \sum \Delta H_f°_{\,reactants},$$

$$\Delta S_{rxn}° = \text{sum of standard entropies} - \text{sum of standard entropies}$$
$$\text{of \textbf{products}} \quad \text{of \textbf{reactants}}$$

$$= \sum S°_{\,products} - \sum S°_{\,reactants},$$

where each standard free energy of formation, $\Delta G_f°$, standard enthalpy of formation, $\Delta H_f°$, or standard entropy, $S°$, must be multiplied by the coefficient of the substance in the chemical equation.

You now have two methods of calculating $\Delta G_{rxn}°$, $\Delta H_{rxn}°$, and $\Delta S_{rxn}°$ for a reaction. One method is from the data in Appendix E or similar tables. The second method is from the Gibbs free energy equation, $\Delta G° = \Delta H° - T\Delta S°$, when you know the temperature and two of the three quantities, $\Delta G_{rxn}°$, $\Delta H_{rxn}°$, and $\Delta S_{rxn}°$. The choice usually depends on which data are available to you for that reaction (see Practice Problems 2, 3, 5 and 6 at the end of this chapter).

SAMPLE PROBLEM 17.7 A

▶▶ For the following reaction at 25 °C and standard conditions,

$$H_2C{=}CH_2(g) + 3O_2(g) \rightarrow 2CO_2(g) + 2H_2O(g),$$

calculate $\Delta G°$ using the standard free energies of formation in Appendix E. Compare your results with the calculation of $\Delta G°$ for this reaction using the free energy equation in Sample Problem 17.6 A.

Solution:

To avoid errors, look up the standard free energies of formation for each substance in Appendix E and write them under the formulas in the reaction,

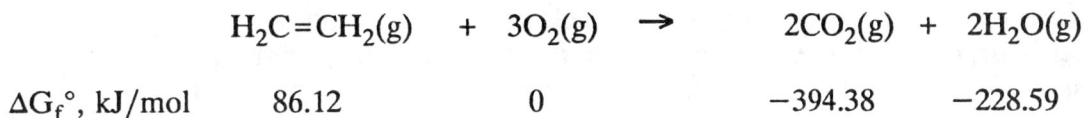

$$H_2C=CH_2(g) \quad + \quad 3O_2(g) \quad \rightarrow \quad 2CO_2(g) \quad + \quad 2H_2O(g)$$

$\Delta G_f°$, kJ/mol $\quad$ 86.12 $\qquad\qquad$ 0 $\qquad\qquad$ −394.38 $\qquad$ −228.59

Now use the equation given in the REVIEW to calculate $\Delta G_{rxn}°$, remembering that each $\Delta G_f°$ must be multiplied by the coefficient of the substance in the chemical equation,

$$\Delta G_{rxn}° = \sum \Delta G_f°\,_{products} - \sum \Delta G_f°\,_{reactants}$$

$$= [2 \text{ mol } CO_2(-394.38 \text{ kJ/mol } CO_2) + 2 \text{ mol } H_2O(-228.59 \text{ kJ/mol } H_2O)]$$

$$- [1 \text{ mol } H_2C=CH_2(86.12 \text{ kJ/mol } H_2C=CH_2) + 3 \text{ mol } O_2(0 \text{ kJ/mol } O_2)]$$

$$= [-788.76 \text{ kJ} - 457.18 \text{ kJ}] - [86.12 \text{ kJ}]$$

$$= -1332.06 \text{ kJ}$$

This value of $\Delta G_{rxn}°$ is very close to the value of $\Delta G_{rxn}°$, −1308.26 kJ, calculated for this same reaction from the free energy equation in Sample Problem 17.6.

SECTION 17.8 ESTIMATION OF $\Delta G°$ AT DIFFERENT TEMPERATURES
SECTION 17.9 ESTIMATION OF TEMPERATURE AT WHICH DIRECTION OF SPONTANEOUS CHANGE REVERSES

After completing this section, you will be able to do the following:

■ Calculate $\Delta G°$ for a reaction at temperatures other than 25 °C;

■ Estimate the temperature at which the direction of spontaneous change reverses.

REVIEW

If we assume that $\Delta H_f°$ and $S°$ do not change with temperature, then the values of $\Delta H_{rxn}°$ and $\Delta S_{rxn}°$ calculated at 25 °C do not change with temperature; they can be used to estimate $\Delta G_{rxn}°$ at temperatures other than 25 °C with the Gibbs free energy equation, $\Delta G = \Delta H - T\Delta S$ (Sample Problem 17.8 A).

Earlier in this chapter, we saw that if both ΔH and ΔS have the same sign, temperature determines the direction of spontaneous change (sample Problem 17.6 A, part c). That is, the free energy equation can be used to estimate the temperature at which a physical or chemical change that is spontaneous in one direction becomes spontaneous in the other direction. This temperature is the one at which a process is at equilibrium; **at equilibrium, ΔG° is zero**. Going from above to below that equilibrium temperature, the direction of spontaneity changes. The melting point and boiling point of a substance are temperatures at which a process is at equilibrium. When a substance melts, solid and liquid are in equilibrium; when a substance boils, liquid and vapor are in equilibrium. When a process is at equilibrium, ΔG° is zero and the free energy equation simplifies to

$$\Delta G° = 0 = \Delta H° - T\Delta S°$$

Solving for the temperature at which the direction of spontaneity changes (Sample Problem 17.9 A), we obtain

$$T = \frac{\Delta H°}{\Delta S°}$$

At the melting point and boiling point of a substance, the ΔH and ΔS are often given the subscripts **fusion** and **vaporization** to indicate the equilibrium present. Thus, at the melting point of a substance, ΔH_{fus} and ΔS_{fus} are used; at the boiling point, ΔH_{vap} and ΔS_{vap} are used. If the boiling point and ΔH_{vap} are known for a substance, the same equation (ΔG° = 0) can be used to find the ΔS_{vap} (Sample Problem 17.9B). A similar process is used to find ΔS_{fus}.

SAMPLE PROBLEM 17.8 A

▶▶ Estimate ΔG° for the following reaction at 225 °C

$$H_2C=CH_2(g) + 3O_2(g) \rightarrow 2CO_2(g) + 2H_2O(g)$$

Solution:

To estimate ΔG° at various temperatures, you need to know ΔH° and ΔS° at 25 °C and assume they are temperature independent. The values of ΔH° and ΔS° for this reaction at 25 °C (from Sample Problem 17.6 A) are −1317.14 kJ and −29.8 J/K, respectively. Substitute these values (ΔS° converted to kJ) and the temperature of 498 K into the free energy equation to calculate ΔG° at 498 K,

$$\Delta G° = \Delta H° + T\Delta S°$$

$$= (-1317.14 \text{ kJ}) - (498 \text{ K})(-0.0298 \text{ kJ/K})$$

$$= -1317.14 \text{ kJ} + 14.8 \text{ kJ} = -1302.3 \text{ kJ}$$

NOTE: The value of $\Delta G°$ at 298 K for this same reaction is -1308.26 kJ (Sample Problem 17.6 A, part d).

SAMPLE PROBLEM 17.9 A

▶▶ Estimate the normal boiling point of hydrogen cyanide, HCN.

Solution:

At the boiling point of any substance, liquid is in equilibrium with vapor and $\Delta G°$ equals zero. You can estimate the boiling point (which is the temperature at which the direction of spontaneous change reverses) from the free energy equation if you know $\Delta H°$ and $\Delta S°$ at the boiling point. The assumption will always be that $\Delta H°$ and $\Delta S°$ calculated at 25 °C will not change with temperature. Use the data from Appendix E to calculate $\Delta H°$ and $\Delta S°$ for the physical change of boiling HCN

	HCN(l)	→	HCN(g)
$\Delta H_f°$, kJ	105		131
S°, J/K•mol	112.8		201.8

$$\Delta H_{rxn}° = \sum \Delta H_f°_{products} - \sum \Delta H_f°_{reactants}$$

$$= [1 \text{ mol HCN(g)}(131 \text{ kJ/mol HCN(g)})] - [1 \text{ mol HCN(l)}(105 \text{ kJ/mol HCN(l)})]$$

$$= 26 \text{ kJ}$$

$$\Delta S°_{rxn}° = \sum S°_{products} - \sum S°_{reactants}$$

$$= [1 \text{ mol HCN(g)}(201.8 \text{ J/K•mol HCN(g)})] - [1 \text{ mol HCN(l)}(112.8 \text{ J/K•mol HCN(l)})]$$

$$= 89.0 \text{ J/K}$$

At equilibrium, $\Delta G° = 0$, and the free energy equation becomes

$$T = \frac{\Delta H°}{\Delta S°}$$

The estimated boiling point is

$$T = \frac{\Delta H°}{\Delta S°} = \frac{26 \text{ kJ}}{0.0890 \text{ kJ/K}} = 2.9 \times 10^2 \text{ K} = 290 \text{ K}$$

or, converting to °C,

$$\text{Boiling point} = 290 \text{ K} - 273 = 20 \text{ °C}$$

NOTE: Recall that 290 K is accurate only to the tens place, which explains why the answer is 20 °C rather than 17 °C.

SAMPLE PROBLEM 17.9 B

▶ ▶ For ethanol, CH_3CH_2OH, the normal boiling point is 76.8 °C and $\Delta H_{vap} = 42.33$ kJ/mol. Calculate ΔS_{vap} for ethyl alcohol.

Solution:

At the boiling point, $\Delta G° = 0$ and the simplified free energy equation can be solved for T (sample Problem 17.9 A), $\Delta S°$ (this problem), or $\Delta H°$, providing that the other two quantities are known. In this case, solving the free energy equation at the boiling point for $\Delta S°$, which more specifically is ΔS_{vap}, gives

$$\Delta G° = 0 = \Delta H° - T\Delta S°$$

$$\Delta S_{vap} = \frac{\Delta H_{vap}}{T}$$

Substituting in the values of ΔH_{vap} and the boiling point in K, one can calculate ΔS_{vap}.

$$\Delta S_{vap} = \frac{\Delta H_{vap}}{T} = \frac{42.33 \text{ kJ/mol}}{350.0 \text{ K}} = 0.1209 \text{ kJ/K} \cdot \text{mol} = 120.9 \text{ J/K} \cdot \text{mol}$$

CHECK: Does this answer make sense? Yes, because the entropy of the gaseous product is 120.9 J/K•mol higher than the liquid reactant. The molecules of a gas are always more disordered than the same molecules as a liquid.

SECTION 17.10   CALCULATION OF ΔG FOR NON-STANDARD CONDITIONS

After completing this section, you will be able to do the following:

- ■ Calculate ΔG at non-standard conditions.

REVIEW

Many reactions, particularly those involving gases, are carried out at pressures other than standard pressure (partial pressure of 1 atm) or **non-standard** conditions. The free energy change accompanying a reaction carried out under **non-standard** conditions, ΔG (**no** superscript), can be calculated from the equation

$$\Delta G = \Delta G° + RT \ln Q,$$

where ΔG° is the standard free energy change, R is the gas constant (8.315 J/K•mol or 0.008 315 kJ/K•mol), T is the Kelvin temperature, ln is the base e logarithm (ln key on calculators), and Q is the reaction quotient (Section 14.4). The units of ΔG, ΔG°, and R must be consistent; that is, all must be J/mol or all must be kJ/mol. The term "RT ln Q" indicates how far the conditions of the reaction are from standard conditions. If a reaction contains all gases and they all have a partial pressure of one, then Q=1, ln(1)=0, and (since all gases would be at standard pressure) ΔG = ΔG°; in other words, you are at standard conditions. Only when the partial pressures of gases are not one will the reaction be occurring under **non-standard conditions**, in which case ΔG ≠ ΔG°.

SAMPLE PROBLEM 17.10 A

▶▶ If ΔG° = 92.0 kJ for the following reaction, find the value of ΔG at 25 °C when p_{H_2O} = 100.0 atm, p_{CO} = 2.0 atm, and p_{H_2} = 2.0 atm.

$$C(s) + H_2O(g) \rightarrow CO(g) + H_2(g)$$

Solution:

You are trying to find the free energy change at **non-standard** conditions, ΔG. The only equation you have available to do that is the one given in the REVIEW

$$\Delta G \;=\; \Delta G^\circ \;+\; RT \ln Q$$

Substitute into the equation the values of ΔG°, R, T (Kelvin), and Q to calculate the value of the free energy change at non-standard conditions, ΔG. Arbitrarily, we will use units of kJ/mol for ΔG° and R, so ΔG will also have units of kJ/mol.

$$\Delta G \;=\; \Delta G^\circ \;+\; RT \ln Q \;=\; \Delta G^\circ \;+\; RT \ln \frac{P_{CO} \cdot P_{H_2}}{P_{H_2O}}$$

$$=\; 92.0 \text{ kJ/mol} \;+\; (0.008\ 315 \text{ kJ/K} \bullet \text{mol})(298 \text{ K}) \ln \frac{(2)(2)}{100}$$

$$=\; 92.0 \text{ kJ/mol} \;+\; (2.48 \text{ kJ/mol}) \ln(0.04)$$

$$=\; 92.0 \text{ kJ/mol} \;+\; (2.48 \text{ kJ/mol})(-3.22)$$

$$=\; 92.0 \text{ kJ/mol} \;+\; (-7.99 \text{ kJ/mol})$$

$$=\; 84.0 \text{ kJ/mol}$$

NOTE: Remember, pure solids do not appear in the reaction quotient, Q. Also, check your math very carefully; the $RT \ln Q$ term is always negative when $Q < 1$.

SECTION 17.11	STANDARD FREE ENERGIES AND EQUILIBRIUM CONSTANTS

After completing this section, you will be able to do the following:

■ Calculate the equilibrium constant from the standard free energy change for a reaction, and vice versa.

REVIEW

At equilibrium, $\Delta G = 0$, $Q = K$, and the equation discussed in Section 17.10 becomes

$$\Delta G \;=\; 0 \;=\; \Delta G^\circ \;+\; RT \ln Q$$

Rearranging to simplify,

$$\Delta G° = -RT \ln K,$$

where K means K_p for gaseous reactions and K_c for reactions that involve solutions. Given the standard free energy change, $\Delta G°$, and T, you can calculate the equilibrium constant K for a reaction, and vice versa.

SAMPLE PROBLEM 17.11 A

▶▶ For the equilibrium reaction

$$PCl_5(g) \rightleftarrows PCl_3(g) + Cl_2(g)$$

$\Delta G° = 37.2$ kJ at 25 °C. Calculate K_p at this temperature.

Solution:

Whenever you are working with a reaction at **equilibrium**, you know that $\Delta G = 0$ and $Q = K$. The only equation that you have to work with that relates $\Delta G°$ and K is the one discussed in the REVIEW. Solving that equation for $\ln K_p$,

$$\ln K_p = \frac{-\Delta G°}{RT}$$

Substitute in the values of $\Delta G°$, R, and T, making sure that the units are consistent, and solve for K_p,

$$\ln K_p = \frac{-\Delta G°}{RT} = \frac{-37.2 \text{ kJ/mol}}{(0.008\ 315 \text{ kJ/K} \bullet \text{mol})(298 \text{ K})} = -15.0$$

Taking the antilog of both sides of the equation ("INV ln x" or "e^x" on calculators),

$$K_p = 3 \times 10^{-7}$$

Test your understanding of the concepts and skills in this chapter by working through the multiple choice questions BEFORE checking the answers. A periodic table will always be available. The ANSWERS TO SELF-TEST QUESTIONS follow immediately after this section.

1. If $\Delta H° = 40.0$ kJ and $\Delta S° = 50$ J/K•mol for a reaction taking place under standard conditions, then the reaction is

   A. spontaneous at all temperatures.
   B. non-spontaneous at all temperatures.
   C. spontaneous only at high temperatures.
   D. spontaneous only at low temperatures.
   E. non-spontaneous only at high temperatures.

2. For which of the following reactions or processes is ΔS negative?

   A. $C_3H_8(g) + 5O_2(g) \rightarrow 3CO_2(g) + 4H_2O(l)$

   B. vaporization of one mole of acetone (C_3H_6O)

   C. $2SO_3(g) \rightarrow O_2(g) + 2SO_2(g)$

   D. mixing 10 mL of acetone with 10 mL of water

   E. $PbCl_2(s) \rightarrow Pb^{2+}(aq) + 2Cl^-(aq)$

3. The only correct statement of the following is that

   A. exothermic reactions are always spontaneous.
   B. at constant T and P, a negative ΔG value requires that ΔS is positive.
   C. a reaction with a negative ΔS value cannot be spontaneous.
   D. free energy, ΔG, is independent of temperature.
   E. at constant T and P, free energy is the maximum energy available to do useful work.

4. Using $\Delta G_f°$ data from Appendix E, the value of $\Delta G°$ for the following reaction is

$$4NH_3(g) + 5O_2(g) \rightarrow 4NO(g) + 6H_2O(l)$$

A. -1142.8 kJ   B. -134.1 kJ   C. -1835.4 kJ   D. -920.2 kJ   E. -1010.8 kJ

5. For which of the following combinations of signs for $\Delta H°$ and $\Delta S°$ is the process spontaneous at all temperatures?

	$\underline{\Delta H°}$	$\underline{\Delta S°}$
A.	$-$	$-$
B.	$-$	$+$
C.	$+$	$-$
D.	$+$	$+$

E. None of these; spontaneity has nothing to do with temperature.

6. Using $\Delta G_f°$ data in Appendix E, the value of $\Delta G°$ for the following reaction is

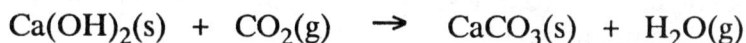

$$Ca(OH)_2(s) + CO_2(g) \rightarrow CaCO_3(s) + H_2O(g)$$

A. -457.26 kJ   B. -64.49 kJ   C. -68.52 kJ   D. $+64.49$ kJ   E. -73.08 kJ

7. For the reaction below, $\Delta G° = 33.1$ kJ, $\Delta H° = 129.3$ kJ, and $\Delta S° = 323$ J/K•mol. What is the kelvin temperature at which the reaction will be at equilibrium when all other conditions are standard, assuming no phase changes?

$$2NaHCO_3(s) \rightarrow Na_2CO_3(s) + CO_2(g) + H_2O(g)$$

A. 315
B. 298
C. 276
D. 4.00×10^2
E. None, because the reaction is spontaneous at all temperatures.

8. The normal boiling point for formic acid is

	$\Delta H_f°$ (kJ/mol)	S° (J/K•mol)
HCOOH(l)	−410	130
HCOOH(g)	−363	249

A. 1.7×10^3 °C   B. 2.7×10^2 °C   C. 3.9×10^2 °C   D. 63 °C   E. 1.2×10^2 °C

9. If $\Delta H° = 126.4$ kJ and $\Delta S° = -74.9$ J/K•mol for the reaction given, estimate the value of $\Delta G°$ for the reaction at 377 °C.

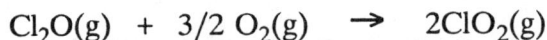

$$Cl_2O(g) + 3/2\ O_2(g) \rightarrow 2ClO_2(g)$$

A. 175.1 kJ   B. 98.2 kJ   C. 51.5 kJ   D. 154.6 kJ   E. 77.7 kJ

10. If $\Delta H° = -194.5$ kJ and $\Delta S° = -32$ J/K•mol for the following equilibrium reaction at 25 °C, the value of K_p at the same temperature is

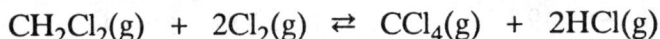

$$CH_2Cl_2(g) + 2Cl_2(g) \rightleftharpoons CCl_4(g) + 2HCl(g)$$

A. 2.7×10^{32} B. 4.7×10^{74} C. 3.7×10^{-33} D. 74.6699   E. 1.01

11. Consider the relationship between the equilibrium constant and the standard free energy change for a reaction at equilibrium. Which of the following statements must be true?

A. If $\Delta G° = +$, then $K > 1$.
B. If $\Delta G° = 0$, then $K = 0$.
C. If $\Delta G° = -$, then $K < 1$.
D. If $\Delta G° = +$, then $K < 1$.
E. If $\Delta G° = -$, then $K = 1$.

ANSWERS TO SELF-TEST QUESTIONS

1. c	5. b	9. a
2. a	6. b	10. a
3. e	7. d	11. d
4. e	8. e	

PRACTICE PROBLEMS

Test your problem-solving skills in this chapter by working through the problems in the space provided **BEFORE** checking the answers. A periodic table will always be available. The ANSWERS TO PRACTICE PROBLEMS follow immediately after this section.

1. Which has the <u>higher</u> entropy?

   (a) $AlCl_3(s)$ or $NaCl(s)$

   (b) solid sugar or sugar dissolved in a cup of coffee

   (c) $CO_2(s)$ (dry ice) at $-78\ °C$ or $CO_2(g)$ at $0\ °C$

   (d) $NH_4Cl(s)$ or $NH_4Cl(aq)$

2. For the following reaction at standard conditions,

$$3H_2(g) + CO(g) \rightarrow CH_4(g) + H_2O(g)$$

   (a) calculate $\Delta G°$, $\Delta H°$, and $\Delta S°$ using the data in Appendix E.

   (b) Is the reaction spontaneous at standard conditions? If not, will changing the temperature in either direction make it spontaneous (and which direction)?

500

3. For the following reaction at standard conditions and 25 °C, $\Delta H_{rxn}° = -6535.1$ kJ and $\Delta S_{rxn}° = -437.9$ J/K,

$$2C_6H_6(l) + 15O_2(g) \rightarrow 12CO_2(g) + 6H_2O(l)$$

(a) calculate $\Delta G°$ at 25 °C (you will not be able to use $\Delta G_f°$ data, since $C_6H_6(l)$ is not listed in Appendix E).

(b) Is the reaction spontaneous at standard conditions? If not, will changing the temperature in either direction make it spontaneous (and which direction)?

4. Which system will have the <u>lower</u> entropy

(a) $Ni(NO_3)_2(aq)$ or $Ni(NO_3)_2(s)$?

(b) 0.40 mol of $CH_4(g)$ at 0.10 atm or 0.40 mol of $CH_4(g)$ at 0.001 atm?

(c) mercury vapor or liquid mercury?

(d) 2.0 mol of $He(g)$ at 0.05 atm or 2.0 mol of $He(g)$ at 5 atm?

5. For the following reaction at standard conditions,

$$2CO_2(g) + 2H_2O(g) \rightarrow HC_2H_3O_2(l) + 2O_2(g)$$

(a) calculate $\Delta G°$, $\Delta H°$, and $\Delta S°$ using the data in Appendix E

(b)   Is the reaction spontaneous at standard conditions?

(c)   Is this reaction spontaneous at (a) high temperatures, (b) low temperatures, (c) all temperatures, or (d) not spontaneous at any temperature? Explain!

6.   For the following reaction at standard conditions,

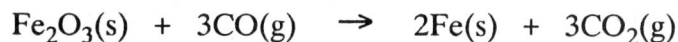

$$Fe_2O_3(s) \ + \ 3CO(g) \ \rightarrow \ 2Fe(s) \ + \ 3CO_2(g)$$

(a)   calculate $\Delta G°$, $\Delta H°$, and $\Delta S°$ using the data in Appendix E; for Fe(s), $S° = 27.3$ J/K•mol.

(b)   Is the reaction spontaneous at standard conditions?

(c)   Is this reaction spontaneous at (a) high temperatures, (b) low temperatures, (c) all temperatures, or (d) not spontaneous at any temperature? Explain!

7.   For ethanol, CH_3CH_2OH, $\Delta H_{vap} = 38.6$ kJ/mol and $\Delta S_{vap} = 110.4$ J/K•mol. What is the normal boiling point of ethanol?

8. For benzene, $C_6\Delta H_6$, $\Delta H_{vap} = 30.8$ kJ/mol and $\Delta S_{vap} = 87.2$ J/K•mol. What is the normal boiling point of benzene?

9. If $\Delta H_{vap} = 32.54$ kJ/mol for carbon tetrachloride, CCl_4, at its boiling point of 75 °C, calculate ΔS_{vap} of carbon tetrachloride.

10. If $\Delta H_{vap} = 31.33$ kJ/mol for chloroform, $CHCl_3$, at its boiling point of 60.0 °C, calculate ΔS_{vap} of chloroform.

11. If $\Delta G° = -234$ kJ for the following reaction, find the value of ΔG at 25 °C when $p_{SO_2} = 1.0 \times 10^{-5}$ atm, $p_{O_3} = 1.0 \times 10^{-4}$ atm, $p_{SO_3} = 1.0 \times 10^{-6}$, and $p_{O_2} = 0.20$ atm.

$$SO_2(g) + O_3(g) \rightarrow SO_3(g) + O_2(g)$$

12. If $K_p = 11.20$ for the following equilibrium reaction at 25 °C, calculate $\Delta G°$ at the same temperature.

$$2NO_2(g) \rightleftarrows N_2O_4(g)$$

13. If $\Delta G° = -142$ kJ for the following equilibrium reaction at 25 °C, calculate K_p at the same temperature.

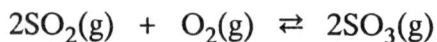

$$2SO_2(g) + O_2(g) \rightleftarrows 2SO_3(g)$$

ANSWERS TO PRACTICE PROBLEMS

1. (a) $AlCl_3(s)$
    (b) sugar dissolved in a cup of coffee
    (c) $CO_2(g)$ at 0 °C
    (d) $NH_4Cl(aq)$

2. (a) $\Delta G_{rxn}° = -142.11$ kJ

    $\Delta H_{rxn}° = -203.25$ kJ

    $\Delta S_{rxn}° = -215.0$ J/K

    (b) $\Delta G_{rxn}°$ is negative, so the reaction is spontaneous.

3. (a) From $\Delta G° = \Delta H° - T\Delta S°$, $\Delta G_{rxn}° = -6405$ kJ.

    (b) $\Delta G_{rxn}°$ is negative, so the reaction is spontaneous.

4. (a) $Ni(NO_3)_2(s)$
    (b) 0.40 mol of $CH_4(g)$ at 0.10 atm
    (c) liquid mercury
    (d) 2.0 mol of He(g) at 5 atm

5. (a) $\Delta H_{rxn}° = 777.9$ kJ

    $\Delta S_{rxn}° = -235$ J/K

    $\Delta G_{rxn}° = 854$ kJ

    (b) $\Delta G_{rxn}°$ is positive, so the reaction is non-spontaneous.

    (c) $\Delta G° = \Delta H° - T\Delta S°$; when $\Delta H_{rxn}°$ is positive and $\Delta S_{rxn}°$ is negative ($-T\Delta S°$ will be positive), $\Delta G_{rxn}°$ will always be positive. This reaction is not spontaneous at any temperature.

6. (a) $\Delta H_{rxn}° = -24.8$ kJ

    $\Delta S_{rxn}° = 14.4$ J/K

    $\Delta G_{rxn}° = -29.1$ kJ

(b) $\Delta G_{rxn}°$ is negative, so the reaction is spontaneous.

(c) $\Delta G° = \Delta H° - T\Delta S°$; when $\Delta H_{rxn}°$ is negative and $\Delta S_{rxn}°$ is positive ($-T\Delta S°$ will be negative), $\Delta G_{rxn}°$ will always be negative. This reaction is spontaneous at all temperatures.

7.   boiling point = 350 K $-$ 273 = 77 °C

8.   boiling point = 353 K $-$ 273 = 8.0×10^1 °C

9. ΔS_{vap} = 0.0935 kJ/K•mol = 93.5 J/K•mol

10. ΔS_{vap} = 0.094 03 kJ/K•mol = 94.03 J/K•mol

11. ΔG = -234 kJ/mol + (0.008 315 kJ/K•mol)(298 K) $\ln(2.0 \times 10^2)$

        = -234 kJ/mol + 13 kJ/mol = -221 kJ/mol

12. $\Delta G°$ = $-RT \ln K_p$ = $-(0.008\ 315$ kJ/K•mol)(298 K) $\ln(11.20)$ = -5.99 kJ/mol

13. $\ln K_p = -\Delta G°/RT$ = 57.3; $K_p = 8 \times 10^{24}$

CHAPTER 18

CHEMICAL KINETICS: A CLOSER LOOK AT REACTION RATES

| SECTION 18.1 | RATES OF REACTIONS |
| SECTION 18.2 | RATE AND CONCENTRATION |

After completing this section, you will be able to do the following:

- Express the rates of a reaction in terms of the rates of disappearance of the reactants and rates of appearance of products;

- Given the rate law for a reaction, give the order with respect to each substance and the overall order, and vice versa.

REVIEW

To determine whether a chemical reaction is feasible (economically, etc.), you need to know if it is spontaneous **and** how fast it takes place. The free energy change, ΔG, indicates the spontaneity of a reaction (Chapter 17). **Chemical kinetics**, the study of how fast chemical reactions take place, is discussed in this chapter. The **rate** at which a reaction occurs is expressed in terms of change in concentration per change in time,

$$\text{rate} = \frac{\Delta \text{conc.}}{\Delta t},$$

where concentration often is in moles per liter (M) and time is typically in minutes (min) or seconds (s).

The rate of a reaction often calculated is the **instantaneous rate** or rate at any point in time, from the beginning of the reaction (**initial rate** at t = 0) to various points in time while the reaction is occurring (e.g., 10 s, 1000 s, 5000 s into the reaction, etc.). The rate of a reaction can be expressed in terms of the rate of **disappearance of a reactant** or the rate of **appearance of a product**. Like quantities of reactants and products, the rates at which different reactants are used up and products formed are related by the stoichiometry of the reaction. Once you know one rate, **you can calculate all the others from it** (Sample Problem 18.1 A).

The mathematical relationship between concentration and rate for a reaction is known as the **rate law,** which can only be determined **experimentally,** NOT from a balanced reaction. For example, when the following reaction occurs at 50°C,

$$NO_2(g) + CO(g) \rightarrow CO_2(g) + NO(g),$$

the rate law has been found experimentally to be

$$rate = k[NO_2]^2,$$

where k is called the **rate constant,** which is constant as long as the temperature is constant, and the square brackets mean molar concentrations. The exponent of each concentration is called the **order of the reaction with respect to that substance.** The **overall order** is the sum of all the orders with respect to each substance (the sum of the exponents). This reaction is **second order** with respect to $NO_2(g)$ and **zero order** with respect to CO(g). The **overall order** is 2 + 0, or **two.** In other words, the rate of this reaction does not depend on the concentration of CO(g). What the rate law for this reaction says is that if the initial concentration of $NO_2(g)$ is doubled, then the initial rate will be 2^2 or four times faster; if the initial concentration of $NO_2(g)$ is tripled, then the rate will be 3^2 or nine times faster. That is, the change in rate is proportional to the square of the change in concentration. For a general reaction,

$$aA + bB + ... \rightarrow cC + dD + ...$$

the rate law determined experimentally usually has the form

$$rate = k[A]^x[B]^y[C]^z ...$$

with each reactant raised to a power (order) that normally is zero or a small positive integer (although orders may be fractions). Note that the order with respect to each substance is not necessarily the same as the coefficient in the balanced reaction. **You cannot predict the rate law (orders) for a reaction from the coefficients in the chemical equation unless you are specifically told it is a single-step reaction.**

Given the rate law for a reaction, you can determine the order with respect to each substance and the overall order, and vice versa (Sample Problem 18.2 A).

SAMPLE PROBLEM 18.1 A

▶▶ If NOCl(g) is reacting (being consumed) at the rate of 1.1×10^{-8} M/min in the following reaction

$$2NOCl(g) \quad \rightarrow \quad 2NO(g) \ + \ Cl_2(g)$$

(a) what is the rate of formation of NO(g)? (b) What is the rate of formation of $Cl_2(g)$?

Solution:

(a)   Since the coefficients of NO and NOCl are identical, the rate of formation of NO must be equal to the rate of disappearance of NOCl, or 1.1×10^{-8} M/min.

(b)   Create a conversion factor from the coefficients in the balanced reaction to find the rate of formation of Cl_2 in M/min (or mol/L•min) from the given rate for the disappearance of NOCl,

$$? \ \frac{mol \ Cl_2}{L \bullet min} \ = \ \frac{1.1 \times 10^{-8} \ \text{mol NOCl}}{L \bullet min} \ \times \ \frac{1 \ mol \ Cl_2}{2 \ \text{mol NOCl}} \ = \ 5.5 \times 10^{-9} \ M/min$$

CHECK: Does this answer make sense? Yes, since the balanced reaction says that when 2 mol of NOCl are consumed, only 1/2 as many moles (1 mol) of Cl_2 are formed. Therefore, the rate of appearance of Cl_2 is 1/2 the rate of disappearance of NOCl.

SAMPLE PROBLEM 18.2 A

▶▶ For the following reaction,

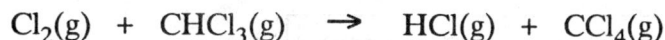

$$Cl_2(g) \ + \ CHCl_3(g) \quad \rightarrow \quad HCl(g) \ + \ CCl_4(g)$$

the experimentally determined rate law is

$$rate \ = \ k[Cl_2]^{1/2}[CHCl_3]$$

Give the order with respect to each substance **and** the overall order.

508

Solution:

The exponent of the concentration of the substance is the order with respect to that substance. The overall order is the sum of the exponents. For this reaction, it is 1/2 order with respect to Cl_2 and first order with respect to $CHCl_3$; the overall order is 1.5 (1/2 + 1). That is, this is a second order reaction.

SAMPLE PROBLEM 18.2 B

▶▶ If the following reaction was found to be second order with respect to A and first order with respect to B, write the rate law for this reaction.

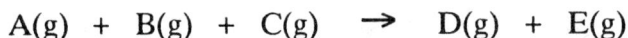

$$A(g) + B(g) + C(g) \rightarrow D(g) + E(g)$$

Solution:

In general, the rate law for a reaction usually contains the concentration of each reactant raised to a power (order). If no order is given with respect to a reactant, then it can be assumed the order is zero with respect to that reactant. For this problem,

$$rate = [A]^x[B]^y[C]^z$$

where x, y and z are the orders. From the information given in the problem, x = 2, y = 1, and z = 0. Substituting this information into the general rate law

$$rate = [A]^2[B][C]^0 = [A]^2[B]$$

where no exponent shown means first-order and any concentration raised to the zero power is always one.

SECTION 18.3 FINDING RATE LAWS

After completing this section, you will be able to do the following:

■ For a given reaction, use the initial rate method to find the rate law and calculate the rate constant and rates of reaction.

REVIEW

Although graphing techniques can be used to find the rate law for a reaction from experimental data, a simpler method is to use the **method of initial rates**. In this method, the order with respect to each reactant is found by changing the concentration of that reactant **only** (while keeping the concentrations of all other reactants constant) and noting the change in rate. This usually results in a simple algebraic equation with one unknown --- the order with respect to that reactant. Then the order with respect to the other reactants is found in exactly the same way. The **key** to this method is to find two sets of experimental data where the concentration of only **one** reactant is changing. Once the rate law is determined, the value of the **rate constant** can be calculated. After the rate law and rate constant are known, then the instantaneous rate of the reaction can be calculated for any set of concentrations (Sample Problem 18.3 A).

SAMPLE PROBLEM 18.3 A

▶▶ The initial rate of the following reaction was measured using different initial concentrations of $(CH_3)_3CBr$ and OH^-

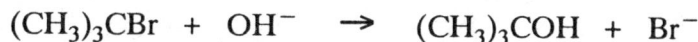

$$(CH_3)_3CBr + OH^- \rightarrow (CH_3)_3COH + Br^-$$

Experiment	$[(CH_3)_3CBr]$	$[OH^-]$	Initial rate, M/s
1	0.10	0.10	0.0010
2	0.20	0.10	0.0020
3	0.30	0.10	0.0030
4	0.10	0.20	0.0010
5	0.10	0.30	0.0010

(a)   What is the rate law for the reaction?
(b)   What is the value of the rate constant, k, with correct units?
(c)   What will be the initial rate of the reaction if the initial concentration of $(CH_3)_3CBr = 0.50$ M and the initial concentration of $OH^- = 0.40$ M?

Solution:

(a)   First write the general form of the rate law for this reaction,

$$\text{rate} = [(CH_3)_3CBr]^x[OH^-]^y$$

You need to determine x and y to find the rate law. Ordinarily, you can find x or y first. Arbitrarily, let's find x, then y. To find x, you need to select two experiments where the concentration of $(CH_3)_3CBr$ changes, but **not** OH^-. You have three choices in this case, all of which are correct. Expts. 1 and 2, 1 and 3, and 2 and 3 show the concentration of $(CH_3)_3CBr$ changing, but **not** OH^-. Again, let's arbitrarily select Expts. 1 and 2. Then substitute this information for each experiment into the general form of the rate law for this reaction

Expt. 1: $0.0010 \text{ M/s} = k(0.10 \text{ M})^x(0.10 \text{ M})^y$

Expt. 2: $0.0020 \text{ M/s} = k(0.20 \text{ M})^x(0.10 \text{ M})^y$

Now divide the equation for Expt. 2 by the equation for Expt. 1 to give

$$\frac{0.0020 \text{ M/s}}{0.0010 \text{ M/s}} = \frac{k(0.20 \text{ M})^x(0.10 \text{ M})^y}{k(0.10 \text{ M})^x(0.10 \text{ M})^y}$$

or

$$2 = (2.0)^x$$

The number 2.0 raised to what power equals 2?

$$x = 1$$

Comparing Expt. 2 with Expt. 1, the initial rate doubled when the $[(CH_3)_3CBr]$ was doubled; in other words, a direct proportion exists. Compare Expts. 1 and 3, and 2 and 3, you will find the same relationship. The rate changes in direct proportion to the change in concentration.

NOTE: To make the math simpler, the general form of the rate law applied to Expt. 2 was divided by the rate law applied to Expt. 1, so you would be working with numbers greater than one. You will be able to quickly do this by always placing the experiment with the higher concentrations in the numerator. If we had divided the data for Expt. 1 by the data from Expt. 2, we would have obtained

$$0.50 = (0.50)^x$$

and again x = 1, but this relationship is perhaps not as easy to see or work with. Also note that when the general form of the rate law for Expt. 2 was divided by the general form of the rate law for Expt. 1, the identical concentrations of OH^- canceled as well as the rate constant k. Even though we don't yet know the value of k, it is constant for this reaction no matter how many experiments are run, as long as the temperature remains constant.

To find y, select two experiments where the concentration of OH^- changes but not the concentration of $(CH_3)_3CBr$. Again, you have three choices, all of which are correct ---Expts. 1 and 4, 1 and 5, and 4 and 5. Arbitrarily, let's choose Expts. 1 and 4, and follow the same procedure as for determining x.

Expt. 1:   0.0010 M/s = k(0.10 M)^x(0.10 M)^y

Expt. 4:   0.0010 M/s = k(0.10 M)^x(0.20 M)^y

As the NOTE indicates above, if we divide the equation for Expt. 4 (higher concentration of OH$^-$) by the equation for Expt. 1, the equation produced will have numbers greater than one,

$$\frac{\cancel{0.0010 \text{ M/s}}}{\cancel{0.0010 \text{ M/s}}} = \frac{k(\cancel{0.10 \text{ M}})^x(0.20 \text{ M})^y}{k(\cancel{0.10 \text{ M}})^x(0.10 \text{ M})^y}$$

or

$$1 = (2.0)^y$$

The number 2.0 raised to what power equals 1? Any number raised to the **zero** power is always one. [If the solution is not obvious, take the logarithms (ln) of both sides of the equation and solve for y.]

$$y = 0$$

NOTE: Zero order means the rate does not depend on the concentration of OH$^-$. Comparing Expts. 1 and 5, and 4 and 5, you find the same thing; the rate is unaffected by any change in concentration of OH$^-$. However, OH$^-$ must be present for this reaction to take place.

The rate law for this reaction is

$$\text{rate} = k[(CH_3)_3CBr]$$

This is a first-order reaction. Normally, a zero-order reactant is not shown in the rate law.

(b)   Only after you know the rate law (orders) can you calculate the rate constant k. Since the rate constant is constant for this reaction (as long as the temperature does not change), you may select **any** experiment to determine the value of k. Choose an experiment where the numbers are as simple as possible to avoid errors. For example, if we choose Expt. 1, solve the rate law for k and substitute into the equation the data for Expt. 1,

$$k = \frac{\text{rate}}{[(CH_3)_3CBr]} = \frac{(0.0010 \text{ M/s})}{(0.10 \text{ M})} = 0.010 \text{ s}^{-1}$$

The units of the rate constant k are always t^{-1} (or 1/time) for first-order reactions. The units of k will depend on the overall order of a reaction.

(c) Once you know the rate law and the value of the rate constant k, you can calculate the initial rate of this reaction for any concentrations given. Substitute the initial concentrations given into the rate law and calculate the rate,

$$\text{rate} = 0.010 \text{ s}^{-1}[(CH_3)_3CBr]$$

$$= 0.010 \text{ s}^{-1}(0.50 \text{ M}) = 0.0050 \text{ M/s or } 0.0050 \text{ M} \bullet \text{s}^{-1}$$

NOTE: While the units of the rate constant k vary with the overall order of a reaction, the units of the rate of a reaction (no matter what the overall order is) are **ALWAYS** change in concentration per change in time, or usually M/t ($M \bullet t^{-1}$). Also, note that the rate of this reaction is independent of the concentration of OH^-.

CHECK: Does this rate for part (c) make sense? Yes. For Expt. 3, the $[(CH_3)_3CBr] = 0.30$ and the rate = 0.0030 M/s. A concentration of 0.50 M for $(CH_3)_3CBr$ is higher and the rate must be proportionally higher.

SECTION 18.4   FIRST-ORDER REACTIONS

After completing this section, you will be able to do the following:

■ For a first-order reaction, calculate the concentration of reactant remaining after a given amount of time and the length of time required to reach a given concentration.

REVIEW

The common overall reaction orders are zero, first, and second order. Each of these common orders has a rate law (which relates **rate and concentration**) and a very useful form called the integrated rate law (which relates **time and concentration**) that can be used to calculate the concentration of a reactant remaining after a given amount of time, and vice versa (see Table 18.3 in the textbook). Our discussion and applications will be limited to **first-order reactions only,** which include many medical applications and all nuclear decay processes.

For a **first-order reaction** of substance A, the rate law and integrated rate law are

$$\text{rate} = k[A]$$

$$\ln[A]_t = -kt + \ln[A]_0 \quad \text{or} \quad \ln \frac{[A]_t}{[A]_0} = -kt,$$

where $[A]_t$ is the concentration of reactant A at any time t, $[A]_0$ is the initial concentration of A, and k is the rate constant. The **half-life** of a reaction, $t_{1/2}$, is the time required for one half of the quantity of reactant originally present to react. For **first-order** reactions (not zero or second order), the half-life does **not** depend on the concentration of the reactant; that is, $t_{1/2}$ is a constant. For a first-order reaction, $t_{1/2}$ is the time at which the concentration of A is 1/2 of the original concentration of A, or $[A]_t = 1/2[A]_0$. If you substitute this information into either form of the integrated rate law above and solve for $t_{1/2}$,

$$t_{1/2} = \frac{0.693}{k}$$

Since half-lives can be readily measured for first-order reactions, this equation allows you to calculate the rate constant k, with which you can perform a variety of concentration-time applications with the first-order integrated rate law.

SAMPLE PROBLEM 18.4 A

▶▶ For the following first-order reaction, $t_{1/2} = 1140$ s at 45 °C.

$$2N_2O_5(g) \rightarrow 2N_2O_4(g) + O_2(g)$$

(a) What is the value of the rate constant k?
(b) If the initial concentration of N_2O_5 is 6.94×10^{-2} M, how long will it take for the concentration of N_2O_5 to reach 5.00×10^{-4} M at 45 °C?
(c) What will the concentration of N_2O_5 be after 1455 s?

Solution:

(a) Use the half-life equation for a first-order reaction given in the REVIEW and solve for the rate constant k,

$$k = \frac{0.693}{t_{1/2}} = \frac{0.693}{1140 \text{ s}} = 6.08 \times 10^{-4} \text{ s}^{-1}$$

(b) For concentration-time applications, use the integrated rate law for first-order reactions given in the REVIEW. When you are solving for time, you will find that the form of the integrated rate law with the logarithm of the ratio of concentrations easier to use.

$$\ln \frac{[N_2O_5]_t}{[N_2O_5]_0} = -kt$$

Substituting in the values for $[N_2O_5]_t$, $[N_2O_5]_0$, and k,

$$\ln \frac{[N_2O_5]_t}{[N_2O_5]_0} = \ln \frac{(5.00 \times 10^{-4})}{(6.94 \times 10^{-2})} = -4.933 = -(6.08 \times 10^{-4} \text{ s}^{-1})t$$

$$t = \frac{-4.933}{-(6.08 \times 10^{-4} \text{ s}^{-1})} = 8.11 \times 10^3 \text{ s}$$

NOTE: To simplify matters, divide the two concentrations **first**, then find the natural logarithm, ln x, of just one number.

(c) Again, this is a concentration-time application; use the integrated rate law for first-order reactions. When solving for one of the two concentrations, you will find it easier to use the other form of the integrated rate law than the one used in part (b).

$$\ln[N_2O_5]_t = -kt + \ln[N_2O_5]_0$$

Substitute in the given information and solve first for $\ln[N_2O_5]$

$$\ln[N_2O_5]_t = -kt + \ln[N_2O_5]_0 = -(6.08 \times 10^{-4} \text{ s}^{-1})(1455 \text{ s}) + \ln(6.94 \times 10^{-2} \text{ M})$$

$$= -0.885 + (-2.668) = -3.553$$

Taking the antilog of both sides of the equation

$$[N_2O_5] = 2.86 \times 10^{-2} \text{ M}$$

CHECK: Does this answer make sense? Yes, since after one half-life (1140 s), the concentration of N_2O_5 would be 1/2 of 6.94×10^{-2} M, or 3.47×10^{-2} M, and after two half-lives (2280 s), the concentration of N_2O_5 would be 1/4 of the initial concentration, or 1.74×10^{-2} M. The time of 1455 s is between 1140 s and 2280 s, and the calculated concentration is between 3.47×10^{-2} M and 1.74×10^{-2} M.

SECTION 18.9   RATE AND TEMPERATURE

After completing this section, you will be able to do the following:

■ Calculate the activation energy, E_a, for a reaction from the values of the rate constant at two different temperatures.

REVIEW

The relationship between the **rate constant** (and therefore rate) and **temperature** is that the rate constant **increases exponentially** as the temperature increases according to the Arrhenius equation

$$k = Ae^{-E_a/RT},$$

where A is a constant characteristic of the reaction, e is the base of the natural logarithm system, R is the gas constant 8.315 J/K•mol, T is the kelvin temperature, and E_a is a constant called the **activation energy** of the reaction. The **activation energy, E_a,** of a reaction is the **minimum energy** that the reactant molecules must have in order to react and form products. If you know the rate constant, k, at two or more temperatures, you can graph this information to obtain the value of the activation energy, E_a (see Practice Problem 18.19 in the textbook). Or, by knowing the value of the rate constant, k, at two temperatures, you can calculate the value of the activation energy, E_a, by using the following form of the Arrhenius equation given above

$$\ln \left(\frac{k_2}{k_1} \right) = \frac{E_a}{R} \left(\frac{1}{T_1} - \frac{1}{T_2} \right),$$

where k_1 is the value of the rate constant at temperature T_1, and k_2 is the value of the rate constant at temperature T_2 (Sample Problem 18.9 A).

SAMPLE PROBLEM 18.9 A

▸▸ For a certain reaction, $k = 6.08 \times 10^{-4}$ s^{-1} at 45.0 °C and $k = 3.35 \times 10^{-3}$ s^{-1} at 60.0 °C. What is the value of the activation energy, E_a for this reaction?

Solution:

Make sure the rate constant and the temperature at which it was measured have the **same** subscript (either 1 or 2, it doesn't matter). Arbitrarily assigning subscripts in the order the information was given in the problem,

$$k_1 = 6.08 \times 10^{-4} \text{ s}^{-1} \text{ at } T_1 = (45.0 + 273.2) = 318.2 \text{ K}$$

$$k_2 = 3.35 \times 10^{-3} \text{ s}^{-1} \text{ at } T_2 = (60.0 + 273.2) = 333.2 \text{ K}$$

Substituting this information and R = 8.315 J/K•mol into the equation given in the REVIEW,

$$\ln\left[\frac{3.35 \times 10^{-3}}{6.08 \times 10^{-4}}\right] = \frac{E_a}{8.315 \text{ J/K} \bullet \text{mol}} \left[\frac{1}{318.2 \text{ K}} - \frac{1}{333.2 \text{ K}}\right]$$

Since there a number of places to make math errors, the individual steps are shown leading to the calculation of E_a,

$$\ln(5.51) = \frac{E_a}{8.315 \text{ J/K} \bullet \text{mol}} \times (1.41 \times 10^{-4} \text{ K}^{-1})$$

$$1.707 = 1.70 \times 10^{-5} E_a$$

$$E_a = 1.00 \times 10^5 \text{ J/mol} = 1.00 \times 10^2 \text{ kJ/mol}$$

NOTE: You will probably find it easier to simplify each side of the equation as shown, rather than solve the equation for E_a initially.

SECTION 8.11   REACTION MECHANISMS

After completing this section, you will be able to do the following:

- Determine the rate law from the mechanism for a reaction.

REVIEW

Many reactions do not occur in a single step but require two or more steps between the reactants and products. The series of steps by which a reaction occurs is called the **reaction mechanism** or pathway for that reaction. Each step in a mechanism is an **elementary process**; this means that the order of each reactant in the rate law for that step is the coefficient of that reactant in the equation for that step. **This is the only time that you can write a rate law from a balanced reaction --- when you know or are told it is a single step or elementary process.** An overall reaction can occur no faster than the **slowest** step. Every multiple-step reaction has one **slowest or rate-determining step (RDS)**. Species that are formed in one step and used up in another step are called **intermediates**.

For some reactions, two or more mechanisms are possible and remain valid until experiments determine which one is correct. For a mechanism to be considered as a possibility for a reaction, two criteria must be met:

- the steps must add up to the overall reaction;

- the rate law for the **slowest** step **IS** the rate law for the overall reaction.

Given a proposed mechanism for a reaction, you should be able to determine the rate law for that reaction (Sample Problem 18.11 A).

SAMPLE PROBLEM 18.11 A

▸▸ At low pressure, the formation of gaseous propene, $CH_2=CHCH_3$, from gaseous cyclopropane, C_3H_6, is believed to occur by the following mechanism

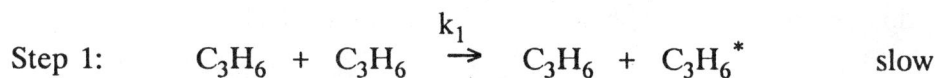

$$\text{Step 1:} \quad C_3H_6 + C_3H_6 \xrightarrow{k_1} C_3H_6 + C_3H_6{}^* \qquad \text{slow}$$

$$\text{Step 2:} \quad C_3H_6{}^* \xrightarrow{k_2} CH_2CH=CH_3 \qquad \text{fast}$$

(where $C_3H_6{}^*$ is an excited cyclopropane molecule; k_1 and k_2 are the rate constants for steps 1 and 2, respectively)

(a) Write the equation for the overall reaction.
(b) Write the rate law for each of the steps.
(c) Write the rate law predicted by the mechanism for the overall reaction.
(d) Which species is an intermediate?

Solution:

(a) Add the steps together to obtain the equation for the overall reaction. In the first step in this problem, two cyclopropane molecules collide to create one cyclopropane molecule in an excited state. The equation for the overall reaction is

$$C_3H_6 \rightarrow CH_2CH=CH_3$$

(b) For steps in a mechanism (which are always elementary processes), the order of each reactant is the coefficient of that reactant in the equation for that step

$$\text{Step 1:} \quad \text{rate} = k_1[C_3H_6]^2$$

$$\text{Step 2:} \quad \text{rate} = k_2[C_3H_6{}^*]$$

(c)   The rate law for the overall reaction is the rate law for the slowest step, which is Step 1.

$$\text{rate} = k_1[C_3H_6]^2 = k[C_3H_6]^2$$

(d)   Intermediates are **formed** in one step and used up in a later step; intermediates do **not** appear in the overall reaction. For this problem, $C_3H_6^*$ is the intermediate, since it is formed in Step 1 and consumed in Step 2.

SECTION 18.13   RATE CONSTANTS AND EQUILIBRIUM CONSTANTS

After completing this section, you will be able to do the following:

■ Write the equilibrium constant expression for a reaction from the rate laws for the forward and reverse reactions;

■ Given the value of the equilibrium constant and the value of the rate constant for the forward or reverse reaction, calculate the value of the other rate constant.

REVIEW

All of the rate laws that we have been writing so far were for the forward reaction. For those reactions that establish equilibrium, we can write a rate law for the reverse reaction as well. Because the rate of the forward reaction is equal to the rate of the reverse reaction at equilibrium, it can be shown that the equilibrium constant, K, is equal to the quotient of the **rate constants**,

$$K = \frac{k_f}{k_r},$$

where k_f and k_r are the rate constants for the forward and reverse reactions, respectively. Remember, CAPITAL K is the equilibrium constant and small k is a rate constant. Sample Problem 18.13 A will illustrate the relationship between K and k_f and k_r for a reaction.

SAMPLE PROBLEM 18.13 A

▶ ▶ Carbon dioxide (CO_2) and hydrogen gas (H_2) react and establish equilibrium with water vapor (H_2O) and carbon monoxide (CO). (a) What is the equilibrium constant expression for the reaction if $\text{rate}_f = k_f[CO_2][H_2]$ and $\text{rate}_r = k_r[H_2O][CO]$? (b) If K = 1.58 and $k_f = 0.15$ $M^{-1}s^{-1}$ at a certain temperature, what is the value of k_r at the same temperature?

(a)   At equilibrium, the forward and reverse rates are equal.

$$rate_f = rate_r$$

Substituting the forward and reverse rate laws into this equation,

$$k_f[CO_2][H_2] = k_r[H_2O][CO]$$

Since the equilibrium constant, K, is equal to k_f/k_r, solve the equation for k_f/k_r to find the equilibrium constant expression for this reaction,

$$\frac{k_f}{k_r} = \frac{[H_2O][CO]}{[CO_2][H_2]} = K$$

(b)   Solve the equation, $K = k_f/k_r$, for k_r and substitute in the information given to find k_r,

$$k_r = \frac{k_f}{K} = \frac{0.15 \text{ M}^{-1}\text{s}^{-1}}{1.58} = 0.095 \text{ M}^{-1}\text{s}^{-1}$$

NOTE: At equilibrium, the forward and reverse **rates** are **equal** but not necessarily the forward and reverse rate **constants**.

SELF-TEST QUESTIONS

Test your understanding of the concepts and skills in this chapter by working through the multiple choice questions BEFORE checking the answers. A periodic table will always be available. The ANSWERS TO SELF-TEST QUESTIONS follow immediately after this section.

1. Formaldehyde, an eye irritant, is formed by the reaction of ozone with ethylene in smog.

$$CH_2=CH_2(g) + O_3(g) \rightleftarrows 2CH_2O(g) + 1/2\,O_2(g)$$

The initial rate of this reaction was measured using different initial concentrations of $CH_2=CH_2$ and O_3.

Experiment	$[CH_2=CH_2]$	$[O_3]$	Initial rate, M/s
1	1.0×10^{-8}	0.50×10^{-7}	1.0×10^{-12}
2	1.0×10^{-8}	1.5×10^{-7}	3.0×10^{-12}
3	2.0×10^{-8}	1.0×10^{-7}	4.0×10^{-12}

The rate law is

A. rate $= k[O_3]$

B. rate $= k[CH_2=CH_2]^2$

C. rate $= k[CH_2=CH_2][O_3]^3$

D. rate $= k[CH_2=CH_2][O_3]$

E. rate $= k[CH_2=CH_2]^2[O_3]$

2. For the following reaction, the rate law is: rate $= k[NO_2][F_2]$

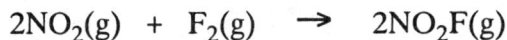

$$2NO_2(g) + F_2(g) \rightarrow 2NO_2F(g)$$

If the initial rate of this reaction is 9.6×10^{-4} M/s when the initial concentration of NO_2 is 0.0060 M and F_2 is 0.0040 M, the value of the rate constant, k, is

A. $2.3 \times 10^{-8}\ M^{-1}s^{-1}$

B. $6.7 \times 10^3\ M^{-1}s^{-1}$

C. $0.025\ M^{-1}s^{-1}$

D. $2.0 \times 10^1\ M^{-1}s^{-1}$

E. $4.0 \times 10^1\ M^{-1}s^{-1}$

3. For the reaction of ammonium ion with nitrous acid, the net ionic reaction is

$$NH_4^+(aq) + HNO_2(aq) \rightarrow N_2(g) + 2H_2O(l) + H^+(aq)$$

The rate law for this reaction is

A. rate = $k[H^+]$
B. rate = $k[NH_4^+][HNO_2]$
C. rate = $k[NH_4^+][HNO_2]/[N_2][H_2O]^2[H^+]$
D. rate = $k[HNO_2]/[H^+]$
E. cannot be determined from the information given.

4. A reaction with a rate law of rate = $k[A]$, has a rate constant of 1.2×10^{-2} s^{-1}. If the initial concentration of A is 2.0 M, what is the concentration of A after 200.0 seconds?

A. 6.0×10^{-3} M   B. 1.0 M   C. 1.7 M   D. 0.18 M   E. 0.55 M

5. What is the half-life of a first-order reaction having a rate constant of 25 min^{-1}?

A. 0.29 s   B. 750 s   C. 1040 s   D. 0.0227 s   E. 1.66 s

6. The initial rate of the following reaction was measured using different initial concentrations of BrO_3^-, Br^-, and H^+

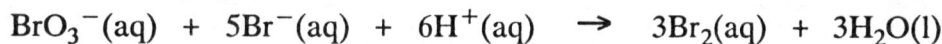

$$BrO_3^-(aq) + 5Br^-(aq) + 6H^+(aq) \rightarrow 3Br_2(aq) + 3H_2O(l)$$

Experiment	$[BrO_3^-]$	$[Br^-]$	$[H^+]$	Initial rate, M/s
1	0.10	0.10	0.10	1.2×10^{-3}
2	0.20	0.10	0.10	2.4×10^{-3}
3	0.10	0.30	0.10	3.6×10^{-3}
4	0.20	0.10	0.15	5.4×10^{-3}

The rate law for this reaction is

A. rate = $k[BrO_3^-][Br^-][H^+]$
B. rate = $k[BrO_3^-]^2[Br^-]^3[H^+]^3$
C. rate = $k[BrO_3^-][Br^-][H^+]^2$
D. rate = $k[BrO_3^-][Br^-]^5[H^+]^6$
E. rate = $k[BrO_3^-][Br^-]^2[H^+]^3$

7. At a certain temperature, $k = 2.8 \ M^{-2}s^{-1}$ for the reaction given below, which was found to be first-order with respect to bromine and second order with respect to chlorine. What is the rate of the reaction when the concentration of bromine and chlorine are 0.20 M and 0.30 M, respectively?

A. 0.30 M/s B. 160 M/s C. 0.050 M/s D. 2.8 M/s E. 0.034 M/s

8. Nitrogen monoxide and hydrogen react at 826 °C according to the following reaction,

$$2NO(g) \ + \ 2H_2(g) \ \rightarrow \ N_2(g) \ + \ 2H_2O(g)$$

for which the rate law is rate = $k[NO]^2[H_2]$. Therefore the reaction

A. must proceed by a multi-step mechanism.
B. may proceed by a multi-step mechanism.
C. must proceed by a single-step mechanism.
D. may proceed by a single-step mechanism.
E. may proceed by either a single-step or a multi-step mechanism.

ANSWERS TO SELF-TEST QUESTIONS

1. d 4. d 7. c
2. e 5. e 8. a
3. e 6. c

PRACTICE PROBLEMS

Test your problem-solving skills in this chapter by working through the problems in the space provided **BEFORE** checking the answers. A periodic table will always be available. The ANSWERS TO PRACTICE PROBLEMS follow immediately after this section.

1. If methane, CH_4, burns at a rate of 0.16 M/s according to the following reaction

$$CH_4(g) + 2O_2(g) \rightarrow CO_2(g) + 2H_2O(g)$$

at what rate is (a) O_2 being consumed? (b) CO_2 being formed? (c) water being formed?

2. If ammonia, NH_3, is being formed at the rate of 6.7×10^{-2} M/min according to the following reaction

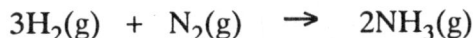

$$3H_2(g) + N_2(g) \rightarrow 2NH_3(g)$$

at what rate is (a) H_2 being consumed? (b) N_2 being consumed?

3. At 400 °C, $k = 1.6 \times 10^{-6}$ s^{-1} for the first-order decomposition of cyclopropane to propylene.

    (a) What is the half-life of this reaction?

    (b) How long will it take for an initial concentration of 0.20 M cyclopropane to decrease to 0.15 M at 400 °C?

    (c) What is the initial rate of the reaction if the initial concentration of cyclopropane is 0.25 M?

4.   At a certain temperature, k = 0.163 $M^{-1}s^{-1}$ for the reaction

$$2ICl(g) + H_2(g) \rightarrow I_2(g) + 2HCl(g)$$

and the rate law is rate = $k[ICl][H_2]$. What is the initial rate of the reaction when the initial concentration of ICl is 0.25 M and the concentration of H_2 is 0.50 M?

5.   The initial rate of the following reaction was measured using different initial concentrations of A and B

$$2A(g) + B(g) \rightarrow C(g) + 2D(g)$$

Experiment	[A]	[B]	Initial rate, M/s
1	0.010	0.010	1.20×10^{-3}
2	0.020	0.010	2.40×10^{-3}
3	0.030	0.010	3.60×10^{-3}
4	0.030	0.020	1.44×10^{-2}

(a)   What is the rate law for the reaction?

(b)   What is the value of the rate constant with proper units?

6.   Ethyl chloride, C_2H_5Cl, decomposes with first-order kinetics at 700 K. If an initial concentration of 0.200 M ethyl chloride decreases to 0.150 M in 60.0 minutes,

(a)   what is the value of the rate constant k?

(b)   What is the half-life of this reaction?

7.   The initial rate of the following reaction was measured using different initial concentrations of NO_2 and O_3

$$NO_2(g) + O_3(g) \rightarrow NO_3(g) + O_2(g)$$

Experiment	$[NO_2]$	$[O_3]$	Initial rate, M/s
1	5.0×10^{-5}	1.0×10^{-5}	0.022
2	5.0×10^{-5}	2.0×10^{-5}	0.044
3	2.5×10^{-5}	2.0×10^{-5}	0.022
4	2.5×10^{-5}	4.0×10^{-5}	0.044

(a)   What is the rate law for the reaction?

(b)   What is the value of the rate constant with proper units?

8.   At 95 °C, $k = 1.8 \times 10^{-3}$ s^{-1} for a certain reaction. When the temperature was raised to 100 °C, the rate constant increased to $k = 4.1 \times 10^{-2}$ s^{-1}. What is the value of the activation energy, E_a, for this reaction?

9. The decomposition of hydrogen peroxide, H_2O_2, is catalyzed by iodide ion and believed to have the mechanism

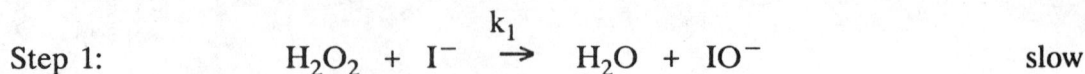

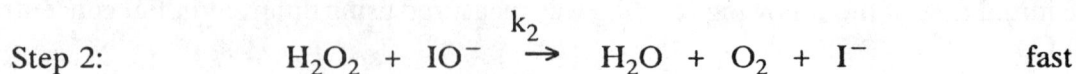

Step 1: $H_2O_2 + I^- \xrightarrow{k_1} H_2O + IO^-$ slow

Step 2: $H_2O_2 + IO^- \xrightarrow{k_2} H_2O + O_2 + I^-$ fast

(a) Write the equation for the overall reaction.

(b) Write the rate law for each of the steps.

(c) Write the rate law predicted by the mechanism for the overall reaction.

(d) Which species is an intermediate?

ANSWERS TO PRACTICE PROBLEMS

1. (a) rate of reaction = -0.32 M O_2/s
   (b) rate of reaction = 0.16 M CO_2/s
   (c) rate of reaction = 0.32 M H_2O/s.

2. (a) rate of reaction = -1.0×10^{-1} M H_2/min
   (b) rate of reaction = -3.4×10^{-2} M N_2/min

3. (a) $t_{1/2} = 4.3 \times 10^5$ s
   (b) Use first-order integrated rate law; $t = 1.8 \times 10^5$ s
   (c) Rate = k[cyclopropane] = $(1.6 \times 10^{-6} \text{ s}^{-1})(0.25 \text{ M}) = 4.0 \times 10^{-7}$ M/s

4. rate = 0.020 M/s

5.   (a) rate $= k[A][B]^2$

        (b) $k = 1.2 \times 10^3\ M^{-2}s^{-1}$

6.   (a) Use the first-order integrated rate law; $k = 4.79 \times 10^{-3}\ min^{-1}$

        (b) $t_{1/2} = 145$ min

7.   (a) rate $= k[NO_2][O_3]$

        (b) $k = 4.4 \times 10^7\ M^{-1}s^{-1}$

8. $\ln\left[\dfrac{4.1 \times 10^{-2}}{1.8 \times 10^{-3}}\right] = \dfrac{E_a}{8.315\ J/K\bullet mol}\left[\dfrac{1}{368\ K} - \dfrac{1}{373\ K}\right]$

        $E_a = 7 \times 10^5\ J/mol = 7 \times 10^2\ kJ/mol$

9.   (a) $2H_2O_2\ \rightarrow\ 2H_2O + O_2$

        (b) Step 1: rate $= k_1[H_2O_2][I^-]$ Step 2: rate $= k_2[H_2O_2][IO^-]$

        (c) Rate $= k_1[H_2O_2][I^-] = k[H_2O_2][I^-]$

        (d) Intermediate is IO^-.

CHAPTER 19

ELECTROCHEMISTRY

After completing this section, you will be able to do the following:

■ Distinguish between voltaic and electrolytic electrochemical cells;

■ Make a cell diagram for a voltaic cell.

REVIEW

Electrochemistry deals with (a) the use of **spontaneous** redox reactions to supply electrical energy in a **voltaic or galvanic cell,** and (b) the use of electrical energy to make **non-spontaneous** redox reactions take place in **electrolytic cells.** Both kinds of electrochemical cells contain two electrodes (often a metal) that are connected externally by an ammeter (galvanic) or battery (electrolytic) and inserted into a solution(s) containing the reactants for the redox reaction. The **oxidation** half-reaction **always** occurs at the electrode called the **anode,** and the **reduction** half-reaction **always** occurs at the electrode called the **cathode** (see Section 11.4 for definitions and examples of oxidation and reduction; see Section 11.7 for half-reactions). The **sign** of the electrode depends on the kind of cell, voltaic or electrolytic. To keep the conventions correct for electrochemical cells, a summary of the conventions for the **anode** in both kinds of cells is provided in the following table

ELECTROCHEMICAL CELLS	
Voltaic	Electrolytic
S pontaneous	
L eft	**L** eft
O xidation	**O** xidation
A node	**A** node
N egative	**P** ositive

The **only difference in conventions** between the two kinds of cells is the sign of the electrodes. Although showing the anode on the left side of a schematic drawing is arbitrary, it is a good habit to develop to avoid confusion (see the schematic drawings in Figure 19.3 for a voltaic cell and Figure 19.20 for an electrolytic cell, both in the textbook). The rest of the conventions are mandatory. **Oxidation** always occurs at the **anode** and **reduction** always occurs at the **cathode**, in **ALL** electrochemical cells. [A memory device for that last statement is "an ox" and "red cat".] The acronym **"SLOAN"** quickly reminds you of all the conventions for the anode of spontaneous (voltaic) cells; the other electrode must be the cathode on the right side, where reduction always occurs and the sign is positive. For electrolytic cells, remember that the only difference in the conventions of the anode is the sign (as an alternative to the less memorable "LOAP"). Two additional pieces of information with suggested memory devices that apply to **both** kinds of electrochemical cells:

- in the external circuit, electrons flow from **anode** to **cathode** (a → c);
- in solution, **anions** always migrate to the **anode** (a → a) and **cations** always migrate to the **cathode** (c → c). [It is the movement of the ions in the solutions that conducts the current and completes the circuit for both kinds of cells.]

For **voltaic** cells, **cell diagrams** rather than schematic drawings are often used to describe the cell. For example, aluminum metal spontaneously reduces Ag^+ in aqueous solution according to the redox reaction

$$Al(s) + Ag^+(aq) \rightarrow Al^{3+}(aq) + Ag(s)$$

and the short-hand **cell diagram** for this reaction is

$$Al(s) \mid Al^{3+}(aq) \| Ag^+(aq) \mid Ag(s),$$

where the **anode** is **always** written on the **left** side and the cathode on the right side. A single vertical line shows a boundary between phases, such as the boundary where the surface of the aluminum electrode meets the solution containing the aluminum ion (such as from aluminum nitrate). A double vertical line shows a salt bridge or other barrier between solutions that allows ions but not the bulk solution to flow through.

SAMPLE PROBLEM 19.1 A

▶ ▶ Make a cell diagram for the voltaic cell that uses the spontaneous reaction between magnesium metal and zinc nitrate solution to form metallic zinc and magnesium nitrate solution.

Solution:

First, write out the net ionic equation for the reaction given,

$$Mg(s) + Zn^{2+}(aq) \rightarrow Mg^{2+}(aq) + Zn(s)$$

Magnesium metal is undergoing oxidation, which always occurs at the anode and is written on the **left side** (remember "SLOAN"?). Following the procedure discussed in the REVIEW, we have

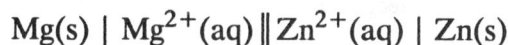

$$Mg(s) \mid Mg^{2+}(aq) \parallel Zn^{2+}(aq) \mid Zn(s)$$

SECTION 19.2   STANDARD POTENTIALS

After completing this section, you will be able to do the following:

■ Calculate a standard cell potential from standard half-cell reduction potentials;

■ Calculate a standard half-cell reduction potential from a measured standard cell potential and one other standard half-cell reduction potential.

REVIEW

The difference in electric potential energy between half-cells is called the **electromotive force, E,** of the cell or, more commonly, the **cell potential** or **cell voltage**. The **cell potential** is the driving force that pushes electrons through an external wire from one electrode to another. One **volt** is the **emf** (electromotive force) required to give one joule of energy to a charge of one coulomb.

$$1 \text{ volt} = \frac{1 \text{ joule}}{\text{coulomb}} \quad \text{or} \quad 1 \text{ V} = \frac{1 \text{ J}}{\text{C}}$$

The cell potential for a voltaic cell is measured with a voltmeter in the external circuit; the voltmeter reads the potential difference between the two half-cells. The **cell potential, E,** measured under **standard conditions** (the same as for thermodynamics, 1 M for solutions and partial pressures of 1 atm for gases) is called the **standard cell potential, E°**. It represents the driving force for a reaction when all materials are in their standard states. Measurement of the potential of a single half-cell is impossible (you cannot have oxidation occurring without reduction); only differences in potential between two half-cells can be measured. The only way, then, that a half-cell potential can be determined is if the potential of the other half-cell is zero. By international agreement, the standard half-cell potential for the following reaction has been defined as exactly **zero**.

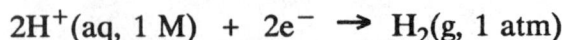

$$2H^+(aq, 1 \text{ M}) + 2e^- \rightarrow H_2(g, 1 \text{ atm})$$

and the half-cell containing this half-reaction is called the **standard hydrogen electrode (SHE)**. The standard potential for this half-reaction, $E°$, is **zero** no matter whether 1 M H^+ is being reduced or hydrogen gas at 1 atm is being oxidized. If one of the half-cells in a cell is the standard hydrogen electrode with its standard potential of **zero**, then the standard cell potential of the other half-cell will be equal to the measured **cell potential, E**, which is the difference between the reduction potentials of the two half-cells

$$E°_{cell} = E°_{reduction\ half-cell} - E°_{oxidation\ half-cell},$$

where the negative sign changes a standard reduction potential into a standard **oxidation potential** for the oxidation half-reaction. For example, the standard potential, $E°$, for the **reduction** of Cu^+ is 0.52 V.

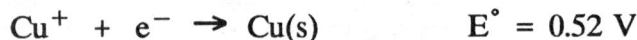

$$Cu^+ + e^- \rightarrow Cu(s) \qquad E° = 0.52 \text{ V}$$

The standard potential for the reverse reaction (**oxidation**) is -0.52 V.

$$Cu(s) \rightarrow Cu^+ + e^- \qquad E° = -0.52 \text{ V}$$

By this method, a series of **standard reduction potentials** can be determined for a variety of substances. By convention, **the emf or potential of voltaic (spontaneous) cells is positive**. That is, for a **spontaneous reaction**, E has a + sign (and ΔG has a − sign). The standard cell potential is an intensive property (i.e., the cell potential does not depend on the size of the cell). Thus, standard potentials for half-cells are **not** multiplied by any factor (such as are needed to add two half-reactions together to produce an overall balanced reaction) when the cell potential, $E°_{cell}$, is calculated.

Table 19.1 in the textbook lists the standard potentials for a number of common **reduction** half-reactions in **order of decreasing value**, called **standard reduction potentials** (placing a negative sign in front of the standard reduction potentials changes them to standard oxidation potentials for the reverse reaction). In Table 19.1, all the species in the left-hand column (to the left of the reaction arrow) are **oxidizing agents**, arranged from the best, $F_2(g)$, to the poorest, Li^+. The species in the right hand-column (to the right of the reaction arrow) are **reducing agents** (for the reverse reactions), arranged from the best reducing agent, Li(s) (with standard oxidation potential of +3.05 V), to the poorest, F^- (standard oxidation potential of -2.87 V). Qualitatively, you can predict that under standard conditions, any species to the **left of the arrows** will (spontaneously) react with any species to the **right of the arrows** that is **lower** in the table. For example, Ag^+ will spontaneously react with Fe^{2+}, H_2O_2, I^-, Cu(s), Cu^+, H_2, Fe(s), etc., but **not**

NO(g), Br^-, $H_2O(l)$, Cr^{3+}, Cl^-, etc. This **diagonal relationship** (from upper left to lower right in Table 19.1) works equally well starting with a reducing agent in the right-hand column; under standard conditions, any species in the right-hand column will spontaneously react with any species to the left of the arrows that is **higher** in the table. For example, Zn(s) will spontaneously react with Cr^{3+}, Fe^{2+}, Fe^{3+}, H^+, Cu^{2+}, etc., but not Cr^{2+}, Al^{3+}, Mg^{2+}, etc. Any species in an intermediate oxidation state, like Fe^{2+}, Cu^+ and Cr^{2+}, can act as either an oxidizing or reducing agent.

SAMPLE PROBLEM 19.2 A

▶▶ Calculate the standard cell potential, E°, for the following reactions and indicate whether each reaction is spontaneous or not.

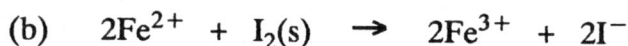

(a) $2Cr(s) + 3Cu^{2+} \rightarrow 2Cr^{3+} + 3Cu(s)$

(b) $2Fe^{2+} + I_2(s) \rightarrow 2Fe^{3+} + 2I^-$

Solution:

(a) The standard cell potential is found from the equation given in the REVIEW,

$$E°_{cell} = E°_{reduction\ half-cell} - E°_{oxidation\ half-cell}$$

Identify the reduction and oxidation half-reactions and their standard potentials from Table 19.1. For reduction, $E°_{Cu^{2+}/Cu(s)} = 0.34$ V. For oxidation, the half-reaction is $Cr(s) \rightarrow Cr^{3+} + 3e^-$, but you want the **reduction potential** to use in the above equation, or $E°_{Cr^{3+}/Cr(s)} = -0.74$ V. There is a standard reduction potential (SRP) listed in Table 19.1 for Cr^{3+}/Cr^{2+} and $Cr^{3+}/Cr(s)$; make sure that you select the correct one.

Now substitute the SRP's into the equation and solve for the E°,

$$E°_{cell} = E°_{Cu^{2+}/Cu(s)} - E°_{Cr^{3+}/Cr(s)} = 0.34\ V - (-0.74\ V) = 1.08\ V$$

The reaction is spontaneous, because E° is +.

(b) Follow the same procedure as in part (a). For reduction, $E°_{I_2(s)/I^-} = 0.54$ V. For oxidation, the half-reaction is $Fe^{2+} \rightarrow Fe^{3+} + e^-$, but you want the **reduction potential** to use in the above equation, or $E°_{Fe^{3+}/Fe^{2+}} = 0.77$ V. Again, there is more than one SRP listed for Fe^{3+} in Table 19.1; make sure that you select the correct one.

Now substitute the SRP's into the equation and solve for the E°,

$$E°_{cell} = E°_{I_2(s)/I^-} - E°_{Fe^{3+}/Fe^{2+}} = 0.54\ V - (0.77\ V) = -0.23\ V$$

The reaction is non-spontaneous, because E° is negative.

SAMPLE PROBLEM 19.2 B

▶▶ The standard cell potential for the reaction

$$2Ag^+ + Pb(s) \rightarrow 2Ag(s) + Pb^{2+}$$

is +0.93 V at 25 °C. What is E° for the half-reaction $Pb^{2+} + 2e^- \rightarrow Pb(s)$ at 25 °C?

Solution:

From Table 19.1, $E°_{Ag^+/Ag(s)}$ = 0.80 V for reduction. Substitute this value and the $E°_{cell}$ value into the equation for calculating a standard cell potential given in the REVIEW,

$$E°_{cell} = E°_{reduction\ half\text{-}cell} - E°_{oxidation\ half\text{-}cell}$$

$$0.93\ V = 0.80\ V - E°_{oxidation\ half\text{-}cell}$$

Solving for $E°_{oxidation\ half\text{-}cell}$,

$$E°_{oxidation\ half\text{-}cell} = -0.13\ V$$

The standard reduction potential for the half-reaction $Pb^{2+} + 2e^- \rightarrow Pb(s)$ is -0.13 V.

NOTE: The standard **reduction potential** for the half-reaction occurring in the oxidation half-cell is $E°_{oxidation\ half\text{-}cell}$.

SECTION 19.3	EFFECT OF CONCENTRATION ON CELL POTENTIAL

After completing this section, you will be able to do the following:

■ Calculate the cell potential at non-standard conditions using the Nernst equation.

REVIEW

Previously (Section 17.10), we discussed the calculation of the free energy change at nonstandard conditions, ΔG, and its relationship to the free energy change at standard conditions, $\Delta G°$. In a very similar manner, the cell potential measured at non-standard conditions (other than 1 M for solutions and 1 atm for gases), E, is related to the standard cell potential, E°, in the so-called Nernst equation,

$$E \; = \; E° \; - \; \frac{RT}{nF} \; \ln Q,$$

where E is the potential of the cell under nonstandard conditions, E° is the standard potential of the cell, R is the gas constant, 8.315 J/K•mol, T is the kelvin temperature, Q is the reaction quotient, n is the number of moles of electrons transferred (which must be the same number for both oxidation and reduction), and F is the Faraday constant, 96 485 coulombs/mol. The **Faraday constant** is the charge on 1 mol of electrons. For simplification at 25 °C, RT/F becomes 0.0257 V and the Nernst equation becomes

$$E \; = \; E° \; - \; \frac{0.0257 \text{ V}}{n} \; \ln Q$$

At standard conditions, $Q = 1$, $\ln Q = 0$, and, as you should expect the cell potential $E = E°$. The term $-(RT/nF) \ln Q$ indicates how far the reaction conditions are from standard conditions. If the conditions of the reaction are close to standard conditions, then there will be very little difference between E and E°. Given an electrochemical cell at nonstandard conditions of concentrations of solutions and/or pressures of gases, you can calculate Q, determine n from the balanced reaction, and calculate E° from Table 19.1. Then substitute these values into the Nernst equation and solve the equation for E°.

SAMPLE PROBLEM 19.3 A

▸▸ What is the potential, E, of the following cell at 25 °C?

$$\text{Pt} \mid H_2(g, \text{ 1.0 atm}) \mid H^+(aq, \text{ 0.30 M}) \| Ag^+(aq, \text{ 0.020 M}) \mid Ag(s)$$

Solution:

To write the equation for the cell, remember that the anode is always written on the **left** side of the cell diagram (to the left of the vertical parallel lines) and the cathode is always written on the right side of the cell diagram. Interpreting the cell diagram from left to right, H_2 gas is oxidized to H^+ and Ag^+ is reduced to silver metal. Write down and balance the reaction for what we just stated in the previous sentence, or look up the balanced half-reactions in Table 19.1 and add them together so that the same number of electrons are gained in reduction as are lost in oxidation (see Section 11.7 for balancing by the half-reaction method). The equation for the cell reaction is

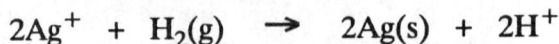

$$2Ag^+ + H_2(g) \rightarrow 2Ag(s) + 2H^+$$

You will need the Nernst equation to calculate the nonstandard potential of any cell

$$E = E° - \frac{0.0257 \text{ V}}{n} \ln Q$$

$E° = 0.80$ V $(Ag^+/Ag) - 0.00$ V $(H^+/H_2) = 0.80$ V. For either oxidation or reduction, n = 2; that is, there are 2 electrons given up by two hydrogen atoms that are transferred to and gained by the two silver ions. Q is written from the balanced reaction and calculated to be

$$Q = \frac{[H^+]^2}{[Ag^+]^2 p_{H_2}} = \frac{(0.30)^2}{(0.020)^2(1.0)} = 2.3 \times 10^2$$

Substituting the values for E°, n, and Q into the Nernst equation,

$$E = 0.80 \text{ V} - \frac{0.0257 \text{ V}}{2} \ln (2.3 \times 10^2)$$

$$= 0.80 \text{ V} - 0.0699 \text{ V} = 0.73 \text{ V}$$

SECTION 19.4   FREE ENERGY AND CELL POTENTIAL

After completing this section, you will be able to do the following:

■ For a reaction at equilibrium, calculate $\Delta G°$ from E°, and vice versa.

REVIEW

Earlier (Section 17.11) we saw that for a system at equilibrium, the nonstandard free energy change, ΔG, is zero and Q = K, which gave the following relationship between the standard free energy change, $\Delta G°$, and the equilibrium constant, K,

$$\Delta G° = -RT \ln K$$

Under those same equilibrium conditions, the non-standard cell potential, E, is zero and Q = K, which when substituted into the Nernst equation yields the relationship between the standard cell potential, E°,

and the equilibrium constant, K,

$$E° = \frac{RT}{nF} \ln K$$

To obtain the relationship between $\Delta G°$ and E°, solve this equation for $-RT\ln K$ (which is $\Delta G°$), which gives the equation

$$\Delta G° = -nFE°$$

Knowing the value of E°, you can calculate K, and vice versa, from the second equation; knowing the value of E°, you can calculate $\Delta G°$, and vice versa, from the last equation.

In summary, the following table indicates how Q, ΔG, and E have been used to predict the direction of spontaneous change,

Direction of Spontaneous Change	Q	ΔG	E
Forward direction	<K	−	+
At equilibrium	=K	0	0
Reverse direction	>K	+	−

SAMPLE PROBLEM 19.4A

▶▶ For the following reaction at equilibrium, calculate (a) E°, (b) $\Delta G°$, (c) K, and (d) indicate whether the reaction is spontaneous or not.

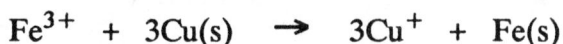

$$Fe^{3+} + 3Cu(s) \rightarrow 3Cu^+ + Fe(s)$$

Solution:

(a) Assuming the only data available are from Table 19.1, use the same procedure as in Sample Problem 19.2 A. That is, look up the standard reduction potentials for Fe^{3+}/Fe(s) and Cu^+/Cu(s) in Table 19.1 and substitute them into the equation for calculating the standard cell potential,

$$E°_{cell} = E°_{reduction\ half-cell} - E°_{oxidation\ half-cell}$$

$$= E°_{Fe^{3+}/Fe(s)} - E°_{Cu^+/Cu(s)}$$

$$= (-0.04\ V) - (0.52\ V) = -0.56\ V$$

(b) Knowing E° from part (a) = −0.56 V = −0.56 J/C (Section 19.2 REVIEW), calculate ΔG° from the last equation given in the REVIEW above, where n = 3 from the reaction given and F = 9.65 x 10⁴ C/mol (Section 19.3)

$$\Delta G° = -nFE° = -(3 \text{ mol})(9.65 \times 10^4 \text{ C/mol})(-0.56 \text{ J/C})$$

$$= 1.6 \times 10^5 \text{ J} = 1.6 \times 10^2 \text{ kJ}$$

(c) Now knowing both E° and ΔG° for this reaction at equilibrium, you can calculate K from either the first or second equation given in the REVIEW. Using the second equation where RT/F = 0.0257 V (Section 19.3) and solving for ln K,

$$E° = \frac{RT}{nF} \ln K = \frac{0.0257 \text{ V}}{n} \ln K$$

$$\ln K = \frac{nE°}{0.0257 \text{ V}} = \frac{(3)(-0.56 \text{ V})}{0.0257 \text{ V}} = -65$$

Taking the antilog of both sides of the equation,

$$K \approx 10^{-29}$$

(d) This reaction is non-spontaneous, because E° is − and ΔG° is +.

SAMPLE PROBLEM 19.4 B

▶▶ If ΔG° = −4.96 x 10² kJ for the following reaction, calculate E°.

$$Sn^{2+} + 2Na(s) \rightarrow Sn(s) + 2Na^+$$

Solution:

The last equation in the REVIEW is the only one that contains the two quantities you are working with in this problem,

$$\Delta G° = -nFE°$$

Solving the equation for E° and substituting in the values of ΔG° = −4.96 x 10² kJ = −4.96 x 10⁵ J, n = 2, F = 9.65 x 10⁴ C/mol,

$$E° = -\frac{\Delta G°}{nF} = \frac{-4.96 \times 10^5 \text{ J}}{(2 \text{ mol})(9.65 \times 10^4 \text{ C/mol})} = -2.57 \text{ J/C} = -2.57 \text{ V}$$

NOTE: ΔG° must be in joules (not kilojoules) for the answer to have the unit of volts.

SECTION 19.7   ELECTROLYTIC CELLS

After completing this section, you will be able to do the following:

■ Predict the anode and cathode reactions in an electrolytic cell.

REVIEW

As discussed earlier (Section 19.1), there are two kinds of **electrochemical cells**: **voltaic** (spontaneous) and **electrolytic** (non-spontaneous). [Be careful! The term "electrochemical" includes both kinds of cells, spontaneous and non-spontaneous; the similar sounding and appearing term "electrolytic" means a non-spontaneous cell only.] The process of causing a non-spontaneous chemical reaction to take place by means of electrical energy is called **electrolysis**. The minimum voltage necessary to bring about electrolysis is the calculated potential for the cell from Table 19.1, although the actual voltage required is always greater than this minimum voltage due to the resistance of the cell, the kinetics of the reaction, etc.

For purposes of predicting which reaction will occur (particularly first) at each electrode in an electrolytic cell, the **three** types of reactions that must be considered are oxidation or reduction of the

1. solute;
2. solvent (ordinarily water);
3. electrode.

Of the possible half-reactions derived by considering these three possibilities, the ones with the **most positive (or least negative)** potentials will usually take place first when energy is supplied to these non-spontaneous cells. The common solvent water can be both oxidized and reduced. When water is oxidized, $E° = -1.23$ V (Table 19.1) which is for $[H^+] = 1$ M; in a neutral solution, $[H^+] = 1 \times 10^{-7}$ M and from the Nernst equation $E = -0.82$ V for the oxidation of water. When water is reduced, $E° = -0.83$ V (Table 19.1), which is for $[OH^-] = 1$ M; in a neutral solution, $[OH^-] = 1 \times 10^{-7}$ M and (from the Nernst equation) $E = -0.42$ V for the reduction of water. When a pure liquid (melt) is electrolyzed, then only the ions of the liquid and the electrode need to be considered (not water).

Given an electrolytic cell, you should be able to predict which half-reactions will occur (first) at the anode and cathode. Your predictions will be correct most of the time, but in some instances, the actual half-reactions will be different, but unpredictable, due to the factors indicated above.

SAMPLE PROBLEM 19.7 A

▶▶ Predict the anode, cathode, and overall reactions using inert electrodes for the electrolysis of (a) NaI(aq) and (b) NaI(l).

Solution:

(a) Using only the data in Table 19.1, what half-reactions need to be considered at the anode and cathode?

Anode (Oxidation)

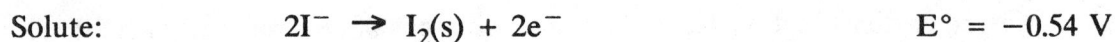

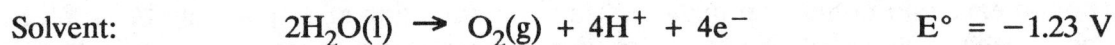

Solute:	$2I^- \rightarrow I_2(s) + 2e^-$	$E° = -0.54$ V
Solvent:	$2H_2O(l) \rightarrow O_2(g) + 4H^+ + 4e^-$	$E° = -1.23$ V
Electrode:	None (inert)	

Cathode (Reduction)

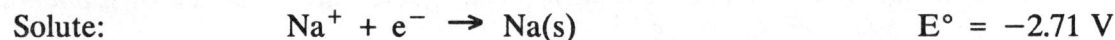

Solute:	$Na^+ + e^- \rightarrow Na(s)$	$E° = -2.71$ V
Solvent:	$2H_2O(l) + 2e^- \rightarrow H_2(g) + 2OH^-$	$E° = -0.83$ V
Electrode:	None (inert)	

As a voltage is applied to this cell, the half-reaction with the **least** negative potential will be predicted to take place. At the anode, a voltage of at least 0.54 V must be applied for I^- to be oxidized, and a higher voltage of at least 1.23 V must be applied to oxidize water. Therefore, at the anode, we would predict that the solute will be oxidized.

Anode (Oxidation)

Solute:	$2I^- \rightarrow I_2(s) + 2e^-$	$E° = -0.54$ V

At the cathode, far less voltage will need to be applied to reduce water (0.83 V) than Na^+ (2.71 V). Therefore, at the cathode, we would predict that the solvent will be reduced.

<u>Cathode (Reduction)</u>

Solvent: $\qquad 2H_2O(l) + 2e^- \rightarrow H_2(g) + 2OH^- \qquad E° = -0.83\ V$

The overall reaction that we would predict is

$$2I^- + 2H_2O(l) \rightarrow I_2(s) + H_2(g) + 2OH^-$$

(b)   In liquid or molten NaI, there are only Na^+ and I^- ions in the liquid state (no solvent). And with inert electrodes, there is only one possible half-reaction at the anode and cathode.

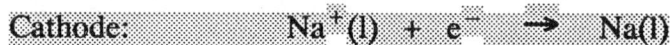

Anode: $\qquad 2I^-(l) \rightarrow I_2(l) + 2e^-$

Cathode: $\qquad Na^+(l) + e^- \rightarrow Na(l)$

Overall: $\qquad 2Na^+(l) + 2I^-(l) \rightarrow 2Na(l) + I_2(l)$

SECTION 19.8   STOICHIOMETRY OF ELECTROCHEMICAL REACTIONS

After completing this section, you will be able to do the following:

■   Given two of the three quantities required to carry out an electrochemical reaction --- mass, current, and time --- calculate the third quantity.

REVIEW

The mass of any substance consumed or deposited at an electrode is directly proportional to the number of Faradays available divided by the number of Faradays required to form or consume the molar mass of the substance

$$\text{mass, g} = \text{\#F available x MM/\#F required,}$$

where the number of Faradays available is the product of the time in seconds times the current in coulombs per second (1 A = 1 C/s) divided by the number of coulombs in one Faraday

$$\text{\#F available} = s \times \frac{\text{\textcrC}}{s} \times \frac{1 \text{ F}}{9.65 \times 10^4 \text{ \textcrC}}$$

The number of Faradays required to form or consume the molar mass of the substance comes from the half-reaction of interest. For example, if nickel is plated out from a Ni^{2+} solution, the half-reaction

$$Ni^{2+} + 2e^- \rightarrow Ni(s)$$

tells us that **1 mol** of nickel metal always requires **2** mol of electrons or **two Faradays** (the charge on one mole of electrons is the definition of one Faraday). Since the molar mass of nickel is 58.69 g/mol Ni, **2 F are always required to produce or consume 1 mole or 58.69 g, of Ni(s).**

Given two of the three quantities involved in an electrolysis cell (current in amps, time, and mass), you should be able to calculate the third quantity.

SAMPLE PROBLEM 19.8 A

▶ ▶ How many grams of magnesium metal can be produced from a Mg^{2+} solution in 1.00 h with a current of 106 A?

Solution:

Using the equation given in the REVIEW

$$\text{mass, g} = \text{\#F available} \times \text{MM}/\text{\#F required}$$

First find the number of Faradays available, then the number of Faradays required to produce the molar mass of magnesium.

Use the second equation in the REVIEW to calculate the #F available for the current and time specified, remembering that 1 A = 1 C/s (which means 106 A = 106 C/s),

$$\text{\#F available} = s \times \frac{\text{\textcrC}}{s} \times \frac{1 \text{ F}}{9.65 \times 10^4 \text{ \textcrC}}$$

$$= 1.00 \text{ h} \times \frac{3600 \text{ s}}{1 \text{ h}} \times \frac{106 \text{ \textcrC}}{s} \times \frac{1 \text{ F}}{9.65 \times 10^4 \text{ \textcrC}} = 3.95 \text{ °F}$$

To find the number of Faradays required to form the molar mass of magnesium, use the atomic mass of magnesium from the inside front cover (24.3050) and the half-reaction for forming magnesium metal,

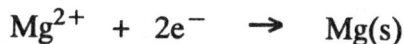

$$Mg^{2+} + 2e^- \rightarrow Mg(s)$$

Two Faradays (2 mol of electrons) are always required to form or consume 1 mol or 24.3050 g of Mg.

Now the information can be substituted into the first equation to calculate the mass of magnesium metal formed.

$$mass, g = \#F \text{ available} \times MM/\#F \text{ required}$$

$$? \text{ g Mg(s)} = 3.95 \text{ F} \times \frac{24.3050 \text{ g Mg}}{2 \text{ F}} = 48.0 \text{ g Mg}$$

CHECK: Does this answer make sense? Yes, because 2 F are required to produce 24.3050 g of Mg metal, so the 3.95 F available will produce almost twice the molar mass, or 48.0 g Mg.

SAMPLE PROBLEM 19.8 B

▶▶ How many hours would it take to produce 25.0 g of chromium metal from a solution of Cr^{3+} with a current of 1.25 A?

Solution:

The number of Faradays that are used (available) for any electrolysis depends on the current and length of time the experiment is run. Since you are looking for time, you will need to find the number of Faradays used (available) first, then the time. The half-reaction of interest in this case is

$$Cr^{3+} + 3e^- \rightarrow Cr(s)$$

which tells us that three Faradays (3 mol of electrons) are always required to form 1 mol or 51.9961 g of Cr metal. From this information and the mass of chromium metal produced, we can calculate the number of Faradays used (available)

$$? \text{ F used} = 25.0 \text{ g Cr} \times \frac{3 \text{ F}}{51.9961 \text{ g Cr}} = 1.44 \text{ F}$$

The time can be found by using the definition of the Faraday in coulombs and the amperage in C/s,

$$? \text{ s } = 1.44 \text{ F } \times \frac{9.65 \times 10^4 \text{ C}}{1 \text{ F}} \times \frac{1 \text{ s}}{1.25 \text{ C}} = 1.11 \times 10^5 \text{ s} \times \frac{1 \text{ h}}{3600 \text{ s}} = 30.9 \text{ h}$$

SELF-TEST QUESTIONS

Test your understanding of the concepts and skills in this chapter by working through the multiple choice questions BEFORE checking the answers. A periodic table will always be available. The ANSWERS TO SELF-TEST QUESTIONS follow immediately after this section.

1. At standard conditions, which choice has the oxidizing agents in order of increasing strength?

    A. $F_2 < Cr_2O_7^{2-} < Zn^{2+} < Ag^+ < H^+$

    B. $H^+ < Cr_2O_7^{2-} < Ag^+ < Zn^{2+} < F_2$

    C. $Zn^{2+} < H^+ < Ag^+ < Cr_2O_7^{2-} < F_2$

    D. $Ag^+ < Zn^{2+} < F_2 < Cr_2O_7^{2-} < H^+$

    E. $F_2 < Cr_2O_7^{2-} < Ag^+ < H^+ < Zn^{2+}$

2. Which of the following reactions in aqueous solution under standard conditions is spontaneous?

    A. $Br_2(l) + 2Cl^- \rightarrow 2Br^- + Cl_2(g)$

    B. $Cu^{2+} + 2Ag(s) \rightarrow Cu(s) + 2Ag^+$

    C. $3Fe^{2+} + Cr^{3+} \rightarrow 3Fe^{3+} + Cr(s)$

    D. $Al(s) + NO_3^- + 4H^+ \rightarrow Al^{3+} + NO(g) + 2H_2O(l)$

    E. More than one of the above is correct.

3. At standard conditions, the strongest oxidizing agent is

A. Cr^{3+} B. Cl_2 C. Ag^+ D. Fe^{2+} E. Mg^{2+}

4. $E°_{cell}$ for the following reaction is

A. 1.31 V   B. 0.30 V   C. 0.060 V   D. 1.18 V   E. 0.46 V

5. For the following voltaic cell, what is the potential, E, at 25 °C if $[NO_3^-] = [H^+] = [Cu^{2+}] = 0.0100$ M and NO(g) = 0.0100 atm?

$$2NO_3^- + 8H^+ + 3Cu(s) \rightarrow 3Cu^{2+} + 2NO(g) + 4H_2O(l)$$

A. 0.52 V   B. 0.57 V   C. 0.03 V   D. 0.72 V   E. 6.1 V

6. A solution is 1.0 M in each of the following ions: Zn^{2+}, Ag^+, and Cu^{2+}. The correct order in which they will be plated out on the cathode (first on left; last on right) as the solution is electrolyzed will be

A. Ag, Cu, Zn
B. Zn, Ag, Cu
C. Zn, Cu, Ag
D. Ag, Zn, Cu
E. Cu, Ag, Zn

7. If pure liquid lanthanum bromide, $LaBr_3$, is electrolyzed with inert electrodes, the products will be

	Anode	Cathode
A.	$O_2(g)$	La(l)
B.	$Br_2(l)$	La(l)
C.	$O_2(g)$	$H_2(g)$
D.	La(l)	$Br_2(l)$
E.	$H_2(g)$	$Br_2(l)$

8. The half-reaction that will occur at the inert anode in the electrolysis of an aqueous 1.0 M $NiSO_4$ solution is

A. $Ni^{2+} + 2e^- \rightarrow Ni(s)$

B. $2H_2O(l) + 2e^- \rightarrow H_2(g) + 2OH^-$

C. $H_2(g) \rightarrow 2H^+ + 2e^-$

D. $2H_2O(l) \rightarrow O_2(g) + 4H^+ + 4e^-$

E. $Ni(s) \rightarrow Ni^{2+} + 2e^-$

9. The mass of nickel that can be electroplated from a solution of $NiSO_4$ by using a current of 0.200 A for 30.0 m will be

A. 1.82×10^{-3} g   B. 0.218 g   C. 0.109 g   D. 0.436 g   E. 36.0 g

10. How long would it take to electroplate 0.100 g of gold on a jewelry item used as the cathode in an electrolysis cell containing $AuCl_4^-$ if a current of 0.500 A is used?

A. 33 s   B. 57 900 s   C. 98 s   D. 73 s   E. 294 s

ANSWERS TO SELF-TEST QUESTIONS

1. c	4. b	7. b
2. d	5. a	8. d
3. b	6. a	9. c
		10. e

PRACTICE PROBLEMS

Test your problem-solving skills in this chapter by working through the problems in the space provided **BEFORE** checking the answers. A periodic table will always be available. The ANSWERS TO PRACTICE PROBLEMS follow immediately after this section.

1. Make a cell diagram for the voltaic cell that uses the spontaneous reaction between calcium metal and iron(III) nitrate solution to form metallic iron and calcium nitrate solution.

2. Which species in each pair is the stronger reducing agent, under standard conditions?

    (a) $Cu(s)$ or $Zn(s)$

    (b) $Br^-(aq)$ or $Ag(s)$

    (c) $Fe(s)$ or $Cr(s)$

    (d) $Ag(s)$ or $Cu(s)$

3. Calculate $E°_{cell}$ for the following reactions

    (a) $3H_2O_2(aq) + 2Cr^{3+} + H_2O(l) \rightarrow Cr_2O_7^{2-} + 8H^+$

    (b) $3MnO_4^- + 4H^+ + 5NO(g) \rightarrow 3Mn^{2+} + 2H_2O(l) + 5NO_3^-$

    (c) $3Cu^{2+} + 2NO(g) + 4H_2O(l) \rightarrow 3Cu(s) + 2NO_3^- + 8H^+$

4. Which species in each pair is the stronger oxidizing agent, under standard conditions?

    (a) $NO_3^-(aq)$ or $Br_2(l)$

    (b) $Cu^{2+}(aq)$ or $Ca^{2+}(aq)$

    (c) $Cl_2(g)$ or $Cr_2O_7^{2-}(aq)$

    (d) $I_2(s)$ or $Fe^{2+}(aq)$

548

5.   What is the potential, E, of the following cell at 25 °C?

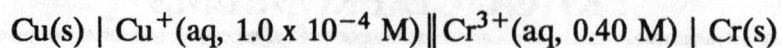

$$Cu(s) \mid Cu^+(aq, 1.0 \times 10^{-4} \text{ M}) \| Cr^{3+}(aq, 0.40 \text{ M}) \mid Cr(s)$$

6.   What is the potential, E, of the following cell at 25 °C?

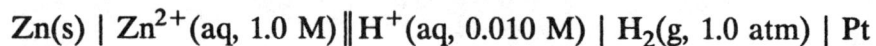

$$Zn(s) \mid Zn^{2+}(aq, 1.0 \text{ M}) \| H^+(aq, 0.010 \text{ M}) \mid H_2(g, 1.0 \text{ atm}) \mid Pt$$

7.   For the following reaction at equilibrium, calculate (a) E°, (b) ΔG°, (c) K, and (d) indicate whether the reaction is spontaneous or not.

$$2H_2O(l) + 2Cl_2(g) \rightarrow O_2(g) + 4Cl^- + 4H^+$$

8. For the following reaction at equilibrium, calculate (a) E°, (b) $\Delta G°$, (c) K, and (d) indicate whether the reaction is spontaneous or not. For $Ni^{2+} + 2e^- \rightarrow Ni(s)$, E° = -0.23 V.

$$2Cr^{2+} + Ni^{2+} \rightarrow 2Cr^{3+} + Ni(s)$$

9. For the following reaction at equilibrium, calculate (a) E°, (b) $\Delta G°$, (c) K, and (d) indicate whether the reaction is spontaneous or not. For $Au^{3+} + 3e^- \rightarrow Au(s)$, E° = 1.50 V.

$$2Au(s) + 3Cl_2(g) \rightarrow 2Au^{3+} + 6Cl^-$$

10. How many hours would it take to produce 7.50 g of Pt(s) from a solution of Pt^{2+} using a current of 0.300 A?

550

11. How many grams of iron can be produced from an Fe^{2+} solution in 10.0 h with a current of 1.44 A?

ANSWERS TO PRACTICE PROBLEMS

1. $Ca(s) \mid Ca^{2+}(aq) \parallel Fe^{3+}(aq) \mid Fe(s)$

2. (a) Zn   (b) Ag   (c) Cr   (d) Cu

3. (a) 0.38 V   (b) 0.55 V   (c) −0.62 V

4. (a) Br_2 (b) Cu^{2+} (c) Cl_2 (d) I_2

5. $3Cu(s) + Cr^{3+} \rightarrow 3Cu^+ + Cr(s)$

$E° = -1.26$ V; n = 3; Q = 2.5×10^{-12}; E = −1.03 V

6. $Zn(s) + 2H^+ \rightarrow Zn^{2+} + H_2(g)$

$E° = 0.76$ V; n = 2; Q = 1.0×10^4; E = 0.64 V

7. (a) $E° = 0.13$ V
(b) $\Delta G° = -5.0 \times 10^4$ J $= -5.0 \times 10^1$ kJ
(c) $K = 6 \times 10^8$
(d) Spontaneous ($E°$ is +; $\Delta G°$ is −)

8. (a) $E° = 0.19$ V
    (b) $\Delta G° = -3.7 \times 10^4$ J $= -3.7 \times 10^1$ kJ
    (c) $K = 3 \times 10^6$
    (d) Spontaneous ($E°$ is $+$; $\Delta G°$ is $-$)

9. (a) $E° = -0.14$ V
    (b) $\Delta G° = 8.1 \times 10^4$ J $= 8.1 \times 10^1$ kJ
    (c) $K = 6 \times 10^{-15}$
    (d) Non-spontaneous ($E°$ is $-$; $\Delta G°$ is $+$)

10. 6.87 h

11. 15.0 g Fe

CHAPTER 20

NUCLEAR CHEMISTRY

SECTION 20.1 RADIOACTIVE DECAY PROCESSES

After completing this section, you will be able to do the following:

■ Complete and balance nuclear reactions.

REVIEW

The chemical reactions studied so far involve a rearrangement of the valence electrons (transfer or sharing) around the nucleus of an atom, with **no change** in the arrangement or number of protons and neutrons in the nucleus. The area of chemistry that deals with a change in the number of protons and/or neutrons in the nucleus is called **nuclear chemistry**. The change in the number of protons and/or neutrons within the nucleus of an atom is independent of whether the atom is chemically bonded to other atoms in a compound or an atom of a free element. Three types of rays are given off when nuclei undergo radioactive decay: **alpha rays**, which are He^{2+} ions; **beta rays**, which are electrons from **inside the nucleus**; and **gamma rays**, which have even greater energy and penetrating power than x-rays. In addition, a **positron** or positive electron, a **proton** or hydrogen ion, and a neutron are additional emission products in a number of nuclear reactions. A summary of these major decay emissions is given in the following table, where the **subscript** indicates the atomic number or nuclear charge and the **superscript** is the mass number or number or protons and neutrons (see Section 2.11 for interpretation of sub- and superscripts).

Emission	Symbols	
Alpha ray	α	^4_2He
Beta ray	β^-	$^0_{-1}\text{e}$
Gamma ray	γ	$^0_0\gamma$
Positron	β^+	$^0_{+1}\text{e}$
Proton	H	^1_1H
Neutron	n	^1_0n

Nuclear reactions, like chemical reactions, are described by balanced reactions or equations. In **nuclear equations**, the total of the mass numbers (superscripts) must be equal on each side of the reaction, and the total of the nuclear charges (subscripts) must also be equal on each side of the reaction. In nuclear equations (unlike chemical equations), the number of atoms of each element is not necessarily the same on each side of the reaction; every time the nuclear charge (subscript) changes, so does the identity of the element. [Remember, the subscript in this notation is the unique atomic number or nuclear charge for each element or nuclear particle.] You will need to know the symbols with sub- and superscripts for the nuclear emissions shown in the table above because you will be expected to **complete** nuclear reactions by balancing the sub- and superscripts.

SAMPLE PROBLEM 20.1 A

▶ ▶ Write balanced nuclear reactions for the following, including subscript, superscript, and symbol for the missing species.

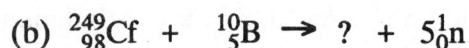

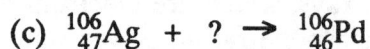

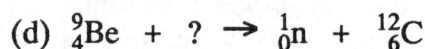

(a) $^{81}_{37}\text{Rb} \rightarrow \; ? \; + \; ^0_1\text{e}$

(b) $^{249}_{98}\text{Cf} \; + \; ^{10}_5\text{B} \rightarrow \; ? \; + \; 5^1_0\text{n}$

(c) $^{106}_{47}\text{Ag} \; + \; ? \rightarrow \; ^{106}_{46}\text{Pd}$

(d) $^9_4\text{Be} \; + \; ? \rightarrow \; ^1_0\text{n} \; + \; ^{12}_6\text{C}$

(e) Proton emission by $^{18}_9\text{F}$

(f) $^{209}_{83}\text{Bi} \; + \; ^4_2\text{He} \rightarrow \; ^{210}_{85}\text{At} \; + \; ?$

Solution:

HINT: Calculate the subscript and superscript of the unknown species; determine the identity of the unknown species from the subscript (and superscript for positron).

(a)   Superscripts: 81 on left = 81 on right. Unknown superscript is 81.
            Subscripts: 37 on left = 37 on right. Unknown subscript must be 36.
            Identity: A subscript of 36 is the atomic number of Kr.

Equation: $^{81}_{37}Rb \rightarrow {}^{81}_{36}Kr + {}^{0}_{1}e$

(b)   Superscripts: 259 on left = 259 on right. Unknown superscript is 254.
            Subscripts: 103 on left = 103 on right. Unknown subscript must be 103.
            Identity: A subscript of 103 is the atomic number of Lr.

Equation: $^{249}_{98}Cf + {}^{10}_{5}B \rightarrow {}^{254}_{103}Lr + 5{}^{1}_{0}n$

(c)   Superscripts: 106 on left = 106 on right. Unknown superscript is 0.
            Subscripts: 46 on left = 46 on right. Unknown subscript must be -1.
            Identity: A subscript of -1 can only be an electron.

Equation: $^{106}_{47}Ag + {}^{0}_{-1}e \rightarrow {}^{106}_{46}Pd$

(d)   Superscripts: 13 on left = 13 on right. Unknown superscript is 4.
            Subscripts: 6 on left = 6 on right. Unknown subscript must be 2.
            Identity: A subscript of 2 with a nonzero superscript can only be the element He.

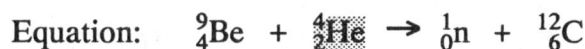

Equation: $^{9}_{4}Be + {}^{4}_{2}He \rightarrow {}^{1}_{0}n + {}^{12}_{6}C$

(e)   Since a proton has the symbol $^{1}_{1}H$,

            Superscripts: 18 on left = 18 on right. Unknown superscript is 17.
            Subscripts: 9 on left = 9 on right. Unknown subscript must be 8.
            Identity: A subscript of 8 is the atomic number of O.

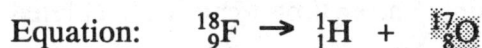

Equation: $^{18}_{9}F \rightarrow {}^{1}_{1}H + {}^{17}_{8}O$

(f)   Superscripts: 213 on left = 213 on right. Unknown superscript is 3.
            Subscripts: 85 on left = 85 on right. Unknown subscript must be 0.
            Identity: A subscript of 0 with a nonzero superscript can only be a neutron; the only way the superscript can be 3 is if there are 3 neutrons.

Equation: $^{209}_{83}Bi + ^{4}_{2}He \rightarrow ^{210}_{85}At + 3^{1}_{0}n$

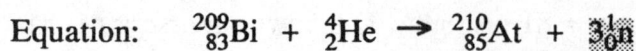

NOTE: You must have a periodic table handy and know the symbols for the major nuclear emissions given in the REVIEW in order to write nuclear equations.

SECTION 20.3   RATES OF NUCLEAR REACTIONS

After completing this section, you will be able to do the following:

■   From the half-life for a radionuclide, calculate the quantity of radionuclide remaining after a given amount of time.

REVIEW

All radioactive decay processes are **first-order** kinetics, for which the rate law, integrated rate law, and half-life equations are (Section 18.4)

First Order Equations		
Rate law	Integrated rate law	Half-life
rate = kN	$\ln \dfrac{N_t}{N_0} = -kt$	$t_{1/2} = \dfrac{0.693}{k}$

where k is the rate constant, N_0 is the amount of radionuclide present at time 0, and N_t is the amount of radionuclide left after time t. As long as the same units are used for both N_t and N_0, N can be expressed in a variety of units, including atoms, counts, mass, moles, and molarity. The number of disintegrations per second for a radionuclide can be calculated from the half-life and rate law (Sample Problem 20.2 in the textbook). The amount of a radionuclide remaining after time t can be calculated from the half-life and integrated rate law (Sample Problem 20.3 A).

SAMPLE PROBLEM 20.3 A

▶▶ The half-life for the radioactive decay of copper-64 is 12.70 hours. Calculate the amount in grams of copper-64 that will be left from a 2.40 g sample after (a) 25.40 h (b) 8.35 h.

Solution:

(a)   In any half-life problem, it is worth a few seconds to see if the time for the decay is a whole-numbered multiple of the half-life or not. If it is, then the problem is much shorter. For example, in this case 25.40 h is exactly two half-lives (2 x 12.70 h). The fraction of any nuclide remaining after n half-lives is

$$\text{fraction remaining} = (1/2)^n = (1/2)^2 = 1/4$$

After two half-lives, 1/4 of the original amount will remain.

$$?\ \text{g copper-64 remaining} = (1/4) \times 2.40\ \text{g Cu-64} = 0.60\ \text{g Cu-64}$$

If the number of half-lives is not a whole number, then the integrated rate law must be used as in part (b).

(b)   A decay time of 8.35 hours is not even one half-life, so the first-order integrated rate law must be used, which requires the rate constant for the radionuclide. Start by using the first order half-life equation to find k, then use the integrated rate law to find the amount remaining after 8.35 h. The rate constant for copper-64 is

$$k = \frac{0.693}{t_{1/2}} = \frac{0.693}{12.70\ \text{h}} = 0.0546\ \text{h}^{-1}$$

Substituting $k = 0.0546$, $N_0 = 2.40$ g, and $t = 8.35$ h into the integrated rate law, solve the equation for N, the number of grams of radioactive copper-64 remaining after 8.35 hours

$$\ln \frac{N_t}{2.40\ \text{g}} = -(0.0546\ \text{h}^{-1})(8.35\ \text{h}) = -0.456$$

Taking the antilog of both sides of the equation

$$\frac{N_t}{2.40\ \text{g}} = 0.634$$

or the fraction (N_t/N_0) of radioactive Cu-64 remaining undecayed after 8.35 hours is 0.6340 or 63.40%. Solving for N_t, the number of grams of undecayed Cu-64 remaining after 8.35 hours,

$$N_t = 1.52\ \text{g Cu-64}$$

CHECK: Does this answer make sense? Yes, because it would take one half-life (12.70 h) for the initial amount to be reduced to 1/2 (1.20 g); since 8.35 hours is not even one half-life, then more than 1/2 (0.6340) of the original amount will be left undecayed (1.52 g).

SECTION 20.4 PREDICTING WHICH TYPE OF RADIOACTIVE DECAY WILL TAKE PLACE

After completing this section, you will be able to do the following:

■ Predict whether a nuclide is stable or unstable;

■ For unstable nuclides, predict the most likely method of radioactive decay.

REVIEW

Whether a nuclide is stable (non-radioactive) or unstable (radioactive) can often be predicted from the ratio of the number of neutrons to the number of protons in the nucleus, n/p. A graph of the number of neutrons vs. the number of protons for all known nuclides (Figure 20.7 in the textbook), reveals a band of stable nuclides (or **band of stability**) in which the neutron to proton ratio, n/p, varies from ~1.0 (for atomic numbers up to 20) to 1.52 (for atomic numbers between 20 and 83). In other words, up to atomic number 20, stable nuclides have approximately one neutron in the nucleus for each proton. Between atomic numbers 20 and 83, more than one neutron per proton is required to maintain stability. **All nuclides above Z = 83 are unstable (radioactive)**, no matter how many neutrons per proton are present.

Nuclides that have a n/p ratio either higher or lower than the stable n/p ratio are radioactive and undergo spontaneous decay until a stable (nonradioactive) n/p ratio is obtained. Up to Z = 84, if the n/p ratio is **above the band of stable nuclides**, stability is achieved by **beta emission**, which **lowers** the n/p ratio; if the n/p ratio is **below the band of stable nuclides**, stability is achieved by **positron emission** (lower Z) and/or **electron capture** (high Z), both of which **increase** the n/p ratio. Of all the stable nuclides, those with an **even** number of protons (atomic number) **and** neutrons are the most numerous; those nuclides with either an **even** number of protons **or** neutrons are the next most abundant group, and the least abundant stable nuclides are those with both an odd number of protons and neutrons.

SAMPLE PROBLEM 20.4 A

▶▶ Without referring to Figure 20.7 in the textbook, predict which of the following nuclides are probably stable or unstable (radioactive). For any radioactive nuclides, predict the type of decay to achieve stability is most likely.

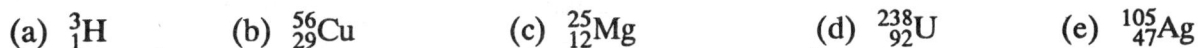

(a) ^3_1H (b) $^{56}_{29}Cu$ (c) $^{25}_{12}Mg$ (d) $^{238}_{92}U$ (e) $^{105}_{47}Ag$

Solution:

(a) **Unstable.** Hydrogen-3 has an atomic mass considerably higher than naturally occurring hydrogen (1.00794) and a n/p ratio of 2.0; this means it contains too many neutrons. Hydrogen-3 must lie above the band of stability and will decay by beta emission.

(b) **Unstable.** Copper-56 has an atomic mass considerably lower than naturally occurring copper (63.546) and a n/p ratio less than 1; this means it contains too few neutrons. Copper-56 must lie below the band of stability and, since it has a lower atomic number, will probably decay by positron emission.

(c) **Stable.** Magnesium-25 has an atomic mass very close to that of naturally occurring magnesium (24.3050), and the n/p ratio is slightly over 1.

(d) **Unstable.** All nuclides with Z > 83 are unstable (radioactive). Uranium-238 is beyond the band of stability and will probably decay by alpha emission.

(e) **Unstable.** Silver-105 has an atomic mass less than that of naturally occurring silver (107.8682); this means it contains too few neutrons. Silver-105 must lie below the band of stability, and since it has a higher atomic number, it will probably decay by electron capture.

SECTION 20.6   ENERGY CHANGES ACCOMPANYING NUCLEAR REACTIONS

After completing this section, you will be able to do the following:

■ Calculate the energy change, ΔE, for a nuclear reaction;

■ Calculate the mass defect, total binding energy, and binding energy per nucleon for a nuclide.

REVIEW

The Law of Conservation of Mass and Energy says that matter cannot be created or destroyed, only changed from one form to another. The conversion of mass to energy and vice versa is quantified by Einstein's famous equation, $E = mc^2$. As usual, we are most interested in the change in total energy, ΔE, for a nuclear reaction, which is calculated from a more useful form of Einstein's equation,

$$\Delta E = \Delta mc^2$$

where ΔE is the energy change in joules, J, if the change in mass, Δm, is in kg and the speed of light, c, is in $m \cdot s^{-1}$ (recall $1\ J = 1\ kg \cdot m^2 \cdot s^{-2}$). The conventions for ΔE are the same as for the other thermodynamic quantities. Nuclear reactions with the mass of the products less than the mass of the reactants will have a **negative** Δm and a **negative** ΔE, and the reaction **spontaneously** converts mass into energy. Similarly, if the mass of the products in a nuclear reaction is greater than the mass of the reactants, Δm and ΔE will be positive and non-spontaneous. Given a nuclear reaction, you should be able to calculate the energy change, ΔE (Sample Problem 20.6 A).

When isolated **nucleons** (protons + neutrons) combine to form a nucleus, the mass of the nucleus is always **smaller** than the sum of the masses of the individual **nucleons** that compose the nucleus (Δm is $-$). This difference between the observed mass of a nucleus and the sum of the masses of the individual nucleons is called the **mass defect** because each nucleon loses a small, but measurable, amount of mass when the nucleons combine to form a nucleus. The **mass defect** is converted into energy, ΔE, which can be calculated from Einstein's equation. The energy required to break a nucleus apart into its nucleons is called the **nuclear binding energy** of the nucleus. That is, when a nucleus is formed from individual nucleons, energy is released (Δm, mass defect, and ΔE are both negative). The energy change for the breaking apart of a nucleus (the reverse of forming the nucleus) is $-\Delta E$, which more specifically is called the nuclear binding energy. Finally, the **binding energy per nucleon** is the nuclear binding energy for the nucleus divided by the number of nucleons in the nucleus (Sample Problem 20.6 B).

SAMPLE PROBLEM 20.6 A

▶ ▶ For the following reaction

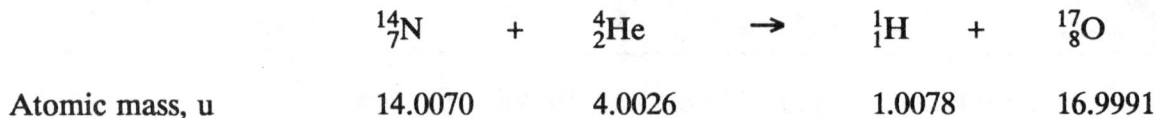

$$^{14}_{7}N\ +\ ^{4}_{2}He\ \rightarrow\ ^{1}_{1}H\ +\ ^{17}_{8}O$$

	$^{14}_{7}N$	$^{4}_{2}He$	$^{1}_{1}H$	$^{17}_{8}O$
Atomic mass, u	14.0070	4.0026	1.0078	16.9991

calculate (a) the change in mass, Δm, and (b) the energy change, ΔE, in kJ/mol.

Solution:

(a) The change in mass (per nucleus), Δm, is the atomic masses of the products minus the atomic masses of the reactants,

$$\Delta m = m_p - m_r = [(1.0078) + (16.9991)] - [(14.0070) + (4.0026)]$$

$$= [18.0069] - [18.0096] = -0.0027 \text{ u}$$

This reaction spontaneously loses 0.0027 u when one nitrogen-14 atom reacts with one helium-4 atom to form one hydrogen-1 atom and one oxygen-17 atom.

(b) The energy change for this reaction, ΔE, is calculated from $\Delta E = \Delta mc^2$. To obtain the energy change in units of kJ/mol, you can save yourself considerable effort by noting carefully that a mass of one u per atom is numerically equivalent to a mass of one g per mol of atoms. That is,

$$1 \text{ u/atom} = 1 \text{ g/mol of atoms}$$

because there are Avogadro's number of u's in 1 g and Avogadro's number of atoms in 1 mol of atoms. Thus, for this problem,

$$\Delta m = -0.0027 \text{ u/atom} = -0.0027 \text{ g/mol of atoms}$$

Substituting $\Delta m = -0.0027$ g/mol and $c = 3.00 \times 10^8$ m•s^{-1} into Einstein's equation and converting g to kg (1.00 g = 0.001 kg) to obtain the energy change in J and kJ, we have

$$\Delta E = \Delta mc^2 = (-0.0027 \text{ g•mol}^{-1})(1.00 \times 10^{-3} \text{ kg•g}^{-1})(3.00 \times 10^8 \text{ m•s}^{-1})^2$$

$$= -2.4 \times 10^{11} \text{ kg•m}^2\text{•s}^{-2}\text{•mol}^{-1} = -2.4 \times 10^{11} \text{ J/mol}$$

$$= -2.4 \times 10^8 \text{ kJ/mol}$$

NOTE: Remember that $1 \text{ kg•m}^2\text{•s}^{-2} = 1$ J.

SAMPLE PROBLEM 20.6 B

►► If the atomic mass of $^{65}_{29}$Cu is 64.9278 u, calculate for this nuclide the (a) mass defect, (b) total binding energy in J, and (c) binding energy per nucleon in J.

Solution:

(a) You will find it easiest to write out a nuclear reaction for the formation of a copper-65 nucleus and calculate Δm (just like in Sample Problem 20.6 A), which is the mass defect. A copper-65 nucleus contains 29 protons and (65 − 29) 36 neutrons. The atomic mass of hydrogen-1, 1.007 825 u, must be used for the proton, and the mass of the neutron, 1.008 665 u, is found on the inside back cover. The nuclear reaction and masses are

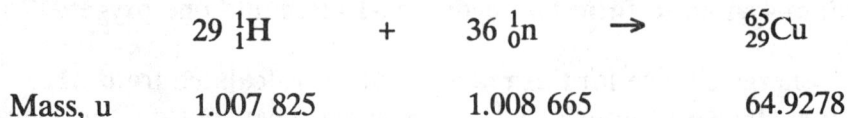

$$29\,{}^{1}_{1}\text{H} \quad + \quad 36\,{}^{1}_{0}\text{n} \quad \rightarrow \quad {}^{65}_{29}\text{Cu}$$

Mass, u   1.007 825   1.008 665   64.9278

The value of Δm, the mass defect, for this nuclear reaction is

$$\Delta m = m_p - m_r = [64.9278] - [29(1.007\ 825) + 36(1.008\ 665)]$$

$$= [64.9278] - [65.538\ 865] = -0.6111\ \text{u}$$

That is, 65 isolated nucleons will lose a total of 0.6111 u when they combine to form a copper-65 nucleus. This mass loss is converted into energy in part (b).

(b) Calculate ΔE for this reaction from Einstein's equation, which will be the energy released when one copper-65 nucleus is formed. Changing the sign of ΔE will give you the energy required to break apart the nucleus (the reverse of the above reaction), which is the **total binding energy**. Because you will be calculating ΔE in J/nucleus, (not J/mol of nuclei), you cannot use the shortcut that we took in Sample Problem 20.6 A. Using Δm = −0.6111 u•nucleus^{-1}, c = 3.00 x 10^8 m•s^{-1}, and the conversion factor for changing mass in u to kg found on the inside back cover

$$1\ \text{u} = 1.660\ 56 \times 10^{-27}\ \text{kg}$$

substitute these values into Einstein's equation to solve for ΔE in J/nucleus

$$\Delta E = \Delta mc^2 = (-0.6111\ \text{u•nucleus}^{-1})(1.660\ 56 \times 10^{-27}\ \text{kg•u}^{-1})(3.00 \times 10^8\ \text{m•s}^{-1})^2$$

$$= -9.13 \times 10^{-11}\ \text{kg•m}^2\text{•s}^{-2}\text{•nucleus}^{-1} = -9.13 \times 10^{-11}\ \text{J/nucleus}$$

The total binding energy = +9.13 x 10^{-11} J/nucleus. That is, it takes 9.13 x 10^{-11} J to break apart one copper-65 nucleus into 29 separate protons and 36 separate neutrons.

(c) The binding energy (BE) per nucleon in J/nucleon, is easily found by dividing the total binding energy by the number of nucleons in copper-65

$$? \ \frac{BE}{nucleon} \ = \ \frac{9.13 \times 10^{-11} \ \text{J}/\text{nucleus}}{65 \ \text{nucleons}/\text{nucleus}} \ = \ 1.40 \times 10^{-12} \ \text{J}/\text{nucleon}$$

SECTION 20.10 USES OF RADIONUCLIDES

After completing this section, you will be able to do the following:

- Calculate the age of an object by means of carbon-14 dating.

REVIEW

The decay of radioactive carbon-14 can be used to date objects containing once-living material. The rate of constantly forming carbon-14 equals the rate of decay of carbon-14, producing a steady-state level of radioactivity of 14.9 disintegrations per minute per gram of total carbon, whether it be in the atmosphere or in the food we eat. When a plant or animal dies, it no longer takes in carbon-14, but the carbon-14 the plant or animal took in when alive continues to decay and the radioactivity due to carbon-14 decreases. From a comparison of the radioactivity of the carbon-14 in an ancient object with the radioactivity of carbon-14 at the steady state (14.9 disintegrations/min/g C) and a knowledge of the half-life of carbon-14 ($t_{1/2}$ = 5730 years), the age of the object can be determined. Since all nuclear decay processes are first-order kinetics, the first-order integrated rate law contains the necessary amount-time relationship for calculating the age of the object by carbon-14 dating.

SAMPLE PROBLEM 20.10 A

▶▶ A buried bone was found to have a carbon-14 radioactivity level that is 30.0% of that in living matter. If the half-life of carbon-14 is 5730 years, how old is the bone?

Solution:

The first order integrated rate law for decay processes contains the time-amount relationship we are concerned with.

$$\ln \frac{[N]_t}{[N]_0} \ = \ -kt$$

Using the first-order half-life equation to find the rate constant, k,

$$k = \frac{0.693}{t_{1/2}} = \frac{0.693}{5730 \text{ y}} = 1.219 \times 10^{-4} \text{ y}^{-1}$$

The ratio N_t/N_0 is the fraction of the activity level remaining after time t, or 0.300. Substituting this information into the integrated rate law, we can solve for the time t

$$\ln \frac{[N]_t}{[N]_0} = \ln 0.300 = -(1.219 \times 10^{-4} \text{ y}^{-1})t$$

$$t = 9.95 \times 10^3 \text{ y}$$

NOTE: Alternatively, you can calculate the values of N_0 and N_t, since the steady-state activity level in carbon-14 is 14.9 disintegrations/min/g C.   Thus, N_0 = 14.9 disintegrations/min/g C, and N_t = 4.47 disintegrations/min/g C, which is 30.0% of the steady-state level.

PRACTICE PROBLEMS

Test your problem-solving skills in this chapter by working through the problems in the space provided BEFORE checking the answers.   A periodic table will always be available.   The ANSWERS TO PRACTICE PROBLEMS follow immediately after this section.

1.   Write balanced nuclear reactions for the following, including subscript, superscript, and symbol.

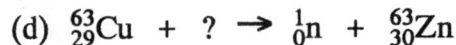

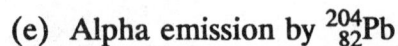

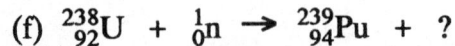

(a) $? \rightarrow {}^{4}_{2}\text{He} + {}^{204}_{85}\text{At}$

(b) ${}^{253}_{99}\text{Es} + {}^{4}_{2}\text{He} \rightarrow ? + {}^{1}_{0}\text{n}$

(c) ${}^{15}_{8}\text{O} \rightarrow ? + {}^{15}_{7}\text{N}$

(d) ${}^{63}_{29}\text{Cu} + ? \rightarrow {}^{1}_{0}\text{n} + {}^{63}_{30}\text{Zn}$

(e)   Alpha emission by ${}^{204}_{82}\text{Pb}$

(f) ${}^{238}_{92}\text{U} + {}^{1}_{0}\text{n} \rightarrow {}^{239}_{94}\text{Pu} + ?$

2. How many grams of calcium-47 ($t_{1/2}$ = 4.536 d) will be left from a 6.82 g sample after (a) 72.00 h? (b) exactly one week?

3. How many grams of cesium-137 ($t_{1/2}$ = 30.17 years) will be left from a 24.75 g sample after 100.0 years?

4. For the following reaction

$$^{6}_{3}Li \quad + \quad ^{1}_{1}H \quad \rightarrow \quad ^{3}_{2}He \quad + \quad ^{4}_{2}He$$

| Atomic Mass, u | 6.015 12 | 1.007 82 | 3.016 03 | 4.002 60 |

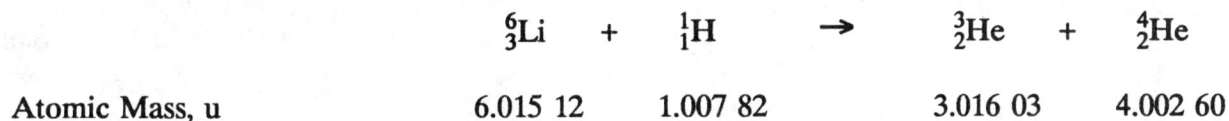

calculate (a) the change in mass, Δm, and (b) the energy change, ΔE, in kJ/mol.

5. For the following reaction

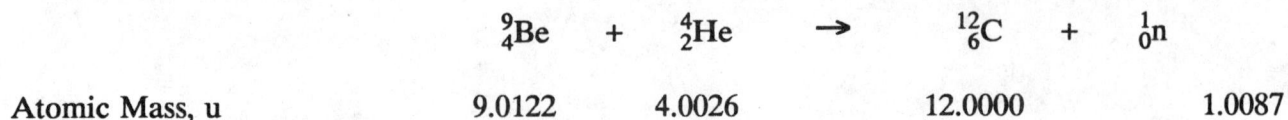

$$\text{}^{9}_{4}\text{Be} \quad + \quad \text{}^{4}_{2}\text{He} \quad \longrightarrow \quad \text{}^{12}_{6}\text{C} \quad + \quad \text{}^{1}_{0}\text{n}$$

Atomic Mass, u   9.0122   4.0026   12.0000   1.0087

calculate (a) the change in mass, Δm, and (b) the energy change, ΔE, in kJ/mol.

6. If the atomic mass of $^{35}_{17}\text{Cl}$ is 34.9689 u, calculate for this nuclide the (a) mass defect, (b) total binding energy in J, and (c) binding energy per nucleon in J. [For the mass of the proton, use the atomic mass of hydrogen-1, 1.007 825 u; see the inside back cover for the mass of the neutron.]

7. If the atomic mass of $^{48}_{22}\text{Ti}$ is 47.9480 u, calculate for this nuclide the (a) mass defect, (b) total binding energy in J, and (c) binding energy per nucleon in J. [For the mass of the proton, use the atomic mass of hydrogen-1, 1.007 825 u; see the inside back cover for the mass of the neutron.]

8. The carbon in an object uncovered in an archaeological expedition has an activity level of 1.7 disintegrations/min/g C. How old is the object?   [For carbon-14, $t_{1/2}$ = 5730 y.]

ANSWERS TO PRACTICE PROBLEMS

1. (a) $^{208}_{87}Fr$ (b) $^{256}_{101}Md$ (c) $^{0}_{1}e$

   (d) $^{1}_{1}H$ (e) $^{204}_{82}Pb \rightarrow ^{4}_{2}He + ^{200}_{80}Hg$ (f) $2\,^{0}_{-1}e$

2. (a) k = 0.1528 d^{-1}; N_t = 4.31 g

   (b) N_t = 2.34 g

3. k = 0.02297 y^{-1}; N_t = 2.49 g

4. (a) Δm = −0.00431 u/atom = −0.00431 g/mol

   (b) ΔE = −3.88 x 10^8 kJ/mol

5. (a) Δm = −0.0061 u/atom = −0.0061 g/mol

   (b) ΔE = −5.5 x 10^8 kJ/mol

6. (a) Δm = −0.3201 u

   (b) ΔE = −4.784 x 10^{-11} J/nucleus;   BE = +4.784 x 10^{-11} J/nucleus

   (c) BE/nucleon = 1.367 x 10^{-12} J/nucleon

7. (a) $\Delta m = -0.4494$ u

    (b) $\Delta E = -6.716 \times 10^{-11}$ J/nucleus; BE $= +6.716 \times 10^{-11}$ J/nucleus

    (c) BE/nucleon $= 1.399 \times 10^{-12}$ J/nucleon

8. $N_0 = 14.9$ disintegrations/min/g C; $N_t = 1.7$ disintegrations/min/g C; $k = 1.21 \times 10^{-4}$ y^{-1}

    $t = 1.80 \times 10^4$ y

CHAPTER 21

A CLOSER LOOK AT HYDROGEN AND ITS COMPOUNDS

SECTION 21.3   PREPARATION OF HYDROGEN

After completing this section, you will be able to do the following:

■ Write equations for the preparation of hydrogen gas.

REVIEW

Industrially, hydrogen gas is produced by the

(a) reaction of hydrocarbons (contain only C and H) with steam, e.g.,

$$C_2H_6(g) \ + \ 2H_2O(g) \ \rightarrow \ 5H_2(g) \ + \ 2CO(g)$$

(b) electrolysis of aqueous sodium chloride

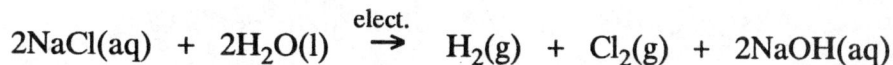

$$2NaCl(aq) \ + \ 2H_2O(l) \ \xrightarrow{elect.} \ H_2(g) \ + \ Cl_2(g) \ + \ 2NaOH(aq)$$

Commercially, large-scale production of a chemical is economically feasible only if there is a cheap source of raw materials (reactants) and a high output in a reasonable amount of time owing to the high cost of manufacturing equipment, storage, and transportation. In addition, if more than one commercial product is obtained (as is the case for reaction (b)), then the process is even more profitable.

In the laboratory, a simpler apparatus and slightly more expensive chemicals are used to illustrate the preparation of various chemicals. In the case of preparing hydrogen gas, usually a moderately active metal (e.g., Zn, Fe, Al, Sn, etc.) and dilute HCl are used,

$$Sn(s) + 2HCl(aq) \rightarrow SnCl_2(aq) + H_2(g)$$

You should be able to write several equations for the preparation of hydrogen gas (Sample Problem 21.3 A).

SAMPLE PROBLEM 21.3 A

▶▶ (a) Write one equation each for the preparation of hydrogen gas industrially and in the laboratory. (b) If both equations are exothermic, will a higher yield of product be favored at higher or lower temperature? Explain.

Solution:

(a) See the discussion in the REVIEW. Hydrogen gas is prepared industrially by the electrolysis of aqueous sodium chloride and by reacting a hydrocarbon with steam. An example of the latter is the reaction of butane gas with steam

$$CH_3CH_2CH_2CH_3(g) + H_2O(g) \rightarrow 9H_2(g) + 4CO(g)$$

In the laboratory, hydrogen gas is produced by the reaction of a moderately active metal with dilute HCl(aq), such as aluminum metal with HCl(aq),

$$Al(s) + 6HCl(aq) \rightarrow 2AlCl_3(aq) + 3H_2(g)$$

(b) If both reactions are exothermic, thermal energy is a product (on the right side of the equation). The equilibrium will be shifted in the forward direction (where the products are) by removing the "product heat," which means lowering the temperature.

SECTION 21.4 CHEMICAL PROPERTIES OF HYDROGEN

After completing this section, you will be able to do the following:

■ Write equations for the preparation and reactions of salt-like hydrides.

REVIEW

Hydrogen gas is colorless, odorless, and tasteless. At **low temperatures**, hydrogen gas is relatively **unreactive**. A balloon filled with two volumes of hydrogen gas to one volume of oxygen gas will remain unreactive until a spark or flame is brought near the mixture; then a spontaneous and exothermic reaction occurs so rapidly that an explosion is observed. Thus at high temperatures, enough energy is available to break the H−H bond, at which point hydrogen reacts vigorously with many metals and nonmetals to form several types of binary (two-element) compounds: salt-like hydrides, molecular hydrides, and metal-like hydrides. We will look at the first type only.

Group IA and IIA elements react with hydrogen gas at high temperatures to produce solid **salt-like** hydrides -- that is, ionic compounds consisting of the Group IA or Group IIA metal ion and the hydride ion, H^- (the names of binary compounds always end with -ide). In other words, at high temperatures, the H−H bond is broken and each H atom accepts an electron from an active Group IA or Group IIA metal. These are the only instances where hydrogen has an oxidation number of -1. For example, when strontium metal reacts with hydrogen gas at high temperatures (indicated by a delta sign, Δ), we have

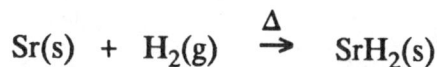

$$Sr(s) \ + \ H_2(g) \ \overset{\Delta}{\rightarrow} \ SrH_2(s)$$

The hydrides are a good source of hydrogen gas, which is easily released upon the addition of water.

$$SrH_2(s) \ + \ 2H_2O(l) \ \rightarrow \ H_2(g) \ + \ Sr(OH)_2(aq)$$

Similar reactions can be written for the other Group IIA elements and the Group IA elements.

SAMPLE PROBLEM 21.4 A

▶▶ (a) Write an equation for the reaction of barium metal with hydrogen gas. (b) Write the molecular, complete ionic, and net ionic equations for the reaction of barium hydride with water. (c) Using the net ionic reaction from part (b), which species acts as a Bronsted-Lowry acid and which species acts as a Bronsted-Lowry base? (d) Using the net ionic reaction from part (b), which element undergoes oxidation, which element undergoes reduction, which species is the oxidizing agent, and which species is the reducing agent?

Solution:

(a)   See the discussion in the REVIEW; the product is solid barium hydride.

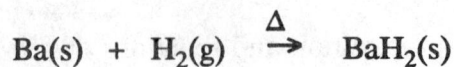

$$Ba(s) + H_2(g) \xrightarrow{\Delta} BaH_2(s)$$

(b)   See the REVIEW; see also Section 4.4 for a discussion on writing net ionic, complete ionic, and molecular equations. Hydrides react with water to produce hydrogen gas and the corresponding hydroxide.

Molecular: $BaH_2(s) + 2H_2O(l) \rightarrow H_2(g) + Ba(OH)_2(aq)$

Complete ionic: $Ba^{2+} + 2H^- + 2H_2O(l) \rightarrow H_2(g) + Ba^{2+} + 2OH^-$

Net ionic: $H^- + H_2O(l) \rightarrow H_2(g) + OH^-$

(c)   A Bronsted-Lowry base accepts a proton (H^+); H^- accepts a proton from the acid water to become a hydrogen molecule. H^- and H_2 are a conjugate acid-base pair that differ in their formulas only by a proton (H^+).

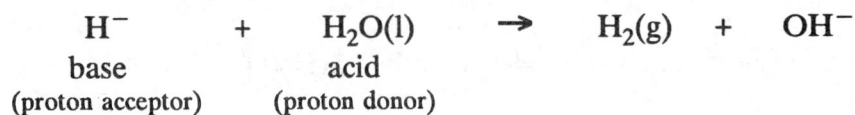

$$H^- + H_2O(l) \rightarrow H_2(g) + OH^-$$

H^-		$H_2O(l)$			
base		acid			
(proton acceptor)		(proton donor)			

(d)   Hydrogen has an oxidation number of $+1$ in all its compounds **except hydrides**, where it is -1. The hydride ion is undergoing oxidation, and one of the two hydrogens in water is undergoing reduction.

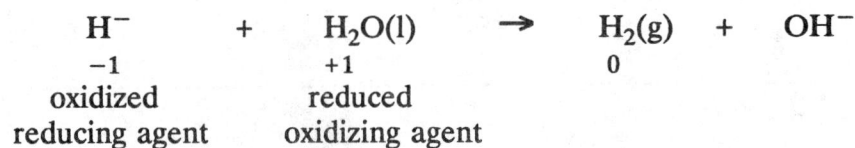

$$H^- + H_2O(l) \rightarrow H_2(g) + OH^-$$

H^-		$H_2O(l)$		$H_2(g)$	
-1		$+1$		0	
oxidized		reduced			
reducing agent		oxidizing agent			

PRACTICE PROBLEMS

Test your problem-solving skills in this chapter by working through the problems in the space provided **BEFORE** checking the answers. A periodic table will always be available. The ANSWERS TO PRACTICE PROBLEMS follow immediately after this section.

1. Write one equation each (not used in Section 21.3 of this study guide or the textbook) for the industrial preparation and laboratory preparation of hydrogen gas.

2. (a) Write an equation for the reaction of potassium metal with hydrogen gas.

   (b) Write the molecular, complete ionic, and net ionic equations for the reaction of potassium hydride with water.

   (c) Using the net ionic reaction from part (b), which species acts as a Bronsted-Lowry acid and which species acts as a Bronsted-Lowry base?

   (d) Using the net ionic reaction from part (b), which element undergoes oxidation, which element undergoes reduction, which species is the oxidizing agent, and which species is the reducing agent?

ANSWERS TO PRACTICE PROBLEMS

1. Industrial: Select another hydrocarbon

$$CH_2=CH_2(g) \ + \ 2H_2O(g) \ \rightarrow \ 4H_2(g) \ + \ 2CO(g)$$

Laboratory: Any moderately active metal and dilute HCl

$$2Fe(s) \ + \ 6HCl(aq) \ \rightarrow \ 3H_2(g) \ + \ 2FeCl_3(aq)$$

2. (a)
$$2K(s) \ + \ H_2(g) \ \xrightarrow{\Delta} \ 2KH(s)$$

(b) Molecular: $$KH(s) \ + \ H_2O(l) \ \rightarrow \ H_2(g) \ + \ KOH(aq)$$

Complete ionic: $$K^+ + H^- + H_2O(l) \ \rightarrow \ H_2(g) + K^+ + OH^-$$

Net ionic: $$H^- + H_2O(l) \ \rightarrow \ H_2(g) \ + \ OH^-$$

(c)
$$\underset{\substack{\text{base} \\ \text{(proton acceptor)}}}{H^-} \ + \ \underset{\substack{\text{acid} \\ \text{(proton donor)}}}{H_2O(l)} \ \rightarrow \ H_2(g) \ + \ OH^-$$

(d)
$$\underset{\substack{-1 \\ \text{oxidized} \\ \text{reducing agent}}}{H^-} \ + \ \underset{\substack{+1 \\ \text{reduced} \\ \text{oxidizing agent}}}{H_2O(l)} \ \rightarrow \ \underset{0}{H_2(g)} \ + \ OH^-$$

CHAPTER 22

A CLOSER LOOK AT ORGANIC CHEMISTRY

SECTION 22.3   STEREOISOMERS

After completing this section, you will be able to do the following:

- Identify stereocenters in an organic compound;

- Define and draw the cis-trans forms of diastereomers.

REVIEW

Isomers are compounds that have the same molecular formula but differ in either the order of attachment of the atoms (structural or constitutional isomers) or the arrangement of the atoms in space (stereoisomers).

- **Structural** or **constitutional isomers** have the same molecular formula but differ in the order of attachment of the atoms, and as a result they have different physical and chemical properties (discussed in Section 22.2 in the textbook).

- **Stereoisomers** have the same molecular formula **and** order of attachment of the atoms, but differ in the arrangement of these atoms in three-dimensional space.

  - If the stereoisomers are non-superimposable mirror images of each other, they are called **enantiomers**.

- If the stereoisomers have a carbon-carbon double bond or ring structure, they are called **diastereomers** (or cis-trans or geometric isomers).

A pair of **enantiomers** exist when a molecule forms a **non-superimposable mirror image** of itself. That is, your left and right hands are mirror images of each other; if you hold your left hand in front of a mirror, you will see your right hand in the mirror. Yet you **cannot** superimpose your right hand on your left, or vice versa; palms up (or down), you cannot have the thumb on top of thumb, index finger on top of index finger, etc. For compounds, a pair of **enantiomers** or mirror images will exist if there is a **stereocenter** present. In organic compounds, this means a carbon atom has **four different atoms or groups of atoms** bonded to it. For example, the carbon atom in the compound CHFClBr has four **different** atoms bonded to it, which makes it a stereocenter; this compound has a non-superimposable mirror image, i.e., a pair of **enantiomers** exist. On the other hand, CH_2ClBr has only three different atoms bonded to the carbon and does **not** have a stereocenter; this compound forms a mirror image, but the mirror image is **superimposable**. When this occurs, the compound and its superimposable mirror image are **identical**, not enantiomers. [Conversely, if you find two or more identical atoms or groups of atoms bonded to a carbon atom, that carbon atom cannot be a stereocenter.] You can readily verify the statements in this section by using toothpicks for bonds and different colored gumdrops or marshmallows for atoms or groups of atoms. Pairs of **enantiomers** have the same physical and chemical properties; they differ only in the direction of rotation of **plane-polarized** light.

A pair of **diastereomers** has a different spatial arrangement of atoms around a **double bond** or in a **ring structure**. Unlike carbon-carbon single bonds, carbon-carbon double bonds do not permit free rotation about the bond; this "freezes" the atoms into two spatial arrangements called -cis and -trans (or geometric) isomers, which are **non-superimposable** on each other. For example, the two kinds of 2-pentene are cis-trans isomers

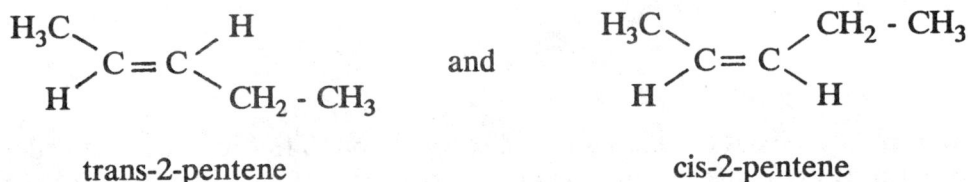

trans-2-pentene   cis-2-pentene

When the groups attached to the double-bonded carbons, in this case the -CH$_3$ and -CH$_2$-CH$_3$ groups (or two H's), are located **"across"** the double bond (across means one group above the double bond and one below), it is the -**trans** isomer. When the same two groups are located on the **same side** of the double bond (either both above or both below the double bond), it is the -**cis** isomer. The two isomers are **not superimposable** on each other. (You cannot rotate the carbon-carbon double bond.) On the other hand, just because a carbon-carbon double bond is present in an organic compound does **not** mean cis-trans isomers always exist. **Cis-trans or geometric isomers** exist only when there are two **different atoms or groups of atoms** attached to **each** double-bonded carbon in the compound. For 2-pentene, the left-hand double-bonded carbon has two different atoms or groups of atoms attached to it (-CH$_3$ and -H) and so does the right hand double-bonded carbon (-CH$_2$-CH$_3$ and -H). If each double-bonded carbon does **not** have two different atoms or groups of atoms attached to it, then any two "cis-trans isomers" that you try to draw are in fact identical; they can be superimposed on top of each other and are not isomers.

SAMPLE PROBLEM 22.3 A

▶▶ Indicate whether the following compounds will have enantiomers (non-superimposable mirror images); if so, mark the stereocenter with an asterisk.

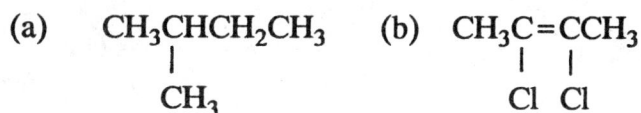

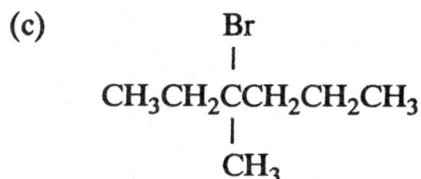

(a) $CH_3CHCH_2CH_3$ (b) $CH_3C=CCH_3$
        |
       CH_3 Cl Cl

(c) Br
                 |
     $CH_3CH_2CCH_2CH_2CH_3$
                 |
                CH_3

Solution:

(a) **No enantiomers**; no stereocenters. Any carbon with two or more identical atoms or groups of atoms cannot be a stereocenter.

(b) **No enantiomers**; no stereocenters. A double-bonded carbon can never have four different atoms or groups of atoms attached to it.

(c) **Enantiomers exist, with the stereocenter marked below.** The four different groups attached to the starred carbon are $-CH_2CH_3$, $-Br$, $-CH_3$, and $-CH_2CH_2CH_3$.

            Br
                 |
    $CH_3CH_2C^*CH_2CH_2CH_3$
                |
               CH_3

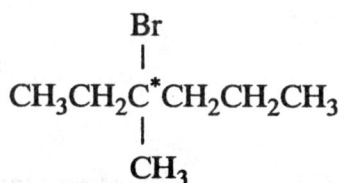

SAMPLE PROBLEM 22.3 B

▶▶ If cis-trans isomers exist, draw the isomers and label them "cis" and "trans".

(a) $Cl_2C=CHCH_3$ (b) $(CH_3)_2C=CH_2$ (c) $ClHC=CHCl$

(a)   Focus on the two double-bonded carbons. Each double-bonded carbon **must** have two different atoms or groups of atoms attached to it in order for cis-trans isomers to exist. The left-hand double-bonded carbon has two identical chlorine atoms attached to it; no cis-trans isomers exist.

(b)   No cis-trans isomers exist. The left-hand double-bonded carbon has two identical -CH_3 groups attached to it, and the right-hand double-bonded carbon has two identical hydrogen atoms bonded to it.

(c)   Cis-trans isomers exist, because each double-bonded carbon has two different atoms or groups of atoms bonded to it, -H and -Cl in each case.

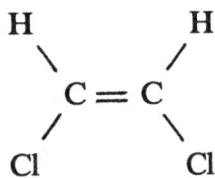

"cis"

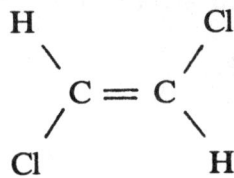

"trans"

SECTION 22.5   ALKANES

After completing this section, you will be able to do the following:

■ Define and identify hydrocarbons, alkanes, and cycloalkanes.

REVIEW

With over nine million organic compounds known, a system of organizing these compounds into families with similar structures or properties becomes necessary. One very important and large family is the **hydrocarbons,** which consist of all the organic compounds that contain only the elements hydrogen and carbon. The subfamilies of hydrocarbons include **alkanes,** which contain only carbon-carbon single bonds; **alkenes,** which contain at least one carbon-carbon double bond; **alkynes,** which contain at least one carbon-carbon triple bond; and **cyclic** compounds which contain carbon-carbon ring structures. This section deals only with **alkanes.**

Alkanes are often called saturated hydrocarbons because they have no multiple bonds and are saturated with the maximum number of hydrogen atoms possible. [Unsaturated hydrocarbons contain one or more multiple bonds.] The general formula of alkanes is C_nH_{2n+2}, where n is an integer. That is, the five-carbon alkane, pentane, contains five carbon atoms and $2n+2$ or 12 hydrogen atoms, C_5H_{12}. The formulas, names, and physical properties of the first ten continuous or straight-chain alkanes are given in Table 22.3 in the textbook. You will need to memorize the names of the first ten alkanes, since they form the basis for naming alkanes (to be discussed in Section 22.6) and many other hydrocarbons as well. **Alkanes** are not very reactive (compared to alkenes, alkynes, etc.) and only commonly undergo combustion, pyrolysis, and halogenation (reaction with F, Cl, or Br).

SAMPLE PROBLEM 22.5 A

▶▶ Which of the following compounds are hydrocarbons? alkanes? cycloalkanes?

$$O$$

(a) CH_3COOH (b) $(CH_3)_2NH$ (c) $H_2C=CH_2$

(d) 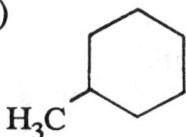   (e) CH_4 (f) $CH3(CH_2)_{16}CH_3$

(g) △   (h) $CHCl_3$

Solution:

Hydrocarbons: c, d, e, f, g   (all contain only C and H, including multiple bonds and rings)

Alkanes: e, f   (a subset of hydrocarbons containing only single C-C bonds and excluding cyclic compounds)

Cycloalkanes: d, g   (a subset of hydrocarbons containing only single C-C bonds in a ring or cyclic structure)

NOTE: a, b, and h are not hydrocarbons, so they cannot possibly be considered as an alkane or cycloalkane.

SECTION 22.6   NOMENCLATURE OF ORGANIC COMPOUNDS

After completing this section, you will be able to do the following:

■ Name an alkane from the structural formula, and vice versa.

REVIEW

The system used for naming organic compounds is the one developed by the International Union of Pure and Applied Chemistry, better known by its acronym, IUPAC. This system of nomenclature is based on the straight or continuous chain alkanes. To name an alkane, you will need to **memorize** the names of the first ten straight-chain or continuous alkanes (which are given in Table 22.3 of the textbook) and carefully follow the rules given below. A detailed explanation of each rule is given in Section 22.6 of the textbook. You must also be able to write the structure from the name of an alkane, which is usually easier to do.

RULES FOR NAMING ALKANES

1. Find the longest continuous chain and name it (memorize the names in Table 22.3).

2. If there are two or more longest chains of equal length, choose the one with the greater number of groups attached to it for the main chain.

3. Number the main chain starting from the end that gives the lower number to the carbon where the first branch is attached.

4. Show the position of attachment of the side chain and its structure as a prefix to the name of the main chain. If two or more identical groups are attached to the main chain, use a prefix (di- for two, tri- for three, etc.) to show how many of the groups there are.

5. If the groups attached to the main chain are different, the side chains are arranged in alphabetical order according to the name of the alkyl group (side chain). The prefixes (di-, tri-, etc.) are **not** included in the alphabetizing of side chains.

6. Hyphens separate numbers from words and commas separate numbers.

SAMPLE PROBLEM 22.6 A

▶ ▶ Name the following alkane

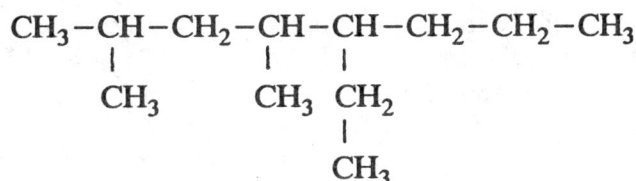

$$CH_3-CH-CH_2-CH-CH-CH_2-CH_2-CH_3$$

$$\begin{array}{cccc} & | & & | & | \\ & CH_3 & & CH_3 & CH_2 \\ & & & & | \\ & & & & CH_3 \end{array}$$

Solution:

Carefully follow the rules given in the REVIEW; see Section 22.6 in the textbook for a detailed explanation of each rule.

Step 1: *Find the longest continuous chain and name it (memorize the names in Table 22.3).*

While there are over a half-dozen continuous chains of varying length, there are only two longest chains of 8 carbons each, which are shaded in the structures shown below. The name of the longest chain is the same as the straight-chain alkane with 8 carbons, octane. Therefore, this compound is some kind of octane. If there were no side chains, the name would be complete.

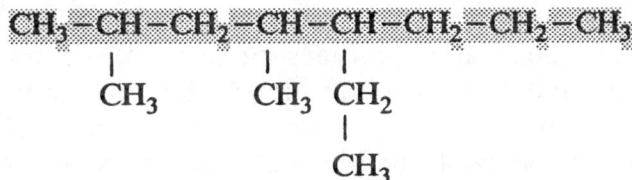

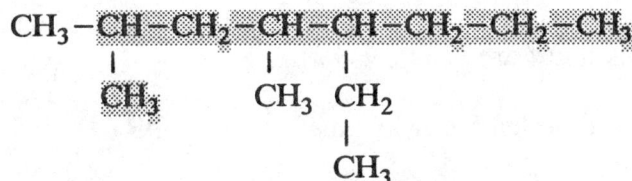

$$CH_3-CH-CH_2-CH-CH-CH_2-CH_2-CH_3$$

$$\begin{array}{cccc} & | & & | & | \\ & CH_3 & & CH_3 & CH_2 \\ & & & & | \\ & & & & CH_3 \end{array}$$

$$CH_3-CH-CH_2-CH-CH-CH_2-CH_2-CH_3$$

$$\begin{array}{cccc} & | & & | & | \\ & CH_3 & & CH_3 & CH_2 \\ & & & & | \\ & & & & CH_3 \end{array}$$

Step 2: *If there are two or more longest chains of equal length, choose the one with the greater number of groups attached to it for the main chain.*

The two longest chains are equivalent; both will have the same three side chains. Either longest chain can be used; both longest chains will give the same correct name for the compound.

Step 3:   *Number the main chain starting from the end that gives the lower number to the carbon where the first branch is attached.*

Arbitrarily, we will use the longest chain straight across and number the carbons 1 to 8 from left to right; numbered this way, the first side chain ($-CH_3$) is attached to carbon 2 of the longest chain. [If you numbered the longest chain (incorrectly) from right to left, the first side chain ($-CH_2-CH_3$) would be attached to carbon 4.] The correctly numbered carbons in the longest chain are

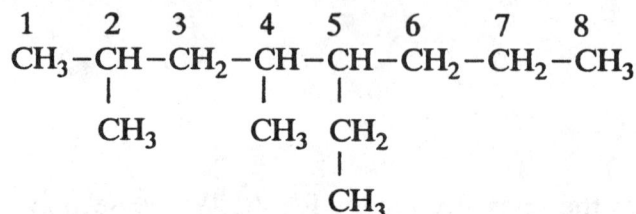

$$
\begin{array}{cccccccc}
1 & 2 & 3 & 4 & 5 & 6 & 7 & 8 \\
CH_3 & -CH & -CH_2 & -CH & -CH & -CH_2 & -CH_2 & -CH_3 \\
 & | & & | & | & & & \\
 & CH_3 & & CH_3 & CH_2 & & & \\
 & & & & | & & & \\
 & & & & CH_3 & & &
\end{array}
$$

Step 4:   *Show the position of attachment of the side chain and its structure as a prefix to the name of the main chain. If two or more identical groups are attached to the main chain, use a prefix (di- for two, tri- for three, etc.) to show how many of the groups there are.*

We have three side chains or branches attached to the longest continuous chain. The two one-carbon branches are called methyl branches, with the name taken from the straight-chain alkane with one carbon (methane) and replacing the -ane with -yl. The methyl branches are attached to carbons 2 and 4 of the longest chain. **We must give the name and location of each side chain in a compound.** The locations are carbons 2 and 4; the location precedes the name of the branch, with a hyphen between the location and the name of the branch. However, we don't use "2-methyl-4-methyl" but "2,4-dimethyl" instead. It conveys the same information, but in a simpler way. Similarly, the two-carbon branch is called an ethyl branch, and since it is located at carbon 5 of the longest chain, we use "5-ethyl" to describe it. The names and locations of the branches precede the name of the longest chain. Back in step 1, we determined that this compound is some kind of octane. Two choices for the name at this point are

     2,4-dimethyl-5-ethyloctane   OR   5-ethyl-2,4-dimethyloctane

Step 5:   *If the groups attached to the main chain are different, the side chains are arranged in alphabetical order according to the name of the alkyl group (side chain). The prefixes (di-, tri-, etc.) are **not** included in the alphabetizing of side chains.*

The side chains are alphabetized, **excluding** prefixes, which indicate the number of identical side chains (di-, tri-, etc.). Since "e" in ethyl comes before "m" in methyl, the **only** correct name for this compound is

5-ethyl-2,4-dimethyloctane

Step 6: *Hyphens separate numbers from words and commas separate numbers.*

We have already incorporated step 6 into the correct name. This step and some of the preceding steps may sound very picky, but with over 10 million organic compounds, you have to be able to locate the name fast and the name must be unique to one compound only.

SAMPLE PROBLEM 22.6 B

▶▶ Write the structure for 2,3,5-trimethylheptane.

Solution:

The last part of the name of a compound is the name of the longest continuous chain; this is a heptane, which means the longest continuous chain has seven carbons in it. The rest of the name says that there are three methyl groups attached to carbons 2, 3 and 5 of the longest continuous chain. Draw out a seven-carbon longest chain and number the chain from right to left, or vice versa, since you will be placing the side chains on the appropriate carbon in the longest chain.

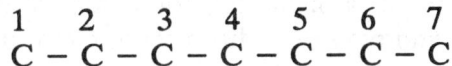

$$
\begin{array}{ccccccc}
1 & 2 & 3 & 4 & 5 & 6 & 7 \\
C - & C - & C - & C - & C - & C - & C
\end{array}
$$

Add the methyl branches to carbons 2, 3, and 5

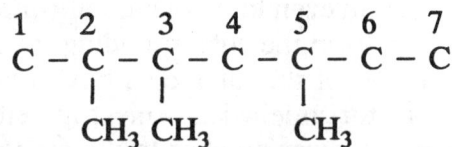

$$
\begin{array}{ccccccc}
1 & 2 & 3 & 4 & 5 & 6 & 7 \\
C - & C - & C - & C - & C - & C - & C \\
 & | & | & & | & & \\
 & CH_3 & CH_3 & & CH_3 & &
\end{array}
$$

Finally, there must be four bonds around each carbon. Count up the carbon-carbon bonds already present, then add sufficient hydrogen atoms until there is a total of four bonds around each carbon.

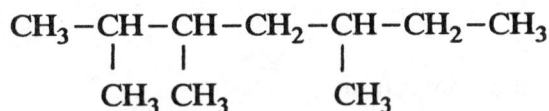

$$
\begin{array}{c}
CH_3-CH-CH-CH_2-CH-CH_2-CH_3 \\
\quad\;\; | \quad\; | \qquad\qquad | \\
\quad\;\; CH_3 \; CH_3 \qquad\; CH_3
\end{array}
$$

This is the correct structure for 2,3,5-trimethylheptane.

NOTE: If you numbered the longest chain from right to left, you will end up with a structure that may look different than this one until you flip it over 180°; it is identical to this one.

SECTION 22.7   FUNCTIONAL GROUPS

After completing this section, you will be able to do the following:

■ Name an alcohol from the structural formula, and vice versa;

■ Identify functional groups in complete and condensed structural formulas.

REVIEW

The classification system that is used for over 10 million organic compounds is based on the fact that when organic compounds take part in a reaction, only certain bonds and organic groups are affected, while the rest of the molecule remains unchanged. The part of the molecule that reacts is called a **functional group**, which serves as the basis for classifying organic compounds. All molecules that have the same functional group react similarly. Given a structural formula for a compound, you should be able to recognize the common functional groups shown in Table 22.4 of the textbook. Be careful, however, since a number of the functional groups look very similar and may only differ from each other by one atom.

Like alkanes, to name a compound under the IUPAC system that contains a functional group, you must find the longest continuous chain that **has the functional group in it or attached to it** (often called the parent or main chain). This has priority over an even longer chain of carbon atoms without the functional group. The name of the main chain is taken from the corresponding straight-chain alkane with the same number of carbon atoms and replacing the "e" of the alkane with with an ending unique to a particular functional group (e.g., -ol for alcohols, -al for aldehydes, -one for ketones, etc). The main chain is numbered starting from the end that gives the lower number to the carbon where the functional group is attached; the location of the functional group is indicated with a number immediately preceding the name of the main chain. The name and location of side chains follows the same procedure as used for alkanes.

SAMPLE PROBLEM 22.7 A

▶▶ What is the name of the compound

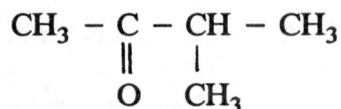

$$CH_3 - \underset{\underset{O}{\|}}{C} - \underset{\underset{CH_3}{|}}{CH} - CH_3$$

Solution:

From Table 22.4 in the textbook, the functional group that has a **carbonyl group** (C=O) with two hydrocarbon groups attached to it is a **ketone**. There are two equivalent longest chains that contain the ketone group, which are shown shaded.

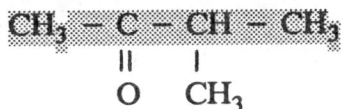

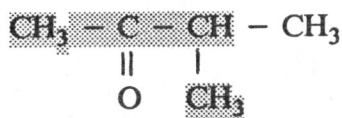

$$CH_3 - C - CH - CH_3$$
$$\underset{O}{\overset{\|}{}} \quad \underset{CH_3}{\overset{|}{}}$$

$$CH_3 - C - CH - CH_3$$
$$\underset{O}{\overset{\|}{}} \quad \underset{CH_3}{\overset{|}{}}$$

The name of the main chain is derived from the four-carbon straight-chain alkane, butane, with the "e" replaced by "one", making this compound a butanone. Arbitrarily selecting the main chain straight across, carbon 1 is on the left side so that the ketone group has the lowest number at carbon 2.

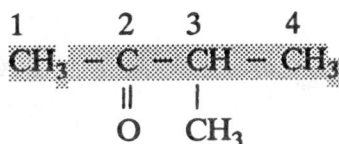

$$\overset{1}{CH_3} - \overset{2}{C} - \overset{3}{CH} - \overset{4}{CH_3}$$
$$\underset{O}{\overset{\|}{}} \quad \underset{CH_3}{\overset{|}{}}$$

The location of the functional group must be given; so far, this is a 2-butanone. Finally, the one carbon methyl branch is attached to carbon 3 of the main chain. The complete name for this compound is 3-methyl-2-butanone.

SAMPLE PROBLEM 22.7 B

▶ ▶ Draw the structure for 2,2-dimethyl-3-pentanone.

Solution:

From the "3-pentanone" part of the name, you should be able to interpret this compound as having a five-carbon main chain with a ketone group at carbon 3 (a double-bonded oxygen attached to carbon 3). Draw out the five-carbon skeleton chain, number it from either direction (which you can do when placing the branches at their proper location last), locate the double-bonded oxygen at carbon 3, and place the two methyl branches on carbon 2,

$$
\begin{array}{c}
CH_3 \\
1 \quad\; 2 \mid \;\; 3 \quad\; 4 \qquad 5 \\
CH_3 - C - C - CH_2 - CH_3 \\
\mid \;\;\; \parallel \\
CH_3 \; O
\end{array}
$$

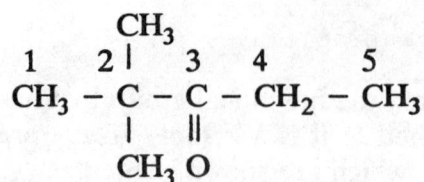

SAMPLE PROBLEM 22.7 C

▶▶ Identify the functional group(s) in the following compound

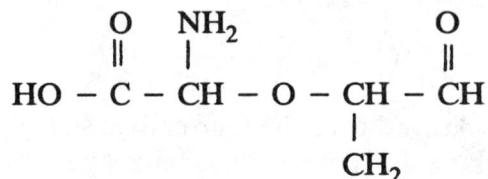

$$
\begin{array}{c}
O \quad\;\; NH_2 \qquad\qquad\quad O \\
\parallel \quad\;\; \mid \qquad\qquad\qquad\; \parallel \\
HO - C - CH - O - CH - CH \\
\mid \\
CH_2
\end{array}
$$

Solution:

Using Table 22.4 in the textbook and starting from the left-hand end of the compound,

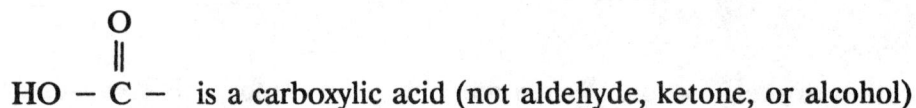

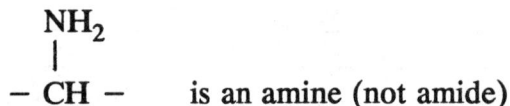

$$
\begin{array}{c}
O \\
\parallel \\
HO - C -
\end{array}
$$ is a carboxylic acid (not aldehyde, ketone, or alcohol)

$$
\begin{array}{c}
NH_2 \\
\mid \\
- CH -
\end{array}
$$ is an amine (not amide)

$C - O - C$ is an ether

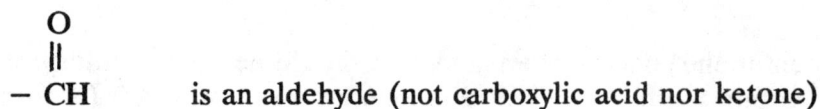

$$
\begin{array}{c}
O \\
\parallel \\
- CH
\end{array}
$$ is an aldehyde (not carboxylic acid nor ketone)

SECTION 22.10   STRUCTURE DETERMINATION

After completing this section, you will be able to do the following:

- ■ Predict the number of multiple bonds and/or rings in a compound from its molecular formula;

- ■ Determine the structure from hydrogenation reactions.

REVIEW

Both the physical and chemical properties of compounds depend on their structure. To determine the structure of a compound, the molecular formula is determined first by mass spectrometry or from the percent composition of the compound and its molecular mass (Sections 3.9-3.11). From the molecular formula, the number of multiple bonds and/or rings in the compound can be reasoned out using the general formulas for alkanes, alkenes, alkynes, cycloalkanes, and cycloalkenes.

	General formula	Example	Molecular formula	Number of H's less than alkane
Alkane	C_nH_{2n+2}	$CH_3CH_2CH_3$	C_3H_8	0
Alkene	C_nH_{2n}	$CH_3CH=CH_2$	C_3H_6	2
Cycloalkane	C_nH_{2n}	$H_2C \overset{CH_2}{\diagup\;\diagdown} CH_2$	C_3H_6	2
Alkyne	C_nH_{2n-2}	$CH_3C\equiv CH$	C_3H_4	4
Cycloalkene	C_nH_{2n-2}		C_4H_6	4

That is, when a single bond in an alkane is replaced by a double bond (making it an alkene) or an alkane becomes a cycloalkane, the number of hydrogens **decreases by two**. Similarly, when a single bond in an alkane is replaced by a triple bond, when two single bonds are replaced by two double bonds, or when the alkane forms two rings bonded together, the number of hydrogens **decreases by four**. From the given molecular formula for a compound, you can determine its general formula in the table above and predict the possibilities for multiple bonds and rings.

Results from hydrogenation reactions (reacting a hydrocarbon with hydrogen gas) will allow you to determine the number of multiple bonds, since hydrogen gas reacts with multiple bonds but not rings at room temperature. Furthermore, it takes 1.0 mol of hydrogen gas to react with and break one double bond, 2.0 mol of H_2 gas to break one triple or two double bonds, etc. By knowing the number of moles of hydrogen gas that reacted with one mole of the unknown compound, you can determine the number of multiple bonds present and reduce the number of possibilities for structures of the compound.

SAMPLE PROBLEM 22.10 A

▶▶ An organic compound of unknown structure has the molecular formula, C_6H_{12}. (a) How many multiple bonds and/or rings are there in the compound? (b) Write structures for two possible structural (constitutional) isomers.

Solution:

(a)   This compound has six carbons. The formula for an alkane with six carbons is C_6H_{14}. The unknown compound (C_6H_{12}) has two fewer hydrogens than C_6H_{14}. Therefore, two fewer hydrogens can be explained by either one double bond or one ring.

(b)   One possible double-bonded structure and the ring structure are

$$CH_2 = CH(CH_2)_3CH_3$$

SAMPLE PROBLEM 22.10 B

▶▶ When the unknown organic compound in Sample Problem 22.10 A, C_6H_{12}, was subjected to catalytic hydrogenation at room temperature, no reaction occurred. Write the structure(s) for possible structural (constitutional) isomers.

Solution:

As pointed out in the REVIEW, multiple bonds react with hydrogen gas at room temperature to produce alkanes, but ring structures do not. Since no reaction occurred, the double-bonded isomers are eliminated; the only choice left for the structure of C_6H_{12} is the ring structure

This ring structure is known as cyclohexane.

PRACTICE PROBLEMS

Test your problem solving skills in this chapter by working through the problems in the space provided **BEFORE** checking with the answers. A periodic table will always be available. The ANSWERS TO PRACTICE PROBLEMS follows immediately after this section.

1. Indicate whether the following compounds will have enantiomers (non-superimposable mirror images); if so, mark the stereocenter with an asterisk.

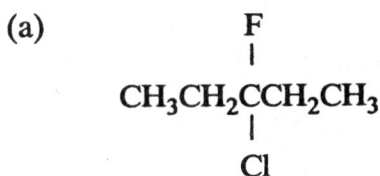

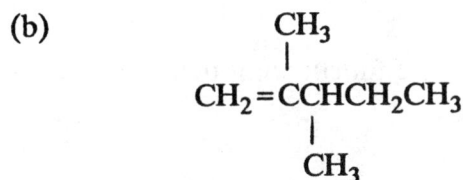

(a)
$$\begin{array}{c} F \\ | \\ CH_3CH_2CCH_2CH_3 \\ | \\ Cl \end{array}$$

(b)
$$\begin{array}{c} CH_3 \\ | \\ CH_2{=}CCHCH_2CH_3 \\ | \\ CH_3 \end{array}$$

2. If cis-trans isomers exist, draw the isomers and label them "cis" and "trans".

(a) $(CH_3)_2C{=}C(CH_3)_2$

(b) $CH_3CH{=}CHCH_2CH_3$

3. Which of the following compounds are hydrocarbons? alkanes? cycloalkanes?

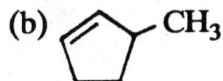

(a) HCOOH

(b) —CH$_3$

(c) $(CH_3)_3N$

(d) H_2CO_3

(e) $HC{\equiv}CCH_2CH_3$

(f)

(g) C_2H_5OH

(h) $CH_3C(CH_3)_2CH_2CH_3$

(i) $CH_2{=}CHCH{=}CHCH_3$

4. Name the following alkanes

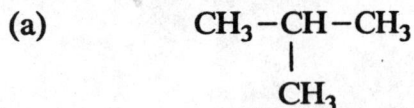

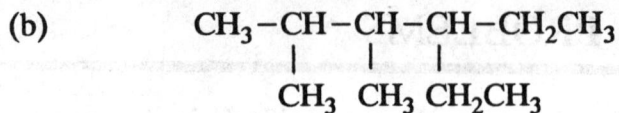

   (a)
   $$CH_3-CH-CH_3$$
   $$|$$
   $$CH_3$$

   (b)
   $$CH_3-CH-CH-CH-CH_2CH_3$$
   $$\quad\quad\ |\quad\ \ |\quad\ |$$
   $$\quad\quad CH_3\ \ CH_3\ CH_2CH_3$$

5. Write the structures for

   (a) 3-ethylhexane

   (b) 2,2,3,3-tetramethylbutane

6. Identify the functional group(s) in each of the following compounds

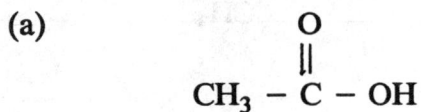

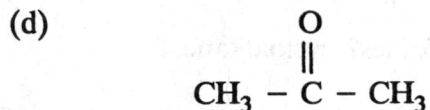

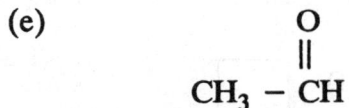

   (a)
   $$O$$
   $$\|$$
   $$CH_3 - C - OH$$

   (b) $CH_2=CH_2$

   (c) CH_3-O-CH_3

   (d)
   $$O$$
   $$\|$$
   $$CH_3 - C - CH_3$$

   (e)
   $$O$$
   $$\|$$
   $$CH_3 - CH$$

ANSWERS TO PRACTICE PROBLEMS

1. (a) No stereocenter; no enantiomers.

   (b) Enantiomers exist; stereocenter marked with an asterisk.

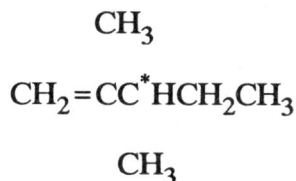

$$CH_3$$
$$CH_2=CC^*HCH_2CH_3$$
$$CH_3$$

2. (a) No cis-trans isomers; each double-bonded carbon must have two different atoms or groups of atoms attached to it.

   (b) Cis-trans isomers do exist; each double-bonded carbon has two different atoms or groups of atoms attached to it. This compound is 2-pentene, whose cis-trans isomers are shown in the REVIEW.

3. Hydrocarbons: b, e, f, h, i

   Alkanes: h

   Cycloalkanes: f

4. (a) 2-methylpropane

   (b) 4-ethyl-2,3-dimethylhexane

5. (a) $CH_3-CH_2-CH-CH_2-CH_2-CH_3$
   $$CH_2CH_3$$

   (b) $$CH_3\ CH_3$$
   $$CH_3 - C - C - CH_3$$
   $$CH_3\ CH_3$$

6.   (a)   carboxylic acid
        (b)   alkene
        (c)   ether
        (d)   ketone
        (e)   aldehyde

CHAPTER 23

A CLOSER LOOK AT INORGANIC CHEMISTRY: NONMETALS AND METALLOIDS AND THEIR COMPOUNDS

SECTION 23.2   CARBON

After completing this section, you will be able to do the following:

■ Complete and balance selected reactions of carbon and its inorganic compounds.

REVIEW

In order to complete and balance reactions in general, it is best to learn **type reactions** (rather than each and every specific example) that occur which can be applied to many examples. One **type reaction** that we have discussed a number of times is the reaction of an acid with a base to produce a salt and water

$$acid \ + \ base \ \rightarrow \ salt \ + \ water$$

This **type reaction** can be used to complete and balance dozens of reactions regardless of the formula of the acid or base. Although not all **type reactions** are this general, nevertheless these **type reactions** are one of the best ways to summarize thousands of specific reactions.

Following is a summary of the more important **type reactions of carbon and its inorganic compounds** that will be useful for completing and balancing the reactions given in the Sample Problems:

1.   At high temperatures, both diamond and graphite react with a number of elements and compounds.

594

2. Carbon monoxide is a reducing agent for producing metals from their ores.
3. On a laboratory scale, carbon dioxide is made by treating a carbonate with a dilute acid.
4. Neutralization of HCN with a base.

SAMPLE PROBLEM 23.2 A

▶▶ Write molecular equations for each of the following reactions:

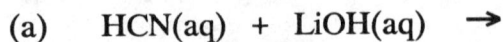

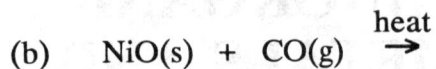

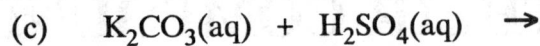

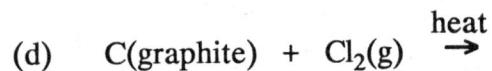

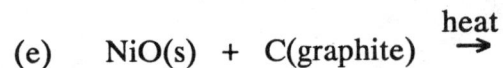

(a) $HCN(aq) + LiOH(aq) \rightarrow$

(b) $NiO(s) + CO(g) \xrightarrow{heat}$

(c) $K_2CO_3(aq) + H_2SO_4(aq) \rightarrow$

(d) $C(graphite) + Cl_2(g) \xrightarrow{heat}$

(e) $NiO(s) + C(graphite) \xrightarrow{heat}$

Solution:

HINT: Learn the type reactions in the REVIEW.

(a) Type reaction #4.

   $HCN(aq) + LiOH(aq) \rightarrow LiCN(aq) + H_2O(l)$

(b) Type reaction #2.

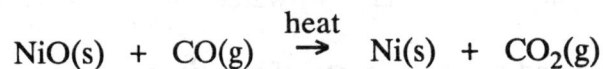

   $NiO(s) + CO(g) \xrightarrow{heat} Ni(s) + CO_2(g)$

(c) Type reaction #3.

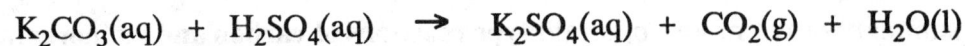

   $K_2CO_3(aq) + H_2SO_4(aq) \rightarrow K_2SO_4(aq) + CO_2(g) + H_2O(l)$

(d) Type reaction #1

$$C(graphite) + 2Cl_2(g) \xrightarrow{heat} CCl_4(g)$$

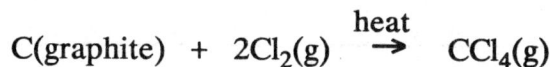

(e)   Type reaction #1.

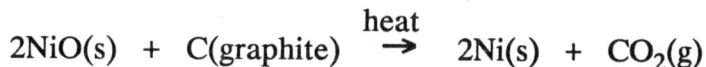

$$2NiO(s) + C(graphite) \xrightarrow{heat} 2Ni(s) + CO_2(g)$$

SECTION 23.4   PHOSPHORUS

After completing this section, you will be able to do the following:

■ Complete and balance selected reactions of phosphorus and its compounds.

REVIEW

In order to complete and balance reactions in general, it is best to learn **type reactions** (rather than each and every specific example) that occur which can be applied to many examples. One **type reaction** that we have discussed a number of times is the reaction of an acid with a base to produce a salt and water

$$acid + base \rightarrow salt + water$$

This **type reaction** can be used to complete and balance dozens of reactions regardless of the formula of the acid or base. Although not all **type reactions** are this general, nevertheless these **type reactions** are one of the best ways to summarize thousands of specific reactions.

Following is a summary of the more important **type reactions of phosphorus and its compounds** that will be useful for completing and balancing the reactions given in the Sample Problems:

1. White phosphorus reacts spontaneously with the halogens (F, Cl, Br, I) at room temperature and with S and Group IA metals on warming, forming a binary compound in each case.
2. Excess H_3PO_4 reacts with a carbonate to produce the dihydrogen phosphate salt; H_3PO_4 reacts with excess carbonate to produce the monohydrogen phosphate salt; H_3PO_4 reacts with strong base to produce the phosphate salt.
3. Non-metal oxide reacts with water forms an acid.
4. Non-metal burns to form an oxide.

SAMPLE PROBLEM 23.4 A

▸▸ Write molecular equations for each of the following reactions:

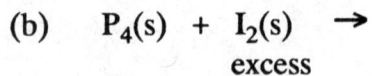

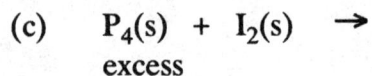

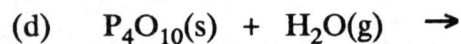

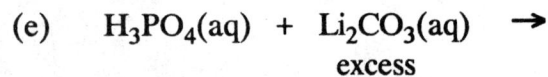

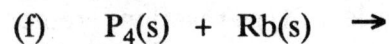

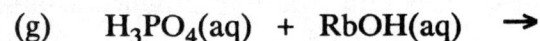

(a) $P_4(l) + O_2(g) \rightarrow$

(b) $P_4(s) + I_2(s) \rightarrow$
        excess

(c) $P_4(s) + I_2(s) \rightarrow$
        excess

(d) $P_4O_{10}(s) + H_2O(g) \rightarrow$

(e) $H_3PO_4(aq) + Li_2CO_3(aq) \rightarrow$
                            excess

(f) $P_4(s) + Rb(s) \rightarrow$

(g) $H_3PO_4(aq) + RbOH(aq) \rightarrow$

Solution:

(a)   Type reaction #4.

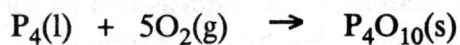

        $P_4(l) + 5O_2(g) \rightarrow P_4O_{10}(s)$

(b)   Type reaction #1.

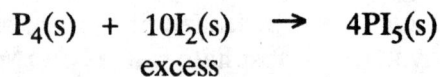

        $P_4(s) + 10I_2(s) \rightarrow 4PI_5(s)$
                    excess

(c)   Type reaction #1.

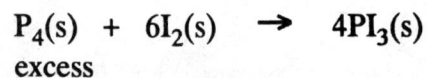

        $P_4(s) + 6I_2(s) \rightarrow 4PI_3(s)$
        excess

(d)   Type reaction #3.

$$P_4O_{10}(s) + 6H_2O(g) \rightarrow 4H_3PO_4(aq)$$

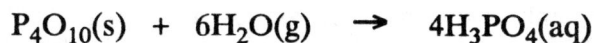

(e) Type reaction #2.

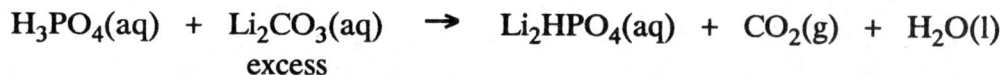

$$H_3PO_4(aq) + Li_2CO_3(aq) \rightarrow Li_2HPO_4(aq) + CO_2(g) + H_2O(l)$$
$$\text{excess}$$

(f) Type reaction #1.

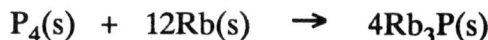

$$P_4(s) + 12Rb(s) \rightarrow 4Rb_3P(s)$$

(g) Type reaction #2.

$$H_3PO_4(aq) + 3RbOH(aq) \rightarrow Rb_3PO_4(aq) + 3H_2O(l)$$

NOTE: In reactions (b) and (c), the reactant in excess has a higher atom ratio in the product; excess I_2 gives a 5:1 ratio of I:P in PI_5 rather than 3:1 in PI_3, and excess P produces a higher 1:3 ratio of P:I in PI_3 rather than the 1:5 ratio in PI_5.

SECTION 23.5 NITROGEN

After completing this section, you will be able to do the following:

■ Complete and balance selected reactions of nitrogen and its compounds.

REVIEW

In order to complete and balance reactions in general, it is best to learn **type reactions** (rather than each and every specific example) that occur which can be applied to many examples. One **type reaction** that we have discussed a number of times is the reaction of an acid with a base to produce a salt and water

$$\text{acid} + \text{base} \rightarrow \text{salt} + \text{water}$$

This **type reaction** can be used to complete and balance dozens of reactions regardless of the formula of the acid or base. Although not all **type reactions** are this general, nevertheless these **type reactions** are one of the best ways to summarize thousands of specific reactions.

Following is a summary of the more important **type reactions of nitrogen and its compounds** that will be useful for completing and balancing the reactions given in the Sample Problems:

1. At high temperatures, nitrogen combines with many metals to form a binary nitride.
2. Ammonia is made in the laboratory by treating an ammonium salt with a strong base.
3. Ammonia reacts with an acid to produce a salt.
4. Ammonium salts decompose on heating.
5. Many metals are oxidized by HNO_3, with the nitrogen-containing product depending primarily on the concentration of the HNO_3; 16 M HNO_3 favors $NO_2(g)$ as a product, 6 M HNO_3 favors $NO(g)$, 3 M HNO_3 favors $N_2O(g)$, 1 M HNO_3 favors $N_2(g)$, and 0.1 M HNO_3 favors NH_4NO_3.

Nitrogen also exhibits more oxidation states in its compounds than any other element. There are 10 oxidation states of nitrogen, ranging from +5 to −3, including −1/3.

SAMPLE PROBLEM 23.5 A

▸▸ Give the formula of at least one molecule or ion that exhibits the odd oxidation states of nitrogen.

Solution:

+5: HNO_3, N_2O_5

+3: HNO_2, N_2O_3

+1: N_2O

−1: NH_2OH

−3: NH_3, NH_4Cl

SAMPLE PROBLEM 23.5 B

▸▸ Write molecular equations for each of the following reactions:

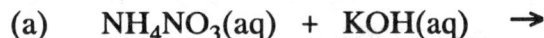

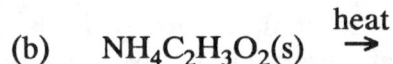

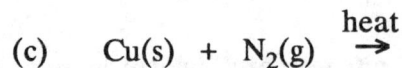

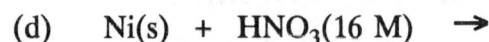

(a) $NH_4NO_3(aq)$ + $KOH(aq)$ $\rightarrow$

(b) $NH_4C_2H_3O_2(s)$ $\xrightarrow{\text{heat}}$

(c) $Cu(s)$ + $N_2(g)$ $\xrightarrow{\text{heat}}$

(d) $Ni(s)$ + $HNO_3(16\ M)$ $\rightarrow$

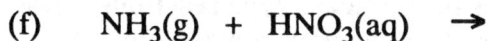

(e) $Ni(s) + HNO_3(6\text{ M}) \rightarrow$

(f) $NH_3(g) + HNO_3(aq) \rightarrow$

Solution:

(a)   Type reaction #2.

   $NH_4NO_3(aq) + KOH(aq) \rightarrow KNO_3(aq) + NH_3(g) + H_2O(l)$

(b)   Type reaction #4.

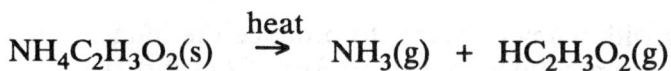

   $NH_4C_2H_3O_2(s) \overset{heat}{\rightarrow} NH_3(g) + HC_2H_3O_2(g)$

(c)   Type reaction #1.

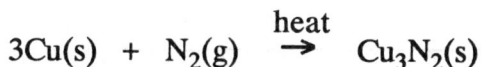

   $3Cu(s) + N_2(g) \overset{heat}{\rightarrow} Cu_3N_2(s)$

(d)   Type reaction #5.

   $Ni(s) + 4HNO_3(16\text{ M}) \rightarrow Ni(NO_3)_2(aq) + 2NO_2(g) + 2H_2O(l)$

(e)   Type reaction #5.

   $3Ni(s) + 8HNO_3(6\text{ M}) \rightarrow 3Ni(NO_3)_2(aq) + 2NO(g) + 4H_2O(l)$

(f)   Type reaction #3.

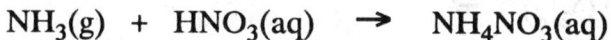

   $NH_3(g) + HNO_3(aq) \rightarrow NH_4NO_3(aq)$

SECTION 23.6   OXYGEN

After completing this section, you will be able to do the following:

■ Complete and balance selected reactions of oxygen and its compounds.

REVIEW

In order to complete and balance reactions in general, it is best to learn **type reactions** (rather than each and every specific example) that occur which can be applied to many examples. One **type reaction** that we have discussed a number of times is the reaction of an acid with a base to produce a salt and water

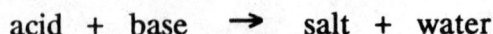

$$acid + base \rightarrow salt + water$$

This **type reaction** can be used to complete and balance dozens of reactions regardless of the formula of the acid or base. Although not all **type reactions** are this general, nevertheless these **type reactions** are one of the best ways to summarize thousands of specific reactions.

Following is a summary of the more important **type reactions of oxygen and its compounds** that will be useful for completing and balancing the reactions given in the Sample Problems:

1. Oxygen combines directly with many elements to produce the binary oxide.
2. Oxygen reacts with organic compounds (burns) to form CO_2 and H_2O.
3. Oxygen reacts with inorganic compounds to form one or more oxides only if the compound contains an element that forms compounds with oxygen and is **not** in its highest oxidation state.

SAMPLE PROBLEM 23.6 A

▶▶ Write molecular equations for each of the following reactions:

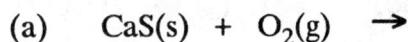

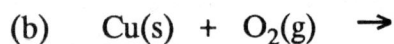

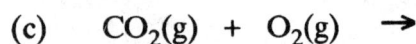

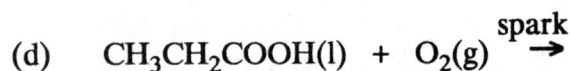

(a) $CaS(s) + O_2(g) \rightarrow$

(b) $Cu(s) + O_2(g) \rightarrow$

(c) $CO_2(g) + O_2(g) \rightarrow$

(d) $CH_3CH_2COOH(l) + O_2(g) \xrightarrow{spark}$

Solution:

(a) Type reaction #3.

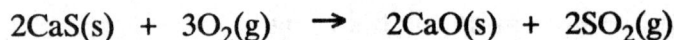

$$2CaS(s) + 3O_2(g) \rightarrow 2CaO(s) + 2SO_2(g)$$

(b) Type reaction #1.

$$2Cu(s) + O_2(g) \rightarrow 2CuO(s)$$

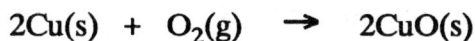

(c)   Type reaction #3.

$$CO_2(g) + O_2(g) \rightarrow \text{No reaction} \qquad \text{[carbon is in its maximum oxidation state]}$$

(d)   Type reaction #2.

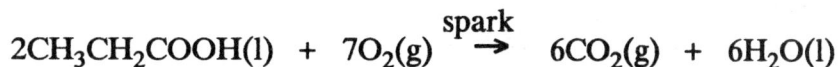

$$2CH_3CH_2COOH(l) + 7O_2(g) \xrightarrow{\text{spark}} 6CO_2(g) + 6H_2O(l)$$

SECTION 23.8   THE HALOGENS - FLUORINE, CHLORINE, BROMINE, AND IODINE

After completing this section, you will be able to do the following:

■ Complete and balance selected reactions of the halogens and their compounds.

REVIEW

In order to complete and balance reactions in general, it is best to learn **type reactions** (rather than each and every specific example) that occur which can be applied to many examples. One **type reaction** that we have discussed a number of times is the reaction of an acid with a base to produce a salt and water

$$\text{acid} + \text{base} \rightarrow \text{salt} + \text{water}$$

This **type reaction** can be used to complete and balance dozens of reactions regardless of the formula of the acid or base. Although not all **type reactions** are this general, nevertheless these **type reactions** are one of the best ways to summarize thousands of specific reactions.

Following is a summary of the more important **type reactions of the halogens and their compounds** that will be useful for completing and balancing the reactions given in the Sample Problems:

1. Fluorine forms binary fluorides with all elements except He, Ne, and Ar.
2. HCl can be prepared from a chloride salt and concentrated sulfuric acid; HBr and HI can be prepared from a bromide or iodide salt and concentrated phosphoric acid.
3. Aqueous solutions of HCl, HBr, and HI undergo the usual reactions with bases and active metals.
4. Cl_2 is a stronger oxidizing agent than Br_2, which is stronger than I_2; this means that Cl_2 will

oxidize bromides to bromine and iodides to iodine, and Br_2 will oxidize iodides to iodine.

SAMPLE PROBLEM 23.8 A

▸▸ Write molecular equations for each of the following reactions:

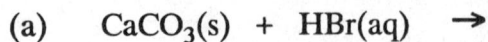

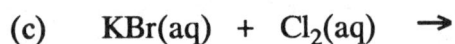

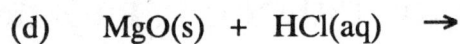

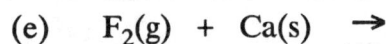

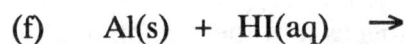

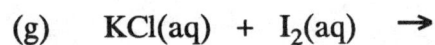

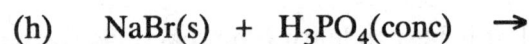

(a) $CaCO_3(s) + HBr(aq) \rightarrow$

(b) $KCl(s) + H_2SO_4(conc) \rightarrow$

(c) $KBr(aq) + Cl_2(aq) \rightarrow$

(d) $MgO(s) + HCl(aq) \rightarrow$

(e) $F_2(g) + Ca(s) \rightarrow$

(f) $Al(s) + HI(aq) \rightarrow$

(g) $KCl(aq) + I_2(aq) \rightarrow$

(h) $NaBr(s) + H_3PO_4(conc) \rightarrow$

Solution:

(a)   Type reaction #3.

$$CaCO_3(s) + 2HBr(aq) \rightarrow CaBr_2(aq) + CO_2(g) + H_2O(l)$$

(b)   Type reaction #2.

$$2KCl(s) + H_2SO_4(conc) \rightarrow 2HCl(g) + K_2SO_4(s)$$

(c)   Type reaction #4.

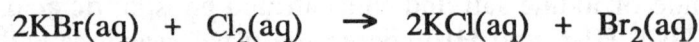

$$2KBr(aq) + Cl_2(aq) \rightarrow 2KCl(aq) + Br_2(aq)$$

(d)   Type reaction #3.

$$MgO(s) \; + \; 2HCl(aq) \; \rightarrow \; MgCl_2(aq) \; + \; H_2O(l)$$

(e)   Type reaction #1.

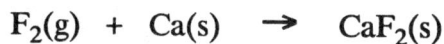

$$F_2(g) \; + \; Ca(s) \; \rightarrow \; CaF_2(s)$$

(f)   Type reaction #3.

$$2Al(s) \; + \; 6HI(aq) \; \rightarrow \; 2AlI_3(aq) \; + \; 3H_2(g)$$

(g)   Type reaction #4.

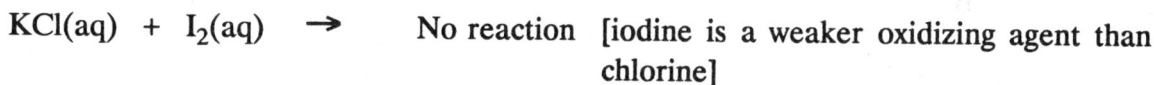

$$KCl(aq) \; + \; I_2(aq) \; \rightarrow \;$$ No reaction   [iodine is a weaker oxidizing agent than chlorine]

(h)   Type reaction #2.

$$3NaBr(s) \; + \; H_3PO_4(conc) \; \rightarrow \; 3HBr(l) \; + \; Na_3PO_4(s)$$

PRACTICE PROBLEMS

Test your problem-solving skills in this chapter by working through the problems in the space provided **BEFORE** checking the answers. A periodic table will always be available. The ANSWERS TO PRACTICE PROBLEMS follow immediately after this section.

1.   Write molecular equations for each of the following reactions:

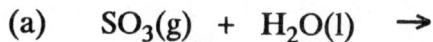

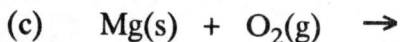

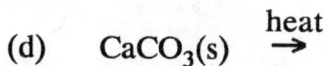

(a) $SO_3(g) \; + \; H_2O(l) \; \rightarrow$

(b) $Li(s) \; + \; S(s) \; \rightarrow$

(c) $Mg(s) \; + \; O_2(g) \; \rightarrow$

(d) $CaCO_3(s) \; \overset{heat}{\rightarrow}$

(e) $P_4O_{10}(s) + H_2O(l) \rightarrow$

(f) $Rb(s) + H_2O(l) \rightarrow$

(g) $Ca(s) + Br_2(l) \rightarrow$

(h) $K(s) + S(s) \rightarrow$

(i) $NH_3(g) + HCl(g) \rightarrow$

(j) $Br_2(l) + H_2(g) \rightarrow$

(k) $Na(s) + H_2O(l) \rightarrow$

(l) $HNO_3(aq) + Ca(OH)_2(s) \rightarrow$

(m) $Ca(s) + N_2(g) \rightarrow$

(n) $Ca(s) + Cl_2(g) \rightarrow$

(o) $N_2(g) + H_2(g) \rightarrow$

(p) $HNO_3(aq) + Na_2CO_3(s) \rightarrow$

(q) $H_2(g) + I_2(s) \rightarrow$

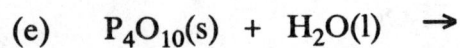

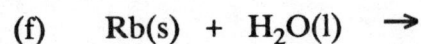

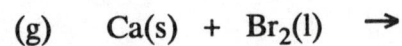

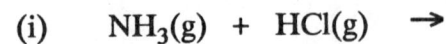

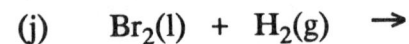

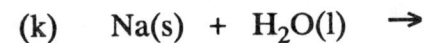

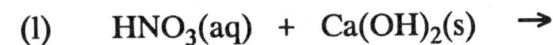

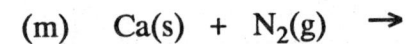

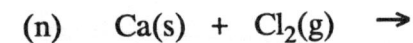

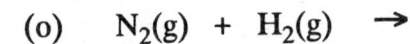

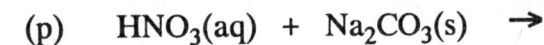

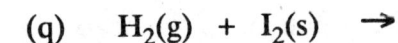

2.   Give the names of the following compounds:

(a) SrO

(b) As_2S_3

(c) Li_2O

(d) $PbCl_2$

(e) H_2O_2

(f)　　N_2O_3

(g)　　MgS

(h)　　HNO_2

(i)　　$SnCl_4$

(j)　　Sb_2S_3

(k)　　HI

(l)　　H_2SO_3

(m)　　CaO

(n)　　NO_2

(o)　　BaO

(p)　　N_2O_5

(q)　　CaS

ANSWERS TO PRACTICE PROBLEMS

1.　　(a)　　$SO_3(g) + H_2O(l) \rightarrow H_2SO_4(aq)$

　　　(b)　　$2Li(s) + S(s) \rightarrow Li_2S(s)$

　　　(c)　　$2Mg(s) + O_2(g) \rightarrow 2MgO(s)$

　　　(d)　　$CaCO_3(s) \xrightarrow{heat} CaO(s) + CO_2(g)$

　　　(e)　　$P_4O_{10}(s) + 6H_2O(l) \rightarrow 4H_3PO_4(aq)$

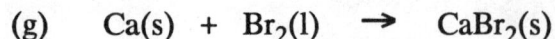

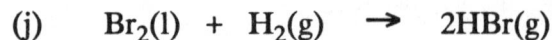

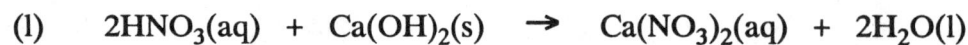

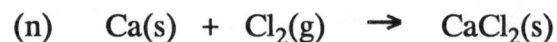

(f) $2Rb(s) + 2H_2O(l) \rightarrow 2RbOH(aq) + H_2(g)$

(g) $Ca(s) + Br_2(l) \rightarrow CaBr_2(s)$

(h) $2K(s) + S(s) \rightarrow K_2S(s)$

(i) $NH_3(g) + HCl(g) \rightarrow NH_4Cl(s)$

(j) $Br_2(l) + H_2(g) \rightarrow 2HBr(g)$

(k) $2Na(s) + 2H_2O(l) \rightarrow 2NaOH(aq) + H_2(g)$

(l) $2HNO_3(aq) + Ca(OH)_2(s) \rightarrow Ca(NO_3)_2(aq) + 2H_2O(l)$

(m) $3Ca(s) + N_2(g) \rightarrow Ca_3N_2(s)$

(n) $Ca(s) + Cl_2(g) \rightarrow CaCl_2(s)$

(o) $N_2(g) + 3H_2(g) \rightarrow 2NH_3(g)$

(p) $2HNO_3(aq) + Na_2CO_3(s) \rightarrow 2NaNO_3(aq) + CO_2(g) + H_2O(l)$

(q) $H_2(g) + I_2(s) \rightarrow 2HI(g)$

2.   (a) SrO strontium oxide

     (b) As_2S_3 arsenic(III) sulfide

     (c) Li_2O lithium oxide

     (d) $PbCl_2$ lead(II) chloride

     (e) H_2O_2 hydrogen peroxide

     (f) N_2O_3 dinitrogen trioxide

     (g) MgS magnesium sulfide

(h) HNO_2 nitrous acid

(i) $SnCl_4$ tin(IV) chloride

(j) Sb_2S_3 antimony(III) sulfide

(k)   HI   hydrogen iodide (or hydriodic acid if aqueous)

(l) H_2SO_3 hydrogen sulfite (or sulfurous acid if aqueous)

(m)   CaO   calcium oxide

(n) NO_2 nitrogen dioxide

(o)   BaO   barium oxide

(p) N_2O_5 dinitrogen pentoxide

(q)   CaS   calcium sulfide

CHAPTER 24

A CLOSER LOOK AT INORGANIC CHEMISTRY: METALS AND THEIR COMPOUNDS

SECTION 24.3   THE ALKALI METALS (GROUP IA)

After completing this section, you will be able to do the following:

- Complete and balance selected reactions of the alkali metals.

REVIEW

In order to complete and balance reactions in general, it is best to learn **type reactions** (rather than each and every specific example) that occur, which can be applied to many examples. One **type reaction** that we have discussed a number of times is the reaction of an acid with a base to produce a salt and water

$$acid \ + \ base \ \rightarrow \ salt \ + \ water$$

This **type reaction** can be used to complete and balance dozens of reactions regardless of the formula of the acid or base. Although not all **type reactions** are this general, nevertheless these **type reactions** are one of the best ways to summarize thousands of specific reactions.

Following is a summary of the more important **type reactions of the alkali metals and their compounds** that will be useful for completing and balancing the reactions given in the Sample Problems:

1.   All alkali metals react with halogens to form salts;

609

610

2. All alkali metals react with hydrogen to form hydrides;
3. Lithium metal reacts with oxygen to form the oxide, sodium metal reacts with oxygen to form the peroxide, and the rest of the alkali metals react with oxygen to form superoxides;
4. All alkali metals react with water to form hydrogen and the corresponding base.

SAMPLE PROBLEM 24.3 A

▶▶ Write molecular equations for each of the following reactions:

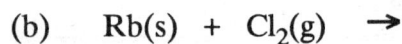

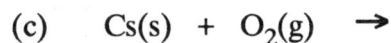

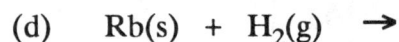

(a) $Na(s) + H_2O(l) \rightarrow$

(b) $Rb(s) + Cl_2(g) \rightarrow$

(c) $Cs(s) + O_2(g) \rightarrow$

(d) $Rb(s) + H_2(g) \rightarrow$

Solution:

HINT: Learn the type reactions in the REVIEW.

(a)   Type reaction #4.

$$2Na(s) + 2H_2O(l) \rightarrow 2NaOH(aq) + H_2(g)$$

(b)   Type reaction #1.

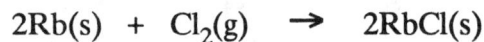

$$2Rb(s) + Cl_2(g) \rightarrow 2RbCl(s)$$

(c)   Type reaction #3.

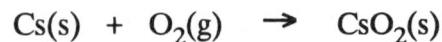

$$Cs(s) + O_2(g) \rightarrow CsO_2(s)$$

(d)   Type reaction #2.

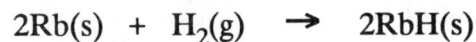

$$2Rb(s) + H_2(g) \rightarrow 2RbH(s)$$

SECTION 24.4 THE ALKALINE EARTH METALS (GROUP IIA)

After completing this section, you will be able to do the following:

■ Complete and balance selected reactions of the alkaline earth metals.

REVIEW

In order to complete and balance reactions in general, it is best to learn **type reactions** (rather than each and every specific example) that occur, which can be applied to many examples. One **type reaction** that we have discussed a number of times is the reaction of an acid with a base to produce a salt and water

$$acid \; + \; base \; \rightarrow \; salt \; + \; water$$

This **type reaction** can be used to complete and balance dozens of reactions regardless of the formula of the acid or base. Although not all **type reactions** are this general, nevertheless these **type reactions** are one of the best ways to summarize thousands of specific reactions.

Following is a summary of the more important **type reactions of the alkaline earth metals and their compounds** that will be useful for completing and balancing the reactions given in the Sample Problems:

1. Mg, Ca, Sr and Ba oxides and hydroxides are bases that dissolve in acids only; Be oxide and hydroxide are amphoteric and dissolve in both acid and base;
2. The solubility of the hydroxides in water varies from insoluble for magnesium to slightly soluble for calcium and strontium to soluble for barium;
3. The metals burn in air forming the oxides;
4. The metal carbonate decompose on heating to form carbon dioxide and the corresponding oxide.

SAMPLE PROBLEM 24.4 A

▶▶ Write molecular equations for each of the following reactions:

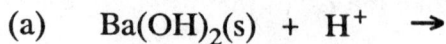

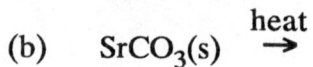

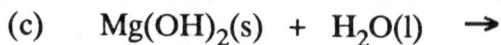

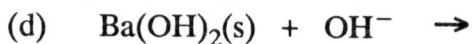

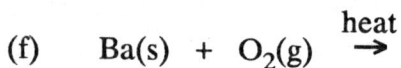

    (a) $Ba(OH)_2(s) + H^+$ →

    (b) $SrCO_3(s) \xrightarrow{heat}$

    (c) $Mg(OH)_2(s) + H_2O(l)$ →

    (d) $Ba(OH)_2(s) + OH^-$ →

    (e) $BeO(s) + OH^- + H_2O(l)$ →

    (f) $Ba(s) + O_2(g) \xrightarrow{heat}$

Solution:

NOTE: Learn the type reactions in the **REVIEW.**

    (a)   Type reaction #1.

$$Ba(OH)_2(s) + 2H^+ \rightarrow Ba^{2+} + 2H_2O(l)$$

    (b)   Type reaction #4.

$$SrCO_3(s) \xrightarrow{heat} SrO(s) + CO_2(g)$$

    (c)   Type reaction #2.

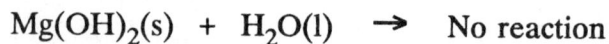

$$Mg(OH)_2(s) + H_2O(l) \rightarrow \text{No reaction}$$

    (d)   Type reaction #1.

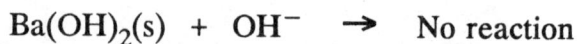

$$Ba(OH)_2(s) + OH^- \rightarrow \text{No reaction}$$

(e)   Type reaction #1.

$$BeO(s) + 2OH^- + H_2O(l) \rightarrow Be(OH)_4^{2-}$$

(f)   Type reaction #3.

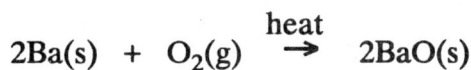

$$2Ba(s) + O_2(g) \xrightarrow{heat} 2BaO(s)$$

SECTION 24.5   GROUP IIIA

After completing this section, you will be able to do the following:

■ Complete and balance selected reactions of the Group IIIA metals.

REVIEW

In order to complete and balance reactions in general, it is best to learn **type reactions** (rather than each and every specific example) that occur, which can be applied to many examples. One **type reaction** that we have discussed a number of times is the reaction of an acid with a base to produce a salt and water

$$acid + base \rightarrow salt + water$$

This **type reaction** can be used to complete and balance dozens of reactions regardless of the formula of the acid or base. Although not all **type reactions** are this general, nevertheless these **type reactions** are one of the best ways to summarize thousands of specific reactions.

Following is a summary of the more important **type reactions of the Group IIIA metals and their compounds** that will be useful for completing and balancing the reactions given in the Sample Problems:

1. Aluminum is prepared by the reduction of aluminum chloride with an alkali metal;
2. Aluminum is used as a reducing agent to produce Co, Cr, and Mn from their oxides;
3. Aluminum, gallium, and indium metals, oxides, and hydroxides are amphoteric (dissolve in both acid and base);
4. Aluminum reacts with halogens to produce the corresponding halide.

SAMPLE PROBLEM 24.5 A

▶▶ Write molecular equations for each of the following reactions:

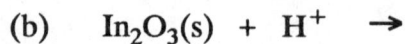

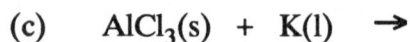

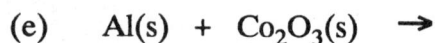

(a) $Al(s) + Cl_2(g) \rightarrow$

(b) $In_2O_3(s) + H^+ \rightarrow$

(c) $AlCl_3(s) + K(l) \rightarrow$

(d) $In_2O_3(s) + OH^- + H_2O(l) \rightarrow$

(e) $Al(s) + Co_2O_3(s) \rightarrow$

Solution:

NOTE: Learn the type reactions in the REVIEW.

(a)   Type reaction #4.

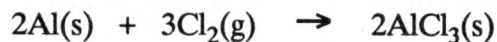

$$2Al(s) + 3Cl_2(g) \rightarrow 2AlCl_3(s)$$

(b)   Type reaction #3.

$$In_2O_3(s) + 6H^+ \rightarrow 2In^{3+} + 3H_2O(l)$$

(c)   Type reaction #1.

$$AlCl_3(s) + 3K(l) \rightarrow Al(s) + 3KCl(s)$$

(d)   Type reaction #3.

$$In_2O_3(s) + 2OH^- + 3H_2O(l) \rightarrow 2In(OH)_4{}^-$$

(e)   Type reaction #2.

$$2Al(s) + Co_2O_3(s) \rightarrow 2Co(s) + Al_2O_3(s)$$

SECTION 24.7   OBSERVATIONS OF COMPLEXES

After completing this section, you will be able to do the following:

■ In transition metal complexes, identify the central atom and determine its coordination number and oxidation number.

REVIEW

A **coordination compound** most commonly consists of a transition metal **complex ion** and enough **free ions** of opposite charge so that the compound is neutral, or a neutral complex species. The formula of the **complex ion** is shown within square brackets with the **central atom** written first, followed by the ionic and neutral ligands **in that order**. **Ligands** are the ionic and neutral species covalently bonded to the central atom. The **free ions** are **not** covalently bonded to the central atom and only function as counter ions (of opposite charge) to make the compound electrically neutral. For example, $[PtCl_2(NH_3)_4]Cl_2$ is a coordination compound which consists of the complex ion and free ions,

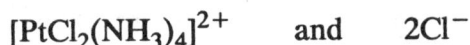

$$[PtCl_2(NH_3)_4]^{2+} \quad \text{and} \quad 2Cl^-$$

The **complex ion** consists of the **central atom**, Pt, which is the **first** element inside the brackets, followed by the ionic ($2Cl^-$) and neutral ($4NH_3$) ligands. The **coordination number** of the central atom is the total **number of bonds** formed between the central atom and the ligands. In this complex ion, the coordination number (CN) of platinum is **6** ($2Cl^- + 4NH_3$). [For our purposes in this section, the total number of ionic and neutral ligands is the coordination number; however, there are other ligands known where each ligand forms two or more bonds to the central atom.] The sum of the charges of all the species in the complex ion gives the charge on the complex ion. By knowing the charge on the complex ion, you can determine the **oxidation number** of the central atom. To find the oxidation number of Pt in this complex ion, we need to know the charge on each of the ligands. Chloride, Cl^-, is its usual $1-$ and NH_3 is neutral or zero. (Be careful! NH_3 is zero overall and NH_4^+ is $1+$). Substituting these oxidation numbers into the complex ion and letting x equal the oxidation number of Pt,

sum of the charges of all species in the complex ion = charge on complex ion

$$[x + 2(-1) + 4(0)] = 2+$$

and solving the equation for x

$$[x - 2] = +2 \quad \text{or} \quad x = +4.$$

The oxidation number of platinum in this complex ion is $+4$. For charges on ligands, you will need to review the charges on the common monatomic anions (Section 1.9) and realize that when you see the formula of a compound as a ligand (e.g., NH_3, H_2O, CO, etc.), the overall charge is zero. Although "ionic" ligands like chloride act as a Lewis base and furnish a pair of valence electrons to covalently bond to the central atom, we use the ionic charge (or oxidation number) to determine the oxidation number of the central atom.

SAMPLE PROBLEM 24.7 A

▸▸ In the following coordination compounds, indicate the (a) formula and charge of the complex ion, and (b) the central atom of the complex ion, and its coordination number and oxidation number.

$$[CrCl(NH_3)_5]Cl_2 \qquad\qquad K[AuCl_4] \qquad\qquad [Co(NH_3)_4(H_2O)_2]Br_3$$

Solution:

$$[CrCl(NH_3)_5]Cl_2$$

(a) The complex ion is the species inside the brackets, with a charge equal and opposite to the total charge of the free ions. Since the total charge of the free ions is -2, the formula of the complex ion is

$$[CrCl(NH_3)_5]^{2+}$$

(b) The central atom is chromium, Cr, the first element appearing in the complex ion.

The coordination number is 6, since there are a total of six ligands.

The oxidation number of Cr is $+3$, since NH_3 is zero, Cl^- is -1, and the sum of all the oxidation numbers must add up to the charge on the complex ion, $2+$,

$$[Cr + 1(-1) + 5(0)] = 2+.$$

$$K[AuCl_4]$$

(a) Follow the same procedure used for the previous compound. Since the total charge of the free ion is $1+$, the formula of the complex ion is

$$[AuCl_4]^-$$

(b) The central atom is gold, Au, the first element appearing in the complex ion.

The coordination number is 4, since there are a total of four ligands.

The oxidation number of Au is +3, since Cl^- is -1 and the sum of all the oxidation numbers must add up to the charge on the complex ion, $1-$,

$$[Au + 4(-1)] = 1-.$$

$[Co(NH_3)_4(H_2O)_2]Br_3$

(a) Follow the same procedure used for the previous compound. Since the total charge of the free ions is $3-$, the formula of the complex ion is

$$[Co(NH_3)_4(H_2O)_2]^{3+}$$

(b) The central atom is cobalt, Co, the first element appearing in the complex ion.

The coordination number is 6, since there are a total of six ligands.

The oxidation number of Co is +3, since both NH_3 and H_2O have an overall charge of zero and the sum of all the oxidation numbers must add up to the charge on the complex ion, $3+$,

$$[Co + 4(0) + 2(0)] = 3+.$$

SECTION 24.11   NOMENCLATURE OF COMPLEXES

After completing this section, you will be able to do the following:

■ Name a coordination compound from its formula, and vice versa.

REVIEW

To write the **name** of a coordination compound **from the formula**:

1. The ligands are named first in alphabetical order (regardless of whether they are ionic or neutral), with the prefixes di-, tri-, etc., used to indicate the number of each kind of ligand (the prefixes are **not** used for alphabetizing); the names of **anionic** ligands end in **-o**, NH_3 is called **ammine** with two "m's," and H_2O is called **aqua** (see Table 24.6 in the textbook for names of common ligands).

2. Name the central atom next followed by its oxidation state in roman numerals (see Sample

Problem 24.11 A). If the complex ion has a **negative** charge, the suffix -ate is added to the name **unless** the name of the central atom ends in -ium, -um, -ese, or -en, in which case the -ate **replaces** those endings (see Table 24.7 in the textbook for the names of common anionic complex ions). Note that the Latin name of the central atom is often used instead of the English name.

3. Steps 1 and 2 are one word. Then give the name of the free ion(s) as a separate word (no prefix indicating the number of free ions is needed). The **cation is always named first** and the anion is named second, regardless of whether the complex ion has a positive or negative charge.

To write the **formula** of a coordination compound from the **name**:

1. The cation and anion are always separated by a space, regardless of whether the complex ion is positively or negatively charged. Write the formula of the cation first; if the cation is the complex ion, inside the brackets write the symbol for the central atom, followed by the ionic and neutral ligands, in that order.

2. From the oxidation number of the central atom and the charges on the ligands, determine the charge on the complex ion. This will allow you to determine the number of free ions needed to balance the charge on the complex ion.

SAMPLE PROBLEM 24.11 A

▶▶ Name the compound $K_3[Mn(CN)_6]$.

Solution:

Follow the procedure in the REVIEW. The first word of the two-word name is always the name of the cation, potassium.

The second word is always the name of the anion, which in this case is the complex ion. The name of this complex ion starts with the number and name of each ligand in alphabetical order, followed by the name and oxidation number of the central atom. Since there are 6 cyano- ligands and the central atom is manganese, we start by writing

hexacyanomanganese(?)

Since the three K^+ ions have a total charge of $3+$, the charge on the complex anion must be $3-$, which means the name of the central atom ends with **-ate** replacing -ese, or manganate. Since CN^- has a charge of $1-$ (Table 24.6), the oxidation number of manganese must be $+3$

(see Sample Problem 24.7 A for procedure). The name of the complex anion is

hexacyanomanganate(III)

The complete name of the compound, $K_3[Mn(CN)_6]$, is

potassium hexacyanomanganate(III)

SAMPLE PROBLEM 24.11 B

▶▶ Write the formula for tetraaquadicyanoiron(III) nitrate.

Solution:

The first word of the two-word name is the cation, which in this compound is the complex ion. The second word is the anion.

The formula of the complex ion starts with the symbol for the central atom, which is always the last part of the name (in this case iron, Fe). Next comes the formula with subscripts of the ionic ligands (cyano- is CN^-) followed by the neutral ligands (aqua is H_2O). At this point for the complex cation, we have

$$[Fe(CN)_2(H_2O)_4]^?$$

The roman numeral in the name tells you that the oxidation number of iron is $+3$. Since cyano- is -1 and water is neutral, the charge on the complex ion must be $1+$; only one free nitrate ion will be required to balance the charge of the complex ion. The formula for tetraaquadicyanoiron(III) nitrate is

$$[Fe(CN)_2(H_2O)_4]NO_3$$

SECTION 24.13   BONDING IN COMPLEXES

After completing this section, you will be able to do the following:

- ■ Sketch crystal field d-orbital splitting diagrams for octahedral complexes.

REVIEW

Since color is one of the most characteristic properties of complexes, crystal field theory (CFT) is needed to explain the colors as well as their magnetic properties. According to **crystal field theory**, electrostatic attraction between the positive charge on the central metal ion and the negative charges on the ligands **bonds** the ligands to the metal ion. The theory applied to octahedral-shaped complexes is discussed in this section.

In an isolated gaseous metal ion, its five d orbitals are degenerate, i.e., they have the same size, shape, and energy. However, as six ligands (negatively charged or the negative end of polar molecules) approach the metal ion along its x, y, and z axes to form an octahedral complex, the repulsion between the negative charges on the ligands and the negative charges on the electrons present in the d orbitals of the metal ion **raises the energy of all five d orbitals**. However, the **two d orbitals** that lie **along the axes** are closer to the incoming ligands (resulting in stronger repulsions) and as a result have their energies **raised more** than the energies of the other **three d orbitals** that lie **between** the axes. Thus, the five d orbitals are split into a set of **two high-energy orbitals** and a set of **three low-energy orbitals** (although both sets are higher in energy than in the absence of the ligands). The difference in energy between the two sets of orbitals is called the **crystal field splitting, Δ**.

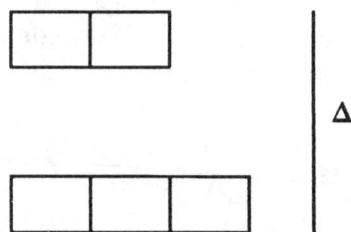

The strength of the crystal field splitting depends on the charge or polarity of the ligand. As electrons are added to the d orbitals, less energy will be required to place the fourth electron in one of the high-energy orbitals than to put a second electron into one of the low-energy orbitals and form a pair (which also requires energy), if the splitting is small (Δ is small). This electron configuration is called **weak field** (small Δ) or **high spin** (maximum unpaired d electrons).

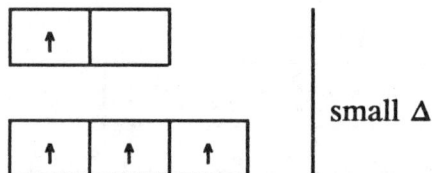

If the crystal field splitting is large (Δ is large), less energy will be required to place the fourth electron into one of the low-energy orbitals and form a pair than to put it in one of the high-energy orbitals. This electron configuration is called **strong field** (large Δ) or **low spin** (minimum unpaired d electrons).

```
┌──────┬──────┐
│      │      │
└──────┴──────┘
```
large Δ

```
┌──────┬──────┬──────┐
│  ↑↓  │  ↑   │  ↑   │
└──────┴──────┴──────┘
```

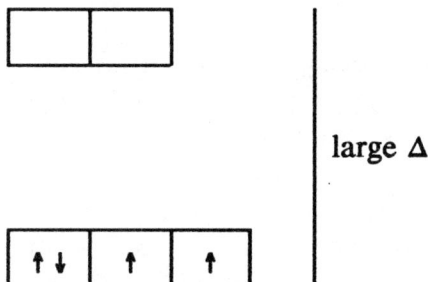

Given the number of unpaired electrons in a complex from other experiments, you can predict whether the complex is high spin (many unpaired electrons) or low spin (few, if any, unpaired electrons).

SAMPLE PROBLEM 24.13 A

▶▶ The complex ion $[CoCl_6]^{3-}$ is diamagnetic. Sketch the d-orbital splitting diagram.

Solution:

With six ligands, the complex ion is probably octahedral. The electron configuration of the cobalt atom is [Ar] $3d^74s^2$; the electron configuration of Co^{3+} is [Ar] $3d^6$, so this is a d^6 ion. The fact that this complex is diamagnetic means that there are no unpaired electrons and it must be a low spin or strong field complex (large splitting between the two sets of d orbitals). Placing six electrons in the low-spin orbital diagram, we have

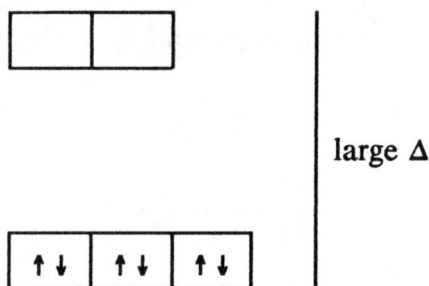

```
┌──────┬──────┐
│      │      │
└──────┴──────┘
```
large Δ

```
┌──────┬──────┬──────┐
│  ↑↓  │  ↑↓  │  ↑↓  │
└──────┴──────┴──────┘
```

SAMPLE PROBLEM 24.13 B

▶▶ The complex ion $[CrCl_6]^{3-}$ has three unpaired electrons. Is this complex a high-spin or low-spin complex. Sketch the d-orbital diagram.

In this octahedral complex, the electron configuration of the chromium atom is [Ar] $3d^44s^2$; the electron configuration of Cr^{3+} is [Ar] $3d^3$, so this is a d^3 ion. With three unpaired electrons, you might expect the ion to be high spin (small splitting between sets of d orbitals). However, there are no high-spin/low-spin complexes when there are **three or less d electrons** in the metal ion; the electrons are all unpaired. The choice of high spin or low spin doesn't occur until the fourth electron needs to be placed in an orbital.

Similarly, when the metal ion has 8, 9, or 10 d electrons, there is no difference in the number of unpaired electrons in the high-spin vs. the low-spin configuration. A d^8 metal ion has two unpaired electrons, a d^9 metal ion has one unpaired electron, and a d^{10} metal ion has zero unpaired electrons.

The complex ion $[CrCl_6]^{3-}$ does not form high- and low-spin octahedral complexes.

SELF-TEST QUESTIONS

Test your understanding of the concepts and skills in this chapter by working through the multiple choice questions BEFORE checking the answers. A periodic table will always be available. The ANSWERS TO SELF-TEST QUESTIONS follow immediately after this section.

1. The only transition metal ion with the correct ground state electron configuration is

   A. Ti^{2+} [Ar] $3d^24s^2$

   B. Cr^{3+} [Ar] $3d^54s^1$

   C. Co^{3+} [Ar] $3d^4$

   D. Sc^{3+} [Ar] $3d^44s^2$

   E. Cu^{2+} [Ar] $3d^9$

2. The only coordination compound named correctly is

    A. $[Mn(H_2O)_6]Cl_2$ manganese chloride hexahydrate

    B. $[FeBr_2(NH_3)_4]Br$ tetraammoniairontribromide

    C. $K_2[NiCl_4]$ potassium nickel(IV) chloride

    D. $[Cu(NH_3)_4]SO_4$ tetraamminecopper(II) sulfate

    E. $[Co(NH_3)_6](NO_3)_2$ cobalt(III)hexaammine nitrate

3. If the central atom in a complex ion has a coordination number of six, the expected geometry of the ion is

    A. octahedral
    B. linear
    C. trigonal bipyramidal
    D. tetrahedral
    E. trigonal planar

4. The alkaline earth element is

    A. Li   B. Ti   C. U   D. Ca   E. P

5. The most reactive element with water at room temperature is

    A. Fe   B. N   C. K   D. Zn   E. Ar

6. If CN^- is a strong field ligand, the number of unpaired electrons in the complex ion, $[Fe(CN)_6]^{3-}$, is

    A. 5   B. 1   C. 2   D. 3   E. 4

7. The electron configuration of the ground state Cr^{3+} ion is

   A. $[Ar]\ 4d^{1}4s^{2}$

   B. $[Ar]\ 3d^{1}4s^{2}$

   C. $[Ar]\ 3d^{4}4s^{2}$

   D. $[Ar]\ 3d^{5}4s^{1}$

   E. $[Ar]\ 3d^{3}$

8. The choice that contains the elements that spontaneously form peroxides or superoxides when exposed to oxygen is the

   A. lanthanide metals
   B. metalloids
   C. noble gases
   D. alkali metals
   E. halogens

9. The compound that gives a basic solution when dissolved in water is

   A. Li_2O B. CO_2 C. NO_2 D. SO_3 E. P_4O_{10}

10. The only compound named correctly is

   A. $[Co(H_2O)_6]SO_4$ hexaaquacobalt sulfate

   B. $[CrCl_2(H_2O)_4]^{+}$ tetraaquadichlorochromium(III) ion

   C. $K_3[Fe(CN)_6]$ potassium hexacyanoiron(III)

   D. $[Pt(NH_3)_2Cl_2]$ diamminedichloropotassium(II)

   E. $Na[FeCl_4]$ sodium tetrachloriron

11. The correct formula for tetraamminedichlorocobalt(III) bromide is

   A. $[CoCl_2(NH_3)_4Br]^{3+}$

   B. $[CoCl_2(N_4H_2)]Br$

C. $[CoCl_2(NH_3)_4]Br$

D. $[Co_3Cl_2(NH_3)_4Br]$

E. $[CoCl_2NH_3)_4]Br_3$

ANSWERS TO SELF-TEST QUESTIONS

1. e	5. c	9. a
2. d	6. b	10. b
3. a	7. e	11. c
4. d	8. d	

PRACTICE PROBLEMS

Test your problem-solving skills in this chapter by working through the problems in the space provided **BEFORE** checking the answers. A periodic table will always be available. The ANSWERS TO PRACTICE PROBLEMS follow immediately after this section.

1. [See Chapter 23, Practice Problem #1, for completing and balancing reactions involving metals from Groups IA, IIA, and IIIA.]

2. [See Chapter 23, Practice Problem #2, for naming compounds containing metals from Groups IA, IIA, and IIIA.]

3. Write the correct ground-state electron configuration for

    (a) Cu

    (b) Cr^{2+}

    (c) V

    (d) Co^{3+}

(e) Ni

(f) V^{3+}

(g) Cr

(h) Cu^+

4. Write the formula for each of the following:

(a) diaquadicyanoplatinum(IV) sulfate

(b) hexafluorocobaltate(III) ion

(c) potassium diamminetetrabromocobaltate(II)

(d) hexachloronickelate(II) ion

5. Write the name for each of the following:

(a) $Na[CrBr_4(NH_3)_2]$

(b) $[Cr(NH_3)_6](NO_3)_3$

(c) $K[PtBr_3(NH_3)]$

(d) $[CoCl_2(NH_3)_4]Cl$

6. How many <u>unpaired electrons</u> are present in each of the following complex ions?

(a) $[Mn(CN)_6]^{3-}$ high spin

(b) $[Co(NH_3)_6]^{2+}$ low spin

(c) $[Cr(H_2O)_6]^{2+}$ low spin

(d) $[Fe(NH_3)_6]^{3+}$ high spin

ANSWERS TO PRACTICE PROBLEMS

1.   [See Chapter 23, #1.]

2.   [See Chapter 23, #2.]

3.   (a) $[Ar]\ 3d^{10}4s^1$

      (b) $[Ar]\ 3d^4$

      (c) $[Ar]\ 3d^34s^2$

      (d) $[Ar]\ 3d^6$

      (e) $[Ar]\ 3d^84s^2$

      (f) $[Ar]\ 3d^2$

      (g) $[Ar]\ 3d^54s^1$

      (h) $[Ar]\ 3d^{10}$

4.   (a) $[Pt(CN)_2(H_2O)_2]SO_4$

      (b) $[CoF_6]^{3-}$

      (c) $K_2[CoBr_4(NH_3)_2]$

      (d) $[NiCl_6]^{4-}$

5. (a)   sodium diamminetetrabromochromate(III)

    (b)   hexaamminechromium(III) nitrate

    (c)   potassium amminetribromoplatinate(II)

    (d)   tetraamminedichlorocobalt(III) chloride

6. (a) Mn^{3+} is a d^4 ion; high-spin configuration means 4 unpaired electrons.

    (b) Co^{2+} is a d^7 ion; low-spin configuration means 1 unpaired electron.

    (c) Cr^{2+} is a d^4 ion; low-spin configuration means 2 unpaired electrons.

    (d) Fe^{3+} is a d^5 ion; high-spin configuration means 5 unpaired electrons.